Nuclear Structure Physics

Nuclear Structure Physics

Edited by

Amritanshu Shukla, PhD
Department of Physics,
Rajiv Gandhi Institute of Petroleum Technology
Jais, Uttar Pradesh, India

Suresh Kumar Patra, PhD
Institute of Physics
Bhubaneshwar, India
and
Homi Bhabha National School
Mumbai, India

CRC Press
Taylor & Francis Group
Boca Raton London New York

CRC Press is an imprint of the
Taylor & Francis Group, an **informa** business

First edition published 2021
by CRC Press
6000 Broken Sound Parkway NW, Suite 300, Boca Raton, FL 33487-2742

and by CRC Press
4 Park Square, Milton Park, Abingdon, Oxon OX14 4RN

© 2021 Taylor & Francis Group, LLC

CRC Press is an imprint of Taylor & Francis Group, an Informa business

ISBN: 978-0-367-25610-4 (hbk)
ISBN: 978-0-429-28864-7 (ebk)

Typeset in Palatino
by codeMantra

Contents

Preface

How neutrons and protons can be joined together to form a nuclei? At most how many neutrons and protons can be held together to form a nuclei? This fundamental riddle seeks to be answered yet. What we have learnt so far and has also been testified with extremely sophisticated experiments is that the nucleus of an atom is held together by the strong nuclear force that binds together protons and neutrons. Most importantly, this 'glue' – strong nuclear force, which is the strongest of the four fundamental forces – acts only between nuclear particles and over very short – typically nuclear – distances where this nuclear force is much more powerful than the electrostatic repulsion between protons. Also, this 'glue' is slightly stronger when acting between a proton and a neutron than between either pair of like particles. As a result, atoms are usually stable so long as the numbers of protons and neutrons are not too uneven. If this balance isn't right, atoms can split apart through radioactive decay or nuclear fission. If an atom gets too heavy with neutrons or protons, extra nucleons simply won't stick at all — not even for an instant. Nuclear physicists have long been trying to map out where this boundary of stability line – termed to be the 'drip line' – exists, as nuclei beyond this point are like oversized droplets that drip small fragments. Similarly, there are theoretical speculations that there could lie a region of very stable nuclei across the unstable ones at the extremes of mass, i.e. n the superheavy region of the nuclear landscape. Key to all these questions in knowing the exact form of nuclear force, this is yet to be figured out, making nuclear physics a forbidding subject, even for highly trained physicists. It is important to mention here that nuclear physics research began with the investigation of a handful of nuclei (few tens), but today, this number has grown many folds (few thousands) due to tremendous efforts and contributions made in last 100 years.

A systematic theoretical treatment of nuclei provides a simple lab of complex nuclei, which could provide access to those areas of nuclear landscape, which is still beyond experimental reach. In a way, it is an extraordinarily exciting time for researchers in nuclear structure physics due to the advance tools available to experimentalists as well as theoretical researchers. On the one hand, there has been a great progress in detector techniques helping in the detection of nuclei lying on the extremes of mass and isospin; on the other hand, involvement of supercomputing facilities has led to the development of ab initio-based theoretical approaches increasing the accuracy in the study of many body systems phenomenally. The book chapters cover many interesting topics to be seen through the window of nuclear structure. These include mean field model and recent parametrization, magic nuclei, superheavy nuclei, bubble nuclei, and also topics related to weak interaction. These chapters not only discuss the potential of the subject to impact other current research areas such as mapping the astrophysical p-process in nucleosynthesis, and the exotic one- and two-proton decay modes but also showcase the applicability and success of the different nuclear effective interaction parameters near the drip line, where hints for level reordering have already been seen, and where one can test the isospin dependence of the interaction. Overall, the chapters written by highly experienced and well-known researchers/experts from academia may serve as a standard, up-to-date research reference book for the postgraduate and graduate students, and researchers working in related fields.

We would like to convey our appreciation to all the contributors as well as our academic collaborators, who have been always a source of inspiration to work in this field of research.

Our special thanks are also due to Ms. Aastha Sharma, Ms. Shikha Garg, Mr. Varun Gopal, and the entire editorial team members from CRC publishers for their kind support and great efforts in bringing the book to fruition. Last but not the least, editors are sincerely thankful to their family members, whose unending support and affection has always been a driving force in making any professional assignment a success!

Amritanshu Shukla
Suresh Kumar Patra

Editors

Dr. Amritanshu Shukla earned his Master's degree in Physics from University of Lucknow (1999) and his Ph.D. in Physics from IIT Kharagpur (2005), and did his postdoctoral work at Institute of Physics Bhubaneswar (under the Department of Atomic Energy, Govt. of India); University of North Carolina Chapel Hill, USA; University of Rome, Italy; and Physical Research Laboratory Ahmedabad (under the Department of Space, Govt. of India). Currently, he is an Associate Professor in Physics and also the Head of Division at Rajiv Gandhi Institute of Petroleum Technology (RGIPT) (set up through an Act of Parliament by the Ministry of Petroleum & Natural Gas, Govt. of India as an "Institute of National Importance" on the lines of IITs). The institute is co-promoted as an energy domain-specific institute by six leading PSUs (Public sector undertakings)—ONGC Ltd., IOCL, OIL, GAIL, BPCL, and HPCL in association with OIDB. The mission of the institute is to promote energy self-sufficiency in the country through its teaching and R&D efforts.

His research interests include theoretical physics, nuclear physics, and physics of renewable energy systems. He has published about 100 research articles in various international journals and conference proceedings of national and international repute. Currently, he is working on the different problems in nuclear structure and nuclear reactions with a special focus on bubble nuclei, drip-line nuclei, and superheavy nuclei. He is also supervising several research students, actively involved in a number of sponsored research projects, and having several national as well as international active research collaborations from India and abroad on the topics of his research interests.

Prof. Suresh Kumar Patra is a well-known Indian Theoretical Nuclear Physicist. He earned his Ph.D. (Theoretical Nuclear Physics) from Institute of Physics, India. After his Ph.D., he has done postdoctoral studies at Tohoku University (Japan), J.W. Goethe University (Germany), Chung Yuan Christian University (Taiwan), University of Barcelona (Spain), and University of Surrey (U.K.). He also had short visits to other reputed universities for collaborative research work and to give lectures. Currently, he is a Professor at Institute of Physics, Bhubaneswar.

Prof. Patra has a wide-ranging expertise in various areas of nuclear physics, such as nuclear structure, nuclear reaction, and nuclear astrophysics. He has contributed considerably to the structure of exotic and superheavy nuclei, nucleus–nucleus reaction for both beta-stable and beta-unstable nuclei, and structure of neutron stars and gravitational waves from their merger. He has published about 170 research articles in international peer-reviewed journals, more than 13 book chapters, and about 160 contributions to various national/international symposiums/workshops and is the co-editor of several conference proceedings. He has also served as the organizer/advisory member of various national/international symposiums/workshops and the referee of various journals. He has guided more than 16 Ph.D. and 8 postdoctoral scholars.

Contributors

Bijay Kumar Agrawal
Theory Division
Saha Institute of Nuclear Physics
Kolkata, India
and
Homi Bhabha National Institute
Anushakti Nagar
Mumbai, India

A. N. Antanov
Institute of Nuclear Research and Nuclear
    Energy
Bulgarian Academy of Sciences
Sofia, Bulgaria

Naftali Auerbach
School of Physics and Astronomy
Tel Aviv University
Tel Aviv, Israel
and
Department of Physics and Astronomy
National Superconducting
Cyclotron Laboratory/Facility for Rare
    Isotope Beams, Michigan
State University
East Lansing, Michigan

Awanish Bajpeyi
Division of Humanities and Basic Sciences
Rajiv Gandhi Institute of Petroleum
    Technology
Jais, Uttar Pradesh, India

M. Bhuyam
Instituto Technológico de Aeronáutica
São José dos Campos
São Paulo, Brazil

Dennis Bonatsos
Institute of Nuclear and Particle Physics
National Center for Scientific Research
    "Demokritos"
Aghia Paraskevi, Attiki, Greece

B. V. Carlson
Instituto Technológico de Aeronáutica
São José dos Campos
São Paulo, Brazil

M. Dutra
Departamento de física
Instito Technólogico de Aeronáutica
DCTA, Säo José Dos Campos
São Paulo, Brazil

M. K. Gaidarov
Institute of Nuclear Research and Nuclear
    Energy
Bulgarian Academy of Sciences
Sofia, Bulgaria

J. Gellanki
Department of Physics and Astrophysics
University of Delhi
Delhi, India

Neha Grover
School of Physics and Materials Science
Thapar Institute of Engineering and
    Technology
Patiala, India

E. Moya de Guerra
Departamento de Fisica Atomica, Molecular
    y Nuclear
Facultad de Ciencias Fisicas
Universidad Complutense de Madrid
Madrid, Spain

D. Jain
Department of Physics
Mata Gujri College
Fatehgarh, Punjab, India

D. N. Kadrev
Institute of Nuclear Research and Nuclear
　Energy
Bulgarian Academy of Sciences
Sofia, Bulgaria

Mandeep Kaur
Department of Physics
Guru Granth Sahib World University
Fatehgarh Sahib
Punjab, India

Manpreet Kaur
Institute of Physics
Bhubaneshwar, India

Bharat Kumar
Institute of Physics
Sachivalaya Marg
Bhubaneshwar, India
and
Homi Bhabha National School
Training School Complex
Anushakti Nagar
Mumbai, India

Raj Kumar
School of Physics and Material Science
Thapar Institute of Engineering and
　Technology
Patiyala, Punjab, India

Bui Minh Loc
School of Physics and Astronomy
Tel Aviv University
Tel Aviv, Israel
and
Department of Physics
Ho Chi Minh City University of Education
Ho Chi Minh City, Vietnam

O. Lourenço
Departamento de física
Instito Technólogico de Aeronáutica
DCTA, São José Dos Campos
São Paulo, Brazil

Bhoomika Maheshwari
Department of Physics
Faculty of Science
University of Malaya
Kuala Lumpur, Malaysia

Tuhin Malik
Department of Physics
BITS-Pilani
K. K. Birla Campus
Goa, India

S. K. Mandal
Department of Physics and Astrophysics
University of Delhi
Delhi, India

Andriana Martinou
Institute of Nuclear and Particle Physics
National Center for Scientific Research
　"Demokritos"
Aghia Paraskevi
Attiki, Greece

K. C. Naik
Department of Physics
Siksha 'O' Anusnadhan Deemed to be
　University
Bhubaneshwar, India

R. N. Panda
Department of Physics
Siksha 'O' Anusnadhan Deemed to be
　University
Bhubaneshwar, India

Suresh Kumar Patra
Institute of Physics
Sachivalaya Marg
Bhubaneshwar, India
and
Homi Bhabha National School
Training School Complex
Anushakti Nagar, Mumbai, India

V. Safoora
School of Pure and Applied Physics
Kannur University
Swami Anandatheertha Campus
Payyanur, Kerala, India

K. P. Santhosh
School of Pure and Applied Physics
Kannur University
Swami Anandatheertha Campus
Payyanur, Kerala, India

P. Sarriguren
Instituto de Estructura de la Materia
IEM-CSIC, Serrano 123
Madrid, Spain

Manoj K. Sharma
School of Physics and Materials Science
Thapar Institute of Engineering and
    Technology
Patiala, India

Amritanshu Shukla
Department of Physics
Rajiv Gandhi Institute of Petroleum
    Technology
Jais, Uttar Pradesh, India

BirBikram Singh
Department of Physics
Guru Granth Sahib World University
Fatehgarh Sahib
Punjab, India

Ajeet Singh
Division of Humanities and Basic Sciences
Rajiv Gandhi Institute of Petroleum
    Technology
Jais, Uttar Pradesh, India

P. D. Stevenson
Department of Physics
University of Surrey
Guildford, United Kingdom

Akhilesh Kumar Yadav
Division of Humanities and Basic Sciences
Rajiv Gandhi Institute of Petroleum
    Technology
Jais, Uttar Pradesh, India

Bladimir Zelevinsky
Department of Physics and Astronomy
National Superconducting
Cyclotron Laboratory/Facility for Rare
Isotope Beams, Michigan
State University
East Lansing, Michigan

1

Magic Numbers of Cylindrical Symmetry

Andriana Martinou and Dennis Bonatsos

National Centre for Scientific Research "Demokritos"

CONTENTS

1.1 Introduction

Magic numbers were one of the first discoveries in nuclear structure [1]. It was realized that nuclei formed by particular numbers of protons (Z) and neutrons (N) were extremely stable. This experimental discovery led to the concept of a nuclear closed shell [2], in analogy to atomic closed shells. As a result, the shell model was introduced as the basic microscopic model for the description of atomic nuclei, its magic numbers for either protons or neutrons being 2, 8, 20, 28, 50, 82 [3–7]. The next magic number for neutrons is known to be 126 [8], while the one for protons remains elusive and is currently the subject of intensive research in superheavy elements [9–12].

At the infancy of nuclear structure, it has also been realized that excitation energies of the atomic nucleus can be arranged into bands [13] with increasing angular momentum J. Soon thereafter, it was realized that energy bands near closed shells would increase almost linearly with angular momentum, i.e. $E(J) = AJ$, therefore called vibrational or spherical, while energy bands near the middle of the nuclear valence shells would increase almost following the rigid rotator expression $E(J) = AJ(J+1)$, therefore called rotational or deformed [14]. From the macroscopic point of view, these bands have been described in the framework of the Bohr Hamiltonian [15], soon extended into the Bohr Mottelson model [16,17], in terms of the collective deformation parameters β and γ, of which the former describes the degree of deviation of the nuclear shape from sphericity, while the latter describes its departure from spherical symmetry towards cylindrical symmetry realized through axial prolate shapes (elongated like the ball of American football) or axial oblate (pancake-like) shapes, or even towards purely triaxial shapes [18,19], in which the cylindrical symmetry is also broken. It should be noticed that the Bohr Mottelson model [15–17,20,21] involves five dimensions, the two collective variables β and γ plus the three Euler angles describing the orientation of the atomic nucleus in space. In addition to the angular momentum J, nuclear bands are additionally characterized by the projection of the angular momentum on the z-axis of the coordinate system attached to the nucleus [14,22], which is called the intrinsic set of coordinates. In the case of even nuclei, the symbol K is used for this quantum number. Bands with $K = 0$ contain the angular momenta $J = 0, 2, 4, 6, 8, \ldots$. The lowest one is the band built on the ground state (which is always zero for even-even nuclei, i.e. nuclei with an even number of protons and an even number of neutrons), called the ground state band. For the rest of $K = 0$ bands, the term β bands is used. In contrast, bands with $K = 2$ contain the angular momenta $J = 2, 3, 4, 5, 6, \ldots$, and are called γ bands [14,22].

Symmetries that use the language and techniques of group theory [23–29] have also been used for the description of atomic nuclei [30–36]. It has been realized [37–42] that the Bohr Hamiltonian possesses a U(5) symmetry, having an O(5) subalgebra, which in turn has an SO(3) subalgebra. The SO(3) subalgebra is needed within all symmetries describing atomic nuclei, since the quantum number determining its irreducible representations is the angular momentum J. Therefore, SO(3) has to be present, so that the labeling of energy levels by the angular momentum quantum number J is possible. U(5) is counting the number of excitation quanta present, while the irreducible representations of O(5) are characterized by the seniority quantum number [37–40] corresponding to the number of nucleon pairs coupled to nonzero angular momentum. The O(5) symmetry imposes certain degeneracies in the spectrum. For example, the first excited $J = 2$ state has to be degenerated with the first $J = 4$ state, the first excited $J = 4$ and $J = 3$ states have to be degenerated to the first $J = 6$ state, and so on (see Table I and Figure 3 of Ref. [43] for further details). Since the seniority quantum number appears to be near closed shells [7], O(5) is a symmetry expected to appear there.

For deformed nuclei, the SU(3) symmetry, possessing an SO(3) subalgebra, has been used. In addition to the three components of the angular momentum, forming the SO(3) subalgebra, it also contains the five components of the quadrupole operator. Its relevance to nuclear spectra has been demonstrated by Elliott [44–47] and will be further discussed in Section 1.4.

An algebraic model incorporating U(5) and SU(3) as special subcases is the Interacting Boson Model [30–34,36] possessing an overall U(6) symmetry. In addition to the U(5) [48] and SU(3) [49] subalgebras, it also contains an O(6) [50] subalgebra, which in turn contains O(5) and SO(3) subalgebras. The O(6) symmetry [50] turns out to be applicable to γ-unstable nuclei, i.e. to nuclei which can change easily their axial or triaxial character at the expense of no energy. Since the O(6) symmetry contains the same O(5) and SO(3) subalgebras as the U(5) symmetry, they exhibit the same set of degeneracies in their bands. In the Interacting Boson Model, the nucleus is approximated by a set of bosons corresponding to correlated valence proton pairs and valence neutron pairs, i.e. nucleon pairs outside closed shells. But symmetries appear also in the fermionic description of the nucleus [35], as we shall see below.

It should be noticed that the study of the three-dimensional harmonic oscillator (3D-HO) in spherical coordinates leads to shell model algebras possessing SU(3) subalgebras [23,51,52]. We shall use the spectroscopic notation, in which angular momenta $J = 0, 1, 2, 3, 4, 5, 6, \ldots$ correspond to the symbols s, p, d, f, g, h, i, $\ldots$ respectively. The p shell closes a U(3) algebra; the sd shell closes a U(6) algebra; the pf, sdg, pfh, and sdgi shells close U(10), U(15), U(21), and U(28) algebras, respectively; and all of them possess SU(3) and

SO(3) subalgebras [52]. The s shell in the beginning of this series possesses a U(1) algebra. Since each value of the angular momentum can accommodate 2(2J+1) nucleons of the same type (protons or neutrons), the factor of two due to the two possible orientations of spin, one can easily work out the magic numbers appearing in the 3D-HO, which are 2, 8, 20, 40, 70, 112, 168, ...

One can remark that the shell model magic numbers and the 3D-HO magic numbers are identical up to 20. This means that up to the sd shell, the SU(3) symmetry of the 3D-HO oscillator is present, as first realized by Elliott [44–47]. The agreement between the shell model magic numbers and the 3D-HO magic numbers is destroyed beyond the sd shell by the spin-orbit interaction, which is a relativistic effect which has to be introduced by hand in nonrelativistic shell model Hamiltonians in order to establish agreement with the experimentally observed magic numbers [3–7]. Various approximation methods have been developed over the years, attempting to reestablish the SU(3) symmetry beyond the sd shell. Pseudo-SU(3) [53–57] and quasi-SU(3) [58,59] were the earliest ones, followed recently by proxy-SU(3) [60–62], to be considered in more detail in Section 1.5. The search for approximate SU(3) symmetries beyond the sd shell has been stimulated by the tremendous amount of existing analytical and mathematical work in SU(3), which in this way would become available for use in the nuclear structure framework.

Following a different path, deformed nuclear shapes have been described since a very early stage within an elementary shell model, using a three-dimensional (3D) anisotropic harmonic oscillator with cylindrical symmetry (occurring when two of the three frequencies are equal to each other), to which the spin-orbit interaction has been added. This is the Nilsson model [63,64], which over the years has been extremely useful in classifying and understanding a huge bulk of experimental data [65,66]. The Nilsson Hamiltonian is diagonalized in a set of states which become exact eigenstates only at large deformations, called the asymptotic basis wave functions [67–69], which are characterized by the quantum numbers $K[\mathcal{N}n_z\Lambda]$, where $\mathcal{N}$ is the number of oscillator quanta, n_z is the number of quanta along the cylindrical symmetry axis, Λ and K are the projections of the orbital angular momentum and the total angular momentum, respectively, along the same axis. These quantum numbers remain rather good even at intermediate deformation values [64]. Ben Mottelson [66] has remarked that the asymptotic quantum numbers of the Nilsson model can be seen as a generalization of Elliott's SU(3), applicable to heavy deformed nuclei.

A nontrivial question arising above is the following one: SU(3) is known to be the symmetry of the 3D isotropic harmonic oscillator [23,51]. How is one allowed to use it in the case of the deformed (anisotropic) harmonic oscillator with cylindrical symmetry [70]? This is a formidable mathematical problem, which has already been addressed in the literature in several different ways, based on the fact that the z-component of the angular momentum operators remains intact by the transition from spherical to cylindrical symmetry [71–76]. In particular, using quantum group techniques [77–79], it has been shown how the irreducible representations of the anisotropic HO with rational ratios of frequencies can be constructed, leaving the z-component of the angular momentum operator unchanged [74]. Furthermore, Smirnov and collaborators have shown how the isotropic HO can be transformed into the anisotropic oscillator through the use of a dilatation operator [80]. The resurrection of the SU(3) symmetry of the isotropic HO for large anisotropic deformations has been proved numerically in [81].

We have described above some basic nuclear shapes, namely, spherical, γ-unstable, axially deformed, and the symmetries corresponding to them, viz. U(5), O(6), and SU(3), respectively. There are also nuclei exhibiting behavior intermediate among these symmetries [82]. Considering various series of isotopes, one can see a gradual development from one symmetry into another. If at some point this transition becomes abrupt, we say that we have a shape/phase transition [83–86]. Such an abrupt change is seen in the transition from vibrational (U(5)) to γ-unstable (O(6)) nuclei, called the E(5) critical point symmetry [87–91]. Another case appears in the transition from spherical (U(5)) to axially deformed prolate (SU(3)) nuclei, called the X(5) shape phase transition [92–95]. According to the Ehrenfest classification [84], E(5) is a second-order phase transition, while X(5) is a first-order one, i.e. it is more abrupt. As a consequence, it is much easier to locate experimental examples for X(5) than for E(5) (see the review articles [96–98] for specific experimental manifestations of these critical point symmetries). Extensive literature exists [86] on shape phase transitions, within both the Bohr collective framework and the Interacting Boson Model approach.

A peculiar situation, which might be related to the concept of shape phase transitions [99,100], occurs in several even-even nuclei, in which the ground-state band is accompanied by another $K = 0$ band, which lies close in energy but possesses a radically different structure [101,102]. This effect, which also appears in odd nuclei [103], is called shape coexistence and is attracting recently wide interest. In particular, shape coexistence is known to appear in certain areas of the nuclear chart, while it seems to be absent in others (see Figure 8 of [102]). The arise of the additional coexisting $K = 0$ band is usually attributed to two-particle–two-hole (and, more generally, n-particle–n-hole) excitations across a nuclear closed shell [101,102]. Within the Bohr Hamiltonian approach, the two $K = 0$ coexisting bands are considered to live within two different minima of a sextic potential [104,105]. *However, no explanation has been provided yet for the borders of the regions, within which shape coexistence is observed.*

The main purpose of the present chapter is to show a path towards an explanation of shape coexistence in terms of a mechanism based on the two different sets of magic numbers (shell model, 3D-HO) mentioned above, taking advantage in parallel of the exact SU(3) symmetry present in the 3D-HO shells and the approximate proxy-SU(3) symmetry present in the shell model. In addition to predicting (free of any free parameters) the borders of the regions of coexistence, this method provides also specific predictions for regions of coexistence of a prolate with an oblate band, or of a prolate band with another prolate band of significantly different deformation. The collective model deformation parameters characterizing each band are also predicted by the theory, without involving any free parameters.

1.2   The Origin of QQ Interaction

A nucleus can exhibit collective features only if there is a kind of interaction between distant nucleons. So, there must be a long-range effective potential to unite the whole shell, or even different shells. The most important term of this long-range potential is proportional to the quadrupole interaction.

Suppose there is a kind of two-body central force. The Taylor expansion of it shall be [47]:

$$V^{(2)} = \sum_{i<i'} V\left(\frac{r_{ii'}}{\alpha}\right) = \sum_{i<i'} \left(\xi_0 + \xi_2 \frac{r_{ii'}^2}{\alpha^2} + \xi_4 \frac{r_{ii'}^4}{\alpha^4} + ...\right), \tag{1.1}$$

where α is a range parameter.

The first term is dominant for a very long-range potential, which for an A-body problem is approximately $\xi_0[A(A-1)/2]$. The second term simply verifies that the average potential of all nucleons can be represented by an isotropic harmonic oscillator potential. The third term is [47]:

$$\sum_{i<i'} r_{ii'}^4 = \sum_{i<i'} \left[r_i^4 + r_{i'}^4 + \frac{8}{3} r_i^2 r_{i'}^2 - 4(r_i^2 + r_{i'}^2) r_i r_{i'} \cos\theta_{ii'} + \frac{4}{3} r_i^2 r_{i'}^2 P_2(\cos\theta_{ii'}) \right]. \tag{1.2}$$

The first three terms in the above expansion do not contribute any splitting in the energy degeneracy problem. Therefore, the only significant term is $\frac{4}{3} r_i^2 r_{i'}^2 P_2(\cos\theta_{ii'})$, where P_2 is the Legendre polynomial. This important term can be written as [47]:

$$r_i^2 r_{i'}^2 P_2(\cos\theta_{ii'}) = (r_i^2 Y_{20}(\theta_i \phi_i)) \cdot (r_{i'}^2 Y_{20}(\theta_{i'} \phi_{i'})), \tag{1.3}$$

where Y_{20} is the spherical harmonic with $l = 2, m = 0$.

But the spherical harmonics are related to the collective quadrupole moment. The collective operators have the following relation with the space variables [47]:

$$r_i^2 Y_{2m}(\theta_i \phi_i) = \frac{b^2}{4} \sqrt{\frac{5}{\pi}} q_m^{(c)}, \tag{1.4}$$

where b is the oscillator length parameter with $b = \sqrt{\frac{\hbar}{m\omega}}$. The difference between q and Q is that the first stands for one nucleon, while the capital for the whole nucleus.

So, it happens that [47]:

$$r_i^2 r_{i'}^2 P_2(\cos\theta_{ii'}) = \left(\frac{b^2}{4} \sqrt{\frac{5}{\pi}} \right)^2 q_0^{(c)} q_0^{(c)}. \tag{1.5}$$

This result proves that any central potential contains a long-range effective QQ interaction.

1.3 The Nilsson Model

The Nilsson model is a microscopic nuclear model, in the spirit of shell model, applicable in deformed nuclei. The following will be a review of Eqs. (2)–(7) of [63]. The Hamiltonian for the nuclear coordinate system is:

$$H = H_0 + u_{ls} \hbar\omega_0(\delta) \Lambda \cdot \Sigma + u_{ll} \hbar\omega_0(\delta) \Lambda^2, \tag{1.6}$$

$$H_0 = -\frac{\hbar^2}{2m} \nabla'^2 + \frac{m}{2} (\omega_x^2 x'^2 + \omega_y^2 y'^2 + \omega_z^2 z'^2). \tag{1.7}$$

The frequencies are set to be:

$$\omega_x^2 = \omega_y^2 = \omega_0^2(\delta) \left(1 + \frac{2}{3}\delta \right), \tag{1.8}$$

$$\omega_z^2 = \omega_0^2(\delta) \left(1 - \frac{4}{3}\delta \right). \tag{1.9}$$

The volume conservation restriction leads to the relation:

$$\omega_0(\delta) = \tilde{\omega}_0 \left(1 - \frac{4}{3}\delta^2 - \frac{16}{27}\delta^3 \right)^{-1/6}. \tag{1.10}$$

The deformation parameter δ is connected with the deformation β of the Bohr and Mottelson model as:

$$\delta \approx \frac{3}{2} \sqrt{\frac{5}{4\pi}} \beta \approx 0.95\beta. \tag{1.11}$$

The dimensionless coordinate system can be created by setting:

$$x = \sqrt{\frac{m\tilde{\omega}_0}{\hbar}}x', \qquad y = \sqrt{\frac{m\tilde{\omega}_0}{\hbar}}y', \qquad z = \sqrt{\frac{m\tilde{\omega}_0}{\hbar}}z'. \tag{1.12}$$

Then, the spatial part of the Hamiltonian becomes:

$$H_0 = \tilde{H}_0 + H_\delta, \tag{1.13}$$

$$\tilde{H}_0 = \hbar\omega_0(\delta)\frac{1}{2}(-\nabla^2 + r^2), \tag{1.14}$$

$$H_\delta = -\delta\hbar\omega_0(\delta)\frac{4}{3}\sqrt{\frac{\pi}{5}}r^2 Y_{20}. \tag{1.15}$$

Therefore, the spatial part of the Hamiltonian can be interpreted as an isotropic 3D harmonic oscillator with frequency $\omega_0(\delta)$ plus a distortion to the isotropic field H_δ.

This distortion contains the term $r^2 Y_{20}$, which (recall Eq. (1.4)) is the $q_0^{(c)}$ component of the quadrupole moment of the nucleon. So:

$$H_\delta = -\delta\hbar\omega_0(\delta)\frac{b^2}{3}q_0^{(c)}. \tag{1.16}$$

This term is the quadrupole interaction of the nucleon with the mean quadrupole field of all the others. It is noticeable that in the Nilsson model, only the q_0 component appears, while the other four $(q_{\pm 2}, q_{\pm 1})$ are neglected.

The non-vanishing matrix elements of H_δ are among the Nilsson states with [63]:

$$\Lambda = \Lambda', \qquad \Sigma = \Sigma', \qquad l = l', l' \pm 2, \qquad \mathcal{N} = \mathcal{N}', \mathcal{N}' \pm 2, \tag{1.17}$$

where Λ, Σ are the projections of the angular momentum and spin, respectively; l is the angular momentum; and N is the total number of quanta. Obviously, the $q^{(c)}$ interplays with orbitals with the same parity, since they have $\mathcal{N}, \mathcal{N} \pm 2$.

The full Nilsson model Hamiltonian can therefore be written as:

$$H = \tilde{H}_0 + u_{ls}\hbar\omega_0(\delta)\Lambda \cdot \Sigma + u_{ll}\hbar\omega_0(\delta)\Lambda^2 - \delta\hbar\omega_0(\delta)\frac{4}{3}\sqrt{\frac{\pi}{5}}r^2 Y_{20}. \tag{1.18}$$

1.3.1   The Nilsson Basis

The asymptotic wave functions of Nilsson orbitals are labeled usually as $K[\mathcal{N}n_z\Lambda]$ [64]. These wave functions occur when the deformation ϵ is large enough. An elegant method for the derivation of the asymptotic wave functions is presented in [64]. In the following, emphasis will be placed on an equivalent notation $|n_z rs\Sigma\rangle$ of the asymptotic wave functions, which highlights the number of quanta in the $x - y$ plane [64].

The Nilsson problem is solved in the cylindrical coordinate system. The quanta on the $x - y$ plane are being created by the $R^\dagger, R, S^\dagger, S$ operators [106] which are exactly the same as the $a_+^\dagger, a_+, a_-^\dagger, a_-$ of Section 1.4:

$$R^\dagger = a_+^\dagger, \qquad R = a_+, \qquad S^\dagger = a_-^\dagger, \qquad S = a_-. \tag{1.19}$$

It is valid that:

$$n_\perp = r + s = n_x + n_y, \tag{1.20}$$

where $R^\dagger R |r\rangle = r|r\rangle$ and $S^\dagger S |s\rangle = s|s\rangle$. The actions of the creation and annihilation operators on the relevant states are [107]:

$$R\,|r\rangle = \sqrt{r}\,|r-1\rangle\,, \quad R^\dagger\,|r\rangle = \sqrt{r+1}\,|r+1\rangle\,, \quad S\,|s\rangle = \sqrt{s}\,|s-1\rangle\,,$$
$$S^\dagger\,|s\rangle = \sqrt{s+1}\,|s+1\rangle\,. \tag{1.21}$$

In addition, the absolute value of the projection of orbital angular momentum in the Nilsson model is [64]:

$$\Lambda = r - s. \tag{1.22}$$

Lastly, for the projection of total angular momentum K and the projection of spin Σ, the equation [64]:

$$K = \Lambda + \Sigma, \tag{1.23}$$

applies.

With these tools:

$$r + s = \mathcal{N} - n_z, \tag{1.24}$$
$$r - s = \Lambda, \tag{1.25}$$
$$\Sigma = K - \Lambda, \tag{1.26}$$

one can transform the Nilsson orbitals between the two notations:

$$K[\mathcal{N}n_z\Lambda] \rightarrow |n_z r s \Sigma\rangle\,. \tag{1.27}$$

Some transformations are given in Table 1.1.

TABLE 1.1
Nilsson Orbitals in Two Different Notations

Shell model	$K[\mathcal{N}n_z\Lambda]$	$|n_z r s \Sigma\rangle$
$1g_{9/2}$	9/2[404]	$|040+\rangle$
	7/2[413]	$|130+\rangle$
	5/2[422]	$|220+\rangle$
	3/2[431]	$|310+\rangle$
	1/2[440]	$|400+\rangle$
$2p_{1/2}$	1/2[301]	$|021-\rangle$
$1f_{5/2}$	5/2[303]	$|030-\rangle$
	3/2[301]	$|021+\rangle$
	1/2[310]	$|111+\rangle$
$2p_{3/2}$	3/2[312]	$|120-\rangle$
	1/2[321]	$|210-\rangle$
$1f_{7/2}$	7/2[303]	$|030+\rangle$
	5/2[312]	$|120+\rangle$
	3/2[321]	$|210+\rangle$
	1/2[330]	$|300+\rangle$
$1d_{3/2}$	3/2[202]	$|020-\rangle$
	1/2[200]	$|011+\rangle$
$2s_{1/2}$	1/2[211]	$|110-\rangle$
$1d_{5/2}$	5/2[202]	$|020+\rangle$
	3/2[211]	$|110+\rangle$
	1/2[220]	$|200+\rangle$
$1p_{1/2}$	1/2[101]	$|010-\rangle$
$1p_{3/2}$	3/2[101]	$|010+\rangle$
	1/2[110]	$|100+\rangle$
$1s_{1/2}$	1/2[000]	$|000+\rangle$

1.4   The Elliott SU(3)

The majority of physicists nowadays are familiar with the nuclear shell model, but few have been engaged in the Elliott SU(3) symmetry, which was first introduced in [44–46]. *Briefly this symmetry is the fulfillment of the shell model. While the shell model treats single-particle states, the Elliott SU(3) teaches us, how to couple the valence nucleons towards the derivation of the collective nuclear states.* Moreover, the Elliott SU(3) has a remarkable and unique property: it is a fermionic, collective, nuclear model. As a consequence, the collective states emerge in accordance with the Pauli principle.

The nucleus consists of Z protons and N neutrons. The mass number is $A = Z + N$. All these nucleons create a mean field potential, which participates in the 3D isotropic, harmonic oscillator Hamiltonian:

$$H = \sum_{i=1}^{A} H_i = \sum_{i=1}^{A} \left(\frac{p_i^2}{2m} + \frac{1}{2} m \omega_i^2 r_i^2 \right). \tag{1.28}$$

The single-particle eigenstates of:

$$H_i = \frac{p_i^2}{2m} + \frac{1}{2} m \omega_i^2 r_i^2, \tag{1.29}$$

are labeled by the total number of oscillator quanta $\mathcal{N}_i$ in the three spacial dimensions.

The single-particle, isotropic, harmonic oscillator Hamiltonian can be transformed in the second quantization with the use of the quanta annihilation and creation operators in each Cartesian axis $a_j, a_j^\dagger$, with $j = x, y, z$. The Hamiltonian of the ith particle in the x direction is $H_{x,i} = \frac{p_{x,i}^2}{2m} + \frac{1}{2} m \omega_i^2 x_i^2$. The eigenstates are $|n_x\rangle$. The action of the $a_x, a_x^\dagger$ on the eigenstates is [107]:

$$a_x |n_x\rangle = \sqrt{n_x} |n_x - 1\rangle, \qquad a_x^\dagger |n_x\rangle = \sqrt{n_x + 1} |n_x + 1\rangle. \tag{1.30}$$

By changing the script x to y or z, one gets the relevant actions in the other two Cartesian directions. In addition, the number operators are:

$$a_j^\dagger a_j |n_j\rangle = n_j |n_j\rangle. \tag{1.31}$$

The above operators being boson (quanta) operators follow the well-known commutation relations:

$$[a_j^\dagger, a_{j'}^\dagger] = [a_j, a_{j'}] = 0, \qquad [a_j, a_{j'}^\dagger] = \delta_{jj'}, \quad \text{for } , j, j' = x, y, z. \tag{1.32}$$

Thus, the single-particle 3D Hamiltonian can be transformed as [107]:

$$H_i = \frac{p_i^2}{2m} + \frac{1}{2} m \omega^2 r_i^2 = \left(a_x a_x^\dagger + a_y a_y^\dagger + a_z a_z^\dagger + \frac{3}{2} \right) \hbar \omega_i, \tag{1.33}$$

with eigenvalues:

$$E_i = \left(n_x + n_y + n_z + \frac{3}{2} \right) \hbar \omega_i = \left(\mathcal{N}_i + \frac{3}{2} \right) \hbar \omega_i. \tag{1.34}$$

Now the problem can be turned in cylindrical coordinates, which are more suitable for axially deformed nuclei. In order to distinguish the $x - y$ plane from the z-axis, the left and right quanta operators are defined as [108]:

$$a_\pm = \frac{a_x \mp i a_y}{\sqrt{2}}, \tag{1.35}$$

$$a_\pm^\dagger = \frac{a_x^\dagger \pm i a_y^\dagger}{\sqrt{2}}. \tag{1.36}$$

The operators $a_+, a_+^\dagger$ are called right-hand operators, because they seem to destroy/create a circular quantum that rotates in right-hand side in the $x - y$ plane [108]. For the same reason, the other two operators are called left-hand operators. The operators in the z-axis are renamed:

$$a_0 = a_z, \qquad a_0^\dagger = a_z^\dagger. \tag{1.37}$$

Now the Hamiltonian is [109]:

$$H_i = \hbar\omega_i \left(a_+^\dagger a_+ + a_-^\dagger a_- + a_0^\dagger a_0 + \frac{3}{2} \right). \tag{1.38}$$

With these ingredients, one can define eight operators which are the generators of SU(3) [109]:

$$l_0 = a_+^\dagger a_+ - a_-^\dagger a_-, \tag{1.39}$$

$$l_\pm = \mp(a_0^\dagger a_\mp - a_\mp^\dagger a_0), \tag{1.40}$$

$$q_{\pm 2} = -\sqrt{6} a_\pm^\dagger a_\mp, \tag{1.41}$$

$$q_{\pm 1} = \mp\sqrt{3}(a_0^\dagger a_\mp + a_\pm^\dagger a_0), \tag{1.42}$$

$$q_0 = 2a_0^\dagger a_0 - a_+^\dagger a_+ - a_-^\dagger a_-. \tag{1.43}$$

The first three operators $(l_0, l_\pm)$ form the SU(2) algebra of angular momentum, while the last five are the five components of the single-particle quadrupole operator.

The eight of them close the commutation relations of the SU(3) algebra [109].

$$[l_0, l_\pm] = \pm l_\pm, \tag{1.44}$$

$$[l_+, l_-] = 2l_0, \tag{1.45}$$

$$[l_0, q_m] = m q_m, \tag{1.46}$$

$$[l_\pm, q_m] = \sqrt{6 - m(m \pm 1)} q_{m\pm 1}, \tag{1.47}$$

$$[q_0, q_{\pm 1}] = \pm 3\sqrt{3} l_\pm, \tag{1.48}$$

$$[q_1, q_{-1}] = -3l_0, \tag{1.49}$$

$$[q_2, q_{-2}] = 6l_0, \tag{1.50}$$

$$[q_{\pm 2}, q_{\mp 1}] = \pm 3\sqrt{2} l_\pm, \tag{1.51}$$

$$[q_0, q_{\pm 2}] = [q_{\pm 1}, q_{\pm 2}] = 0. \tag{1.52}$$

The commutator of two operators from the set (1.39)–(1.43) results to one operator from the set (1.39)–(1.43) again. Thus, the eight operators of (1.39)–(1.43) close the SU(3) algebra.

In simple words, the Elliott SU(3) consists of the three angular momentum operators and of the five components of the quadrupole operator. The eight generators of SU(3) are certain linear combinations of the boson operators $a_j^\dagger, a_j$. The 3D isotropic harmonic oscillator Hamiltonian is invariant under U(3) and SU(3) transformations. The $\frac{(\mathcal{N}+1)(\mathcal{N}+2)}{2}$ orbitals of a harmonic oscillator shell with $\mathcal{N}$ total number of quanta have SU(3) symmetry.

1.4.1 Derivation of the Highest Weight irrep

In the Elliott SU(3) model, the calculation of every observable is based on the Elliott quantum numbers (λ, μ) [44]. The highest weight (h.w.) irrep describes the ground-state properties of the nucleus. The method for the calculation of the h.w. (λ, μ) has been presented in steps at [110,111].

For the h.w. irreps, a simple method will be presented here.

1. The Cartesian states $|n_z, n_x, n_y\rangle$ are vectors of a $U\left(\frac{(\mathcal{N}+1)(\mathcal{N}+2)}{2}\right)$ algebra. The degenerate 3D isotropic harmonic oscillator orbitals are ordered as:

$$|n_z, n_x, n_y\rangle = |n,0,0\rangle, \quad |n-1,1,0\rangle, \quad |n-1,0,1\rangle, \quad ..., \quad |0,0,n\rangle. \quad (1.53)$$

   For total quanta $\mathcal{N} = 2$, there are $\frac{(\mathcal{N}+1)(\mathcal{N}+2)}{2} = 6$ orbitals. One has to order these six orbitals as follows:

$$|1\rangle = |2,0,0\rangle, \quad |2\rangle = |1,1,0\rangle, \quad |3\rangle = |1,0,1\rangle, \quad |4\rangle = |0,2,0\rangle, \quad |5\rangle = |0,1,1\rangle,$$
$$|6\rangle = |0,0,2\rangle. \quad (1.54)$$

2. Then, at most two protons or neutrons are placed in each orbital $|n_z, n_x, n_y\rangle$ with the order of Eq. (1.54). For instance, $^{31}_{12}\text{Mg}_{19}$ has $19 - 8 = 11$ neutrons in the 8–20 shell. Two neutrons are placed in each of the $|1\rangle, ..., |5\rangle$ orbitals, while the last neutron occupies the $|6\rangle$.

3. Afterwards, the summations of the quanta in each Cartesian axis are calculated:

$$\sum_{i=1}^{Z_{val}, N_{val}} n_{zi}, \qquad \sum_{i=1}^{Z_{val}, N_{val}} n_{xi}, \qquad \sum_{i=1}^{Z_{val}, N_{val}} n_{yi}. \quad (1.55)$$

   For the 11 neutrons in $U(6)$, the above expressions give $8, 8, 6$, respectively.

4. The results of the summations are placed in decreasing order, and these are the quantum numbers $[f_1, f_2, f_3]$ of the U(3) algebra. For our example, $[f_1, f_2, f_3] = [8, 8, 6]$.

5. The λ and μ are:

$$\lambda = f_1 - f_2, \qquad \mu = f_2 - f_3, \quad (1.56)$$

which leads to $(\lambda, \mu) = (0, 2)$ for 11 particles in U(6).

1.4.2 The Collective Operators

The single-particle operators can be extended to the many-particle ones, by summing over all valence particles [109]:

$$L = \sum_{i}^{A_{val}} l_i, \quad (1.57)$$

$$Q_m = \sum_{i}^{A_{val}} q_{im}. \quad (1.58)$$

The former one is the angular momentum, while the latter is the m component of the algebraic quadrupole moment for the whole shell. These many-particle operators satisfy the same commutation relations with their relevant single ones, and therefore, they close an SU(3) algebra.

The simplest SU(3) Hamiltonian of the many nucleon problem, which is suitable for a collective 0^+ state, is [112]:

$$H = \sum_i^A H_i - \frac{1}{2}\chi Q \cdot Q = H_0 - \frac{1}{2}\chi Q \cdot Q, \qquad (1.59)$$

where χ is the strength of the algebraic quadrupole-quadrupole (QQ) interaction and if i and i' are two distinct valence nucleons:

$$QQ = \sum_{i,i'}^{A_{val}} \sum_{m=-2}^{2} (-1)^m q_{mi}q_{-mi'} = \sum_{m=-2}^{2} (-1)^m Q_m Q_{-m}. \qquad (1.60)$$

The SU(3) algebra has a second- and a third-order Casimir operator [113]:

$$C_2 = \lambda^2 + \mu^2 + \lambda\mu + 3(\lambda + \mu), \qquad (1.61)$$

$$C_3 = \frac{2}{9}(\lambda^3 - \mu^3) + \frac{1}{3}\lambda\mu(\lambda - \mu) + (\lambda + 2)(2\lambda + \mu). \qquad (1.62)$$

The Casimir operators commute with all the generators of the algebra. The QQ interaction is measured via the C_2 as [113]:

$$QQ = 4C_2 - 3L^2. \qquad (1.63)$$

The 3D isotropic harmonic oscillator Hamiltonian H_0 has dimensionless eigenvalues (i.e. $\hbar\omega_i = 1$):

$$N_0 = \sum_{i=1}^{A} \left(\mathcal{N}_i + \frac{3}{2}\right) \qquad (1.64)$$

The strength χ is measured by the equation [114]:

$$\frac{\chi}{2} = \frac{\hbar\omega}{4N_0}, \qquad (1.65)$$

where $\hbar\omega$ is measured in MeV.

The deformation variables β, γ follow the formulas [115]:

$$\beta^2 = \frac{4\pi}{5(A\bar{r}^2)^2}(C_2 + 3), \qquad (1.66)$$

$$\gamma = \tan^{-1}\frac{\sqrt{3}(\mu + 1)}{2\lambda + \mu + 3}, \qquad (1.67)$$

where A is the mass number and the dimensionless mean square radius is $\bar{r}^2 = 0.87^2 A^{1/3}$ [116]. The β^2 may be multiplied by the scaling factor $(A/S)^2$, where S is the size of the proton and neutron valence shell [61]:

$$\beta^2 = \frac{4\pi}{5(S\bar{r}^2)^2}(C_2 + 3). \qquad (1.68)$$

1.4.3 The SU(3) → SU(2) × U(1) Decomposition

Most nuclei have quadrupole deformation. Such nuclei look like an ellipsoid with cylindrical symmetry. The cylinder is divided into the $x - y$ plane and the symmetry z-axis. If the ellipsoid has elongation across one axis (z), the shape is prolate, while if the elongation is across two axes (x, y), it is oblate.

The algebra SU(3) is decomposed into two subalgebras SU(2) and U(1) [45]. The former represents rotations in the $x - y$ plane, while the latter involves the z-axis. The U(1) is characterized by the operator q_0, which has eigenvalues $q_0 = 2n_z - n_\perp$, with $n = n_z + n_\perp$.

The SU(2) algebra has the following generators [45]:

$$u_\pm = \mp \left(\frac{1}{2\sqrt{3}} \right) q_{\pm 2}, \qquad u_0 = \frac{1}{2} l_0, \tag{1.69}$$

which satisfy the commutation relations:

$$[u_\pm, u_0] = \mp u_\pm, \qquad [u_+, u_-] = -u_0. \tag{1.70}$$

The SU(2) multiplets are characterized by a quantum number of j-type and one of m_j-type [113]. In the present algebraic structure, the m_j-type quantum number is the eigenvalue of u_0 and the j-type, let it be named Λ_E (Λ of Elliott), is

$$\Lambda_E = \frac{1}{2}(n_x + n_y) = \frac{1}{2} n_\perp. \tag{1.71}$$

Since:

$$u_0 = \frac{l_0}{2}, \tag{1.72}$$

and the possible values of u_0 are $-\Lambda_E, -\Lambda_E + 1, \ldots, \Lambda_E$ obeying the angular momentum algebra, obviously:

$$l_0 = -2\Lambda_E, -2\Lambda_E + 2, \ldots, 2\Lambda_E. \tag{1.73}$$

The basis of the $x - y$ plane is labeled as $|\Lambda_E, u_0\rangle$. These vectors with the use of (1.71) and (1.72) are more conveniently named $|n_\perp, l_0\rangle$.

As a consequence, the SU(2) subalgebra of the Elliott SU(3) depends solely on the quanta on the $x - y$ plane. The physical quantity of the SU(2) is the projection of the orbital angular momentum (l_0), while the $q_{\pm 2}$ are useful as ladder operators. The number of quanta in the z-axis is isolated in the U(1) subalgebra (see Figure 1.1). This decomposition of the Elliott SU(3) suits perfectly in nuclei with quadrupole deformation.

1.5   Proxy-SU(3) Symmetry

A nucleus consists of Z protons and N neutrons. The nucleons interact mainly via the strong nuclear force. Since the exact formula for this force is not yet known, one can approach the result of all the inter-nucleon interactions as in Section 1.2. The main result of all the inter-nucleon interactions is the creation of a mean field potential, which is described by a 3D isotropic harmonic oscillator potential. This mean field creates nuclear shells, which consist of orbitals with a common number of oscillator quanta $\mathcal{N}$. The magic numbers of these shells are the 3D harmonic oscillator (HO) magic numbers: 2, 8, 20, 40, 70, 112, ...

Unfortunately the mean field is not the only interaction. There exists a single-particle spin-orbit (SO) term ls, which destroys the harmonic oscillator shells. The ls interaction forces the orbitals to move in energy. As an example, the 20–40 harmonic oscillator shell becomes 28–50. The mechanism of this procedure is the following:

(a)   The harmonic oscillator shell 20–40 consists of the shell model orbitals $1f_{7/2}, 2p_{3/2}, 1f_{5/2}, 2p_{1/2}$. The orbital $1g_{9/2}$ is part of the harmonic oscillator shell 40–70. The orbitals are listed in Table 1.2.

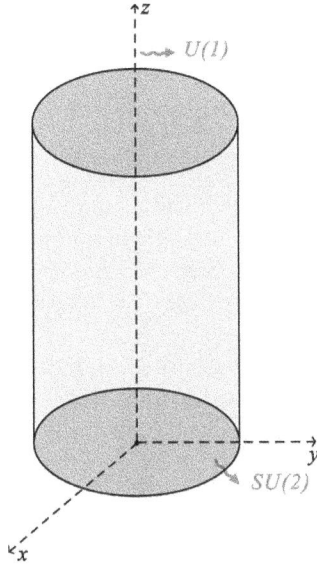

FIGURE 1.1
Cylindrical symmetry. The $SU(3)$ algebra is decomposed into an $SU(2)$, which describes the $x - y$ plane, and a $U(1)$, which refers to the z-axis. This decomposition suits perfectly to a prolate or oblate nuclear shape.

TABLE 1.2
Nilsson Orbitals for the Nucleons between 20 and 50

Shell model	$K[\mathcal{N}n_z\Lambda]$	$|n_z rs\Sigma\rangle$
$1g_{9/2}$	$9/2[404]$	$|040+\rangle$
	$7/2[413]$	$|130+\rangle$
	$5/2[422]$	$|220+\rangle$
	$3/2[431]$	$|310+\rangle$
	$1/2[440]$	$|400+\rangle$
$2p_{1/2}$	$1/2[301]$	$|021-\rangle$
$1f_{5/2}$	$5/2[303]$	$|030-\rangle$
	$3/2[301]$	$|021+\rangle$
	$1/2[310]$	$|111+\rangle$
$2p_{3/2}$	$3/2[312]$	$|120-\rangle$
	$1/2[321]$	$|210-\rangle$
$1f_{7/2}$	$7/2[303]$	$|030+\rangle$
	$5/2[312]$	$|120+\rangle$
	$3/2[321]$	$|210+\rangle$
	$1/2[330]$	$|300+\rangle$

Due to the ls interaction, the $1f_{7/2}$ orbital is excluded from the shell under discussion, while the $1g_{9/2}$ orbital is included. Thus, the harmonic oscillator shell 20–40 becomes the nuclear shell 28–50. The 28-50 shell consists of some orbitals with $\mathcal{N} = 3$ and some orbitals with $\mathcal{N} = 4$ quanta [14]. Therefore, the Elliott SU(3) symmetry is not valid in the 28–50 shell.

(b) The *ls* interaction lowers the energy of the $1f_{7/2}$ orbital so much, that this orbital is no longer included in the shell under discussion.

(c) Instead, the $1g_{9/2}$ from the 40–70 HO shell is lowered and becomes part of the shell under discussion.

To resume, the harmonic oscillator shell 20–40 loses the $1f_{7/2}$ orbital and gains instead the $1g_{9/2}$ orbital. This new shell lies between magic numbers 28 and 50.

Thus after, the ls interaction is applied, each harmonic oscillator shell with $\mathcal{N}$ quanta loses the orbital with total angular momentum $j = \mathcal{N} + \frac{1}{2}$ and gains the orbital with $\mathcal{N} + 1$ quanta and $j' = \mathcal{N} + 1 + \frac{1}{2}$.

Proxy-SU(3) was introduced in [60] in order to describe a nuclear shell which consists of some orbitals with $\mathcal{N}$ quanta and some orbitals with $\mathcal{N} + 1$ quanta. Nuclear shells above 28 protons or neutrons are of this type. Since such shells do not match with an exact SU(3) symmetry, an approximation is necessary in order to make the algebraic tools applicable. The proxy-SU(3) replacement of orbitals is justified in Table 1.3 and described in Table 1.4 for the 28–50 shell.

The proxy-SU(3) approximation affects only the orbitals with $\mathcal{N} + 1$ quanta (intruder orbitals), by erasing one quantum from the z-axis of each intruder orbital. The normal parity orbitals remain intact.

1.5.1 The Exact Symmetry Behind Proxy-SU(3)

Proxy-SU(3) has been very successful in predicting for various isotopic chains, without any parameters, the β and γ deformations of each nucleus, as well as the prolate over oblate dominance along with the prolate-oblate transition [61]. This success relies on the h.w. irrep and on the fact that there is an exact symmetry behind proxy-SU(3).

TABLE 1.3

The $1f_{7/2}$ Orbital has been Excluded from the Harmonic Oscillator Shell 20–40 due to the *ls* Interaction, while the $1g_{9/2}$ Orbital has Intruded in It

Shell Model	$|n_z r s \Sigma\rangle$	Shell Model	$|n_z r s \Sigma\rangle$
$1g_{9/2}$	$|040+\rangle$		X
	$|130+\rangle$	$1f_{7/2}$	$|030+\rangle$
	$|220+\rangle$		$|120+\rangle$
	$|310+\rangle$		$|210+\rangle$
	$|400+\rangle$		$|300+\rangle$

The $1f_{7/2}$ and $1g_{9/2}$ orbitals consist of Nilsson orbitals with $\mathcal{N} = 3$ and $\mathcal{N} = 4$, respectively, as seen in Table 1.2. These orbitals are listed here using the $|n_z r s \Sigma\rangle$ notation. All of them have a common projection of spin Σ. Each of the orbitals of $1f_{7/2}$ has identical distribution of quanta in the $x-y$ plane ($|rs\rangle$) with an orbital of $1g_{9/2}$. Such orbitals appear in the same line in the table and are depicted in the same color. The orbitals with the same color differ by one quantum in the z-axis. Therefore, the operators $q_{\pm 2}, l_0$ of the SU(2) subalgebra of SU(3) (Section 1.4.3) give exactly the same eigenvalues when acting on the colored orbitals. As a consequence, the SU(2) operators cannot discriminate among the two sets of the colored orbitals [117].

TABLE 1.4
The Replacement of the Orbitals in Proxy-SU(3)

	Before		After	
Shell model	$K[\mathcal{N}n_z\Lambda]$	$\lvert n_z rs\Sigma\rangle$	$K[\mathcal{N}n_z\Lambda]$	$\lvert n_z rs\Sigma\rangle$
$1g_{9/2}$	$9/2[404]$	$\lvert 040+\rangle$	X	X
	$7/2[413]$	$\lvert 130+\rangle$	$7/2[303]$	$\lvert 030+\rangle$
	$5/2[422]$	$\lvert 220+\rangle$	$5/2[312]$	$\lvert 120+\rangle$
	$3/2[431]$	$\lvert 310+\rangle$	$3/2[321]$	$\lvert 210+\rangle$
	$1/2[440]$	$\lvert 400+\rangle$	$1/2[330]$	$\lvert 300+\rangle$
$2p_{1/2}$	$1/2[301]$	$\lvert 021-\rangle$		
$1f_{5/2}$	$5/2[303]$	$\lvert 030-\rangle$		
	$3/2[301]$	$\lvert 021-\rangle$		
	$1/2[310]$	$\lvert 111+\rangle$		
$2p_{3/2}$	$3/2[312]$	$\lvert 120-\rangle$		
	$1/2[321]$	$\lvert 210-\rangle$		

One quantum from the z-axis is erased from each intruder orbital, while the n_x, n_y (or, equivalently, the r, s) of the intruders and the normal parity orbitals remain intact. The intruder with the highest projection of total angular momentum (K in Nilsson notation) is excluded. The new shell (after the approximation) is a complete harmonic oscillator shell; thus, it has an SU(3) symmetry.

The total wave function of a single-particle state is the product of a spatial and a spinor part. The spatial wave function is characterized by the number of quanta in the z-axis ($\lvert n_z\rangle$) and in the $x-y$ plane ($\lvert rs\rangle$). In a full harmonic oscillator shell with $\mathcal{N}$ quanta, the n_z gets values:

$$n_z = \mathcal{N}, \mathcal{N}-1, \mathcal{N}-2, ..., 0. \tag{1.74}$$

The quanta in the $x-y$ plane are, respectively:

$$r+s = 0, 1, 2, ..., \mathcal{N}, \tag{1.75}$$

with the restriction [64]:

$$r \geq s. \tag{1.76}$$

Thus, the following combinations are valid:

$$(n_z = \mathcal{N}, \quad r = s = 0),$$
$$(n_z = \mathcal{N}-1, \quad r = 1, \quad s = 0),$$
$$(n_z = \mathcal{N}-2, \quad r = 2, \quad s = 0),$$
$$(n_z = \mathcal{N}-2, \quad r = 1, \quad s = 1),$$
$$...$$

As an example, the orbitals of the "Before" column of Table 1.4, excluding the $9/2[404]$ orbital, contain all the r, s values of a full harmonic oscillator shell with $\mathcal{N} = 3$ quanta.

Consequently, a nuclear shell, such as 28–50, 50–82, and 82–126, after excluding the Nilsson orbital with the maximum value of K, has the SU(2) subalgebra of the Elliott SU(3). The proxy-SU(3) approximation is not affecting the number of quanta in the $x-y$ plane. Thus, the orbitals of the proxy-SU(3) shell preserve the exact SU(2) symmetry.

1.6   Magic Numbers Below 28

In Section 1.5, it has been described how the spin-orbit interaction creates the nuclear shell 28–50. By the same procedure, the shells 50–82 and 82–126 are created. But a significant spin-orbit term in the Hamiltonian can create magic numbers below 28 particles, as described in Table 1.5, which we are going to call, following Ref. [8], SO-like magic numbers. The SO-like magic numbers, that will arise, are:

$$\textit{SO-like magic numbers below 28}: 2, 6, 14. \tag{1.77}$$

Actually, the SO-like magic number 14 is already well established experimentally [8].

Clearly, the above SO-like magic numbers have not emerged in this work through the study of energy gaps, but through symmetry considerations. The sets of orbitals among the mentioned SO-like magic numbers, after excluding the Nilsson orbital with maximum K, preserve the SU(2) symmetry of Elliott.

As seen in Table 1.5, the SO-like shells 6–14 and 14–28 have a proxy-SU(3) symmetry. These shells can be treated as in [60,61], to calculate the deformation variables (β, γ). The shell 14–28 is useful for the neutrons of the Mg isotopes, while the 6–14 can be applied in the neutrons of the Be isotopes, to predict the parity inversion at ^{11}Be.

1.7   Magic Numbers → Shape Coexistence → Inversion of States

The Mg isotopic chain is the best example that demonstrates how the SO-like magic numbers below 28 may lead to right predictions about deformation β, shape coexistence and inversion

TABLE 1.5

A Significant Spin-Orbit Interaction may Create the SO-Like Shells 2–6 and 6–14

Shell model	Before $K[\mathcal{N}n_z\Lambda]$	After $K[\mathcal{N}n_z\Lambda]$	SO-like magic number
$1f_{7/2}$	7/2[303]	X	28
	5/2[312]	5/2[202]	
	3/2[321]	3/2[211]	
	1/2[330]	1/2[220]	
$1d_{3/2}$	3/2[202]	3/2[202]	
	1/2[200]	1/2[200]	
$2s_{1/2}$	1/2[211]	1/2[211]	
$1d_{5/2}$	5/2[202]	X	14
	3/2[211]	3/2[101]	
	1/2[220]	1/2[110]	
$1p_{1/2}$	1/2[101]	1/2[101]	
$1p_{3/2}$	3/2[101]	X	6
	1/2[110]	1/2[000]	
$1s_{1/2}$	1/2[000]	1/2[000]	2

Proxy-SU(3) can be applied in these shells. The column "Before" refers to the original nuclear shell and the column "After" to the proxy-SU(3) shell [117].

of states. A bulk of experimental work [118–123] has established that the Mg isotopic chain exhibits both shape coexistence and inversion of states. A shell model mechanism that uses particle-hole excitations is the state-of-the-art theoretical approach for these phenomena in Mg [124–126].

Detailed calculations will be presented for $^{26-40}_{12}\text{Mg}_{14-28}$ within the proxy-SU(3) framework. In the following, a mechanism, which was initially proposed in [127], will be used to explain why:

(a) $N = 20$ is not a magic number for the ground state of $^{32}_{12}\text{Mg}_{20}$,

(b) shape coexistence occurs among neutron numbers $N = 18 - 20$ for this isotopic chain, and

(c) inversion of states appears at $^{31}_{12}\text{Mg}_{19}$.

The isotonic chains with $N \approx 20$ possess values of the deformation β which exhibit a minimum at $Z = 14$ [128]. This provides experimental support for the findings of the previous section, in which 14 was found to be an SO-like magic number, for protons in the present case. The 12 protons of Mg lie in the 6–14 shell, and in proxy-SU(3) have $(\lambda, \mu)_p = (0, 0)$.

The neutrons of $^{26-40}_{12}\text{Mg}_{14-28}$ lie either in the SO-like shell 14–28 or in the harmonic oscillator shells 8–20 and 20–40. The h.w. SU(3) irreps for these shells are listed in Table 1.6. Since protons do not contribute at all, the deformation formula becomes:

$$\beta^2 = \frac{4\pi}{5(S_n \bar{r}^2)^2}(C_2 + 3),\tag{1.78}$$

where $\bar{r}^2 = 0.87^2 N^{1/3}$ [116], $C_2 = \lambda_n^2 + \mu_n^2 + \lambda_n \mu_n + 3(\lambda_n + \mu_n)$ (where the subscript n stands for neutrons) and $S_n = 12$ for the 8–20 and the 14–28 proxy-SU(3) shell, while $S_n = 20$ for the 20–40 shell. The theoretical predictions for β for the Mg isotopic chain for the spin-orbit like magic numbers and the harmonic oscillator magic numbers are presented in Figure 1.2.

Shape coexistence is a phenomenon where the nucleus exhibits a ground-state band with a certain type of deformation (prolate for instance) and a slightly excited band with another type of deformation (oblate for example) [101–103]. The idea proposed in [117,127] is that shape coexistence is the consequence of two sets of magic numbers, namely the SO-like 6, 14, 28, 50, 82, 126 magic numbers and the harmonic oscillator 2, 8, 20, 40, 70, 112 magic numbers:

TABLE 1.6
The h.w. SU(3) Irreps [61] for 13 to 28 Nucleons, Listed in the Case in Which the SO-Like Magic Numbers 14 and 28 and the Proxy-SU(3) Symmetry Are Taken into Account, as well as in the Case in Which the Harmonic Oscillator Magic Numbers 8, 20, and 40 and the HO SU(3) Symmetry Are Considered

Even Particle Number	$(\lambda, \mu)_{SO}$	$(\lambda, \mu)_{ho}$	Odd Particle Number	$(\lambda, \mu)_{SO}$	$(\lambda, \mu)_{ho}$
14	(0, 0)	(6, 0)	13	(0, 0)	(5, 1)
16	(4, 0)	(2, 4)	15	(2, 0)	(4, 2)
18	(4, 2)	(0, 4)	17	(4, 1)	(1, 4)
20	(6, 0)	(0, 0)	19	(5, 1)	(0, 2)
22	(2, 4)	(6, 0)	21	(4, 2)	(3, 0)
24	(0, 4)	(8, 2)	23	(1, 4)	(7, 1)
26	(0, 0)	(12, 0)	25	(0, 2)	(10, 1)
28	(0, 0)	(10, 4)	27	(0, 0)	(11, 2)

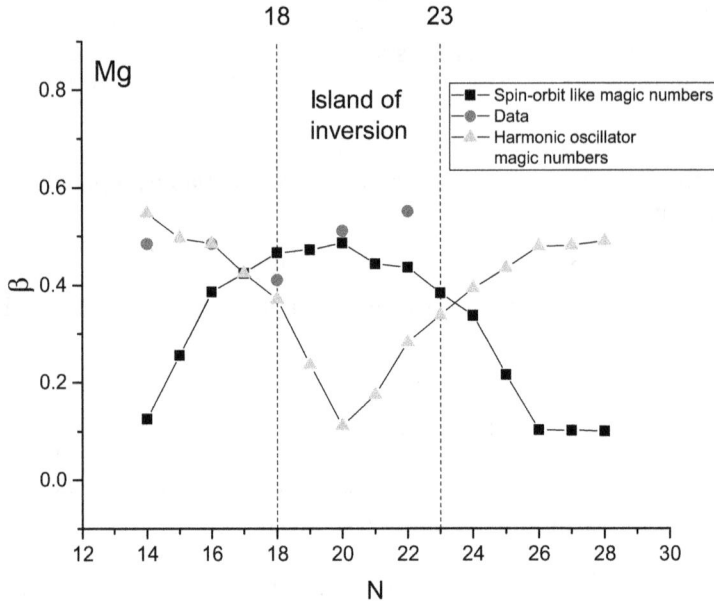

FIGURE 1.2

Shape coexistence begins when the deformation obtained with the harmonic oscillator magic numbers is less than the deformation calculated with the SO-like magic numbers, *i.e.* $\beta_{ho} \leq \beta_{SO}$ and usually stops at a harmonic oscillator magic number. The mechanism of [111,117, 127] predicts that the nuclei $^{30-32}_{12}\mathrm{Mg}_{18-20}$ are candidates for shape coexistence. Data for β have been taken from [129]. The data indicate that the neutrons of the Mg isotopes with neutron number $N \leq 17$ follow the harmonic oscillator magic numbers and do not exhibit shape coexistence. As a consequence, when $N \leq 17$, there is a unique ground state obeying the rules of the 8–20 HO shell. But as soon as shape coexistence begins, $N > 17$, there may be two states low lying in energy. As shown in Eq. (1.82), in general the state with the higher deformation is the ground state. As seen from the comparison to the data [128], the ground state is derived from the SO-like shell 14-28. The excited state is derived in general from the coupling of the 8–20 HO shell with the 14–28 SO-like shell. To summarize, when $N > 17$ the ground state follows the 14–28 SO-like shell, while when $N < 17$ the ground state follows the 8–20 HO shell. This may be the cause of the inversion of states, which begins at $^{30}_{12}\mathrm{Mg}_{18}$ [118].

When the nuclear deformation with the harmonic oscillator magic numbers becomes less than the deformation with the SO-like magic numbers:

$$\beta_{ho} \leq \beta_{SO}, \tag{1.79}$$

then shape coexistence is likely to appear.

The ground state of such nuclei derives from the SO-like magic numbers, because this shell possesses the maximum deformation and QQ interaction, as we can see using the Hamiltonian of Eq. (1.59). Indeed, schematically one has

$$H_{ho} = H_0 - \frac{\chi}{2} QQ_{ho}, \tag{1.80}$$

$$H_{SO} = H_0 - \frac{\chi}{2} QQ_{SO}, \tag{1.81}$$

$$QQ_{ho} \leq QQ_{SO} \Rightarrow H_{SO} \leq H_{ho}. \tag{1.82}$$

The excited $K = 0^+$ state of nuclei with shape coexistence derives from the coupling of the harmonic oscillator shell 8–20 with the SO-like shell 14–28. In the review articles [101,102], one can see that the experimentally known examples of shape coexistence in several series of isotopes appear below and stop at a harmonic oscillator magic number. For example, in the Hg series of isotopes, shape coexistence appears below and stops at $N = 112$, as seen in Figure 10 of Ref. [102]). In the present case, this means that shape coexistence should appear below and stop at $^{32}_{12}\text{Mg}_{20}$.

The inversion of states is a side effect of shape coexistence. One can divide the Mg isotopic chain in two halves:

(a) the first half contains the isotopes with neutron numbers $N \leq 17$ with $\beta_{ho} \geq \beta_{SO}$, and

(b) the second contains the isotopes with $17 < N < 23$ with $\beta_{ho} < \beta_{SO}$.

The neutrons of the isotopes of the first half clearly follow the harmonic oscillator magic numbers 8–20 (see Figure 1.2). But the isotopes of the second half may exhibit shape coexistence. Such isotopes have two low-lying states: a ground-state and a slightly excited one.

While the ground state of Mg isotopes with $N \leq 17$ followed the 8–20 HO shell, now the ground state of isotopes with $N > 17$ follows the rules of the 14–28 SO-like shell. The inversion of states may be caused by a change of magic numbers.

The cause of the inversion of states may be hidden in the collective nuclear features instead of the single-particle ones. The Elliott SU(3) model contains all the necessary techniques, which can reveal the J^π of the ground state and of the excited state of even-odd nuclei, which lie within the islands of inversion.

1.8   Conclusions

A mechanism is proposed for shape coexistence and inversion of states [111,117,127]. The mechanism involves two sets of magic numbers: the spin-orbit (SO) like magic numbers 6, 14, 28, 50, 82, 126 and the harmonic oscillator magic numbers 2, 8 20, 40, 70, 112. Shape coexistence arises when $\beta_{SO} \geq \beta_{ho}$ and stops at a harmonic oscillator closure. The state with the maximum deformation lies lower in energy. As a consequence, within the islands of shape coexistence shown in Figure 8 of Ref. [102], the ground state is derived by the SO-like magic numbers, while the coexisting excited 0^+ state is derived from the coupling of the harmonic oscillator with the SO-like magic numbers. The inversion of states is a side effect of shape coexistence. The proposed mechanism predicts without any parameters the borders of the islands of shape coexistence appearing at Figure 8 of [102]. This mechanism can be applied in every mass region to predict the borders of the islands of shape coexistence and all the relevant nuclear observables.

Acknowledgements

Financial support by the Greek State Scholarships Foundation (IKY) and the European Union within the MIS 5033021 action is gratefully acknowledged.

References

[1] Elsasser, W. 1934. On the principle of Pauli in the nuclei - III. *J. Phys. Radium* 5: 635–639.

[2] Goeppert-Mayer, M. 1948. On Closed Shells in Nuclei. *Phys. Rev.* 74: 235–239.

[3] Goeppert-Mayer, M. (1949) 1969. On Closed Shells in Nuclei II. *Phys. Rev.* 75: 1969–1970.

[4] Haxel, O., J.H.D. Jensen, and H.E. Suess. 1949. On the "Magic Numbers" in Nuclear Structure. *Phys. Rev.* 75: 1766.

[5] Goeppert-Mayer, M. and J.H.D. Jensen. 1955. *Elementary Theory of Nuclear Shell Structure.* New York: Wiley.

[6] Heyde, K. 1990. *The nuclear shell model.* Berlin: Springer.

[7] Talmi, I. 1993. *Simple models of complex nuclei: The shell model and interacting boson model.* Chur: Harwood.

[8] Sorlin, O. and M.-G. Porquet. 2008. Nuclear magic numbers: New features far from stability. *Prog. Part. Nucl. Phys.* 61: 602–673.

[9] Oganessian, Y.T., V.K. Utyonkov, Yu.V. Lobanov et al. 2006. Synthesis of the isotopes of elements 118 and 116 in the ^{249}Cf and ^{245}Cm $+$ ^{48}Ca fusion reactions. *Phys. Rev. C* 74: 044602.

[10] Oganessian, Y.T., F.Sh. Abdullin, C. Alexander et al. 2012. Production and Decay of the Heaviest Nuclei ^{293,294}117 and ^{294}118. *Phys. Rev. Lett.* 109: 162501.

[11] Agbemawa, S.E., A.V. Afanasev, T. Nakatsukasa, and P. Ring. 2015. Covariant density functional theory: Reexamining the structure of superheavy nuclei. *Phys. Rev. C* 92: 054310.

[12] Prassa, V., T. Nikšić, G.A. Lalazissis, and D. Vretenar. 2012. Relativistic energy density functional description of shape transitions in superheavy nuclei. *Phys. Rev. C* 86: 024317.

[13] Scharff-Goldhaber, G. 1955. System of even-even nuclei. *Phys. Rev.* 98: 212–214.

[14] Casten, R.F. 1990. *Nuclear structure from a simple perspective.* Oxford: Oxford University Press.

[15] Bohr, A. 1952. The coupling of the Nuclear Surface Oscillations to the motion of individual nucleons. *Mat. Fys. Medd. K. Dan. Vidensk. Selsk.* 26: no. 14.

[16] Bohr, A., and B.R. Mottelson. 1969. *Nuclear Structure Vol. I: Single particle motion.* New York: Benjamin.

[17] Bohr, A., and B.R. Mottelson. 1975. *Nuclear Structure Vol. II: Nuclear deformations.* New York: Benjamin.

[18] Davydov, A.S., and G.F. Filippov. 1958. Rotational states in even atomic nuclei. *Nucl. Phys.* 8: 237–249.

[19] Meyer-ter-Vehn, J. 1975. Collective model description of transitional odd-A nuclei. *Nucl. Phys. A* 249: 111–140.

[20] Eisenberg, J.M., and W. Greiner. 1975. *Nuclear Theory Vol. I: Nuclear Models.* Amsterdam: North-Holland.

[21] Próchniak, L., and S.G. Rohozinski. 2009. Quadrupole collective states within the Bohr collective Hamiltonian. *J. Phys. G: Nucl. Part. Phys.* 36: 123101.

[22] Greiner, W., and J.A. Maruhn. 1996. *Nuclear Models.* Berlin: Springer.

[23] Wybourne, B.G. 1974. *Classical groups for physicists.* New York: Wiley.

[24] Gilmore, R. 1974. *Lie groups, Lie algebras, and some of their applicatons.* New York: Wiley.

[25] Elliott, J.P., and P.G. Dawber. 1979. *Symmetry in physics Vol. 1: Principles and simple applications.* London: Macmillan.

[26] Elliott, J. P., and P. G. Dawber. 1979. *Symmetry in physics Vol. 2: Further applications.* London: Macmillan.

[27] Tung, W.-K. 1985. *Group theory in physics.* Singapore: World Scientific.

[28] Chen, J.-Q., J. Ping and F. Wang. 2002. *Group representation theory for physicists.* Singapore: World Scientific.

[29] Iachello, F. 2006. *Lie algebras and applications.* Berlin: Springer.

[30] Iachello, F., and A. Arima. 1987. *The Interacting Boson Model.* Cambridge: Cambridge University Press.

[31] Iachello, F., and P. Van Isacker. 1991. *The interacting boson-fermion model.* Cambridge: Cambridge University Press.

[32] Frank, A., and P. Van Isacker. 1994. *Symmetry methods in molecules and nuclei.* New York: Wiley.

[33] Bonatsos, D. 1988. *Interacting boson models of nuclear structure.* Oxford: Clarendon.

[34] Frank, A., J. Jolie, and P. Van Isacker. 2009. *Symmetries in atomic nuclei: From isospin to supersymmetry.* Berlin: Springer.

[35] Rowe, D.J., and J.L. Wood. 2010. *Fundamentals of nuclear models: Foundational models.* Singapore: World Scientific.

[36] Castaños, O., E. Chacón, A. Frank, and M. Moshinsky. 1979. Group theory of the Interacting Boson model of the nucleus. *J. Math. Phys.* 20: 35–44.

[37] Wilets, L., and M. Jean. 1956. Surface oscillations in even-even nuclei. *Phys. Rev.* 102: 788–796.

[38] Rakavy, G. 1957. The classification of states of surface vibrations. *Nucl. Phys.* 4: 289–294.

[39] Bès, D. R. 1959. The γ-dependent part of the wave functions representing γ-unstable surface vibrations. *Nucl. Phys.* 10: 373–385.

[40] Corrigan, T.M., F.J. Margetan, and S.A. Williams. 1976. Exact solution of the quadrupole surface vibration Hamiltonian in body-fixed coordinates. *Phys. Rev. C* 14: 2279–2296.

[41] Chacón, E., M. Moshinsky, and R.T. Sharp. 1976. U(5)⊃O(5)⊃O(3) and the exact solution for the problem of quadrupole vibrations of the nucleus. *J. Math. Phys.* 17: 668–676.

[42] Chacón, E., and M. Moshinsky. 1977. Group theory of the collective model of the nucleus. *J. Math. Phys.* 18: 870–880.

[43] Bonatsos, D., D. Lenis, N. Minkov et al. 2004. Sequence of potentials interpolating between the U(5) and E(5) symmetries. *Phys. Rev. C* 69: 044316.

[44] Elliott, J.P. 1958. Collective motion in the nuclear shell model I. Classification schemes for states of mixed configurations. *Proc. Roy. Soc. Ser. A* 245: 128–145.

[45] Elliott, J.P. 1958. Collective motion in the nuclear shell model II. The introduction of intrinsic wave-functions. *Proc. Roy. Soc. Ser. A* 245: 562–581.

[46] Elliott, J.P., and M. Harvey. 1963. Collective motion in the nuclear shell model III. The calculation of spectra. *Proc. Roy. Soc. Ser. A* 272: 557–577.

[47] Harvey, M. 1968. The Nuclear SU(3) Model. *In Andvances in Nuclear Physics* Vol 1, ed. M. Baranger, and E. Vogt, 67–180, New York: Plenum Press.

[48] Arima, A., and F. Iachello. 1976. Interacting Boson Model of collective states I. The vibrational limit. *Ann. Phys.* 99: 253–317.

[49] Arima, A., and F. Iachello. 1978. Interacting Boson Model of collective nuclear states I. The rotational limit. *Ann. Phys.* 111: 201–238.

[50] Arima, A., and F. Iachello. 1979. Interacting Boson Model of collective nuclear states IV. The O(6) limit. *Ann. Phys.* 123: 468–492.

[51] Moshinsky, M., and Yu.F. Smirnov. 1996. *The harmonic oscillator in modern physics.* Amsterdam: Harwood.

[52] Bonatsos, D., and A. Klein. 1986. Exact boson mappings for nuclear neutron (proton) shell-model algebras having SU(3) subalgebras. *Ann. Phys.* 169: 61–103.

[53] Ratna Raju, R.D., J.P. Draayer, and K.T. Hecht. 1973. Search for a coupling scheme in heavy deformed nuclei: The pseudo SU(3) model. *Nucl. Phys. A* 202: 433–466.

[54] Draayer, J.P., K.J. Weeks, and K.T. Hecht. 1982. Strength of the $Q_\pi \cdot Q_\nu$ interaction and the strong-coupled pseudo-SU(3) limit. *Nucl. Phys. A* 381: 1–12.

[55] Draayer, J.P., and K.J. Weeks. 1983. Shell-model description of the low-energy structure of strongly deformed nuclei. *Phys. Rev. Lett.* 51: 1422–1425.

[56] Draayer, J.P., and K.J. Weeks. 1984. Towards a shell model description of the low-energy structure of deformed nuclei I. Even-even systems. *Ann. Phys.* 156: 41–67.

[57] Ginocchio, J.N. 1997. Pseudospin as a relativistic symmetry. *Phys. Rev. Lett.* 78: 436–439.

[58] Zuker, A.P., J. Retamosa, A. Poves, and E. Caurier. 1995. Spherical shell model description of rotational motion. *Phys. Rev. C* 52: R1741-R1745.

[59] Zuker, A. P., A. Poves, F. Nowacki, and S. M. Lenzi. 2015. Nilsson–SU(3) self-consistency in heavy N = Z nuclei. *Phys. Rev. C* 92: 024320.

[60] Bonatsos, D., I. E. Assimakis, N. Minkov et al. 2017. Proxy-SU(3) symmetry in heavy deformed nuclei. *Phys. Rev. C* 95: 064325.

[61] Bonatsos, D., I.E. Assimakis, N. Minkov et al. 2017. Analytic predictions for nuclear shapes, the prolate dominance and the prolate-oblate shape transition in the proxy-SU(3) model. *Phys. Rev. C* 95: 064326.

[62] Bonatsos, D. 2017. Prolate over oblate dominance in deformed nuclei as a consequence of the SU(3) symmetry and the Pauli principle. *Eur. J. Phys. A* 53: 148.

[63] Nilsson, S.G. 1955. Binding states of individual nucleons in strongly deformed nuclei. *Mat. Fys. Medd. K. Dan. Vidensk. Selsk.* 29: no. 16.

[64] Nilsson, S.V. and I. Ragnarsson. 1995. *Shapes and Shells in Nuclear Structure.* Cambridge: Cambridge University Press.

[65] Ragnarsson, I., S.G. Nilsson, and R.K. Sheline. 1978. Shell structure in nuclei. *Phys. Reports* 45: 1–87.

[66] Mottelson, B. 2006. *The Nilsson Model and Sven Gösta Nilsson.* Phys. Scr. T125: editorial.

[67] Rassey, A.J. 1958. Nucleonic binding states in nonspherical nuclei: Asymptotic representation. *Phys. Rev.* 109: 949–957.

[68] Quentin, P., and R. Babinet. 1970. Nilsson's Hamiltonian in the asymptotic basis. *Nucl. Phys. A* 156: 365–384.

[69] Boisson, J.P., and R. Piepenbring. Treatment of a Nilsson potential in the basis of asymptotic quantum numbers. *Nucl. Phys. A* 168: 385–398.

[70] Takahashi, Y. 1975. SU(3) shell model in a deformed harmonic oscillator basis. *Prog. Theor. Phys.* 53: 461–479.

[71] Rosensteel, G., and J.P. Draayer. 1989. Symmetry algebra of the anisotropic harmonic oscillator with commensurate frequencies. *J. Phys. A: Math. Gen.* 22: 1323–1327.

[72] Nazarewicz, W., and J. Dobaczewski. 1992. Dynamical symmetries, multiclustering, and octupole susceptibility in superdeformed and hyperdeformed nuclei. *Phys. Rev. Lett.* 68: 154–157.

[73] Nazarewicz, W., J. Dobaczewski, and P. Van Isacker. Shell model calculations at superdeformed shapes. *AIP Conf. Proc.* 259: 30.

[74] Bonatsos, D., C. Daskaloyannis, P. Kolokotronis, and D. Lenis. 1994. The symmetry algebra of the N-dimensional anisotropic quantum harmonic oscillator with rational ratios of frequencies and the Nilsson model. arXiv: hep-th/9411218.

[75] Sugawara-Tanabe, K., and A. Arima. 1997. New dynamical SU(3) symmetry in deformed nuclei. *Nucl. Phys. A* 619: 88–96.

[76] Arima, A. 1999. Elliott's SU(3) model and its developments in nuclear physics. *J. Phys. G: Nucl. Part. Phys.* 25: 581–588.

[77] Chari, V., and A. Pressley. 1994. *A guide to quantum groups.* Cambridge: Cambridge University Press.

[78] Klimyk, A., and K. Schmüdgen. 1997. *Quantum groups and their representations.* Berlin: Springer.

[79] Bonatsos, D., and C. Daskaloyannis. 1999. Quantum groups and their applications in nuclear physics. *Prog. Part. Nucl. Phys.* 43: 537–618.

[80] Asherova, R.M., Yu.F. Smirnov, V.N. Tolstoy, and A.P. Shustov. 1981. Algebraic approach to the projected deformed oscillator model. *Nucl. Phys. A* 355: 25–44.

[81] Sugawara-Tanabe, K., A. Arima, and N. Yoshida. 1995. Resurrection of the L-S coupling scheme in superdeformation. *Phys. Rev. C* 51: 1809–1818.

[82] Scholten, O., F. Iachello, and A. Arima. 1978. Interacting Boson Model of collective nuclear states III. The transition from SU(5) to SU(3). *Ann. Phys.* 115: 325.

[83] Iachello, F., and N.V. Zamfir. 2004. Quantum phase transitions in mesoscopic systems. *Phys. Rev. Lett.* 92: 212501.

[84] Iachello, F. 2006. Quantum phase transitions in mesoscopic systems. *Int. J. Mod. Phys. B* 20: 2687.

[85] Bonatsos, D., E.A. McCutchan, R.F. Casten, and R.J. Casperson. 2008. Simple empirical order parameter for a first-order quantum phase transition in atomic nuclei. *Phys. Rev. Lett.* 100: 142501.

[86] Cejnar, P., J. Jolie, and R.F. Casten. 2010. Quantum phase transitions in the shapes of atomic nuclei. *Rev. Mod. Phys.* 82: 2155.

[87] Iachello, F. 2000. Dynamic symmetries at the critical point. *Phys. Rev. Lett.* 85: 3580.

[88] Casten, R.F., and N.V. Zamfir. Evidence for a possible E(5) symmetry in ^{134}Ba. *Phys. Rev. Lett.* 85: 3584.

[89] Bonatsos, D., D. Lenis, N. Minkov et al. 2004. Sequence of potentials interpolating between the U(5) and E(5) symmetries. *Phys. Rev. C* 69: 044316.

[90] Rowe, D.J. 2004. Phase transitions and quasidynamical symmetry in nuclear collective models: I. The U(5) to O(6) phase transition in the IBM. *Nucl. Phys. A* 745: 47.

[91] Turner, P.S., and D.J. Rowe. 2005. Phase transitions and quasidynamical symmetry in nuclear collective models. II. The spherical vibrator to gamma-soft rotor transition in an SO(5)-invariant Bohr model. *Nucl. Phys. A* 756: 333.

[92] Iachello, F. 2001 Analytic description of critical point nuclei in a spherical-axially deformed shape phase transition. *Phys. Rev. Lett.* 87: 052502.

[93] Casten R.F., and N.V. Zamfir. 2001. Empirical realization of a critical point description in atomic nuclei. *Phys. Rev. Lett.* 87: 052503.

[94] Bonatsos, D., D. Lenis, N. Minkov et al. 2004 Sequence of potentials lying between the U(5) and E(5) symmetries. *Phys. Rev. C* 69: 014302.

[95] Rosensteel, G., and D.J. Rowe. 2005. Phase transitions and quasi-dynamical symmetry in nuclear collective models. III. The U(5) to SU(3) phase transition in the IBM. *Nucl. Phys. A* 759: 92.

[96] Casten, R.F., and E.A. McCutchan. 2007. Quantum phase transitions and structural evolution in nuclei. *J. Phys. G: Nucl. Part. Phys.* 34: R285.

[97] Bonatsos, D., D. Lenis, and D. Petrellis. 2007. Special solutions of the Bohr Hamiltonian related to shape phase transitions in nuclei. *Rom. Rep. Phys.* 59: 273.

[98] Casten, R.F. 2009. Quantum phase transitions and structural evolution in nuclei. *Prog. Part. Nucl. Phys.* 62: 183.

[99] Heyde, K., J. Jolie, R. Fossion et al. 2004. Phase tansitions versus shape coexistence. *Phys. Rev. C* 69: 054304.

[100] García-Ramos, J.-E., and K. Heyde. 2018. On the nature of the shape coexistence and the quantum phase transition phenomena: lead region and Zr isotopes. arXiv: [nucl-th] 1802.04219.

[101] Wood, J.L., K. Heyde, W. Nazarewicz et al. 1992. Coexistence in even-mass nuclei. *Phys. Rep.* 215: 101–201.

[102] Heyde, K., and J.L. Wood. 2011. Shape coexistence in atomic nuclei. *Rev. Mod. Phys.* 83: 1467–1521.

[103] Heyde, K., P. Van Isacker, M. Waroquier et al. 1983. Coexistence in odd-mass nuclei. *Phys. Rep.* 102: 291–393.

[104] Budaca, R., P. Buganu, and A.I. Budaca. 2019. Geometrical model description of shape coexistence in Se isotopes. *Nucl. Phys. A* 990: 137–148.

[105] Georgoudis, P.E., and A. Leviatan. Aspects of shape coexistence in the geometric collective model of nuclei. arXiv: [nucl-th] 1712.04392.

[106] Mottelson, B.R. and S.G. Nilsson 1959. The shape of the nuclear photo-resonance in deformed nuclei. *Nucl. Phys.* 13: 281–291.

[107] Shankar, R. 1980. *Principles of Quantum Mechanics.* New York and London: Plenum Press.

[108] Cohen-Tannoudji, C., B. Diu and F. Laloe 1991. *Quantum Mechanics Vol. 1.* Paris: Wiley.

[109] Lipkin, H.J. 1965. *Lie Groups for Pedestrians.* Amsterdam: North-Holland Publishing Company.

[110] Draayer, J.P., Y. Leschber, S.C. Park and R. Lopez. 1989. Representations of $U(3)$ in $U(N)$. *Comp. Phys. Commun.* 56: 279–290.

[111] Martinou, A., D. Bonatsos, N. Minkov et al. 2018. Highest weight SU(3) irreducible representations for nuclei with shape coexistence. Proc. of 37-th International Workshop on Nuclear Theory, Nucl. Th. 37. arXiv: 1810.11870 [nucl-th].

[112] Smirnov, Y., N.A. Smirnova and P.V. Isacker. 2000. $SU(3)$ realization of the rigid asymmetric rotor within the interacting boson model. *Phys. Rev. C* 61: 041302(R).

[113] Draayer, J.P. 1993. Fermion Models. In *Contemporary Concepts in Physics Vol 6, Algebraic Approaches to Nuclear Structure,* ed. R.F. Casten, 423–549, New York: Harwood Academic Publishers.

[114] Rowe, D. J., G. Thiamova and J. L. Wood. 2006. Implications of Deformation and Shape Coexistence for the Nuclear Shell Model. *Phys. Rev. Lett.* 97: 202501.

[115] Castaños, O., J.P. Draayer and Y. Leschber. 1988. Shape Variables and the Shell Model. *Z. Phys. A-Atomic Nuclei* 329: 33–43.

[116] Stone, J.R., N.J. Stone, and S. Moszkowski 2014. Incompressibility in finite nuclei and the nuclear matter. *Phys. Rev. C* 89: 044316.

[117] Martinou, A., D. Bonatsos, N. Minkov et al. 2018. Nucleon numbers for nuclei with shape coexistence. *Proceedings of the 27th annual Symposium of the Hellenic Nuclear Physics Society.* arXiv: 1810.11860 [nucl-th].

[118] Scheit, H. 2011. Spectroscopy in and around the island of inversion. *J. Phys. Conf. Ser.* 312: 092010.

[119] Wimmer, K., T. Kröll, R. Krücken et al. 2010. Discovery of the Shape Coexisting 0^+ State in the ^{32}Mg by a Two Neutron Transfer Reaction. *Phys. Rev. Lett.* 105: 252501.

[120] Schwerdtfeger, W., P.G. Thirolf, K. Wimmer et al. 2009. Shape Coexistence Near Neutron Number $N = 20$: First Identification of the $E0$ Decay from the Deformed First Excited $J^\pi = 0^+$ State of ^{30}Mg. *Phys. Rev. Lett.* 103: 012501.

[121] Kowalska, M., D.T. Yordanov, K. Blaum et al. 2008. Nuclear ground-state spins and magnetic moments of ^{27}Mg, ^{29}Mg and ^{31}Mg. *Phys. Rev. C* 77: 034307.

[122] Neyens, G., M. Kowalska, D. Yordanov et al. 2005. Measurement of the Spin and Magnetic Moment of ^{31}Mg: Evidence for a Strongly Deformed Intruder Ground State. *Phys. Rev. Lett.* 94: 022501.

[123] Neyens, G., P. Himpe, D.L. Balabanski et al. 2007. The "sland of inversion" from a nuclear moments perspective and the g factor of ^{35}Si. *Eur. Phys. J. Special Topics* 150: 149–153.

[124] Caurier, E., F. Nowacki and A. Poves. 2014. Merging the islands of inversion at $N = 20$ and $N = 28$. *Phys. Rev. C* 90: 014302.

[125] Gade, A., and S.N. Liddick. 2016. Shape coexistence in neutron-rich nuclei. *J. Phys. G: Nucl. Part. Phys.* 43: 024001.

[126] Poves, A. 2016. Shape coexistence: the shell model view. *J. Phys. G: Nucl. Part. Phys.* 43: 024010.

[127] Martinou, A. 2018. Nucleon-nucleon interaction in stable and unstable nuclei. PhD Thesis. National Technical University of Athens. http://hdl.handle.net/10442/hedi/ 43367

[128] Nudat 2. National Nuclear Data Center. Brookhaven Laboratory.

[129] Pritychenko, B., M. Birch, B. Singh and M. Horoi. 2016. Tables of $E2$ transition probabilities from the first 2^+ states in even-even nuclei. *Atomic Data Nuclear Data Tables* 107: 1–139.

2

Skyrme and Relativistic Mean-Field Models in the
Description of Symmetric, Asymmetric, and Stellar
Nuclear Matter

O. Lourenço and M. Dutra

Instituto Tecnológico de Aeronáutica, DCTA

P. D. Stevenson

University of Surrey

CONTENTS

2.1 Introduction

The deuteron, composed of one proton and one neutron, is the simplest bound-state system used to study the nuclear interaction. Well-established experimental data show that the deuteron binding energy is given by $|B| = 2.22452 \pm 0.00010$ MeV [1], indicating that, actually, this is a weakly bound system, even with the two particles interact through the strongest force of the nature, namely, the nuclear force. A possible treatment for this system consists in the use of the Schrödinger equation for a square well potential taking into account that the typical range of the nuclear interaction is around 1.7 fm. This procedure produces the result of around 35 MeV for the depth of the well, confirming the high strength of the interaction in comparison with $|B|$. Other experimental data also show that the deuteron has total spin 1, positive parity for its ground state (which is a triplet), electric quadrupole moment given by $Q = 2.82 \pm 0.01$ mb, and magnetic moment of $\mu = 0.857406 \pm 0.000001$ mn [1].

From the theoretical point of view, a widely used approach to treat both few and many nucleons systems is the one based on the construction of nuclear potentials, as the one pion exchange potential for instance, in which the free parameters are adjusted in order to reproduce phenomenology and results coming from deuteron physics and unbound systems, studied from scattering theory. Such microscopic calculations are performed by using the Brueckner-Hartree-Fock [2,3] treatment, for instance. Alternatively, a second approach arises by fitting directly some of the many-nucleon observables based on a mean-field approach, allowing the construction of thermodynamic equations of state to study infinite nuclear matter, i.e., a hypothetical isotropic system of an infinite number of nucleons with no boundaries and no Coulomb interaction. From this point of view, nuclei can be interpreted as nuclear matter droplets. Among the mainly nonrelativistic models constructed through the last approach, one can mention the widely known Skyrme model [4,5], in which the effective interactions are point-like. In this model, the nucleons interact with each other only when they are in contact. For the relativistic case, on the other hand, a Lagrangian density is proposed and all thermodynamic quantities are derived from it. In its simplest version, the Walecka model for infinite nuclear matter [6], based on relativistic field theory in a mean-field approach, depends on free parameters fitted to reproduce the nuclear matter bulk properties. The applications of these models extend to different ranges of temperatures and densities. For the zero-temperature regime, the detailed knowledge of the hadronic equations of the state, coming from both relativistic and nonrelativistic models, is very important to the description, for example, of neutron stars, studied in densities up to around six times the nuclear saturation density. Due to the properties of the nuclear interaction, namely, essentially attractive but repulsive at short distances, the central density of a nuclear system at certain density will not increase any further even if more nucleons are added. The density at which this occurs is referred to as the saturation density, denoted by ρ_0.

The properties of neutron stars, such as the mass-radius relation, are directly affected by features presented in hadronic parametrizations [7,8]. Still at the $T = 0$ regime, but for a range of very low densities, neutron matter can be studied in the so-called dilute Fermi gas regime. Applications of this subject with the use of relativistic models are found in Ref. [9]. In the finite temperature regime, in which the hadronic models are generalized to $T \neq 0$ but keeping the adjustment of the free parameters performed at $T = 0$, the phenomenon of phase transitions in nuclear matter takes place. In general, hadronic models exhibit a liquid- to gas-phase transition characterized by regions presenting low (gas-phase) and high (liquid-phase) densities at a temperature range of $T \lesssim 20$ MeV. This is the typical feature presented by the known van der Waals model in a temperature range of $T < T_c$, where T_c is the critical temperature. Studies show that in finite nuclei, T_c is approximately half of the value obtained in infinite nuclear matter [10]. Another general feature observed in the hadronic models is that such a liquid- to gas-phase transition always occurs at densities below the saturation density. In the high-temperature regime ($T \sim 200$ MeV), another kind of transition occurs in nuclear matter, namely, the hadron-quark phase transition. In this case, hadrons lose identity in favor of their fundamental constituents, namely, the quarks. Thus, a phase diagram related to this transition presents both hadronic and quark regions. Actually, one believes that the inverse process occurred in nature at the beginning of universe, i.e., the initial quark-gluon plasma became the whole hadronic structure known today. From the experimental point of view, one tries to recreate such extreme conditions through heavy ion collisions performed in the particle accelerators, such as the Relativistic Heavy Ion Collider (RHIC) or the Large Hadron Collider (LHC), in which the energy involved is high enough to become possible the detection of signals of the hadron-quark phase transition [11]. The results from these experiments are used to validate diverse hadronic models. With these data properly reproduced, the models are then able to make predictions on other quantities related to the colliding system, such as the temperature and the chemical potential of the baryons involved. Also in the theoretical field, the treatment of the hadron-quark phase transition is somewhat delicate, since this phenomenon involves two different degrees of freedom. Thus, the available hadronic models must be used together with quarks ones, i.e., models that somehow are able to describe the phenomenology of the Quantum Chromodynamics (QCD).

In this chapter, we use the Skyrme and the relativistic mean-field (RMF) model at zero temperature regime in the description of infinite nuclear matter. We apply such models to symmetric matter (equal number of protons and neutrons), asymmetric matter, and stellar matter with a direct application of describing some important features of compact astrophysical objects, more specifically, neutron stars.

2.2 Relativistic Mean-Field Models

The Walecka model, proposed in 1974 by J. D. Walecka [12], is the main representative of the relativistic hadronic models. It treats protons and neutrons as fundamental particles interacting with each other through the exchange of scalar (σ) and vector (ω) mesons, which physically represent the attractive and repulsive parts, respectively, of the nuclear interaction. The free parameters of the theory, given by the coupling constants between these mesons and nucleons, are determined by requiring the model reproduces, at zero temperature, well-established quantities related to infinite nuclear matter, namely, binding energy ($B \sim -16$ MeV) and saturation density ($\rho_0 \sim 0.15$ fm^{-3}). The basis of the Walecka theory, also called Quantum Hadrodynamics (QHD), is Quantum Field Theory. Its starting point is a Lorentz-invariant Lagrangian density from which many properties of the system are extracted. In particular, the nuclear potential is given by the Yukawa form [13] given by

$$V(r) = -\frac{g_\sigma}{4\pi}\frac{e^{-m_\sigma r}}{r} + \frac{g_\omega}{4\pi}\frac{e^{-m_\omega r}}{r}, \tag{2.1}$$

taking into account equal numbers of protons and neutrons. The first term corresponds to the meson σ, with mass m_σ, and its negative sign represents the attractive nature of this interaction. The repulsion is represented by the second term, characterized by the meson ω. Notice that the interaction range is essentially given by the mass of each meson. Consequently, the limit of $m_\sigma, m_\omega \to \infty$ leads to a contact potential. In this case, the nucleons would interact with each other via point-like interactions. In Figure 2.1, we show $V(r)$ as a function of the distance between the nucleons, r, for a set of coupling constants and meson masses given by $g_\sigma = 9.6$, $g_\omega = 11.7$, $m_\sigma = 550$ MeV, and $m_\omega = 783$ MeV.

Although the Walecka model is able to reproduce the nuclear matter binding energy and the saturation density, it does not give reasonable values for the incompressibility (K_0) and effective nucleon mass (M^*). In order to circumvent these problems, Boguta and Bodmer [14,15] included in the Walecka model cubic and quartic self-interactions in the scalar field σ, introducing two more coupling constants in the theory which are fitted so as

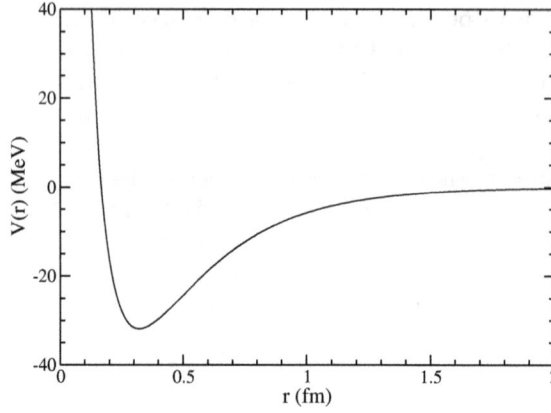

FIGURE 2.1
Nuclear potential generated from Eq. (2.1).

to fix the values of K_0 and M^*. In the same way, more terms can be added to the Boguta-Bodmer model to make it compatible with other observables, such as those related to finite nuclei. Actually, many RMF models and parametrizations have been constructed following this method.

2.2.1   Finite Range Model Description

In the following, we present a more general kind of Lagrangian density describing nucleons interacting with each other through the meson exchange mechanism, namely,

$$\mathcal{L} = \mathcal{L}_{nm} + \mathcal{L}_\sigma + \mathcal{L}_\omega + \mathcal{L}_\rho + \mathcal{L}_{\sigma\omega\rho}, \tag{2.2}$$

where

$$\mathcal{L}_{nm} = \overline{\psi}(i\gamma^\mu\partial_\mu - M)\psi - g_\sigma\sigma\overline{\psi}\psi - g_\omega\overline{\psi}\gamma^\mu\omega_\mu\psi - \frac{g_\rho}{2}\overline{\psi}\gamma^\mu\vec{\rho}_\mu\vec{\tau}\psi, \tag{2.3}$$

$$\mathcal{L}_\sigma = \frac{1}{2}(\partial^\mu\sigma\partial_\mu\sigma - m_\sigma^2\sigma^2) - \frac{A}{3}\sigma^3 - \frac{B}{4}\sigma^4, \tag{2.4}$$

$$\mathcal{L}_\omega = -\frac{1}{4}F^{\mu\nu}F_{\mu\nu} + \frac{1}{2}m_\omega^2\omega_\mu\omega^\mu + \frac{c}{4}(g_\omega^2\omega_\mu\omega^\mu)^2, \tag{2.5}$$

$$\mathcal{L}_\rho = -\frac{1}{4}\vec{B}^{\mu\nu}\vec{B}_{\mu\nu} + \frac{1}{2}m_\rho^2\vec{\rho}_\mu\vec{\rho}^\mu, \tag{2.6}$$

and

$$\mathcal{L}_{\sigma\omega\rho} = -g_\sigma g_\omega^2\sigma\omega_\mu\omega^\mu\left(\alpha_1 - \frac{1}{2}\alpha_1' g_\sigma\sigma\right) - g_\sigma g_\rho^2\sigma\vec{\rho}_\mu\vec{\rho}^\mu\left(\alpha_2 - \frac{1}{2}\alpha_2' g_\sigma\sigma\right)$$
$$+ \frac{1}{2}\alpha_3' g_\omega^2 g_\rho^2\omega_\mu\omega^\mu\vec{\rho}_\mu\vec{\rho}^\mu. \tag{2.7}$$

In this Lagrangian density, $\mathcal{L}_{nm}$ stands for the kinetic part of the nucleons added to the terms representing the interaction between the nucleons and mesons σ, ω, and ρ. The term $\mathcal{L}_j$ represents the free and self-interacting terms of the meson j, for $j = \sigma, \omega$, and ρ. The last term, $\mathcal{L}_{\sigma\omega\rho}$, takes into account crossed interactions between the meson fields. The antisymmetric field tensors $F_{\mu\nu}$ and $\vec{B}_{\mu\nu}$ are given by $F_{\mu\nu} = \partial_\nu\omega_\mu - \partial_\mu\omega_\nu$ and $\vec{B}_{\mu\nu} = \partial_\nu\vec{\rho}_\mu - \partial_\mu\vec{\rho}_\nu$. The nucleon mass is M, and the meson masses are m_j. The nucleon isospin is $\vec{\tau}$ with the arrow in this quantity, and in the field $\vec{\rho}$, indicating that these objects are vectors in isospin space.

Notice that the model described by the Lagrangian density shown in Eqs. (2.2)–(2.7) is much more sophisticated than the Boguta-Bodmer model [14] (found by taking the particular case of $c = \alpha_1 = \alpha_2 = \alpha'_1 = \alpha'_2 = \alpha'_3 = 0$) and the Walecka one ($A = B = c = \alpha_1 = \alpha_2 = \alpha'_1 = \alpha'_2 = \alpha'_3 = 0$) in the sense that self and crossed interactions between the mesons are taken into account. The inclusion of such terms directly affects nuclear matter at zero temperature. For example, the term containing the self-interaction of the meson ω, whose strength is regulated by the c parameter, is important in the high-density regime. One of the applications of hadronic models in this regime is the description of neutron stars. The value of their masses is directly affected by the value of the coupling parameter c [16]. Concerning crossed-interaction terms, it is also known that the interaction between the mesons ρ and ω, for instance, modifies the radius of neutron stars. It also favors softening the symmetry's dependence on energy density [17].

The nucleon and the meson fields presented in Eqs. (2.2)–(2.7) are determined through the Euler-Lagrange equations, given by,

$$\partial_\mu \frac{\partial \mathcal{L}}{\partial(\partial_\mu Q_i)} - \frac{\partial \mathcal{L}}{\partial Q_i} = 0, \tag{2.8}$$

with Q_i being the fields of the theory. The fields satisfy the following nonlinear field equations:

$$(\partial_\mu \partial^\mu + m_\sigma^2)\sigma = -g_\sigma \overline{\psi}\psi - A\sigma^2 - B\sigma^3 - g_\sigma g_\omega^2 \omega_\mu \omega^\mu (\alpha_1 - \alpha'_1 g_\sigma \sigma) - g_\sigma g_\rho^2 \vec{\rho}_\mu \vec{\rho}^\mu (\alpha_2 - \alpha'_2 g_\sigma \sigma), \tag{2.9}$$

$$\partial_\nu F^{\mu\nu} - m_\omega^2 \omega^\mu = -g_\omega \overline{\psi}\gamma^\mu \psi + cg_\omega (g_\omega \omega^\mu)^3 - g_\sigma g_\omega^2 \sigma \omega^\mu (2\alpha_1 - \alpha'_1 g_\sigma \sigma) + \alpha'_3 g_\omega^2 g_\rho^2 \vec{\rho}_\mu \vec{\rho}^\mu \omega^\mu, \tag{2.10}$$

$$\partial_\nu B^{\mu\nu} - m_\rho^2 \vec{\rho}^\mu = -\frac{g_\rho}{2}\overline{\psi}\gamma^\mu \vec{\tau}\psi - g_\sigma g_\rho^2 \sigma \vec{\rho}^\mu (2\alpha_2 - \alpha'_2 g_\sigma \sigma) + \alpha'_3 g_\omega^2 g_\rho^2 \omega_\mu \omega^\mu \vec{\rho}^\mu, \tag{2.11}$$

and

$$[\gamma^\mu(i\partial_\mu - V_\mu) - M^*]\psi = 0. \tag{2.12}$$

One can interpret Eq. (2.12) as a Dirac equation representing a nucleon of effective mass given by $M^* = M + S = M + g_\sigma \sigma$ depending on the scalar field σ and, consequently, on the density since σ depends on ρ, as we will explicitly show later on. From this Dirac equation, one can also define the scalar potential S, namely, a quantity which shifts the nucleon mass. In addition, the nucleon four-momentum is shifted by the quantity $V_\mu = g_\omega \omega_\mu + \frac{g_\rho}{2}\vec{\rho}_\mu \vec{\tau}$. This definition for the nucleon effective mass, also called Dirac mass [18], is purely relativistic. Other definitions of M^*, including a nonrelativistic one, can be found in Ref. [18], for instance. In the relativistic framework, the value of M^* at the saturation density is also related to the spin-orbit splitting in finite nuclei [19].

A widely used approach in relativistic models for the treatment of field equations given by Eqs. (2.9)–(2.12) is the so-called mean-field approximation (MFA) [6,12]. It consists in considering that the system is composed of static and uniform matter in its ground state. Therefore, the fields are replaced by their expectation values, namely,

$$\sigma \to \langle \sigma \rangle \equiv \sigma, \quad \omega_\mu \to \langle \omega_\mu \rangle \equiv \omega_0, \quad \text{and} \quad \vec{\rho}_\mu \to \langle \vec{\rho}_\mu \rangle \equiv \bar{\rho}_0, \tag{2.13}$$

Due to rotational invariance, the spatial components of the four vectors vanish. The source terms have also to be replaced in the form $\overline{\psi}_i\psi_i \to \langle \overline{\psi}_i\psi_i \rangle$. Then, one has

$$\overline{\psi}\psi \to \langle \overline{\psi}\psi \rangle \equiv \rho_s, \quad \overline{\psi}\gamma^\mu \psi \to \langle \overline{\psi}\gamma^\mu \psi \rangle = \langle \overline{\psi}\gamma^0 \psi \rangle \equiv \rho \tag{2.14}$$

and

$$\overline{\psi}\gamma^\mu\vec{\tau}\psi \to \langle\overline{\psi}\gamma^\mu\vec{\tau}\psi\rangle = \langle\overline{\psi}\gamma^0\tau_3\psi\rangle \equiv \rho_3, \tag{2.15}$$

with $\rho_s = \rho_{sp} + \rho_{sn}$ and $\rho = \rho_p + \rho_n$ being the sum of the scalar and vector proton and neutron densities, respectively. Furthermore, the difference between proton and neutron densities is $\rho_3 = \rho_p - \rho_n = (2y-1)\rho$. The third isospin component, τ_3, is equal to 1 for protons and -1 for neutrons, and the proton fraction of the system is defined as $y = \rho_p/\rho$. By taking into account the MFA, the field equations become

$$m_\sigma^2\sigma = -g_\sigma\rho_s - A\sigma^2 - B\sigma^3 - g_\sigma g_\omega^2\omega_0^2(\alpha_1 - \alpha_1' g_\sigma\sigma) - g_\sigma g_\rho^2\bar{\rho}_0^2(\alpha_2 - \alpha_2' g_\sigma\sigma), \tag{2.16}$$

$$m_\omega^2\omega_0 = g_\omega\rho - cg_\omega(g_\omega\omega_0)^3 + g_\sigma g_\omega^2\sigma\omega_0(2\alpha_1 - \alpha_1' g_\sigma\sigma) - \alpha_3' g_\omega^2 g_\rho^2\bar{\rho}_0^2\omega_0, \tag{2.17}$$

$$m_\rho^2\bar{\rho}_0 = \frac{g_\rho}{2}\rho_3 + g_\sigma g_\rho^2\sigma\bar{\rho}_0(2\alpha_2 - \alpha_2' g_\sigma\sigma) - \alpha_3' g_\omega^2 g_\rho^2\bar{\rho}_0\omega_0^2 \tag{2.18}$$

and

$$(i\gamma^\mu\partial_\mu - V_0\gamma^0 - M^*)\psi = 0, \tag{2.19}$$

with

$$V_0 \equiv V = g_\omega\omega_0 + \frac{g_\rho}{2}\bar{\rho}_0\tau_3. \tag{2.20}$$

The scalar potential is computed in the MFA as

$$S = g_\sigma\sigma, \tag{2.21}$$

with σ being the expectation value of the scalar field obtained from the solution of the set of coupled equations given in Eqs. (2.16)–(2.18).

By treating Eq. (2.19) as a Dirac equation, it is possible to write the spinor ψ as a plane wave of the form

$$\psi(\vec{x},t) = u(\vec{k},\lambda)e^{-ik_\mu x^\mu}, \tag{2.22}$$

where $u(\vec{k},s)$ is a spinor of wave vector $\vec{k}$ and spin polarization state λ. The replacement of this quantity in Eq. (2.19) gives

$$(\gamma^\mu k_\mu - V\gamma^0 - M^*)u(\vec{k},\lambda) = 0. \tag{2.23}$$

The product of this equation with the matrix β to the left of both sides and the use of $\alpha^i = \beta\gamma^i$ leads to

$$(\vec{\alpha}\cdot\vec{k} + \beta M^*)u = (k_0 - V)u \equiv E^*u. \tag{2.24}$$

From the structure of this last equation, we conclude that $E^* = \pm\sqrt{k^2 + M^{*2}}$, with the positive (negative) sign related to particles (anti-particles). Since $k_0 = E$, the nucleon energy is then written as

$$E = V \pm \sqrt{k^2 + M^{*2}}, \tag{2.25}$$

i.e., one can verify that the effect of the vector potential V in the relativistic treatment is to shift the nucleon energy.

In order to obtain analytical expressions for the densities given in Eqs. (2.14) and (2.15), we take Eq. (2.24) as

$$u^\dagger(\vec{\alpha}\cdot\vec{k} + \beta M^*) = E^*u^\dagger. \tag{2.26}$$

The product of Eq. (2.24) by $u^\dagger \beta$ (to the left of both sides) and the product of Eq. (2.26) by βu (to the right of both sides) lead to

$$u^\dagger \beta \vec{\alpha} \cdot \vec{k} u + u^\dagger u M^* = u^\dagger \beta u E^* \tag{2.27}$$

and

$$u^\dagger \vec{\alpha} \cdot \vec{k} \beta u + u^\dagger u M^* = u^\dagger \beta u E^*. \tag{2.28}$$

Since $\{\beta, \alpha_i\} = 0$ and $u^\dagger \beta u = u^\dagger \gamma_0 u = \bar{u} u$, the sum of Eqs. (2.27) and (2.28) produces

$$\bar{u} u = \frac{M^*}{E^*} u^\dagger u. \tag{2.29}$$

The analogous procedures with the plane wave spinor for anti-particles given by

$$\psi(\vec{x}, t) = v(\vec{k}, \lambda) e^{ik_\mu x^\mu}, \tag{2.30}$$

lead to

$$\bar{v} v = -\frac{M^*}{E^*} v^\dagger v. \tag{2.31}$$

Now we use the following complete expansion of $\psi(\vec{x}, t)$ and $\psi^\dagger(\vec{x}, t)$

$$\psi(\vec{x}, t) = \frac{1}{(2\pi)^3} \int d^3 k \left[a_{\vec{k}, \lambda} u(\vec{k}, \lambda) e^{-ik_\mu x^\mu} + b_{\vec{k}, \lambda}^\dagger v(\vec{k}, \lambda) e^{ik_\mu x^\mu} \right] \tag{2.32}$$

and

$$\psi^\dagger(\vec{x}, t) = \frac{1}{(2\pi)^3} \int d^3 k \left[a_{\vec{k}, \lambda}^\dagger u^\dagger(\vec{k}, \lambda) e^{ik_\mu x^\mu} + b_{\vec{k}, \lambda} v^\dagger(\vec{k}, \lambda) e^{-ik_\mu x^\mu} \right], \tag{2.33}$$

to compute, for instance, the vector densities

$$\rho_{p,n} = \langle \bar{\psi} \gamma^0 \psi \rangle = \langle 0 | \psi^\dagger \psi | 0 \rangle = \langle 0 | \int d^3 x \, \psi^\dagger \psi \, | 0 \rangle = \langle 0 | \frac{1}{(2\pi)^3} \int d^3 k \left(a_{\vec{k}, \lambda}^\dagger a_{\vec{k}, \lambda} + b_{\vec{k}, \lambda} b_{\vec{k}, \lambda}^\dagger \right) | 0 \rangle$$

$$= \langle 0 | \frac{1}{(2\pi)^3} \int d^3 k \, (a_{\vec{k}, \lambda}^\dagger a_{\vec{k}, \lambda} - b_{\vec{k}, \lambda}^\dagger b_{\vec{k}, \lambda}) | 0 \rangle = \frac{\gamma}{(2\pi)^3} \int d^3 k = \frac{\gamma}{2\pi^2} \int_0^{k_{Fp,n}} k^2 dk$$

$$= \frac{\gamma}{6\pi^2} k_{Fp,n}^3. \tag{2.34}$$

The same procedure is used to determine the scalar densities, namely,

$$\rho_{sp,n} = \langle \bar{\psi} \psi \rangle = \langle 0 | \bar{\psi} \psi | 0 \rangle = \langle 0 | \int d^3 x \, \bar{\psi} \psi \, | 0 \rangle = \langle 0 | \frac{1}{(2\pi)^3} \int d^3 k \frac{M^*}{E^*} \left(a_{\vec{k}, \lambda}^\dagger a_{\vec{k}, \lambda} - b_{\vec{k}, \lambda} b_{\vec{k}, \lambda}^\dagger \right) | 0 \rangle$$

$$= \langle 0 | \frac{1}{(2\pi)^3} \int d^3 k \frac{M^*}{E^*} \left(a_{\vec{k}, \lambda}^\dagger a_{\vec{k}, \lambda} + b_{\vec{k}, \lambda}^\dagger b_{\vec{k}, \lambda} \right) | 0 \rangle$$

$$= \frac{\gamma}{(2\pi)^3} \int d^3 k \frac{M^*}{E^*} = \frac{\gamma M^*}{2\pi^2} \int_0^{k_{Fp,n}} \frac{k^2 dk}{\sqrt{k^2 + M^{*2}}}$$

$$= \frac{\gamma M^* k_{Fp,n}^2}{2\pi^2} \int_0^1 \frac{x^2 dx}{\sqrt{x^2 + z^2}} = \frac{\gamma M^* k_{Fp,n}^2}{2\pi^2} \left[\frac{\sqrt{1 + z^2}}{2} - \frac{z^2}{2} \ln \left(\frac{1 + \sqrt{1 + z^2}}{z} \right) \right], \tag{2.35}$$

where $x = k/k_{Fp,n}$ and $z = M^*/k_{Fp,n}$. The degeneracy factor is $\gamma = 2$ ($\gamma = 4$) for asymmetric (symmetric) matter. The proton (neutron) Fermi momentum is given by k_{Fp} (k_{Fn}).

Finally, the expression for the nucleon effective mass in the MFA can be rewritten by using the field equation for σ as

$$M^* = M - G_\sigma^2 \left[\rho_s + a(\Delta M)^2 + b(\Delta M)^3 + g_\omega^2 \omega_0^2 (\alpha_1 - \alpha_1' \Delta M) + g_\rho^2 \bar{\rho}_0^2 (\alpha_2 - \alpha_2' \Delta M) \right],$$
$$(2.36)$$

with $\Delta M = M^* - M$, $G_\sigma^2 = g_\sigma^2/m_\sigma^2$, $a = A/g_\sigma^3$, and $b = B/g_\sigma^4$. For nonlinear models without crossed interactions between mesonic fields, the effective mass becomes $M^* = M - G_\sigma^2 \left[\rho_s + a(\Delta M)^2 + b(\Delta M)^3 \right]$, and for the Walecka model, M^* is proportional to the scalar density since in this case one has $M^* = M - G_\sigma^2 \rho_s$.

The equations of state of the nonlinear models we will present in the next subsection are determined for each density ρ and proton fraction y from the simultaneous solution of the set of equations given by Eqs. (2.16)–(2.18). Notice that Eq. (2.16) can be replaced by the effective mass self-consistent equation given in Eq. (2.36), since $\sigma = \Delta M/g_\sigma$.

2.2.2   Energy Density and Pressure

All thermodynamics related to the RMF model described before can be constructed within the MFA. The starting point is the determination of the energy-momentum tensor, defined as

$$T_{\mu\nu} = -g_{\mu\nu} \mathcal{L} + \sum_i \frac{\partial \mathcal{L}}{\partial(\partial_\mu Q_i)} \partial_\nu Q_i.$$
$$(2.37)$$

For the RMF model obtained from the MFA, this quantity reads

$$T_{\mu\nu}^{\mathrm{MFA}} = -g_{\mu\nu} \left[-\frac{m_\sigma^2}{2} \sigma^2 - \frac{A}{3} \sigma^3 - \frac{B}{4} \sigma^4 + \frac{m_\omega^2}{2} \omega_0^2 + \frac{c}{4} (g_\omega^2 \omega_0^2)^2 + \frac{1}{2} m_\rho^2 \bar{\rho}_0^2 \right.$$
$$\left. + \frac{1}{2} \alpha_3' g_\omega^2 g_\rho^2 \omega_0^2 \bar{\rho}_0^2 - g_\sigma g_\omega^2 \sigma \omega_0^2 \left(\alpha_1 - \frac{1}{2} \alpha_1' g_\sigma \sigma \right) - g_\sigma g_\rho^2 \sigma \bar{\rho}_0^2 \left(\alpha_2 - \frac{1}{2} \alpha_2' g_\sigma \sigma \right) \right] + i \bar{\psi} \gamma^\mu \partial_\nu \psi.$$

From that, it is possible to determine an expression for the energy density of the model, since they are related through $\mathcal{E} = \langle T_{00}^{\mathrm{MFA}} \rangle$. Therefore, one has

$$\mathcal{E} = \frac{1}{2} m_\sigma^2 \sigma^2 + \frac{A}{3} \sigma^3 + \frac{B}{4} \sigma^4 - \frac{1}{2} m_\omega^2 \omega_0^2 - \frac{c}{4} (g_\omega^2 \omega_0^2)^2 - \frac{1}{2} m_\rho^2 \bar{\rho}_0^2 - \frac{1}{2} \alpha_3' g_\omega^2 g_\rho^2 \omega_0^2 \bar{\rho}_0^2$$
$$+ g_\sigma g_\omega^2 \sigma \omega_0^2 \left(\alpha_1 - \frac{1}{2} \alpha_1' g_\sigma \sigma \right) + g_\sigma g_\rho^2 \sigma \bar{\rho}_0^2 \left(\alpha_2 - \frac{1}{2} \alpha_2' g_\sigma \sigma \right) + i \langle \bar{\psi} \gamma^0 \partial_0 \psi \rangle. \qquad (2.38)$$

However,

$$i \langle \bar{\psi} \gamma^0 \partial_0 \psi \rangle = g_\omega \omega_0 \rho + \frac{g_\rho}{2} \bar{\rho}_0 \rho_3 + \frac{\gamma}{2\pi^2} \sum_{j=p,n} \int_0^{k_{Fj}} k^2 (k^2 + M^*)^{1/2} dk, \qquad (2.39)$$

leading to

$$\mathcal{E}(\rho, y) = \frac{1}{2} m_\sigma^2 \sigma^2 + \frac{A}{3} \sigma^3 + \frac{B}{4} \sigma^4 - \frac{1}{2} m_\omega^2 \omega_0^2 - \frac{c}{4} (g_\omega^2 \omega_0^2)^2 - \frac{1}{2} m_\rho^2 \bar{\rho}_0^2 + g_\omega \omega_0 \rho + \frac{g_\rho}{2} \bar{\rho}_0 \rho_3$$
$$+ g_\sigma g_\omega^2 \sigma \omega_0^2 \left(\alpha_1 - \frac{1}{2} \alpha_1' g_\sigma \sigma \right) + g_\sigma g_\rho^2 \sigma \bar{\rho}_0^2 \left(\alpha_2 - \frac{1}{2} \alpha_2' g_\sigma \sigma \right)$$
$$- \frac{1}{2} \alpha_3' g_\omega^2 g_\rho^2 \omega_0^2 \bar{\rho}_0^2 + \mathcal{E}_{\mathrm{kin}}^p + \mathcal{E}_{\mathrm{kin}}^n, \qquad (2.40)$$

with

$$\mathcal{E}_{kin}^{p,n} = \frac{\gamma}{2\pi^2} \int_0^{k_{F\,p,n}} k^2(k^2 + M^*)^{1/2} dk = \frac{\gamma k_{F\,p,n}^4}{2\pi^2} \int_0^1 x^2(x^2 + z^2)^{1/2} dx$$

$$= \frac{\gamma k_{F\,p,n}^4}{2\pi^2} \left[\left(1 + \frac{z^2}{2}\right) \frac{\sqrt{1+z^2}}{4} - \frac{z^4}{8} \ln\left(\frac{1 + \sqrt{1+z^2}}{z}\right) \right] \quad (2.41)$$

being the kinetic energy of protons and neutrons. From Eq. (2.40), it is possible to recover the particular case of the Boguta-Bodmer model ($c = \alpha_1 = \alpha_2 = \alpha_1' = \alpha_2' = \alpha_3' = 0$), namely,

$$\mathcal{E}(\rho, y) = \frac{1}{2} G_\omega^2 \rho^2 + \frac{1}{8} G_\rho^2 (2y-1)^2 \rho^2 + \frac{(\Delta M)^2}{2 G_\sigma^2} + \frac{a(\Delta M)^3}{3} + \frac{b(\Delta M)^4}{4} + \mathcal{E}_{kin}^p + \mathcal{E}_{kin}^n,$$

$$(2.42)$$

with $G_\omega^2 = g_\omega^2/m_\omega^2$ and $G_\rho^2 = g_\rho^2/m_\rho^2$.

Another important thermodynamical quantity, the pressure, is also obtained from the energy-momentum tensor through the following relation, $P = \langle T_{ii} \rangle /3$. Its expression is given by

$$P = -\frac{1}{2} m_\sigma^2 \sigma^2 - \frac{A}{3} \sigma^3 - \frac{B}{4} \sigma^4 + \frac{1}{2} m_\omega^2 \omega_0^2 + \frac{c}{4} (g_\omega^2 \omega_0^2)^2 + \frac{1}{2} m_\rho^2 \bar{\rho}_0^2 + \frac{1}{2} \alpha_3' g_\omega^2 g_\rho^2 \omega_0^2 \bar{\rho}_0^2$$

$$- g_\sigma g_\omega^2 \sigma \omega_0^2 \left(\alpha_1 - \frac{1}{2} \alpha_1' g_\sigma \sigma\right) - g_\sigma g_\rho^2 \sigma \bar{\rho}_0^2 \left(\alpha_2 - \frac{1}{2} \alpha_2' g_\sigma \sigma\right) + \frac{i}{3} \langle \bar{\psi} \gamma^i \partial_i \psi \rangle, \quad (2.43)$$

with

$$i \langle \bar{\psi} \gamma^i \partial_i \psi \rangle = \frac{\gamma}{2\pi^2} \sum_{j=p,n} \int_0^{k_{Fj}} k^2(k^2 + M^*)^{1/2} dk - \sum_{j=p,n} M^* \rho_{sj}, \quad (2.44)$$

leading to

$$P(\rho, y) = -\frac{1}{2} m_\sigma^2 \sigma^2 - \frac{A}{3} \sigma^3 - \frac{B}{4} \sigma^4 + \frac{1}{2} m_\omega^2 \omega_0^2 + \frac{c}{4} (g_\omega^2 \omega_0^2)^2 + \frac{1}{2} m_\rho^2 \bar{\rho}_0^2 + \frac{1}{2} \alpha_3' g_\omega^2 g_\rho^2 \omega_0^2 \bar{\rho}_0^2$$

$$- g_\sigma g_\omega^2 \sigma \omega_0^2 \left(\alpha_1 - \frac{1}{2} \alpha_1' g_\sigma \sigma\right) - g_\sigma g_\rho^2 \sigma \bar{\rho}_0^2 \left(\alpha_2 - \frac{1}{2} \alpha_2' g_\sigma \sigma\right) + P_{kin}^p + P_{kin}^n, \quad (2.45)$$

in which

$$P_{kin}^{p,n} = \frac{\gamma}{6\pi^2} \int_0^{k_{F\,p,n}} \frac{k^4 dk}{(k^2 + M^*)^{1/2}} = \frac{\gamma k_{F\,p,n}^4}{6\pi^2} \int_0^1 \frac{x^4 dx}{(x^2 + z^2)^{1/2}}$$

$$= \frac{\gamma k_{F\,p,n}^4}{6\pi^2} \left[\left(1 - \frac{3z^2}{2}\right) \frac{\sqrt{1+z^2}}{4} + \frac{3z^4}{8} \ln\left(\frac{1 + \sqrt{1+z^2}}{z}\right) \right] \quad (2.46)$$

are the kinetic pressure of protons and neutrons. For the Boguta-Bodmer model, the pressure reads

$$P(\rho, y) = \frac{1}{2} G_\omega^2 \rho^2 + \frac{1}{8} G_\rho^2 (2y-1)^2 \rho^2 - \frac{(\Delta M)^2}{2 G_\sigma^2} - \frac{a(\Delta M)^3}{3} - \frac{b(\Delta M)^4}{4} + P_{kin}^p + P_{kin}^n.$$

$$(2.47)$$

In the calculations for the energy density and the pressure, we used the Dirac equation Eq. (2.19), the dispersion relation Eq. (2.25), and the plane wave structure of the

Dirac spinor. The vacuum polarization and the contributions of the negative-energy solutions are not explicitly taken into account. This last approach is named as the no-sea approximation (the Dirac sea is disregarded).

In Figure 2.2, we show the energy per nucleon, obtained from Eq. (2.40), of four different parametrizations of the RMF model described by the Lagrangian density shown in Eqs. (2.2)–(2.7). In this figure, we present results of symmetric matter, obtained by making $y = 1/2$ in the field equations (2.16)–(2.18) and in Eq. (2.40). Notice here that the minimum of the curves, defined as the binding energy of the system, is $B \sim -16$ MeV. This is a well-established value reported in the literature. Still regarding the minimum of these curves, it is also possible to relate it with a vanishing pressure. Actually, the energy per volume of a thermodynamical system at zero temperature, as in the case we are treating here, is given by $\mathcal{E} = E/V$. Its density derivative is

$$\frac{\partial \mathcal{E}}{\partial \rho} = \frac{\partial (E/V)}{\partial (N/V)} = \frac{\partial (E/V)}{\partial V} \frac{\partial V}{\partial (N/V)} = -\left(\frac{1}{V}\frac{\partial E}{\partial V} - \frac{E}{V^2}\right)\frac{V^2}{N} = \frac{V}{N}\left(-\frac{\partial E}{\partial V}\right) + \frac{E}{N} = \frac{P + \mathcal{E}}{\rho},$$
(2.48)

since $P = -\dfrac{\partial E}{\partial V}$. At a particular density ρ_0 in which the pressure vanishes, i.e., for $P(\rho_0) = 0$, one has

$$\left.\frac{\partial \mathcal{E}}{\partial \rho}\right|_{\rho=\rho_0} = \frac{\mathcal{E}(\rho_0)}{\rho_0}.$$
(2.49)

Moreover,

$$\frac{\partial (\mathcal{E}/\rho)}{\partial \rho} = \frac{1}{\rho}\frac{\partial \mathcal{E}}{\partial \rho} - \frac{\mathcal{E}}{\rho^2},$$
(2.50)

and due to Eq. (2.49), at $\rho = \rho_0$, one has

$$\left.\frac{\partial (\mathcal{E}/\rho)}{\partial \rho}\right|_{\rho=\rho_0} = \frac{1}{\rho_0}\left.\frac{\partial \mathcal{E}}{\partial \rho}\right|_{\rho=\rho_0} - \frac{\mathcal{E}(\rho_0)}{\rho_0^2} = 0.$$
(2.51)

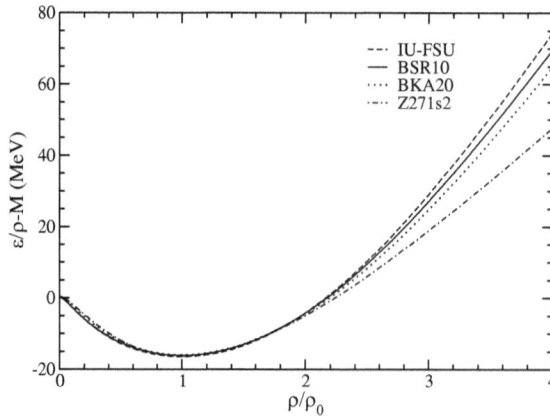

FIGURE 2.2
Energy per nucleon as a function of the density, in units of the saturation density ρ_0, for the following parametrizations of the RMF model: IU-FSU [20], BSR10 [21], BKA20 [22], and Z271s2 [23]. Results for symmetric nuclear matter.

Therefore, the function $\mathcal{E}/\rho$, or equivalently $\mathcal{E}/\rho - M$ (in the relativistic case, one has to subtract the nucleon rest mass in order to correctly compute the energy of the system), has extremes at densities where the pressure vanishes. In other words, use of the condition given in Eq. (2.51) implies the assumption of $P(\rho_0) = 0$.

An important constraint imposed on hadronic parametrizations is its pressure-density dependence. More specifically, this constraint is defined by the limits on the pressure versus density curve at symmetric nuclear matter. The density dependence of the pressure and its curvature can be obtained from analysis of experimental data on the motion of ejected matter in energetic nucleus-nucleus collisions. Measurements of the particle flow in collisions of ^{197}Au nuclei at incident kinetic energy per nucleon varying from about 0.15 to 10 GeV were analyzed in Ref. [24]. The authors extrapolated available data for pressure at about $2\rho_0$ to higher densities, as well as to zero-temperature regime to define the so called flow constraint. In Figure 2.3, the pressure of some parametrizations of the RMF model, namely, BKA22 [22], BKA24 [22], BSR11 [21], BSR12 [21], against the flow constraint is exhibited. In the figure, the same region increased by 20% in its upper limit is also shown. In Ref. [25], the authors extracted the radius r of a 1.4 solar mass neutron star in the range of $10.4 \leqslant r \leqslant 12.9$ km, generating a new constraint that equations of state must satisfy for the mass-radius relation of neutron stars. Their analysis was based on observational data of bursting neutron stars showing photospheric radius expansion, and of transiently accreting neutron stars in quiescence. As a consequence, they also established a new range of validity for the density dependence of the pressure in symmetric nuclear matter, consistent with the previous constraint proposed in Ref. [24] in the lower-pressure region. In the high-pressure region, however, the new constraint is broader than the former. In order to take into account this new phenomenology, the new increasing region of 20% was considered in Ref. [26]. In this reference, the authors collected a set of 263 parametrizations of RMF models, including the one described by Eqs. (2.2)–(2.7). The authors submitted the set to three different sets of constraints, improving the analysis performed in Ref. [27], in which a systematic study was also developed for 240 parametrizations of the nonrelativistic Skyrme model.

Another feature present in the relativistic context is the emergence of the scalar potential S shown in Eq. (2.21). It plays the role of changing the nucleon mass, since an "effective" nucleon mass quite naturally arises in the relativistic treatment, namely, $M^* = M + S$, as we have mentioned before. Figure 2.4 displays the density dependence of this quantity for

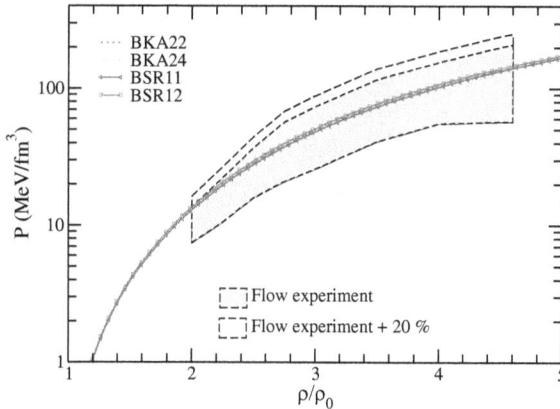

FIGURE 2.3

Pressure as a function of ρ/ρ_0 at symmetric matter for some parametrizations of the RMF mode. Grey band: flow constraint defined in Ref. [24].

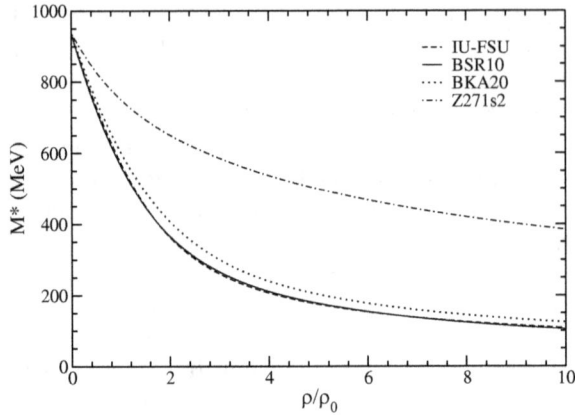

FIGURE 2.4
Nucleon effective mass as a function of the density, in units of ρ_0, for the same parametrizations used in Figure 2.2.

the same parametrizations depicted in Figure 2.1. It is clear from this figure that the effect imposed by the scalar potential in the RMF model is the decrease of the nucleon mass as a function of the density, i.e., the nucleon loses mass in the infinite nuclear matter medium. In particular, the value of M^* at $\rho = \rho_0$ was shown to directly affect the spin-orbit splittings of some finite nuclei as pointed out in Ref. [19]. The authors have shown that Boguta-Bodmer models, in which $0.58 \leqslant m^* \leqslant 0.64$, with $m^* \equiv M^*(\rho_0)/M$, present a good agreement with well-established experimental values for the ^{16}O, ^{40}Ca, and ^{208}Pb nuclei.

2.2.3   Incompressibility

The isothermal compressibility of a system is defined by [28]

$$\chi(\rho) \equiv -\frac{1}{V}\frac{\partial V}{\partial P} = -\frac{N}{V}\frac{\partial(V/N)}{\partial P} = -\rho\frac{\partial(1/\rho)}{\partial P} = -\rho\frac{\partial\rho}{\partial P}\frac{\partial(1/\rho)}{\partial\rho} = -\rho\left(-\frac{1}{\rho^2}\right)\frac{\partial\rho}{\partial P} = \frac{1}{\rho}\frac{\partial\rho}{\partial P}, \tag{2.52}$$

with $\rho = N/V$. From Eq. (2.48), it is possible to write the pressure of the system as

$$P = \rho\frac{\partial\mathcal{E}}{\partial\rho} - \mathcal{E} = \rho^2\frac{\partial(\mathcal{E}/\rho)}{\partial\rho}. \tag{2.53}$$

Therefore, the following relation holds

$$\frac{1}{\chi(\rho)} = \rho\frac{\partial P}{\partial\rho} = \rho\frac{\partial}{\partial\rho}\left[\rho^2\frac{\partial(\mathcal{E}/\rho)}{\partial\rho}\right] = 2\rho^2\frac{\partial(\mathcal{E}/\rho)}{\partial\rho} + \rho^3\frac{\partial^2(\mathcal{E}/\rho)}{\partial\rho^2} = 2P + \rho^3\frac{\partial^2(\mathcal{E}/\rho)}{\partial\rho^2}. \tag{2.54}$$

For infinite symmetric nuclear matter, the incompressibility at the saturation density is defined as [29–32]

$$K_0 \equiv K(\rho_0) = \left[k_F^2\frac{\partial^2(\mathcal{E}/\rho)}{\partial k_F^2}\right]\Bigg|_{k_{F0}} = \left[9\rho^2\frac{\partial^2(\mathcal{E}/\rho)}{\partial\rho^2}\right]\Bigg|_{\rho_0}, \tag{2.55}$$

since $\rho = \gamma k_F^3/(6\pi^2)$, with k_{F0} being the Fermi momentum at the saturation. Thus, the relationship between K_0 and $\chi_0 = \chi(\rho_0)$ is given by

$$K_0 = \frac{9}{\rho_0\chi_0}, \tag{2.56}$$

with the use of Eq. (2.54) evaluated at $\rho = \rho_0$. As mentioned in the previous subsection, at the saturation, one has $P(\rho_0) = 0$. Furthermore, a more general definition for the incompressibility as a function of the density can be extracted for the relation given in Eq. (2.56) as

$$K(\rho) \equiv \frac{9}{\rho\chi(\rho)} = 18\frac{P}{\rho} + 9\rho^2\frac{\partial^2(\mathcal{E}/\rho)}{\partial\rho^2} = 9\left[2\rho\frac{\partial(\mathcal{E}/\rho)}{\partial\rho} + \rho^2\frac{\partial^2(\mathcal{E}/\rho)}{\partial\rho^2}\right], \qquad (2.57)$$

where it was used Eq. (2.54). This is not the only way to write $K(\rho)$. It is possible to use the thermodynamical relations between the pressure and energy density to find out another expressions. For instance, let us analyze the following results:

$$\frac{\partial(\mathcal{E}/\rho)}{\partial\rho} = \frac{1}{\rho}\frac{\partial\mathcal{E}}{\partial\rho} - \frac{\mathcal{E}}{\rho^2} \qquad (2.58)$$

and

$$\frac{\partial^2(\mathcal{E}/\rho)}{\partial\rho^2} = \frac{1}{\rho}\frac{\partial^2\mathcal{E}}{\partial\rho^2} - \frac{2}{\rho^2}\frac{\partial\mathcal{E}}{\partial\rho} + \frac{2\mathcal{E}}{\rho^3}. \qquad (2.59)$$

By multiplying Eq. (2.58) by 2ρ and Eq. (2.59) by ρ^2, one has

$$K(\rho) = 9\left[2\rho\frac{\partial(\mathcal{E}/\rho)}{\partial\rho} + \rho^2\frac{\partial^2(\mathcal{E}/\rho)}{\partial\rho^2}\right] = 9\left[2\frac{\partial\mathcal{E}}{\partial\rho} - \frac{2\mathcal{E}}{\rho} + \rho\frac{\partial^2\mathcal{E}}{\partial\rho^2} - 2\frac{\partial\mathcal{E}}{\partial\rho} + \frac{2\mathcal{E}}{\rho}\right] = 9\rho\frac{\partial^2\mathcal{E}}{\partial\rho^2}. \qquad (2.60)$$

Furthermore, one can still write this expression as

$$K(\rho) = 9\rho\frac{\partial^2\mathcal{E}}{\partial\rho^2} = 9\rho\frac{\partial}{\partial\rho}\left(\frac{\partial\mathcal{E}}{\partial\rho}\right) = 9\rho\frac{\partial}{\partial\rho}\left(\frac{P+\mathcal{E}}{\rho}\right) = 9\left[\frac{\partial(P+\mathcal{E})}{\partial\rho} - \frac{P+\mathcal{E}}{\rho}\right], \qquad (2.61)$$

in which the relation given in Eq. (2.48) was used. Finally, from Eqs. (2.57) and (2.54), one also finds

$$K(\rho) = \frac{9}{\rho\chi(\rho)} = 9\frac{\partial P}{\partial\rho}. \qquad (2.62)$$

For systems at finite temperature, the expressions given in Eqs. (2.57), (2.60), and (2.61) have the same form; however, one needs to replace the energy density by the free energy density $f = F/V$, with $F = E - TS$ (S is the entropy). On the other hand, the expression given in Eq. (2.62) remains unchanged even at finite-temperature regime.

The incompressibility at the saturation density is a quantity of interest in nuclear physics, since it is related to the collective motion of protons and neutrons inside a nucleus [33]. It is experimentally determined from the analysis of the giant monopole resonances in heavy nuclei [34]. Many works establish constraints on K_0, see, for instance, Ref. [26]. As an example, the incompressibility, which defines the stiffness of the EoS, has a value in the range of $K_0 = (240 \pm 10)$ MeV, as found in Refs. [35–37], or even the range of 250 MeV $\leqslant K_0 \leqslant 315$ MeV, more recently obtained in Ref. [7] from a reanalysis of updated data on isoscalar giant monopole resonance energies of Sn and Cd isotopes. Some ranges for K_0 are found through a leptodermous expansion of the finite nucleus incompressibility, with K_0 as one of the terms. However, many works point out to the drawbacks of such a procedure, see, for instance, Refs. [38–40]. The current consensus regarding the value of K_0 is 220 MeV $\leqslant K_0 \leqslant 260$ MeV, as one can see in a very recent review on this subject in Ref. [41], for instance.

Due to the structure of $\mathcal{E}$ and P presented in the previous subsection, it is more convenient to use Eq. (2.61) in order to determine an expression for the incompressibility of the RMF model. We restrict the calculations to the symmetric matter case in which $y = 1/2$ and, consequently, $k_{Fp} = k_{Fn} \equiv k_F$. By taking this condition into account, one has

$$K(\rho) = 9 \left[g_\omega \rho \frac{\partial \omega_0}{\partial \rho} + \frac{k_F^2}{3(k_F^2 + M^{*2})^{1/2}} + \frac{\rho M^*}{(k_F^2 + M^{*2})^{1/2}} \frac{\partial M^*}{\partial \rho} \right], \qquad (2.63)$$

where the relation $\mathcal{E}_{\text{kin}} + P_{\text{kin}} = \rho \sqrt{k_F^2 + M^{*2}}$ was used. The density derivative of the effective mass is

$$\frac{\partial M^*}{\partial \rho} = g_\sigma \frac{\partial \sigma}{\partial \rho} = g_\sigma \frac{a_1 b_2 + a_2 b_3}{a_1 b_1 - a_3 b_3}, \qquad (2.64)$$

where

$$a_1 = m_\omega^2 + 3cg_\omega^4 \omega_0^2 - g_\sigma g_\omega^2 \sigma (2\alpha_1 - \alpha_1' g_\sigma \sigma), \quad a_2 = g_\omega, \quad a_3 = 2 g_\sigma g_\omega^2 \omega_0 (\alpha_1 - \alpha_1' g_\sigma \sigma),$$
$$(2.65)$$

$$b_1 = m_\sigma^2 + 2A\sigma + 3B\sigma^2 - g_\sigma^2 g_\omega^2 \omega_0^2 \alpha_1' + 3g_\sigma^2 \left(\frac{\rho_s}{M^*} - \frac{\rho}{E_F^*} \right), \, b_2 = -\frac{g_\sigma M^*}{E_F^*}, \quad \text{and} \quad b_3 = -a_3,$$
$$(2.66)$$

with $E_F^* = (k_F^2 + M^{*2})^{1/2}$. Moreover, one also has

$$\frac{\partial \omega_0}{\partial \rho} = \frac{a_2 b_1 + a_3 b_2}{a_1 b_1 - a_3 b_3}. \qquad (2.67)$$

The calculation used to obtain $\partial \sigma / \partial \rho$ must take into account the density derivative of the scalar density ρ_s, namely,

$$\frac{\partial \rho_s}{\partial \rho} = \frac{M^*}{E_F^*} + 3g_\sigma \left(\frac{\rho_s}{M^*} - \frac{\rho}{E_F^*} \right) \frac{\partial \sigma}{\partial \rho}, \qquad (2.68)$$

found from the expressions given in Eqs. (2.35) and (2.64).

A straightforward calculation for the Boguta-Bodmer model gives

$$\frac{\partial \sigma}{\partial \rho} = - \frac{g_\sigma M^*}{E_F^* \left[m_\sigma^2 + 2A\sigma + 3B\sigma^2 + 3g_\sigma^2 \left(\frac{\rho_s}{M^*} - \frac{\rho}{E_F^*} \right) \right]} \quad \text{and} \quad \frac{\partial \omega_0}{\partial \rho} = \frac{g_\omega}{m_\omega^2}, \qquad (2.69)$$

used to compute $K(\rho)$ as

$$K(\rho) = 9 G_\omega^2 \rho + \frac{3 k_F^2}{E_F^*} - \frac{9 M^{*2} \rho}{E_F^{*2} \left[\frac{1}{G_\sigma^2} + 2a\Delta M + 3b(\Delta M)^2 + 3 \left(\frac{\rho_s}{M^*} - \frac{\rho}{E_F^*} \right) \right]}, \qquad (2.70)$$

with $b = B/g_\sigma^4$, $\Delta M = M^* - M$, and $G_i^2 = g_i^2 / m_i^2$ for $i = \sigma, \omega$. An interesting feature of parametrizations related to this particular model is the relationship presented between K_0 and $Q_0 \equiv Q(\rho_0)$ (for symmetric matter, $y = 1/2$), with $Q(\rho)$ being the skewness coefficient, given by

$$Q(\rho) = 27 \rho^3 \frac{\partial^3 (\mathcal{E}/\rho)}{\partial \rho^3} = 27 \rho^3 \left[\frac{1}{\rho} \frac{\partial^3 \mathcal{E}}{\partial \rho^3} - \frac{3}{\rho^2} \frac{\partial^2 \mathcal{E}}{\partial \rho^2} + \frac{6}{\rho^3} \frac{\partial \mathcal{E}}{\partial \rho} - \frac{6\mathcal{E}}{\rho^4} \right], \qquad (2.71)$$

In Ref. [42], it was verified that a linear correlation arises between K_0 and Q_0 when one chooses parametrizations with fixed values of m^*. In Figure 2.5a, such a correlation is displayed (panel a) for some Boguta-Bodmer parametrizations, namely, MS2 [16], NLSH [43], NL4 [44], NLRA1 [45], Q1 [46], Hybrid [47], NL3 [43], FAMA1 [48], NL-VT1 [49], NL06 [26], and NLS [50]. In Figure 2.5b, the density dependence of $K(\rho)$ for the same parametrizations is also shown. It is observed that all curves cross each other at a particular "crossing point" at $\rho_c^K = 0.77\rho_0$, with ρ_c^K being the "crossing density" in the $K \times \rho$ curve. This kind of correlation was first verified in Refs. [31,32], where the authors associated the crossing point in the incompressibility function for different nonrelativistic Skyrme models (but not for RMF ones) with the linear correlation between K_0 and Q_0. In Ref. [42], the authors show that this linear correlation holds for the RMF model, more specifically for parametrizations of the Boguta-Bodmer model presenting the same effective mass at the saturation density.

2.2.4 Chemical Potentials

The chemical potential is defined as the energy needed to add (remove) a single particle to (from) the system. In terms of the energy density and nucleon density, μ_i is defined by

$$\mu_i = \frac{\partial \mathcal{E}}{\partial \rho_i},\tag{2.72}$$

for $i = p, n$. By using Eq. (2.40), one can write the expressions for the protons' and neutrons' chemical potentials for the RMF model as

$$\mu_i = c_1 \frac{\partial \sigma}{\partial \rho_i} + c_2 \frac{\partial \omega_0}{\partial \rho_i} + c_3 \frac{\partial \bar{\rho}_0}{\partial \rho_i} + \frac{\partial (\mathcal{E}_{\text{kin}}^p + \mathcal{E}_{\text{kin}}^n)}{\partial \rho_i} + c_4 \frac{\partial \rho}{\partial \rho_i} + c_5 \frac{\partial \rho_3}{\partial \rho_i}\tag{2.73}$$

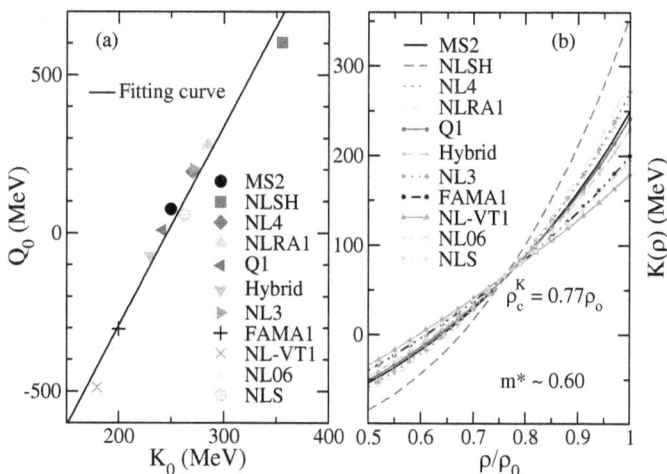

FIGURE 2.5
(a) Q_0 as a function of K_0 for some Boguta-Bodmer parametrizations with $m^* \sim 0.6$.
(b) Incompressibility as a function of ρ/ρ_0 for the same parametrizations. (Figure extracted from Ref. [42].)

with

$$c_1 = m_s^2\sigma + A\sigma^2 + B\sigma^3 + g_\sigma g_\omega^2\omega_0^2(\alpha_1 - \alpha_1' g_\sigma\sigma) + g_\sigma g_\rho^2\bar{\rho}_0^2(\alpha_2 - \alpha_2' g_\sigma\sigma) = -g_\sigma\rho_s, \quad (2.74)$$

$$c_2 = -m_\omega^2\omega_0 - cg_\omega(g_\omega\omega_0)^3 + g_\sigma g_\omega^2\sigma\omega_0(2\alpha_1 - \alpha_1' g_\sigma\sigma) - \alpha_3' g_\omega^2 g_\rho^2\bar{\rho}_0^2\omega_0 + g_\omega\rho = 0, \quad (2.75)$$

$$c_3 = -m_\rho^2\bar{\rho}_0 + g_\sigma g_\rho^2\sigma\bar{\rho}_0(2\alpha_2 - \alpha_2' g_\sigma\sigma) - \alpha_3' g_\omega^2 g_\rho^2\bar{\rho}_0\omega_0^2 + \frac{g_\rho}{2}\rho_3 = 0, \quad (2.76)$$

$$c_4 = g_\omega\omega_0 \quad \text{and} \quad c_5 = \frac{g_\rho}{2}\bar{\rho}_0. \quad (2.77)$$

By using

$$\frac{\partial(\mathcal{E}_{cin}^p + \mathcal{E}_{cin}^n)}{\partial\rho_i} = (k_F^2 + M^{*2})^{1/2} + \rho_s\frac{\partial M^*}{\partial\rho_i}, \quad (2.78)$$

along with the density derivative of the effective nucleon mass, given in Eq. (2.64), and the following derivatives: $\partial\rho/\partial\rho_i = 1$, $\partial\rho_3/\partial\rho_i = \pm 1$ with the positive (negative) sign for protons (neutrons), one can finally find

$$\mu_p = (k_F^2 + M^{*2})^{1/2} + g_\omega\omega_0 + \frac{g_\rho}{2}\bar{\rho}_0 \quad (2.79)$$

and

$$\mu_n = (k_F^2 + M^{*2})^{1/2} + g_\omega\omega_0 - \frac{g_\rho}{2}\bar{\rho}_0. \quad (2.80)$$

The total chemical potential is given in terms of μ_p and μ_n as $\mu = y\mu_p + (1-y)\mu_n$.

2.2.5   Symmetry Energy and Its Derivatives

From the isovector sector of the RMF model, important quantities can be derived such as the symmetry energy, $\mathcal{S}(\rho)$ and its derivatives. This quantity measures the change in binding of the nucleon system as the proton to neutron ratio is changed at a fixed value of the density, $\mathcal{S}(\rho) = E(\rho, 0) - E(\rho, 1/2)$, where $E(\rho, y) = \mathcal{E}/\rho$ is the energy per nucleon. A detailed analysis of the quantity is quite important for understanding many aspects of different isospin asymmetric systems, from astrophysics to finite nuclei. Furthermore, the symmetry energy slope at saturation density provides the dominant contribution to the pressure in neutron stars, as well as affecting the neutron skin thicknesses of heavy nuclei [51,52]. For a recent review regarding the importance of $\mathcal{S}(\rho)$, see Ref. [53]. By considering

$$E(\rho, y) \simeq E(\rho, 1/2) + \mathcal{S}_2(\rho)(1 - 2y)^2 + \mathcal{O}[(1 - 2y)^4], \quad (2.81)$$

one can take $\mathcal{S}(\rho) \simeq \mathcal{S}_2(\rho)$ as a good approximation in order to compute the symmetry energy. Therefore, the symmetry energy can be written as

$$\mathcal{S}(\rho) \simeq \mathcal{S}_2(\rho) = \frac{1}{8}\frac{\partial^2(\mathcal{E}/\rho)}{\partial y^2}\bigg|_{y=\frac{1}{2}}. \quad (2.82)$$

For the RMF model, the calculation of this quantity can be done in the following way: the derivative $\partial\mathcal{E}/\partial\rho_3$ is obtained analogously to the calculation for the chemical potentials performed in the previous subsection. The difference in this procedure is the result given by

$$\frac{\partial(\mathcal{E}_{\text{kin}}^p + \mathcal{E}_{\text{kin}}^n)}{\partial\rho_3} = \frac{1}{2}(E_{Fp}^* - E_{Fn}^*) + \rho_s\frac{\partial M^*}{\partial\rho_3}, \quad (2.83)$$

with $E_{Fp,n}^* = (k_{Fp,n}^2 + M^{*2})^{1/2}$, used in order to obtain

$$\frac{\partial\mathcal{E}}{\partial\rho_3} = \frac{1}{2}(E_{Fp}^* - E_{Fn}^*) + \frac{g_\rho}{2}\bar{\rho}_0. \quad (2.84)$$

From this expression, one finds the second derivative of the energy density as

$$\frac{\partial^2 \mathcal{E}}{\partial \rho_3^2} = \frac{1}{2}\left(\frac{\pi^2}{2E_{Fp}^* k_{Fp}} + \frac{M^*}{E_{Fp}^*}\frac{\partial M^*}{\partial \rho_3} + \frac{\pi^2}{2E_{Fn}^* k_{Fn}} - \frac{M^*}{E_{Fn}^*}\frac{\partial M^*}{\partial \rho_3}\right) + \frac{g_\rho}{2}\frac{\partial \bar{\rho}_0}{\partial \rho_3}. \qquad (2.85)$$

Since the symmetry energy is defined at $y = 1/2$, or equivalently $\rho_3 = 0$, one finally finds

$$\mathcal{S}(\rho) = \frac{k_F^2}{6(k_F^2 + M^{*2})^{1/2}} + \frac{g_\rho^2}{8m_\rho^{*2}}\rho, \qquad (2.86)$$

where $m_\rho^{*2} = m_\rho^2 - g_\sigma g_\rho^2 \sigma(2\alpha_2 - \alpha_2' g_\sigma \sigma) + \alpha_3' g_\omega^2 g_\rho^2 \omega_0^2$ is the effective ρ meson mass. In the case of the Boguta-Bodmer model, this quantity becomes $m_\rho^* = m_\rho$.

From Eq. (2.86), it is possible to determine the slope and curvature of the symmetry energy. These quantities are, respectively, given by

$$L(\rho) \equiv 3\rho\left(\frac{\partial \mathcal{S}}{\partial \rho}\right) = \frac{k_F^2}{3E_F^*} - \frac{k_F^4}{6E_F^{*3}}\left(1 + \frac{2M^* k_F}{\pi^2}\frac{\partial M^*}{\partial \rho}\right) + \frac{3g_\rho^2}{8m_\rho^{*2}}\rho - \frac{3g_\rho^2}{8m_\rho^{*4}}\frac{\partial m_\rho^{*2}}{\partial \rho}\rho^2 \quad (2.87)$$

and

$$\begin{aligned}
K_{\text{sym}}(\rho) \equiv 9\rho^2\left(\frac{\partial^2 \mathcal{S}}{\partial \rho^2}\right) = 9\rho^2\Bigg\{ &-\frac{\pi^2}{12E_F^{*3}k_F}\left(\frac{\pi^2}{k_F} + 2M^*\frac{\partial M^*}{\partial \rho}\right) - \frac{\pi^4}{12E_F^* k_F^4} - \frac{g_\rho^2}{4m_\rho^{*4}}\frac{\partial m_\rho^{*2}}{\partial \rho} \\
&- \left[\frac{\pi^4}{24E_F^{*3}k_F^2} - \frac{k_F \pi^2}{8E_F^{*5}}\left(\frac{\pi^2}{k_F} + 2M^*\frac{\partial M^*}{\partial \rho}\right)\right]\left(1 + \frac{2M^* k_F}{\pi^2}\frac{\partial M^*}{\partial \rho}\right) + \frac{g_\rho^2}{4m_\rho^{*6}}\left(\frac{\partial m_\rho^{*2}}{\partial \rho}\right)^2 \rho \\
&- \frac{k_F \pi^2}{12E_F^{*3}}\left[\frac{M^*}{k_F^2}\frac{\partial M^*}{\partial \rho} + \frac{2k_F}{\pi^2}\left(\frac{\partial M^*}{\partial \rho}\right)^2 + \frac{2k_F M^*}{\pi^2}\frac{\partial^2 M^*}{\partial \rho^2}\right] - \frac{g_\rho^2}{8m_\rho^{*4}}\frac{\partial^2 m_\rho^{*2}}{\partial \rho^2}\rho\Bigg\}. \qquad (2.88)
\end{aligned}$$

In Ref. [54], it was reported that for parametrizations of the Boguta-Bodmer model presenting the same values for m^*, a linear correlation arises between $J \equiv \mathcal{S}(\rho_0)$ and $L_0 \equiv L(\rho_0)$, exactly the same feature concerning the linear relationship for K_0 and Q_0. An example is depicted in Figure 2.6 for the same parametrizations used in Figure 2.5. Such a linear correlation was shown to be directly connected with the crossing point presented in the $\mathcal{S}(\rho)$ curve. In Figure 2.7a, we show the density dependence of $\mathcal{S}(\rho)$ for the RMF parametrizations studied in Ref. [55], namely, FSU-I, FSU-II, FSU-III, FSU-II, FSU-V, and FSUGold. In this figure, it is clear that there is a crossing between the curves at a crossing density of $\rho_c^S = 0.68\rho_0$. Actually, such parametrizations were constructed with this aim. In Figure 2.7b, the linear correlation between the symmetry energy and its slope, both at the saturation density, is displayed for the same parametrizations. For the details regarding the mathematical connection between crossing points and linear correlations for nuclear matter quantities at the saturation density, as presented in Figures 2.5 and 2.7, we address the reader to Ref. [42], in which a broader studied is performed.

The symmetry energy and its slope at the saturation density, namely, J and L_0, were used in Ref. [27] in order to constrain the parametrizations of the Skyrme model. In Ref. [26], their ranges were updated in order to take into account many values obtained from different sources. The ranges used are $25 \text{ MeV} \leqslant J \leqslant 35 \text{ MeV}$ and $25 \text{ MeV} \leqslant L_0 \leqslant 115 \text{ MeV}$. These ranges encompass a set of 28 values extracted from Ref. [56], as one can see in Figure 2.8. In this reference, the authors collected from the literature data obtained from analyses of different terrestrial nuclear experiments and astrophysical observations. They include analyses of isospin diffusion, neutron skins, pygmy dipole resonances, α and β decays, transverse flow, the mass-radius relation of neutron stars, and torsional crust oscillations of neutron stars.

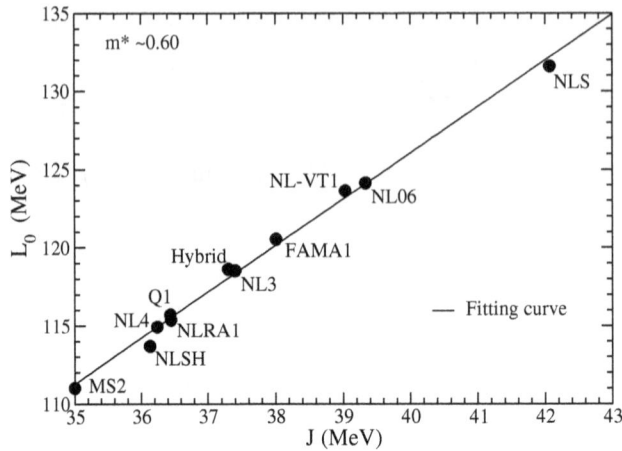

FIGURE 2.6
L_0 as a function of J for the same Boguta-Bodmer parametrizations presented in Figure 2.5 presenting $m^* \sim 0.6$. (Figure extracted from Ref. [54].)

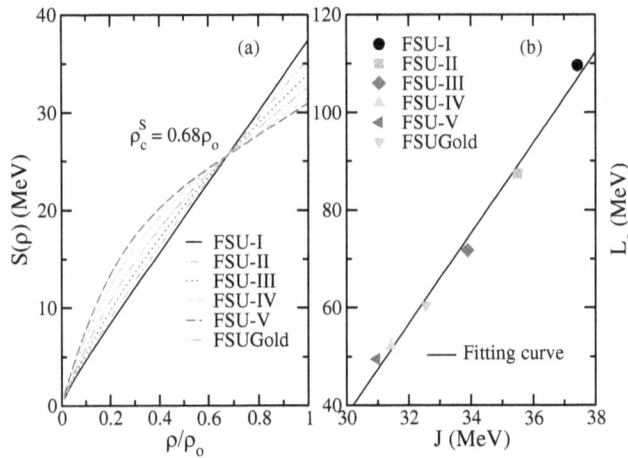

FIGURE 2.7
(a) $\mathcal{S}(\rho)$ as a function of ρ/ρ_0 for some parametrizations of the RMF model described by Eqs. (2.2)–(2.7). (b) Linear correlation between L_0 and J for the same parametrizations. (Figure extracted from Ref. [42].)

2.2.6   Neutron Star Environment

A neutron star (NS) is a very compact object composed not only of neutrons but also of protons and leptons. Different reactions such as the β decay, namely, $n \rightarrow p + e^- + \bar{\nu}_e$ and its inverse process $p + e^- \rightarrow n + \nu_e$, take place in the interior of an NS [8]. For densities in which the electron's chemical potential exceeds the muon mass value, the reactions $e^- \rightarrow \mu^- + \nu_e + \bar{\nu}_\mu$, $p + \mu^- \rightarrow n + \nu_\mu$, and $n \rightarrow p + \mu^- + \bar{\nu}_\mu$ may be energetically allowed. In this case, muons can also emerge. As an approximation, it is very often to disregard the neutrinos since their mean-free path is longer than the radius of the star. By taking these assumptions into account, one can write total energy density and total pressure of the stellar system, respectively, as

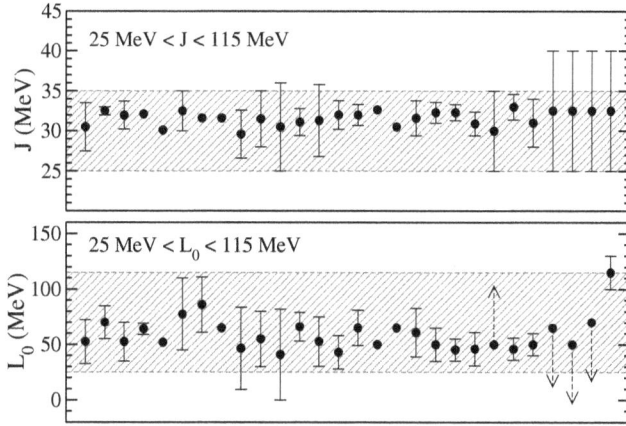

FIGURE 2.8

Ranges of J and L_0 used as constraints in Ref. [26] compared with data extracted from Ref. [56].

$$\mathcal{E}_T(\rho, \rho_e, y) = \mathcal{E}(\rho, y) + \frac{\mu_e^4(\rho_e)}{4\pi^2} + \frac{1}{\pi^2} \int_0^{\sqrt{\mu_\mu^2(\rho_e) - m_\mu^2}} dk\, k^2 (k^2 + m_\mu^2)^{1/2}, \qquad (2.89)$$

and

$$P_T(\rho, \rho_e, y) = P(\rho, y) + \frac{\mu_e^4(\rho_e)}{12\pi^2} + \frac{1}{3\pi^2} \int_0^{\sqrt{\mu_\mu^2(\rho_e) - m_\mu^2}} \frac{dk\, k^4}{(k^2 + m_\mu^2)^{1/2}}, \qquad (2.90)$$

with the muon mass given by $m_\mu = 105.7$ MeV, and massless electrons of density ρ_e and chemical potential μ_e. The energy density and pressure related to the particular hadronic model used to describe the stellar matter are $\mathcal{E}(\rho, y)$ and $P(\rho, y)$, respectively. In the case of the RMF model, these quantities are given in Eqs. (2.40) and (2.45). For the description of the stellar matter composed of protons, neutrons, electrons, and muons, the chemical equilibrium and charge neutrality conditions established in an NS [8] are given, respectively, by

$$\mu_n(\rho, y) - \mu_p(\rho, y) = \mu_e(\rho_e) \qquad (2.91)$$

and

$$\rho_p(\rho, y) - \rho_e = \rho_\mu(\rho_e) \qquad (2.92)$$

with μ_p and μ_n defined in Eqs. (2.79) and (2.80) for the RMF model. Furthermore, one has $\mu_e = (3\pi^2 \rho_e)^{1/3}$, $\rho_\mu = [(\mu_\mu^2 - m_\mu^2)^{3/2}]/(3\pi^2)$, and $\mu_\mu = \mu_e$. Therefore, for each input density ρ, the quantities ρ_e and y are calculated by solving the restrictions for the chemical potentials and densities simultaneously. The output is used to compute $\mathcal{E}_T(\rho, \rho_e, y)$ and $P_T(\rho, \rho_e, y)$ as a function of the density.

The structure of a spherically symmetric and static NS can be studied in terms of the energy density and pressure density by using the Tolman-Oppenheimer-Volkoff (TOV) equations. The general form of the TOV equations [57] is given by,

$$\frac{dp(r)}{dr} = -\frac{[\epsilon(r) + p(r)][m_{\text{NS}}(r) + 4\pi r^3 p(r)]}{r^2 \left[1 - \dfrac{2m_{\text{NS}}(r)}{r}\right]}, \qquad (2.93)$$

and

$$\frac{dm_{\mathrm{NS}}(r)}{dr} = 4\pi r^2 \epsilon(r), \tag{2.94}$$

where the solution is constrained to the following conditions: (i) at the center, $p(0) = p_c$ (central pressure), (ii) $m_{\mathrm{NS}}(0) = 0$ (central mass), and (iii) $\epsilon(0) = \epsilon_c$ (central energy density). At the star surface, in which $r \equiv R$, one has $p(R) = 0$ and $m_{\mathrm{NS}}(R) \equiv M_{\mathrm{NS}}$. These equations represent a star in which pressure coming from hadrons and leptons supports the gravitational pressure, preventing its collapse. In order to solve the TOV equations, one takes $\epsilon = \mathcal{E}_T(\rho, \rho_e, y)$ and $p = P_T(\rho, \rho_e, y)$ as input for Eqs. (2.93) and (2.94). For the description of the NS crust (low-density regime), it is often to use another equation of state that is able to describe the particular structure of such a region, divided into an outer and an inner crust. In terms of densities, the outer crust ranges from about 10^4 g/cm^3 to 4×10^{11} g/cm^3, equivalent to $4 \times 10^{-11}\rho_0 \lesssim \rho \lesssim 1.6 \times 10^{-3}\rho_0$, by taking the typical value of $\rho_0 = 0.15$ fm$^{-3} \simeq 2.5 \times 10^{14}$ g/cm^3. In this region, it becomes energetically favorable for nucleons to cluster into finite nuclei. The so-called drip nuclei are formed as the density increases. The equation of state proposed by Baym, Pethick, and Sutherland, namely, the BPS equation of state, is commonly used to treat this region [58]. Concerning the inner crust, on the other hand, it starts from the neutron drip density of 4×10^{11} g/cm$^3 \simeq 1.6 \times 10^{-3}\rho_0$ up to the density where nuclear matter undergoes a transition to the uniform liquid core, that can be estimated through the thermodynamical method [59–61], for instance. An attempt to describe this region is the use of a polytropic equation of state of the form $p(\epsilon) = A + B\epsilon^{4/3}$ [62–64].

Figure 2.9 shows the mass-radius profile predicted by some parametrizations described in Refs. [21,22], namely, those belonging to the BKA and BSR families. For the construction of these curves, we use an approximation that treats the NS crust only by the BPS equation of state. For this particular case, we use the BPS equation of state for the density range of 0.1581×10^{-10} fm$^{-3} \leqslant \rho \leqslant 0.008907$ fm^{-3}. The figure also displays the bands related to the predictions concerning the mass of the pulsars PSR J1614-2230 with a mass of $(1.97 \pm 0.04)M_\odot$ [65], and the PSR J0348+0432 with a mass of

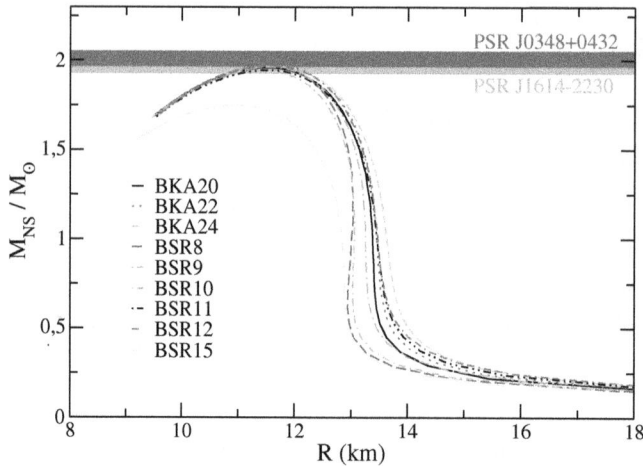

FIGURE 2.9
Neutron star mass-radius profile for some parametrizations of the RMF model, detailed in Refs. [21,22]. Horizontal bands indicate the masses of PSR J1614-2230 [65] (orange) and PSR J038+0432 [66] (blue). (Figure extracted from Ref. [67].)

$(2.01 \pm 0.04)M_\odot$ [66], with $M_\odot$ being the solar mass. After the discovery of these two pulsars, many parameter-dependent models were re-tuned so that they could describe NS masses in these ranges.

Another important constraint imposed to any hadronic model comes from the recent observation of the gravitational wave (GW) emission from the first binary neutron stars merger event, named as GW170817 [68–70]. This event, observed on 17 August 2017 by LIGO and Virgo collaborations, has special importance in nuclear physics since it consists of the emergence of GW from two binary neutron stars. It was possible to establish a relation between the internal structure of the neutron star and the emitted GW through data concerning the quantity called tidal deformability. In a binary NS system, tidal forces originating from the gravitational field induce tidal deformabilities in each companion star analogously to the tides generated on the Earth due to the Moon. Deformations in the stars related to the quadrupole moment generate GW in which the phase evolution depends on the tidal deformability [71–73]. The measured data on the GW170817 event allowed the LIGO/Virgo collaboration to determine constraints on the dimensionless tidal deformabilities for each NS in the binary system, as well as on that one related to a canonical star of mass equal to $1.4M_\odot$. For studies regarding the application of the GW170817 constraint to RMF models, we address the reader to Refs. [74–76], for instance.

2.2.7 Other Kind of Models

The RMF model described by Eqs. (2.2)–(2.7) is not the only relativistic hadronic model available in nuclear physics. Other kind of models are also widely used to describe symmetric, asymmetric, and stellar matters with similar results in comparison with those presented in this section. One of them is a version of the model studied so far, but with the couplings between nucleons and mesons varying with the nuclear medium, i.e., the couplings present specific density dependencies [77]. For this reason, this model is named as "density-dependent" (DD) model. Its Lagrangian density is given by

$$
\begin{aligned}
\mathcal{L}_{\mathrm{DD}} = {} & \overline{\psi}(i\gamma^\mu \partial_\mu - M)\psi + \Gamma_\sigma(\rho)\sigma\overline{\psi}\psi - \Gamma_\omega(\rho)\overline{\psi}\gamma^\mu \omega_\mu \psi - \frac{\Gamma_\rho(\rho)}{2}\overline{\psi}\gamma^\mu \vec{\rho}_\mu \vec{\tau}\psi \\
& + \Gamma_\delta(\rho)\overline{\psi}\vec{\delta}\vec{\tau}\psi + \frac{1}{2}(\partial^\mu \sigma \partial_\mu \sigma - m_\sigma^2 \sigma^2) - \frac{1}{4}F^{\mu\nu}F_{\mu\nu} + \frac{1}{2}m_\omega^2 \omega_\mu \omega^\mu - \frac{1}{4}\vec{B}^{\mu\nu}\vec{B}_{\mu\nu} + \frac{1}{2}m_\rho^2 \vec{\rho}_\mu \vec{\rho}^\mu \\
& + \frac{1}{2}(\partial^\mu \vec{\delta}\partial_\mu \vec{\delta} - m_\delta^2 \vec{\delta}^2),
\end{aligned}
\tag{2.95}
$$

where $\Gamma_i(\rho) = \Gamma_i(\rho_0)f_i(x)$, with

$$
f_i(x) = a_i \frac{1 + b_i(x + d_i)^2}{1 + c_i(x + e_i)^2}
\tag{2.96}
$$

for $i = \sigma, \omega$, and $\Gamma_\rho(\rho) = \Gamma_\rho(\rho_0)e^{-a_\rho(x-1)}$. The ratio of the density over the saturation density is given by $x = \rho/\rho_0$. Some density-dependent parameterizations have couplings different from those of the above equations. In particular, the GDFM model [78] presents the following form for its couplings,

$$
\Gamma_i(\rho) = a_i + (b_i + d_i x^3)e^{-c_i x},
\tag{2.97}
$$

for $i = \sigma, \omega, \rho, \delta$. A correction to the coupling parameter for the meson ω is also taken into account,

$$
\Gamma_{\mathrm{cor}}(\rho) = \Gamma_\omega(\rho) - a_{\mathrm{cor}}e^{-\left(\frac{\rho - \rho_0}{b_{\mathrm{cor}}}\right)^2}.
\tag{2.98}
$$

The DDHδ parametrization has the same coupling parameters as in Eq. (2.96) for the mesons σ and ω, but functions $f_i(x)$ given by [79]

$$f_i(x) = a_i e^{-b_i(x-1)} - c_i(x - d_i), \tag{2.99}$$

for $i = \rho, \delta$.

A consistent thermodynamical description of nuclear matter based on the DD model has to take into account the rearrangement term in the equations of state of the model. It is given by

$$\Sigma_R(\rho) = \frac{\partial \Gamma_\omega}{\partial \rho} \omega_0 \rho + \frac{1}{2} \frac{\partial \Gamma_\rho}{\partial \rho} \bar{\rho}_0 \rho_3 - \frac{\partial \Gamma_\sigma}{\partial \rho} \sigma \rho_s - \frac{\partial \Gamma_\delta}{\partial \rho} \delta \rho_{s3} \tag{2.100}$$

in the mean-field approximation, in which one has $\vec{\delta} \rightarrow\, <\vec{\delta}> \equiv \delta$ (third components of the isospin space vector $\vec{\delta}$).

Another kind of RMF model is the nonlinear point-coupling (NLPC) one [80]. In this model, nucleons interact with each other only through effective point-like interactions, without the meson exchange mechanism. A general Lagrangian density describing this model is

$$\mathcal{L}_{\mathrm{NLPC}} = \bar{\psi}(i\gamma^\mu \partial_\mu - M)\psi - \frac{\alpha_s}{2}(\bar{\psi}\psi)^2 - \frac{\beta_s}{3}(\bar{\psi}\psi)^3 - \frac{\gamma_s}{4}(\bar{\psi}\psi)^4 - \frac{\alpha_v}{2}(\bar{\psi}\gamma^\mu\psi)^2$$
$$- \frac{\gamma_v}{4}(\bar{\psi}\gamma^\mu\psi)^4 - \frac{\alpha_{\mathrm{TV}}}{2}(\bar{\psi}\gamma^\mu\vec{\tau}\psi)^2 - \frac{\gamma_{\mathrm{TV}}}{4}(\bar{\psi}\gamma^\mu\vec{\tau}\psi)^4 - \frac{\alpha_{\mathrm{TS}}}{2}(\bar{\psi}\vec{\tau}\psi)^2$$
$$+ \left[\eta_1 + \eta_2(\bar{\psi}\psi)\right](\bar{\psi}\psi)(\bar{\psi}\gamma^\mu\psi)^2 - \eta_3(\bar{\psi}\psi)(\bar{\psi}\gamma^\mu\vec{\tau}\psi)^2. \tag{2.101}$$

Finally, the density-dependent version of the nonlinear point-coupling model described in Refs. [81,82] presents the following Lagrangian density:

$$\mathcal{L}_{\mathrm{DDNLPC}} = \bar{\psi}(i\gamma^\mu \partial_\mu - M)\psi - \frac{G_s(\rho)}{2}(\bar{\psi}\psi)^2 - \frac{G_V(\rho)}{2}(\bar{\psi}\gamma^\mu\psi)^2 - \frac{G_{\mathrm{TV}}(\rho)}{2}(\bar{\psi}\gamma^\mu\vec{\tau}\psi)^2$$
$$- \frac{G_{\mathrm{TS}}(\rho)}{2}(\bar{\psi}\vec{\tau}\psi)^2 - \frac{D_s(\rho)}{2}(\partial_\nu\bar{\psi}\psi)^2, \tag{2.102}$$

in which the density-dependent couplings are determined from finite-density QCD sum rules and the in-medium chiral perturbation theory. Like the DD finite range model, the thermodynamics of the DDNLPC model also demands the correct implementation of a rearrangement term, analogous to the one obtained in Eq. (2.100). For detailed information on the equations of state of this particular model, and the ones presented in Eqs. (2.95) and (2.101), we address the reader to a very complete review on the subject in Ref. [83].

2.3 The Skyrme Interaction

The Skyrme interaction was proposed as an effective interaction to be used in the nuclear medium [4,5,84,85] under the philosophy that the two-body potential can be written as a contact interaction with a low-momentum expansion, i.e. in the form

$$v_{12} = \delta(\boldsymbol{r}_1 - \boldsymbol{r}_2)v(\boldsymbol{k}', \boldsymbol{k}) \tag{2.103}$$

where $\boldsymbol{k}$ is the operator for the relative wave number $\boldsymbol{k} = \frac{1}{2}i(\boldsymbol{\nabla}_1 - \boldsymbol{\nabla}_2)$ and is placed on the right of the delta function, while $\boldsymbol{k}'$ is the same operator placed on the left. The low-momentum expansion is taken up to quadratic order, and augmented with spin operators, $\boldsymbol{\sigma}$, and $P^\sigma = \frac{1}{2}(1 + \boldsymbol{\sigma}_1 \cdot \boldsymbol{\sigma}_2)$. The form originally given by Skyrme is

$$v(\boldsymbol{k}', \boldsymbol{k}) = t_0(1 + x_0 P^\sigma) + \frac{1}{2}t_1(1 + x_1 P^\sigma)(\boldsymbol{k}'^2 + \boldsymbol{k}^2)$$

$$+ t_2\left(1 + x_2\left(P^\sigma - \frac{4}{5}\right)\right)\boldsymbol{k}' \cdot \boldsymbol{k}$$

$$+ \frac{1}{2}T\left(\boldsymbol{\sigma}_1 \cdot \boldsymbol{k}\boldsymbol{\sigma}_2 \cdot \boldsymbol{k} - \frac{1}{3}\boldsymbol{\sigma}_1 \cdot \boldsymbol{\sigma}_2\boldsymbol{k}^2 + \text{conj.}\right)$$

$$+ \frac{1}{2}U\left(\boldsymbol{\sigma}_1 \cdot \boldsymbol{k}'\boldsymbol{\sigma}_2 \cdot \boldsymbol{k} - \frac{1}{3}\boldsymbol{\sigma}_1 \cdot \boldsymbol{\sigma}_2\boldsymbol{k}' \cdot \boldsymbol{k} + \text{conj.}\right)$$

$$+ V(i(\boldsymbol{\sigma}_1 + \boldsymbol{\sigma}_2) \cdot \boldsymbol{k}' \times \boldsymbol{k}). \tag{2.104}$$

Here, the t_0, t_1, and t_2 terms are the central force, of which the t_1 and t_2 terms are the momentum-dependent component, the T and U terms are the tensor interaction (acting in even and odd channels, respectively), and the V term is the two-body spin orbit force.

This two-body interaction is combined with a three-body term which in its most basic form (as given by Skyrme [5]) reads

$$v_{123} = t_3\delta(\boldsymbol{r}_1 - \boldsymbol{r}_2)\delta(\boldsymbol{r}_3 - \boldsymbol{r}_1). \tag{2.105}$$

This phenomenologically motivated ansatz in functional form requires the free parameters t_0, t_1, t_2, x_0, T, U, V, and t_3, which are to be fitted to data. We will see in this chapter how they can be constrained by nuclear matter properties.

The three-body interaction can be shown to be equivalent [86], when applied in the Hartree-Fock approximation, to the two-body density-dependent interaction:

$$v_3 = \frac{1}{6}t_3(1 + x_3 P^\sigma)\delta(\boldsymbol{r}_1 - \boldsymbol{r}_2)\rho^\alpha\left(\frac{\boldsymbol{r}_1 + \boldsymbol{r}_2}{2}\right) \tag{2.106}$$

with $x_3 = 0$ and $\alpha = 1$. For many implementations of the Skyrme interaction, more general forms with $x_3 \neq 0$ and/or $\alpha \neq 1$ are used. Setting a value for α, which is not a natural number, breaks the derivation of the density-dependent term from a true interaction, but such values of α are frequently used, and we will discuss this further in the context of the nuclear matter incompressibility.

A considerable part of the popularity of the Skyrme interaction comes from its delta function form, which renders it easy to implement and compute in the Hartree-Fock calculations for finite nuclei, with early applications covering spherical ground states [86], deformed ground states [87], deformation-constrained energy curves [88,89], giant resonance states via the generator coordinate method [90], and the extension to time-dependent Hartree-Fock [91]. Continuous developments to the present day are too numerous to give in detail, for which review articles should be consulted [92–97].

2.3.1 Nuclear Matter

As an effective interaction designed for use in-medium and at low momentum, the Skyrme interaction is commonly studied in the mean-field approximation. For the case of nuclear matter, this amounts to supposing the nuclear wave function to consist of a Slater Determinant of single-particle plane wave functions, as discussed by Brueckner [98–100], whose work inspired Skyrme's ideas. We follow that treatment here and give a detailed derivation of symmetric nuclear matter properties with the simplest form of the Skyrme interaction, consisting of the leading order central terms of both the two- and three-body forces:

$$v_{ZR} = t_0\delta(\boldsymbol{r}_1 - \boldsymbol{r}_2) + \frac{1}{6}t_3\delta(\boldsymbol{r}_1 - \boldsymbol{r}_2)\rho\left(\frac{\boldsymbol{r}_1 + \boldsymbol{r}_2}{2}\right). \tag{2.107}$$

For the fuller Skyrme interaction, we quote results. The derivation of these results can be worked out following our scheme here, though most of the literature calculates mean-field expressions first for finite nuclei and carries the results over to infinite nuclear matter [86].

In Eq. (2.107), the subscript in v_{ZR} stands for "zero range" since it omits the momentum-dependent terms which simulate a finite range in the sense of a Taylor expansion. Such a simplified force has sometimes been used in the literature [101–103]. The parameter t_0 is always negative, and t_3 is positive, and this basic version of the Skyrme force is thus analogous to the Yukawa form in Eq. (2.1) for the case of infinitely heavy mesons and point-like interactions.

For the mean-field treatment, we take the single-particle state of a nucleon to be

$$\phi_\lambda(\boldsymbol{r}\sigma\tau) = \frac{1}{\sqrt{\Omega}} e^{i\boldsymbol{k}\cdot\boldsymbol{r}} \chi_\sigma \xi_\tau \tag{2.108}$$

where χ is a spin function and ξ is an isospin function. An arbitrary normalization volume, Ω, is given, which will cancel out in quantities of interest [29].

The number of particles in the system may be written considering the filling of states up to the Fermi momentum in our arbitrary volume as

$$A = \sum_{k\sigma\tau} \theta(k_F - k), \tag{2.109}$$

where θ is the Heaviside step function. Taking the volume to infinity, and hence turning the sum over k into an integral, the expression becomes

$$A = \frac{\Omega}{(2\pi)^3} \sum_{\sigma\tau} \int d^3k \, \theta(k_F - k) \tag{2.110}$$

$$= \frac{4\Omega}{(2\pi)^3} 4\pi \int_0^{k_F} k^2 \, dk \tag{2.111}$$

$$= \frac{2\Omega}{3\pi^2} k_F^3. \tag{2.112}$$

Here, the sum over σ and τ is taken assuming spin- and isospin-saturated matter with a common Fermi momentum between protons and neutrons, giving rise to the factor of 4.

The density is defined as $\rho = A/\Omega$ and is hence

$$\rho = \frac{2k_F^3}{3\pi^2}. \tag{2.113}$$

The kinetic energy is calculated as

$$T = \sum_{k\sigma\tau} \frac{1}{\Omega} \int d^3r \, e^{-i\boldsymbol{k}\cdot\boldsymbol{r}} \chi_\sigma^\dagger \xi_\tau^\dagger \left(-\frac{\hbar^2 \boldsymbol{\nabla}^2}{2m} \right) e^{i\boldsymbol{k}\cdot\boldsymbol{r}} \chi_\sigma \xi_\tau \tag{2.114}$$

$$= \sum_{k\sigma\tau} \frac{\hbar^2}{2m} k^2 \tag{2.115}$$

$$= 4 \frac{\Omega}{(2\pi)^3} \frac{\hbar^2}{2m} 4\pi \int_0^{k_F} k^4 \, dk \tag{2.116}$$

$$= \frac{\Omega \hbar^2 k_F^5}{5\pi^2 m} = \left(\frac{2\Omega k_F^3}{3\pi^2} \right) \frac{3}{5} \frac{\hbar^2 k_F^2}{2m} \tag{2.117}$$

$$= A \frac{3}{5} \frac{\hbar^2 k_F^2}{2m}. \tag{2.118}$$

Hence, we have that the kinetic energy per particle is

$$T/A = \frac{3}{5}\frac{\hbar^2 k_F^2}{2m}, \tag{2.119}$$

which can be related to the density using Eq. (2.113) as

$$T/A = \frac{3}{5}\frac{\hbar^2}{2m}\left(\frac{3\pi}{2}\right)^{2/3}\rho^{2/3}. \tag{2.120}$$

The potential energy is given by

$$V = \frac{1}{2}\sum_{ij}\langle ij|v_{ZR}(1 + P^\sigma P^M P^\tau)|ij\rangle \tag{2.121}$$

$$= \frac{1}{2}\sum_{ij}\langle ij|\left(t_0\delta(\boldsymbol{r}_1 - \boldsymbol{r}_2) + \frac{1}{6}t_3\delta(\boldsymbol{r}_1 - \boldsymbol{r}_2)\rho\left(\frac{\boldsymbol{r}_1 + \boldsymbol{r}_2}{2}\right)\right)(1 + P^\sigma P^M P^\tau)|ij\rangle. \tag{2.122}$$

The inclusion of the operator $(1 + P^\sigma P^M P^\tau)$ allows the direct and exchange terms to be expressed together. Here P^M is the Majorana exchange operator which exchanges the space coordinates of two particles, while P^σ and P^τ exchange the spin and isospin, respectively. Because of the symmetry of the interaction, we have $P^M = 1$. We do not allow proton and neutron states to mix, so P^τ will give a result of zero in the above expression if i and j refer to different isospin species; hence, we write replace P^τ with δ_{τ_i,τ_j} where τ_i and τ_j are the types (proton or neutron) of each particle. For the spin-exchange operator, we make use of the fact that the expectation value of the spin operators disappears when the sum over occupied states includes both the up and down spin states (we are assuming spin-saturated matter here). This means $P^\sigma = \frac{1}{2}(1 + \boldsymbol{\sigma}_1 \cdot \boldsymbol{\sigma}_2)$ can be replaced by $1/2$ inside the expectation value (2.122). The potential energy is hence

$$V = \frac{1}{2}\sum_{k_1\sigma_1\tau_1}\sum_{k_2\sigma_2\tau_2}\frac{1}{\Omega^2}\int\int d^3r_1\,d^3r_2 e^{-ik_1\cdot r_1}e^{-ik_2\cdot r_2}\chi_{\sigma_1}^\dagger(1)\xi_{\tau_1}^\dagger(1)\chi_{\sigma_2}^\dagger(2)\xi_{\tau_2}^\dagger(2) \tag{2.123}$$

$$\cdot\left(t_0\delta(\boldsymbol{r}_1 - \boldsymbol{r}_2) + \frac{1}{6}t_3\delta(\boldsymbol{r}_1 - \boldsymbol{r}_2)\rho\left(\frac{\boldsymbol{r}_1 + \boldsymbol{r}_2}{2}\right)\right)$$

$$\cdot\left(1 - \frac{1}{2}\delta_{\tau_1,\tau_2}\right)e^{ik_1\cdot r_1}e^{ik_2\cdot r_2}\chi_{\sigma_1}(1)\xi_{\tau_1}(1)\chi_{\sigma_2}(2)\xi_{\tau_2}(2). \tag{2.124}$$

Since all explicit spin-dependence has been removed from the interaction, the spin functions cancel each other out, and we can sum over both spin coordinates in the sum to obtain

$$V = 2\sum_{k_1\sigma_1}\sum_{k_2\sigma_2}\frac{1}{\Omega^2}\int\int d^3r_1\,d^3r_2 e^{-ik_1\cdot r_1}e^{-ik_2\cdot r_2}\xi_{\tau_1}^\dagger(1)\xi_{\tau_2}^\dagger(2)$$

$$\cdot\left(t_0\delta(\boldsymbol{r}_1 - \boldsymbol{r}_2) + \frac{1}{6}t_3\delta(\boldsymbol{r}_1 - \boldsymbol{r}_2)\rho\left(\frac{\boldsymbol{r}_1 + \boldsymbol{r}_2}{2}\right)\right) \tag{2.125}$$

$$\cdot(1 - \frac{1}{2}\delta_{\tau_1,\tau_2})e^{ik_1\cdot r_1}e^{ik_2\cdot r_2}\xi_{\tau_1}(1)\xi_{\tau_2}(2). \tag{2.126}$$

The common delta function between the two terms in the interaction can be integrated over, eliminating one of the integrals. Further, though there is a δ_{τ_1,τ_2} operator, this does not affect the operation of the isospinors on each other, and they cancel out, giving

$$V = 2\sum_{k_1\tau_1}\sum_{k_2\tau_2}\frac{1}{\Omega^2}\int d^3r\,e^{-i(k_1+k_2)\cdot r}(t_0 + \frac{1}{6}t_3\rho(\boldsymbol{r}))\cdot(1 - \frac{1}{2}\delta_{\tau_1,\tau_2})e^{i(k_1+k_2)\cdot r}. \tag{2.127}$$

Given that we are dealing only with isospin-symmetric nuclear matter here, and in the sum over occupied states every proton is matched with a neutron, we can write that $\sum_{\tau_1,\tau_2} = 4$ and $\sum_{\tau_1,\tau_2} \delta_{\tau_1,\tau_2} = 2$ and the factor $(1 - \frac{1}{2}\delta_{\tau_1,\tau_2})$ can be replaced with 3. Noting that the plane wave and its conjugate cancel each other, we have

$$V = \sum_{k_1,k_2} \frac{6}{\Omega^2} \int d^3r (t_0 + \frac{1}{6}t_3\rho) = \sum_{k_1,k_2} \frac{6}{\Omega}(t_0 + \frac{1}{6}t_3\rho). \tag{2.128}$$

Taking the sums over wave number to integrals as the volume is taken to infinity, we have

$$V = \frac{6\Omega}{(2\pi)^6} \int_0^{k_F} 4\pi k_1^2 dk_1 \int_0^{k_F} 4\pi k_2^2 (t_0 + \frac{1}{6}t_3\rho) \tag{2.129}$$

$$= \frac{6\Omega}{(2\pi)^6} \frac{16\pi^2}{9} k_F^6 (t_0 + \frac{1}{6}t_3\rho) \tag{2.130}$$

$$= \frac{\Omega}{6\pi^4} k_F^6 (t_0 + \frac{1}{6}t_3\rho). \tag{2.131}$$

Now, replacing the volume with the number of particles through (2.112) $A = (2\Omega k_F^3)/(3\pi^2)$ gives

$$\frac{V}{A} = \frac{3}{12\pi^2} k_F^3 (t_0 + \frac{1}{6}t_3\rho). \tag{2.132}$$

Using the relation (2.113), $\rho = (2k_F^3)/(3\pi^2)$, between ρ and k_F, we have

$$\frac{V}{A} = \frac{3}{8}t_0\rho + \frac{1}{16}t_3\rho^2. \tag{2.133}$$

The full energy per particle is then

$$E/A = \mathcal{E}/\rho = \frac{3}{5}\frac{\hbar^2}{2m}\left(\frac{3\pi}{2}\right)^{\frac{2}{3}}\rho^{\frac{2}{3}} + \frac{3}{8}t_0\rho + \frac{1}{16}t_3\rho^2. \tag{2.134}$$

This equation has two free parameters, given a fixed value for the nucleon mass, and one can fix them by taking, e.g., that the minimum energy occurs at the saturation density. Fixing $E(\rho_0 = 0.16) = -16.0$ MeV and taking a typical value [104] of

$$\frac{\hbar^2}{2m} = 20.73553\,\text{MeV fm}^2 \tag{2.135}$$

give a unique solution for t_0 and t_3 of

$$t_0 = -624.2244193\text{MeV fm}^3, \qquad t_3 = 10852.10393\text{MeV fm}^6. \tag{2.136}$$

For reference, we call this the ZRNM parameter set and plot the energy per particle in symmetric nuclear matter in Figure 2.10. Note that the `Maple` worksheet used to generate the fit may be downloaded by the reader [105].

2.3.2 Incompressibility and the Three-Body Force

The incompressibility for infinite symmetric nuclear matter at the saturation density was derived in the RMF section as (2.55)

$$K_0 \equiv K(\rho_0) = \left[k_F^2 \frac{\partial^2(\mathcal{E}/\rho)}{\partial k_F^2}\right]\Big|_{k_{F0}} = \left[9\rho^2 \frac{\partial^2(\mathcal{E}/\rho)}{\partial\rho^2}\right]\Big|_{\rho_0}. \tag{2.137}$$

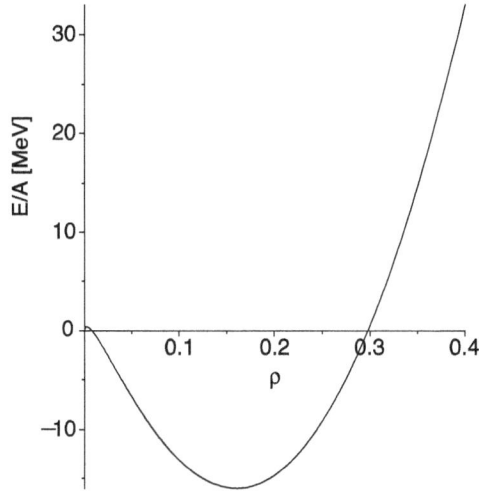

FIGURE 2.10
Energy per particle in symmetric nuclear matter for the ZRNM force parameters defined in (2.136).

In the case of the simplified Skyrme force, this is

$$K_0 = 9\frac{3}{5}\frac{\hbar^2}{2m}\left(\frac{3\pi}{2}\right)^{\frac{2}{3}}\left(\frac{-1}{3}\right)\left(\frac{2}{3}\right)\rho_0^{\frac{2}{3}} + \frac{9}{8}t_3\rho_0^2 \tag{2.138}$$

$$= -\frac{6}{5}\frac{\hbar^2}{2m}\left(\frac{3\pi}{2}\right)^{\frac{2}{3}}\rho_0^{\frac{2}{3}} + \frac{9}{8}t_3\rho_0^2. \tag{2.139}$$

For the ZRNM parameters (2.136), the result is $K_0 = 376.43$ MeV, which is rather higher than the values typically inferred from giant monopole resonance, with a typical value being $K_0 = 230 \pm 30$ MeV [27], though uncertainties in the different methods of extraction of K_0 from data suggest that the range of allowed values may be larger [106].

As seen from (2.139), the value of K_0 depends linearly on t_3 and is independent of t_0. The role of the three-body force in the incompressibility is key, and the use of a value less than unity of the power, α, of the density in the three-body force (2.106), was originally motivated by a desire to reduce the incompressibility of symmetric nuclear matter [107].

Allowing a variable power of the density in the three-body force gives an expression for the energy per particle of symmetric nuclear matter of

$$E/A = \frac{3}{5}\frac{\hbar^2}{2m}\left(\frac{3\pi}{2}\right)^{\frac{2}{3}}\rho^{\frac{2}{3}} + \frac{3}{8}t_0\rho + \frac{1}{16}t_3\rho^{\alpha+1} \tag{2.140}$$

and an incompressibility of [108]

$$K_0 = -\frac{6}{5}\frac{\hbar^2}{2m}\left(\frac{3\pi}{2}\right)^{\frac{2}{3}}\rho_0^{\frac{2}{3}} + \frac{9}{16}\alpha(\alpha+1)t_3\rho_0^{\alpha+1}. \tag{2.141}$$

Now the parameter α strongly alters the incompressibility while allowing the saturation density and energy to still be fitted. Historically fixed fractional values of α were most commonly taken, from which values of K_0 could be explored. Taking this same approach, and adjusting this three-parameter (t_0, t_3, α) force to $\rho_0 = 0.16$ fm^{-3}, $E(\rho_0) = -16.0$ MeV with fixed values of alpha of 1/10, 2/10, 3/10, ..., 1, we produce equations of state as in Figure 2.11.

The incompressibilities are seen to change as the potential shape becomes steeper or stiffer, indicating higher incompressibility, for larger α and shallower or softer, indicating lower incompressibility, for smaller α. The computed values of K_0 for each value of α are found in Table 2.1.

2.3.3   Full-Skyrme Equation of State for Symmetric Nuclear Matter

Going beyond the simplified Skyrme force and lifting the restriction of an equal number of protons and neutrons, one derives the energy per particle in an analogous manner to that of Section 2.3.1.

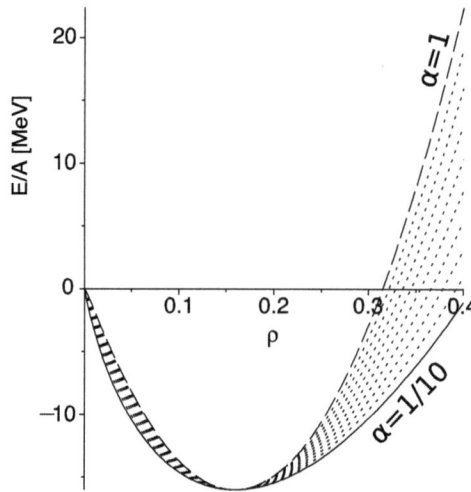

FIGURE 2.11
Energy per particle as a function of density for symmetric nuclear matter with a three-parameter zero-range Skyrme force. The dotted lines between the $\alpha = 1/10$ and $\alpha = 1$ lines are in increments of 1/10. The increasing stiffness of the equation of state as a function of increasing α is seen.

TABLE 2.1
Incompressibilities K_0 as a Function of Power of Density α in Three-Parameter Zero-Range Skyrme Force Fitted to have Energy and Density Saturation of Nuclear Matter at $\rho_0 = 0.16$ fm^{-3} and $E(\rho_0) = -16.0$ MeV

Name	t_0 [MeV fm^3]	t_3 MeV fm$^{3+3\alpha}$	α	K [MeV]
ZRNMα1	−4529.988868	28069.29237	0.1	187.1397998
ZRNMα2	−2582.557248	16857.35643	0.2	208.1720608
ZRNMα3	−1933.413379	13498.52181	0.3	229.2043225
ZRNMα4	−1608.841442	12160.05327	0.4	250.2365838
ZRNMα5	−1414.098281	11684.58969	0.5	271.2688454
ZRNMα6	−1284.269506	11695.53847	0.6	292.3011066
ZRNMα7	−1191.534667	12040.96888	0.7	313.3333680
ZRNMα8	−1121.983539	12654.86420	0.8	334.3656298
ZRNMα9	−1067.888216	13511.17031	0.9	355.3978912
ZRNMαA	−1024.611958	14605.73711	1.0	376.4301526

For a slightly more general version of the original Skyrme force, widely used, with spin-exchange parameters added to each term, and omitting the seldom-used tensor term, a ten-parameter version of the Skyrme interaction is given as [108]

$$V(\boldsymbol{r}_1, \boldsymbol{r}_2) = t_0(1 + x_0 P^\sigma)\delta(\boldsymbol{r}_1 - \boldsymbol{r}_2)$$

$$+ \frac{1}{2}t_1(1 + x + 1P^\sigma)\left[\left(-\frac{1}{2i}(\boldsymbol{\nabla}'_1 - \boldsymbol{\nabla}'_2)\right)^2 \delta(\boldsymbol{r}_1 - \boldsymbol{r}_2) + \delta(\boldsymbol{r}_1 - \boldsymbol{r}_2)\left(\frac{1}{2i}(\boldsymbol{\nabla}_1 - \boldsymbol{\nabla}_2)\right)^2\right]$$

$$+ t_2(1 + x_2 P^\sigma)\left[-\frac{1}{2i}(\boldsymbol{\nabla}'_1 - \boldsymbol{\nabla}'_2) \cdot \delta(\boldsymbol{r}_1 - \boldsymbol{r}_2)\frac{1}{2i}(\boldsymbol{\nabla}_1 - \boldsymbol{\nabla}_2)\right]$$

$$+ \frac{1}{6}t_3(1 + x_3 P^\sigma)\left[\rho((\boldsymbol{r}_1 + \boldsymbol{r}_2)/2)\right]^\alpha \delta(\boldsymbol{r}_1 - \boldsymbol{r}_2)$$

$$+ iW_0\boldsymbol{\sigma} \cdot \left[-\frac{1}{2i}(\boldsymbol{\nabla}'_1 - \boldsymbol{\nabla}'_2) \times \delta(\boldsymbol{r}_1 - \boldsymbol{r}_2)\frac{1}{2i}(\boldsymbol{\nabla}_1 - \boldsymbol{\nabla}_2)\right] \tag{2.142}$$

This version of the Skyrme force, or closely related alternatives, has been fitted many times to data, generating sets of parameters, each of which defines a specific named Skyrme parametrization. There are now hundreds of named parameter sets available in the literature.

For symmetric infinite nuclear matter, this complete version of the Skyrme interaction has an energy per particle that includes a contribution from the momentum-dependent terms as

$$E/A = \frac{3}{5}\frac{\hbar^2}{2m}\left(\frac{3\pi^2}{2}\right)^{\frac{2}{3}}\rho^{\frac{2}{3}} + \frac{3}{8}t_0\rho + \frac{3}{80}\Theta_s\left(\frac{3\pi^2}{2}\right)^{\frac{2}{3}}\rho^{\frac{5}{3}} + \frac{1}{16}t_3\rho^{\alpha+1}, \tag{2.143}$$

where $\Theta_s = 3t_1 + (5 + 4x_2)t_2$ is the coefficient from the momentum-dependent terms of the Skyrme interaction. Note that the three Skyrme parameters here combine in a particular way in the (symmetric nuclear matter) equation of state.

The pressure, whose definitions are given in Section 2.2.2, is given by

$$P = \rho\left(\frac{\hbar^2}{5m}\left(\frac{3\pi^2}{2}\right)^{\frac{2}{3}}\rho^{\frac{2}{3}} + \frac{3}{8}t_0\rho + \frac{1}{16}\Theta_s\left(\frac{3\pi^2}{2}\right)^{\frac{3}{2}}\rho^{\frac{5}{3}} + \frac{1}{16}t_3\alpha(\alpha + 1)\rho^{\sigma+1}\right). \tag{2.144}$$

A range of curves for different Skyrme parameterizations of the dependence of pressure on the density is shown in Figure 2.12. Here, the constraints derived from transverse flow in Au+Au experiments are also shown, as used in Figure 2.3 for comparison with RMF forces. The Skyrme forces, as with the RMF forces, show very similar behavior to each other, and are consistent with experimental constraints.

The incompressibility of symmetric nuclear matter at saturation density with the full Skyrme interaction is

$$K_0 = -\frac{6}{5}\frac{\hbar^2}{2m}\left(\frac{3\pi^2}{2}\right)^{\frac{2}{3}}\rho_0^{\frac{2}{3}} + +\frac{3}{8}\Theta_s\left(\frac{3\pi^2}{2}\right)^{\frac{2}{3}}\rho_0^{\frac{5}{3}} + \frac{9}{16}\alpha(\alpha + 1)t_3\rho_0^{\alpha+1}. \tag{2.145}$$

Earlier, Skyrme forces (e.g., SI, SII [86], SIII, SIV, SV [109]) had incompressibilities in the range of $\simeq$ 300–400 MeV. Most later forces have an incompressibility in the range of $\simeq$ 200–300 MeV [7].

2.3.4 Asymmetric Nuclear Matter

In asymmetric nuclear matter with a neutron fraction $y_p = Z/A$ or asymmetry parameter $I = (N - Z)/A$ parameterized in the asymmetry factor

$$F_m = 2^{m-1}\left(y_p^m + (1 - y_p)^m\right), \qquad F_m = \frac{1}{2}\left((1 + I)^m + (1 - I)^m\right), \tag{2.146}$$

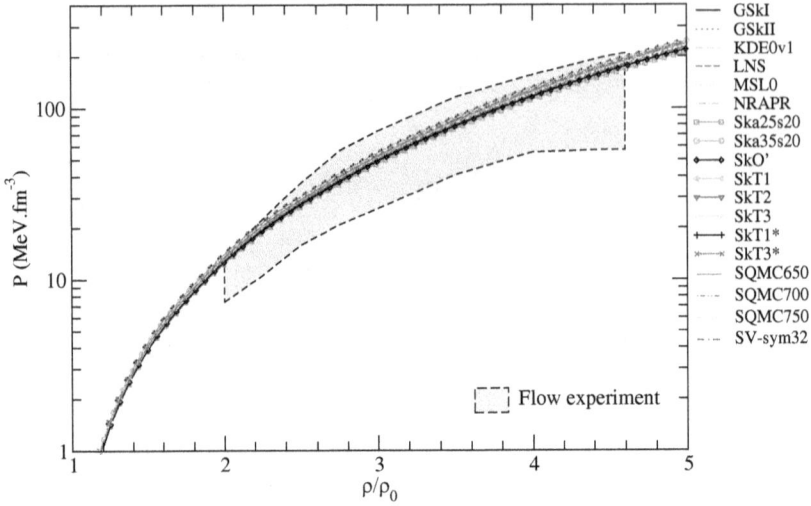

FIGURE 2.12
Pressure as a function of density for a range of different Skyrme parameterizations. Adapted from [27] where each of the named Skyrme parameterizations is cited. Experimental data derived from transverse flow in Au+Au collisions shown in shaded area is from Ref. [24].

the energy per particle is expressed as [108]

$$E/A = \frac{3}{5}\frac{\hbar^2}{2m}\left(\frac{3\pi^2}{2}\right)^{\frac{2}{3}}\rho^{\frac{2}{3}}F_{\frac{5}{3}} + \frac{1}{8}t_0\rho\left(2(x_0+2) - (2x_0+1)F_2\right)$$
$$+ \frac{1}{48}t_3\rho^{\alpha+1}\left(2(x_3+2) - (2x_3+1)F_2\right)$$
$$+ \frac{3}{40}\left(\frac{3\pi^2}{2}\right)^{\frac{2}{3}}\rho^{\frac{5}{3}}\left((t_1(x_1+2) + t_2(x_2+2))F_{\frac{5}{3}} + \frac{1}{2}(t_2(2x_2+1) - t_1(2x_1+1))F_{\frac{8}{3}}\right).$$
$$(2.147)$$

From (2.147), one can write the equation of state for pure neutron matter for which $y_p = 0$

$$\left.\frac{E}{A}\right|_{\text{PNM}} = \frac{3}{5}\frac{\hbar^2}{2m}\left(3\pi^2\right)^{\frac{2}{3}}\rho^{\frac{2}{3}} + \frac{1}{4}\rho t_0(1-x_0) + \frac{1}{24}\rho^{\alpha+1}t_3(1-x_3) + \frac{3}{40}\left(3\pi^2\right)^{\frac{2}{3}}\Theta_n\rho^{\frac{5}{3}}, \quad (2.148)$$

where

$$\Theta_n = t_1(1-x_1) + 3t_2(1+x_2) \quad (2.149)$$

is the combination under which the momentum-dependent term coefficients appear for pure neutron matter.

While most Skyrme parameterizations have very similar properties for the symmetric nuclear matter equation of state across a wide range of densities, the pure neutron matter equation of state is less well constrained. As an example, Figure 2.13 shows the equations of state for symmetric nuclear matter and pure neutron matter for the Skyrme parameterizations TOV-min [110], RD-min [111], and SV-min [112].

These parameterizations were fitted using a procedure which gives theoretical errors in the force parameters, and one can see in the figure the consequent errors in the equations of state. The much better understood equation of state of symmetric nuclear matter is much better constrained by the Skyrme parameterizations than that of pure neutron matter

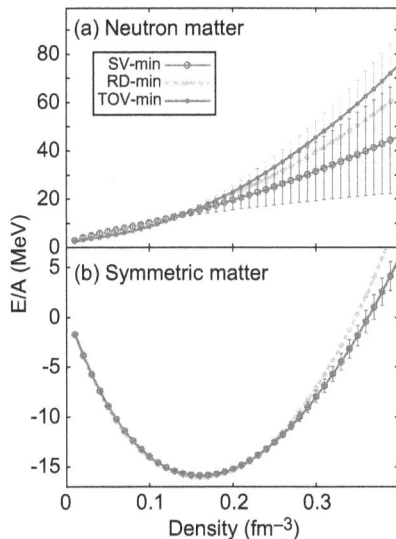

FIGURE 2.13
Energy per particle in pure neutron matter (upper panel) and symmetric nuclear matter (lower panel) for three different Skyrme interactions. (Figure from arXiv version of [110].)

as seen by the error bars in the figure. The correct quantification of theoretical errors in parameter fits has become an important part of the theory of the Skyrme interaction [113–115]. The behavior of neutron-rich nuclear matter is of obvious importance in neutron stars, to be discussed in a later subsection.

The symmetry energy, discussed in Section 2.2.5, measures the energy needed to increase the neutron to proton asymmetry away from a reference value. Defined as (2.82), in the notation we are using in this section for the full Skyrme interaction, we have a straightforward derivation from the expression for E/A as a function of the asymmetry I (through the F_m functions) as

$$a_s = \frac{1}{2} \frac{\partial^2 \frac{E}{A}}{\partial I^2}\bigg|_{I=0} = \frac{1}{3} \frac{\hbar^2}{2m} \left(\frac{3\pi^2}{2}\right)^{\frac{2}{3}} \rho^{\frac{2}{3}} - \frac{1}{8} t_0 (2x_0 + 1)\rho - \frac{1}{24} \left(\frac{3\pi^2}{2}\right)^{\frac{2}{3}}$$

$$\times \Theta_{sym} \rho^{\frac{5}{3}} - \frac{1}{48} t_3 (2x_3 + 1)\rho^{\sigma+1}, \tag{2.150}$$

where the combination of parameters of the momentum-dependent force is

$$\Theta_{sym} = 3t_1 x_1 - t_2 (5 + 4x_2). \tag{2.151}$$

Around saturation density, the available Skyrme parameterizations exhibit a relatively small range of asymmetry energies [7]. Figure 2.14 shows a representative sample of Skyrme interactions and the density dependence of the asymmetry energy as a function of density. This shows the relatively similar behavior between all parameterizations around the saturation density ρ_0, with a marked divergence between the parameterizations at high density.

As mentioned in Section 2.2.5, the symmetry energy at saturation density, J, and its derivative at saturation, L, have been used to constrain the parameters of the Skyrme interaction. There is a particularly strong correlation between the L and J as seen in Figure 2.15. These strong correlations are related to the crossing points of the symmetry energy, as discussed in Section 2.2.5, and the same features hold in the Skyrme interaction.

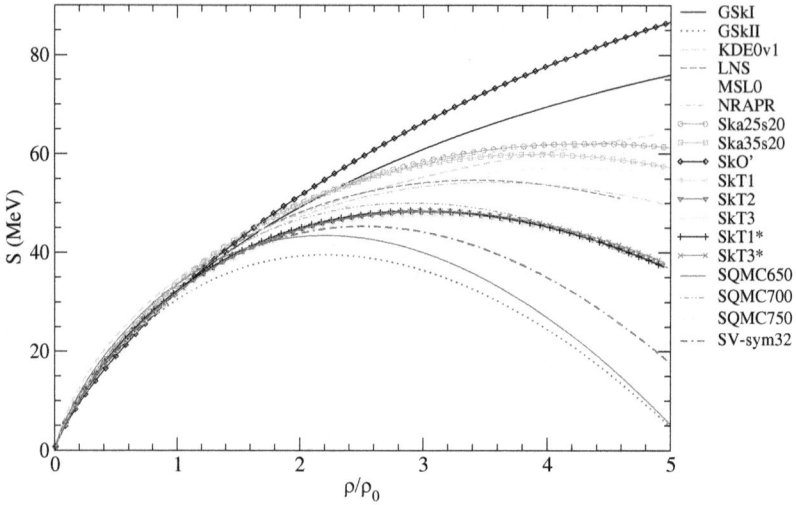

FIGURE 2.14
The asymmetry energy (here labeled S) as a function of density for a selection of the Skyrme interactions. (From Ref. [27].)

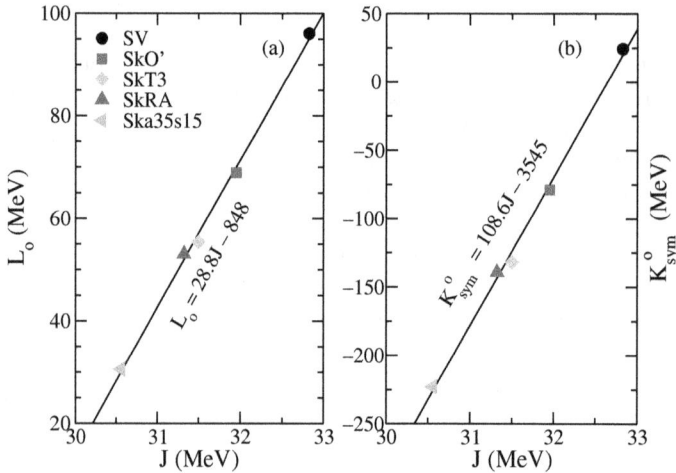

FIGURE 2.15
Left: The correlation between the slope of the symmetry energy at saturation, L, with the value of the symmetry energy at saturation, J for a range of Skyrme parameters. Right: The correlation between J and the second derivative of the symmetry energy K^0_{sym}. (Figure from Ref. [42]).

2.3.5   Neutron Star Environments

Calculations of neutron stars with the Skyrme interactions mirror the theoretical methods discussed in Section 2.2.6. The Tolman-Oppenheimer-Volkoff (TOV) Eq. (2.93) is solved using an equation of state which either fully consists of a Skyrme equation of state at all densities or, more realistically, is a Skyrme equation of state at intermediate densities, matched to appropriate equations of state at high and low densities. As in the RMF case discussed in Section 2.2.6, the ability of Skyrme interactions to generate sufficiently heavy

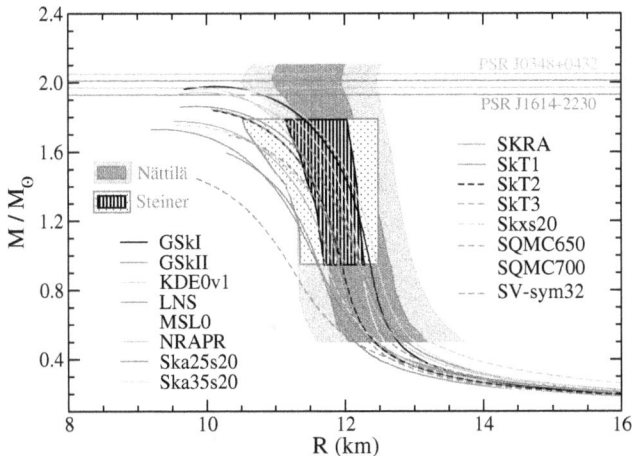

FIGURE 2.16
Radius vs mass plots for different neutron star equations of state derived from a selection of Skyrme parameterizations. (From Ref. [116].)

neutron stars as to match the observed two solar mass pulsars is a strong constraint on Skyrme parameterizations.

As an example of the kinds of predictions made by Skyrme interactions, Figure 2.16 shows mass-radius profiles of allowed (in the sense of being solutions of the TOV equation) neutron stars from various Skyrme-based equations of state. The mass constraints from pulsar observations PSR J1614-2230 [65] and PSR J038+0432 [66] are shown as the two sets of interleaved horizontal lines near two solar masses. Two different sets of constraints are shown in Figure 2.16 for cold neutron star matter, coming from different observational constraints. The first is from data on X-ray bursts, and is labeled "Nättilä" [117]. The second is from a combined analysis of X-ray binaries and X-ray bursts, and is labeled "Steiner" [118]. The range of results show that only a small subset of Skyrme parameterizations are able to reach the highest observed neutron star mass at around 2 solar masses, while most pass through the constraint region of the X-ray binary and X-ray burst observations.

2.3.6 Beyond the Standard Skyrme Model

We end the Skyrme section with a brief list of approaches beyond the standard Skyrme mean-field approach to nuclear matter that has been discussed so far, so that the interested reader can follow up the references with further reading.

Skyrme pointed out even in 1959 [5] that the potential now bearing his name should be considered a kind of *pseudo-potential* thanks to the delta "functions" which are not true functions, but generalized functions, or measures which are well defined only under an integral. The Skyrme interaction can be seen as the lowest orders in a systematic expansion in derivatives of the densities. Exploration of the next orders beyond those provided by Skyrme has been analyzed. This includes the properties of nuclear matter [119].

A more extensive density dependence has been proposed [120], suggesting that multiple fractional powers of the density should be included in a Skyrme-like functional. Many attempts have been made to extend the functional form of the Skyrme interaction to be more general. Extended Skyrme forces which provide momentum and density-dependent terms in combination have been used for nuclear matter and neutron stars successfully [121,122].

Parameterizations in which multiple t_3-like density-dependent terms, each with its own parameter, have been created and used to study nuclear matter and neutron stars [123].

The description of nuclear matter with the Skyrme interaction beyond the mean-field approximation has been studied. With a suitable momentum cut-off, and a refitting of the force parameters to avoid double-counting, a calculation of nuclear matter with explicit many-body correlations is shown to be possible with the Skyrme interaction [124].

2.4   Conclusion

We have presented an overview of the properties of symmetric and asymmetric infinite nuclear matter, and neutron star matter. We have shown the relativistic mean-field and the Skyrme interaction to be powerful and versatile tools to describe the observables and pseudo-observables associated with nuclear matter and neutron stars.

Acknowledgements

PDS acknowledges the financial support from the UK STFC under grant ST/P005314/1. This work is a part of the project INCT-FNA Proc. No. 464898/2014-5 from Brazil, partially supported by the Brazilian agencies Conselho Nacional de Desenvolvimento Científico e Tecnológico (CNPq) under grants 310242/2017-7 and 406958/2018-1 (OL) and 433369/2018-3 (MD), and Fundação de Amparo à Pesquisa do Estado de São Paulo (FAPESP) under thematic project 2013/26258-4 (OL). OL and MD dedicate this chapter to Antonio Delfino Jr, from Federal Fluminense University (UFF), Brazil, who passed away in July 2019.

References

[1] J. M. Eisenberg and W. Greiner, *Nuclear theory, vol. 3*, North-Holland Publishing Company (1972).

[2] P. Ring and P. Schuck, *The nuclear many-body problem*, Springer-Verlag (2000).

[3] H. A. Bethe, *Annu. Rev. Nucl. Sci.* **21**, 93 (1971).

[4] T. H. R. Skyrme, *Phil. Mag.* **1**, 1043–1054 (1956).

[5] T. H. R. Skyrme, *Nucl. Phys.* **9**, 615–634 (1959).

[6] B. D. Serot and J. D. Walecka, *Adv. Nucl. Phys.* **16**, 1 (1986).

[7] J. Rikovska Stone, R. Koncewicz, P. D. Stevenson, and M. R. Strayer, *Phys. Rev. C* **68**, 034324 (2003).

[8] N. K. Glendenning, *Compact stars*, Springer-Verlag (1997); *Compact Stars*, 2nd ed. (Springer, New York, 2000).

[9] J. Piekarewicz, *Phys. Rev. C* **76**, 064310 (2007).

[10] J. B. Natowitz, et al., *Phys. Rev. Lett.* **89**, 212701 (2002).

[11] E. Shuryak, *Rev. Mod. Phys.* **89**, 035001 (2017).

[12] J. D. Walecka, *Ann. Phys.* **83**, 491 (1974).

[13] J. D. Walecka, *Theoretical nuclear and subnuclear physics*, 2ed. Imperial College Press (2004).

[14] J. Boguta and A. R. Bodmer, *Nucl. Phys. A* **292**, 414 (1977).

[15] P.-G. Reinhard, *Phys. Rep.* **52**, 439 (1989).

[16] H. Müller and B. D. Serot, *Nucl. Phys. A* **606**, 508 (1996).

[17] F. J. Fattoyev and J. Piekarewicz, *Phys. Rev. C* **82**, 025805 (2010).

[18] M. Jaminon and C. Mahaux, *Phys. Rev. C* **40**, 354 (1989); *Nucl. Phys. A* **365**, 371 (1981).

[19] R. J. Furnstahl, J. J. Rusnak e B. D. Serot, *Nucl. Phys. A* **632**, 607 (1998).

[20] F. J. Fattoyev, C. J. Horowitz, J. Piekarewicz, and G. Shen, *Phys. Rev. C* **82**, 055803 (2010).

[21] S. K. Dhiman, R. Kumar, and B. K. Agrawal, *Phys. Rev. C* **76**, 045801 (2007).

[22] B. K. Agrawal, *Phys. Rev.* **C 81**, 034323 (2010).

[23] C. J. Horowitz and J. Piekarewicz, *Phys. Rev. C* **66**, 055803 (2002).

[24] P. Danielewicz, R. Lacey, and W. G. Lynch, *Science* **298**, 1592 (2002).

[25] A. W. Steiner, J. M. Lattimer and E. F. Brown, *Astrophys. J. Lett.* **765**, L5 (2013).

[26] M. Dutra, O. Lourenço, S. S. Avancini, B. V. Carlson, A. Delfino, D. P. Menezes, C. Providência, S. Typel, and J. R. Stone, *Phys. Rev. C* **90**, 055203 (2014).

[27] M. Dutra, O. Lourenço, J. S. Sá Martins, A. Delfino, J. R. Stone, and P. D. Stevenson, *Phys. Rev. C* **85**, 035201 (2012).

[28] H. B. Callen, *Thermodynamics and an Introduction to Thermostatics*, 2nd edition (Willey, 1985).

[29] A. L. Fetter and J. D. Walecka, *Quantum Theory of Many-Particle Systems* (McGraw-Hill, New York, 1971).

[30] E. Khan, J. Margueron, G. Colò, K. Hagino, and H. Sagawa, *Phys. Rev. C* **82**, 024322 (2010).

[31] E. Khan, J. Margueron, and I. Vidaña, Phys. *Rev. Lett.* **109**, 092501 (2012).

[32] E. Khan and J. Margueron, *Phys. Rev. C* **88**, 034319 (2013).

[33] W. Greiner and J. Maruhn, *Nuclear models*, Springer (1995).

[34] J. P. Blaizot, *Phys. Rep.* **64**, 171 (1980).

[35] G. Colò, N. Van Giai, J. Meyer, K. Bennaceur, and P. Bonche, *Phys Rev. C* **70**, 024307 (2004).

[36] B. G. Todd-Rutel, and J. Piekarewicz, *Phys. Rev. Lett.* **95**, 122501 (2005).

[37] B. K. Agrawal, S. Shlomo, V. Kim Au, *Phys. Rev. C* **72**, 014310 (2005).

[38] S. Shlomo, and D. H. Youngblood, *Phys. Rev. C* **47**, 529 (1993).

[39] L.-W. Chen, B.-J. Cai, C. M. Ko, B.-A. Li, C. Shen, and J. Xu, *Phys. Rev. C* **80**, 014322 (2009).

[40] J. M. Pearson, N. Chamel, and S. Goriely, *Phys. Rev. C* **82**, 037301 (2010).

[41] U. Garg, and G. Colò, *Prog. Part. Nucl. Phys.* **101**, 55 (2018).

[42] B. M. Santos, M. Dutra, O. Lourenço, and A. Delfino, *Phys. Rev. C* **92**, 015210 (2015).

[43] G. A. Lalazissis, J. König, and P. Ring, *Phys. Rev. C* **55**, 540 (1997).

[44] B. Nerlo-Pomoroska and J. Sykut, *Int. J. Mod. Phys. E* **13**, 75 (2004).

[45] M. Rashdan, *Phys. Rev. C* **63**, 044303 (2001).

[46] R. J. Furnstahl, B. D. Serot, and H. B. Tang, *Nucl. Phys. A* **615**, 441 (1997).

[47] J. Piekarewicz and M. Centelles, *Phys. Rev. C* **79**, 054311 (2009).

[48] J. Piekarewicz, *Phys. Rev. C* **66**, 034305 (2002).

[49] M. Bender, K. Rutz, P. G. Reinhard, J. A. Maruhn, and W. Greiner, *Phys. Rev. C* **60**, 034304 (1999).

[50] P.-G. Reinhard, *Z. Phys. A* **329**, 257 (1988).

[51] C. J. Horowitz, and J. Piekarewicz, *Phys. Rev. Lett.* **86**, 5647 (2001).

[52] M. Bhuyan, B. V. Carlson, S. K. Patra, and S.-G. Zhou, *Phys. Rev. C* **97**, 024322 (2018).

[53] M. Baldo, G.F. Burgio, *Prog. Part. Nucl. Phys.* **91**, 203 (2016).

[54] B. M. Santos, M. Dutra, O. Lourenço, and A. Delfino, *Phys. Rev. C* **90**, 035203 (2014).

[55] B.-J. Cai and L.-W. Chen, *Phys. Rev. C* **85**, 024302 (2012).

[56] B.-A. Li and X. Han, *Phys. Lett. B* **727**, 276 (2013).

[57] R. C. Tolman, Phys. Rev. **55**, 364 (1939); J. R. Oppenheimer and G. M. Volkoff, *Phys. Rev.* **55**, 374 (1939).

[58] G. Baym, C. Pethick, and P. Sutherland, *Astrophys. J.* **170**, 299 (1971).

[59] C. Gonzalez-Boquera, M. Centelles, X. Viñas, and A. Rios, *Phys. Rev. C* **96**, 065806 (2017).

[60] J. Xu, L.-W. Chen, B.-A. Li, and H.-R. Ma, *Astrophys. J.* **697**, 1549 (2009).

[61] C. Gonzalez-Boquera, M. Centelles, X. Viñas, T. R. Routray, arXiv:1904.06566.

[62] J. Carriere, C. Horowitz, and J. Piekarewicz, *Astrophys. J.* **593**, 463 (2003).

[63] J. Piekarewicz, and F. J. Fattoyev, *Phys. Rev. C* **99**, 045802 (2019).

[64] C. Gonzalez-Boquera, M. Centelles, X. Viñas, and L. M. Robledo, *Phys. Lett. B* **779**, 195 (2018).

[65] P. B. Demorest, T. Pennucci, S. M. Ransom, M. S. E. Roberts, and J. W. T. Hessels, *Nature* **467**, 1081 (2010).

[66] J. Antoniadis, P. C. C. Freire, N. Wex, et al., *Science* **340**, 448 (2013).

[67] M. Dutra, O. Lourenço, and D. P. Menezes, *Phys. Rev. C* **93**, 025806 (2016); **94**, 049901(E) (2016).

[68] B. P. Abbott, et al. (The LIGO Scientific Collaboration and the Virgo Collaboration), *Phys. Rev. Lett.* **119**, 161101, (2017).

[69] B. P. Abbott, et al. (The LIGO Scientific Collaboration and the Virgo Collaboration), *Phys. Rev. Lett.* **121**, 161101 (2018).

[70] B. P. Abbott, et al. (LIGO Scientific Collaboration and Virgo Collaboration), *Phys. Rev. X* **9**, 011001 (2019).

[71] T. Hinderer, B. D. Lackey, Ryan N. Lang and J. S. Read, *Phys. Rev. D* **81**, 123016 (2010).

[72] J. S. Read, L. Baiotti, J. D. E. Creighton, J. L. Friedman, B. Giacomazzo, K. Kyutoku, C. Markakis, L. Rezzolla, M. Shibata, and K. Taniguchi, *Phys. Rev. D* **88**, 044042 (2013)

[73] W. Del Pozzo, T. G. F. Li, M. Agathos, and C. Van Den Broeck, *Phys. Rev. Lett.* **111**, 071101 (2013).

[74] O. Lourenço, M. Dutra, C. H. Lenzi, C. V. Flores, and D. P. Menezes, *Phys. Rev. C* **99**, 045202 (2019).

[75] T. Malik, N. Alam, M. Fortin, C. Providência, B. K. Agrawal, T. K. Jha, B. Kumar, and S. K. Patra, *Phys. Rev. C* **98**, 035804 (2018).

[76] R. Nandi, P. Char, and S. Pal, *Phys. Rev. C* **99**, 052802(R) (2019).

[77] S. Typel and H. H. Wolter, *Nucl. Phys. A* **656**, 331 (1999).

[78] P. Gögelein, E. N. E. van Dalen, C. Fuchs, and H. Müther, *Phys. Rev. C* **77**, 025802 (2008).

[79] S. S. Avancini, L. Brito, Ph. Chomaz, D. P. Menezes, and C. Providência, *Phys. Rev. C* **74**, 024317 (2006).

[80] B. A. Nikolaus, T. Hoch, and D. G. Madland, *Phys. Rev. C* **46**, 1757 (1992).

[81] P. Finelli, N. Kaiser, D. Vretenar, W. Weise, *Nucl. Phys. A* **435** 449, (2004).

[82] N. Kaiser, S. Fritsch, W. Weise, *Nucl. Phys. A* **697** 255, (2002).

[83] B.-A. Li, L.-W. Chen, C. M. Ko, *Phys. Rep.* **464**, 113 (2008).

[84] J. S. Bell and T. H. R. Skyrme, *Phil. Mag.* **1**, 1055 (1956).

[85] T. H. R. Skyrme, in Proc. Rehovoth Conf. Nucl. Structure (North Holland Publishing Co., 1958) p. 20.

[86] D. Vautherin and D. M. Brink, *Phys. Rev. C* **5**, 626–647 (1972).

[87] D. Vautherin, *Phys. Rev. C* **7**, 296–316 (1973).

[88] H. Flocard, P. Quentin, A.K. Kerman, and D. Vautherin, *Nucl. Phys. A* **203**, 433–472 (1973).

[89] M. Cailliau, Jocelyne Leterssier, H. Flocard, and P. Quentin, *Phys. Lett. B* **46**, 11–14 (1973).

[90] H. Flocard and D. Vautherin, *Nucl. Phys. A* **264**, 197–220 (1976).

[91] Y. M. Engel, D. M. Brink, K. Goeke, J. Krieger, and D. Vautherin, *Nucl. Phys. A* **249**, 215–238 (1975).

[92] M. Bender, P.-H. Heenen, and P.-G. Reinhard, *Rev. Mod. Phys.* **75**, 121–180 (2003).

[93] K. Bennaceur, P. Bonche, and J. Meyer, *Comptes Rendus Physique* **4**, 555–570 (2003).

[94] J. R. Stone and P.-G. Reinhard, Progress Particle *Nucl. Phys.* **58**, 587–657 (2007).

[95] J. Erler, P. Klüpfel, and P.-G. Reinhard, *J. Phys. G* **38**, 033101 (2011).

[96] Gianluca Colò, Eur. *Phys. J. Plus* **133**, 533 (2018).

[97] P. D. Stevenson and M. C. Barton, *Progress Particle Nucl. Phys.* **104**, 142–164 (2019).

[98] K. A. Brueckner, C. A. Levinson, and H. M. Mahmoud, *Phys. Rev.* **95**, 217–228 (1954).

[99] K. A. Brueckner, *Phys. Rev.* **96**, 508–516 (1954).

[100] K. A. Brueckner, *Phys. Rev.* **97**, 1353–1366 (1955).

[101] P. Bonche, B. Grammaticos, and S. E. Koonin, *Phys. Rev. C* **18**, 2567–2573 (1978).

[102] J.-S. Wu, M. R. Strayer, and M. Baranger, *Phys. Rev. C* **60**, 044302 (1999).

[103] C. I. Pardi and P. D. Stevenson, *Phys. Rev. C* **87**, 014330 (2013).

[104] R. Jodon, M. Bender, K. Bennaceur, and J. Meyer, *Phys. Rev. C* **94**, 024335 (2016).

[105] P. D. Stevenson, *Fitting a zero–range Skyrme force* doi:10.15126/surreydata.9121031

[106] J. R. Stone, N. J. Stone, and S. A. Moszkowski, *Phys. Rev. C* **89**, 044316 (2014).

[107] J. Treiner and H Krivine, *J. Phys. G* **2**, 285–307 (1976).

[108] E. Chabanat, P. Bonche, P. Haensel, J. Meyer, and R. Schaeffer, *Nucl. Phys. A* **627**, 710–746 (1997).

[109] M. Beiner, H. Flocard, Nguyen Van Giai, and P. Quentin, *Nucl. Phys. A* **238**, 29–69 (1975).

[110] J. Erler, C. J. Horowitz, W. Nazarewicz, M. Rafalski, and P. G. Reinhard, *Phys. Rev. C* **87**, 044320 (2013), arXiv:1211.6292.

[111] J. Erler, P. Klüpfel, and P.-G. Reinhard, *Phys. Rev. C* **82**, 044307 (2010).

[112] P. Klüpfel, P.-G. Reinhard, T. J. Bürvenich, and J. A. Maruhn, *Phys. Rev. C* **79**, 034310–23 (2009).

[113] J. Dobaczewski, W. Nazarewicz, and P.-G. Reinhard, *J. Phys. G* **41**, 074001 (2014), arXiv:1402.4657.

[114] N. Schunck, J. D. McDonnell, J. Sarich, S. M. Wild, and D. Higdon, *J. Phys. G* **42**, 034024 (2015).

[115] P.-G. Reinhard, *Phys. Scr.* **91**, 023002 (2016)

[116] O. Lourenço, M. Dutra, C. H. Lenzi, S. K. Biswal, M. Bhuyan, D. P. Menezes, *Eur. Phys. J. A* **56**, 32 (2020)

[117] J. Nättilä, A. W. Steiner, J. J. E. Kajava, V. F.Suleimanov, and J. Poutanen, *Astron. Astrophys.* **591**, A25 (2016)

[118] A. W. Steiner, J. M. Lattimer, and E. F. Brown, *Astro-phys. J.* **722**, 33 (2010).

[119] D. Davesne, A. Pastore, and J. Navarro, *J. Phys. G* **40**, 095104 (2016)

[120] P. Papakonstantinou, T. S. Park, Y. Lim, and C. H. Hyun, Phys. Rev. C 97, 014312 (2018).

[121] N. Chamel, S. Goriely, and J. M. Pearson, *Phys. Rev. C* **80**, 065804 (2006).

[122] Z. Zhang, and L.-W. Chen, *Phys. Rev. C* 94, 064326 (2016)

[123] B. K. Agrawal, S. K. Dhiman, and R. Kumar, *Phys. Rev. C* **73**, 034319 (2006).

[124] C. J. Yang, M. Grasso, X. Roca-Maza, G. Colò, and K. Moghrabi, *Phys. Rev. C* **94**, 034311 (2016).

3

Recent Parameterization in Relativistic Mean-Field Formalism

K.C. Naik and R.N. Panda

Siksha 'O' Anusandhan Deemed to be University

Bharata Kumar and S.K. Patra

Institute of Physics and
Homi Bhabha National Institute

CONTENTS

3.1   Introduction

In the past few decades, the relativistic mean-field (RMF) formalism has been considered one of the most successful theories to describe the properties of finite nuclei and infinite nuclear matter. Using this framework, one can describe the finite nuclei not only for the stability line but also for the nuclei away from the stability valley, including the superheavy region. The RMF theory was originally developed by Walecka [1] from the idea of Dúrr and Teller [2,3] with scalar and vector mesons, which is a simplified form of the technique developed by Miller and Green [4]. The original Walecka model is based on σ and ω mesons, which is known as Quantum Hadrodynamics (QHD)-I and later on included the ρ meson to take care of the proton–neutron asymmetry, i.e. QHD-II. Although this simple $\sigma - \omega$ model is successful to some extent for finite nuclei [5], it gives an unexpectedly large compressibility of $K_\infty = 550 MeV$. To take care of this defect, Boguta and Bodmer [5] included two non-linear terms in the σ-meson coupling, which improves the compressibility to $K_\infty = 200 - 300 \text{MeV}$ and predicts finite nuclei results remarkably [6]. However, this model fails to make the nuclear equation of states (EOSs) softer, a demand for many experimental data including the presently measured mass of the neutron star, which makes it imperative to add some other necessary terms in the RMF Lagrangian.

To see the effect of various terms on finite nuclei and nuclear EOSs, a detailed analysis of these terms is done in Ref. [7]. It is observed that each and every coupling in the Lagrangian is important, at least for one of the properties of either finite nuclei or infinite nuclear matter. Motivated by these effects, a new set of parameters G3 is made based on the effective field theory-motivated relativistic mean-field (E-RMF) Lagrangian, in which all the possible couplings up to some reasonable order are taken into account. It should be noted that there are some parameter sets like G1 and G2 which are already formed based on the E-RMF formalism, in which all possible couplings are included. However, still some crucial couplings such as isoscalar-vector and isovector-vector ($\omega - \rho$) cross-couplings are excluded in the G1 and G2 sets. In addition to this term, the contribution of δ meson to the nuclear phenomena is also ignored in the original E-RMF calculations. Taking the importance of these couplings into consideration, the G3 set [8] is constructed, which is able to reproduce the finite and infinite nuclear properties including the low- and high-density regions. Very recently, a new parameter set IOPB-I [9] is also reported, which may be useful for practical purpose of both finite and infinite nuclear systems [7].

This chapter provides brief information about the progress of E-RMF formalism and the latest parameter sets. Then, the application of force parameters to some of the known properties of finite and infinite nuclear systems is explained. The gravitational tidal deformability λ, which is a crucial parameter in the recently discussed gravitational wave from the binary black hole or binary neutron stars merger, is subsequently discussed.

3.2   The Relativistic Mean-Field Formalism

The RMF theory is a framework for the phenomenological description of nuclei where nucleons are treated as point-like particles and are described by Dirac spinors $\psi^\pm$. These nucleons are interacting through the exchange of mesons which are also point-like particles. The linear scalar field σ produces an attractive force between the nucleons; here, the exchange of vector mesons determines the repulsive part of the interaction. One of the important vector mesons is ω meson which produces the vector field ω^μ. The Coulomb repulsion due to the exchange of photons between the nucleons is produced by the vector field A^μ. The isospin dependence of the nuclear force is also explained by the exchange of ρ meson which gives rise to an isovector-vector field ρ^μ. In order to study the properties of highly asymmetric systems like drip-line nuclei and neutron stars, it is necessary to introduce the effect of the isovector-scalar δ meson.

In this mean-field approximation, the fluctuations in the meson fields are not considered, but their expectation values are used, i.e. $\hat{\sigma} = \langle \sigma \rangle \equiv \sigma$. In order to study the atomic nuclei in general, the E-RMF theory can be used. This is a very successful theory to calculate the nuclear bulk properties such as binding energy, charge radius, and quadruple moment. This theory is a relativistic generalization of the non-RMF effective theory, and the benefit of this theory is that it helps solve basic difficulties such as renormalization and divergence of the systems. The spin–orbit interaction in finite nuclei can properly be analyzed using this formalism, and an additional advantage of this system is that it works better in high-density regions. It gives excellent results for the calculation of bulk properties for both β-stable nuclei and nuclei present throughout the periodic table from proton-rich to neutron-rich regions.

3.2.1 Finite Nuclei

The details of nucleonic-meson E-RMF Lagrangian with δ meson and the cross-coupling of ω and ρ mesons $(W \times R)$ up to the fourth order with the exchange of σ, ω, and ρ mesons and photon A are given in Refs. [7,10–14]. The E-RMF Lagrangian has an infinite number of terms including self- and cross-couplings. This model can be used directly by fitting the coupling constants and some masses of the mesons. The energy density functional $\varepsilon(r)$ corresponding to the E-RMF Lagrangian up to the fourth order is given by [9]

$$\varepsilon(r) = \sum_\alpha \varphi_\alpha^\dagger(r) \left\{ \begin{array}{c} -i\alpha.\nabla + \beta\left[M - \Phi(r) - \tau_3 D(r)\right] + W(r) + \frac{1}{2}\tau_3 R(r) \\ +\frac{1+\tau_3}{2}A(r) - \frac{i\beta\alpha}{2M}.\left(f_\omega \nabla W(r) + \frac{1}{2}f_\rho\tau_3\nabla R(r)\right) \end{array} \right\} \varphi_\alpha(r)$$

$$+ \left(\frac{1}{2} + \frac{k_3}{3!}\frac{\Phi(r)}{M} + \frac{k_4}{4!}\frac{\Phi^2(r)}{M^2}\right)\frac{m_s^2}{g_s^2}\Phi^2(r) - \frac{\varsigma_0}{4!}\frac{1}{g_\omega^2}W^4(r) + \frac{1}{2g_s^2}\left(1 + \alpha_1\frac{\Phi(r)}{M}\right)(\nabla\Phi(r))^2$$

$$\frac{-1}{2g_\omega^2}\left(1 + \alpha_2\frac{\Phi(r)}{M}\right)(\nabla W(r))^2 - \frac{1}{2}\left(1 + \eta_1\frac{\Phi(r)}{M} + \frac{\eta_2}{2}\frac{\Phi^2(r)}{M^2}\right)\frac{m_\omega^2}{g_\omega^2}W^2(r)$$

$$- \frac{1}{2e^2}(\nabla A(r))^2 \frac{-1}{2g_\rho^2}(\nabla R(r))^2 - \frac{1}{2}\left(1 + \eta_\rho\frac{\Phi(r)}{M}\right)\frac{m_\rho^2}{g_\rho^2}R^2(r) - \Lambda_\omega\left(R^2(r) \times W^2(r)\right)$$

$$\frac{+1}{2g_\delta^2}(\nabla D(r))^2 + \frac{1}{2}\frac{m_\delta^2}{g_\delta^2}D^2(r), \tag{3.1}$$

where Φ, W, R, D, and A are the fields; and $g_\sigma, g_\omega, g_\rho, g_\delta$, and $\frac{e^2}{4\pi}$ are the coupling constants; and $m_\sigma, m_\omega, m_\rho$, and m_δ are the masses for σ, ω, ρ, and δ mesons and photon, respectively. The variational principle can be used to solve the field equations for the baryons and mesons. The mesons equations can be obtained by using the equation $\left(\frac{\partial E}{\partial \varphi_i}\right)_{\rho=\text{constant}} = 0$. Using the Lagrange multiplier ε_α, which is the energy eigenvalue of the Dirac equation, one can find the single-particle energy for the nucleons from the relation $\frac{\partial}{\partial \varphi_\alpha^\dagger(r)}\left[\varepsilon(r) - \sum_\alpha \varphi_\alpha^\dagger(r)\varphi_\alpha(r)\right] = 0$, imposing the normalization condition $\sum_\alpha \varphi_\alpha^\dagger(r)\varphi_\alpha(r) = 1$.

A set of coupled equations are obtained as follows:

(a) Dirac equations for nucleons are given by

$$\left\{ \begin{array}{c} -i\alpha.\nabla + \beta\left[M - \Phi(r) - \tau_3 D(r)\right] + W(r) + \frac{1}{2}\tau_3 R(r) + \frac{1+\tau_3}{2}A(r) \\ \frac{-i\beta\alpha}{2M}.\left(f_\omega\nabla W(r) + \frac{1}{2}f_\rho\tau_3\nabla R(r)\right) \end{array} \right\}\varphi_\alpha(r)$$

$$= \varepsilon_\alpha\varphi_\alpha(r). \tag{3.2}$$

(b)   The bosonic equations for the mesons are given by

$$
-\Delta \Phi (r) + m_s^2 \Phi (r) = g_s^2 \rho_s - \left(\frac{k_3}{2!} + \frac{k_4}{3!} \frac{\Phi (r)}{M} \right) \frac{m_s^2}{M} \Phi^2 (r)
$$

$$
+ \frac{g_s^2}{2M} \left(\eta_1 + \eta_2 \frac{\Phi (r)}{M} \right) \frac{m_\omega^2}{g_\omega^2} W^2 (r)
$$

$$
+ \frac{\eta_\rho}{2M} \frac{g_s^2}{g_\rho^2} m_\rho^2 R^2 (r) + \frac{\alpha_1}{2M} \left[(\nabla \Phi (r))^2 + 2\Phi (r) \Delta \Phi (r) \right]
$$

$$
+ \frac{\alpha_2}{2M} \frac{g_s^2}{g_\omega^2} (\nabla W (r))^2 , \tag{3.3}
$$

$$
-\Delta W (r) + m_\omega^2 W (r) = g_\omega^2 \left(\rho (r) - \frac{f_\omega}{2} \rho_T (r) \right) - \left(\eta_1 + \eta_2 \frac{\Phi (r)}{2M} \right) \frac{\Phi (r)}{M} m_\omega^2 W^2 (r)
$$

$$
- \frac{\varsigma_0}{3!} W^3 (r) + \frac{\alpha_2}{M} [\nabla \Phi (r) . \nabla W (r) + \Phi (r) \Delta W (r)]
$$

$$
- 2\Lambda_\omega g_\omega^2 R^2 (r) W (r) , \tag{3.4}
$$

$$
-\Delta R (r) + m_\rho^2 R (r) = \frac{1}{2} g_\rho^2 \left(\rho_3 (r) + \frac{1}{2} f_\rho \rho_{T,3} (r) \right) - \eta_\rho \frac{\Phi (r)}{M} m_\rho^2 R (r)
$$

$$
- 2\Lambda_\omega g_\rho^2 R (r) W^2 (r) , \tag{3.5}
$$

$$
- \Delta A (r) = e^2 \rho_p (r) , \tag{3.6}
$$

and

$$
-\Delta D (r) + m_\delta^2 D (r) = g_\delta^2 \rho_{s,3} (r) , \tag{3.7}
$$

where $\rho (r) = \sum_\alpha \varphi_\alpha^\dagger (r) \varphi_\alpha (r) = \rho_p (r) + \rho_n (r)$ and $\rho_s (r) = \sum_\alpha \varphi_\alpha^\dagger (r) \beta \varphi_\alpha (r) = \rho_{sp} (r) + \rho_{sn} (r)$ are the baryon and scalar densities, respectively, where $\rho_p (r)$ is the proton distribution density and $\rho_n (r)$ is the neutron distribution density. The $\rho_3 (r)$, $\rho_{s3} (r)$, $\rho_p (r)$, $\rho_T (r)$, and $\rho_{T,3} (r)$ are defined in the following:

$$
\rho_3 (r) = \sum_\alpha \varphi_\alpha^\dagger (r) \tau_3 \varphi_\alpha (r) = \rho_p (r) - \rho_n (r) , \tag{3.8}
$$

$$
\rho_{s3} (r) = \sum_\alpha \varphi_\alpha^\dagger (r) \tau_3 \beta \varphi_\alpha (r) = \rho_{sp} (r) - \rho_{sn} (r) , \tag{3.9}
$$

$$
\rho_p (r) = \sum_\alpha \varphi_\alpha^\dagger (r) \left(\frac{1 + \tau_3}{2} \right) \varphi_\alpha (r) , \tag{3.10}
$$

$$
\rho_T (r) = \sum_\alpha \frac{i}{M} \nabla , \tag{3.11}
$$

$$
\rho_{T,3} (r) = \sum_\alpha \frac{i}{M} \nabla . \left[\varphi | \alpha^\dagger (r) \beta \alpha \tau_3 \varphi_\alpha | (r) \right] . \tag{3.12}
$$

The summation used in these equations is all over the occupied states of the nucleons. In a fully relativistic formalism, one has to consider both the negative and positive energy solutions. But, the nucleons and mesons are composite particles, and hence, their vacuum polarization effects in no-sea approximation can be ignored. As a result, the negative-energy states do not have impacts on densities and currents. The effective masses of protons and neutrons are given as $M_p = M - \Phi(r) - D(r)$ and $M_n = M - \Phi(r) + D(r)$. The vector potential is given by

$$V(r) = g_\omega V_0(r) + \frac{1}{2}g_\rho b_0(r) + \frac{(1 - \tau_3)}{2}A_0(r). \tag{3.13}$$

The sets of RMF equations for the nucleon and meson fields can be solved self-consistently to describe the ground-state properties of finite nuclei. An initial guess value of the boson fields is taken, and the Dirac equations are solved. This procedure is repeated until convergence is achieved. After that, one can calculate the total energy of the system as follows:

$$E_{total} = \int \varepsilon(r)\,dr = E_{part} + E_\sigma + E_\omega + E_\rho + E_\delta + E_{\omega\rho} + E_c + E_{pair} + E_{c.m.}, \tag{3.14}$$

where E_{part} is the sum of the single-particle energies of the nucleons and $E_\sigma, E_\omega, E_\rho, E_\delta, E_{\omega\rho}$, and E_c are the contributions of the respective mesons and Coulomb's fields. The pairing energy E_{pair} and the center of mass motion energy $E_{c.m.}$ are incorporated externally. Here, $E_{c.m.} = \frac{3}{4} \times 41A^{-1/3}$ is the non-relativistic approximation of the center of mass energy correction in a harmonic oscillator potential [15]. The E-RMF Lagrangian does not contain the terms like $\varphi_\alpha^\dagger \varphi_\alpha^\dagger$ and $\varphi_\alpha \varphi_\alpha$, which are responsible for pairing correction. If this type of interactions are added in the Lagrangian, then the pairing correlation leads to the violation of the particle number in the mean-field calculations. To take pairing into account, the constant force quasi-BCS approach of Ref. [7] is adopted. In this scheme, the quasi bound states (bounds due to their centrifugal + Coulomb barrier) of the pairing nucleons, i.e. between proton and proton or neutron and neutron, are considered. The calculations are made below and above one harmonic oscillator shell of the Fermi level to ignore the unrealistic pairing in the continuum at highly excited states. When the pairing interaction v_{pain} has non-zero matrix element $\langle \alpha_2 \acute\alpha_2 \vee v_{pair} \vee \alpha_1 \acute\alpha_1 \rangle = -G(\text{force constant})$ with G and m are positive and non-zero values, then the nucleon pairs are under time-reversal symmetry. Here, $\alpha = nljm\rangle$ and $\acute\alpha = nlj - m\rangle$ are the quantum states [7].

3.2.2 Infinite Nuclear Matter

All the gradients of the fields in Eqs. (2.3)–(2.7) vanish for infinite, static, uniform, and isotropic nuclear matter, where we neglect the electromagnetic interaction. Here, energy density and pressure are obtained from the energy-momentum tensor $T_{\mu\nu}$ as follows:

$$T_{\mu\nu} = \sum_i \partial_\nu \phi_i \frac{\partial L}{\partial(\partial^\mu \phi_i)} - g_{\mu\nu}L. \tag{3.15}$$

The zeroth component $<T_{00}>$ of $T_{\mu\nu}$ gives us the energy density, whereas pressure of the system is calculated from the third component $<T_{ii}>$ of $T_{\mu\nu}$.

$$
\begin{aligned}
\varepsilon = {}& \frac{2}{(2\pi)^3}\int d^3k E_i(k) + \rho W + \left(\frac{1}{2} + \frac{k_3}{3!}\frac{\Phi(r)}{M} + \frac{k_4}{4!}\frac{\Phi^2(r)}{M^2}\right)\frac{m_s^2}{g_s^2}\Phi^2(r) \\
& - \frac{1}{2}\left(1 + \eta_1\frac{\Phi(r)}{M} + \frac{\eta_2}{2}\frac{\Phi^2(r)}{M^2}\right)\frac{m_\omega^2}{g_\omega^2}W^2(r) - \frac{\varsigma_0}{4!}\frac{1}{g_\omega^2}W^4(r) + \frac{1}{2}\rho_3 R \\
& - \frac{1}{2}\left(1 + \eta_\rho\frac{\Phi(r)}{M}\right)\frac{m_\rho^2}{g_\rho^2}R^2(r) - \Lambda_\omega\left(R^2(r) \times W^2(r)\right)\frac{+1}{2g_\delta^2}(\nabla D(r))^2 + \frac{1}{2}\frac{m_\delta^2}{g_\delta^2}D^2(r),
\end{aligned} \tag{3.16}
$$

and

$$P = \frac{2}{3\,(2\pi)^3} \int d^3k \frac{k^2}{E_i\,(k)} - \left(\frac{1}{2} + \frac{k_3}{3!}\frac{\Phi\,(r)}{M} + \frac{k_4}{4!}\frac{\Phi^2\,(r)}{M^2}\right)\frac{m_s^2}{g_s^2}\Phi^2\,(r)$$

$$+ \frac{1}{2}\left(1 + \eta_1\frac{\Phi\,(r)}{M} + \frac{\eta_2}{2}\frac{\Phi^2\,(r)}{M^2}\right)\frac{m_\omega^2}{g_\omega^2}W^2\,(r) + \frac{\varsigma_0}{4!}\frac{1}{g_\omega^2}W^4\,(r)$$

$$+ \frac{1}{2}\left(1 + \eta_\rho\frac{\Phi\,(r)}{M}\right)\frac{m_\rho^2}{g_\rho^2}R^2\,(r) + \Lambda_\omega\left(R^2\,(r)\times W^2\,(r)\right) - \frac{1}{2}\frac{m_\delta^2}{g_\delta^2}D^2\,(r) \qquad (3.17)$$

where $E_i\,(k) = \sqrt{k^2 + M_i^2}\,(i = p, n)$ is the energy and k is the Fermi momentum of the nucleon. After getting the EOSs for infinite nuclear matter system, the mass and radius of the isolated static neutron star can be calculated. For this, one has to solve the Tolmann-Oppenheimer-Volkov (TOV) equations. Here, the equations of states are the main inputs. The TOV equations as follows:

$$\frac{dp\,(r)}{dr} = \frac{-\left[\varepsilon\,(r) + p\,(r)\right]\left[M\,(r) + 4\pi r^3 p\,(r)\right]}{r^2\left(1 - \frac{2M(r)}{r}\right)}$$

and

$$\frac{dM\,(r)}{dr} = 4\pi r^2\varepsilon\,(r). \qquad (3.18)$$

3.3   New Parameterization

The fitting of parameters is a crucial process in any of the effective calculations. A large number of parameter sets are designed to calculate the properties of finite nuclei and infinite nuclear matter. Initially, the parameters given by Walecka [5], and later on by Horowitz and Serot [16], are quite successful to describe the properties of finite nuclei and infinite nuclear matter. With the progress of time, the formulation is also used extensively, and one after another, the limitations of the parameter sets are disclosed. For example, as it is mentioned in the Introduction, the linear $\sigma - \omega$ model of Walecka predicts a large value of K_∞ which is rectified by the addition of self-couplings (non-linear terms) to the σ meson. These couplings also include the long-range repulsion of nuclear force and solve the Coester band problem [17]. In this domain, the NL1 [18], NL2 [19], and NL3 [20] forces are very significant, which rectify the shortcoming of the linear models for some time. Although the non-linear models predict the finite nuclei properties remarkably well, they fail to reproduce the heavy-ion collision data of Danielewicz et al. [21], including the soft EOSs. As a result, the Lagrangian density is extended further by the addition of non-linear couplings in the vector mesons, and it is worthy to mention the works of Bodmer [22], Gmuca [23,24], and Toki [25] that gave birth to the sets such as TM1, TM2, and TMA.

Inspired by the E-RMF model, Furnstahl, Serot, and Tang [10,11] extended the RMF Lagrangian to infinite terms with all types of self- and cross-couplings. The naïve dimensional analysis (NDA) and naturalness conditions are imposed to make the Lagrangian finite and workable. Thus, the evolution of G1 and G2 parameter sets is confirmed. In spite of the success of G1 and G2 sets in infinite nuclear matter, like neutron star, the EOSs do not pass through the low-density region in these sets, and the involvement of the additional terms is needed. G3, FSUGold2, FSUGarnet, and IOPB-I are recent examples of such forces.

Construction of parameter sets such as G3 and IOPB-I involves a lot of care [8,9] and also needs to be improved by including the co-variance error analysis, which are some of the future tasks in this direction. The values of these parameters and their nuclear matter properties at saturation for some selected forces are listed in Table 3.1.

The NL3 parameters are obtained with a χ^2 fitting similar to earlier parameterizations [20]. Here the fitting parameters are m_σ, g_σ, g_ω, g_ρ, k_3, and k_4. The masses of ω meson (m_ω) and ρ meson (m_ρ) are taken from the empirical data. These parameters are fitted to reproduce the ground-state properties such as charge radius, total binding energy, and available neutron radius of some of the spherical nuclei. The FSUGarnet set is designed to take into account the theoretical errors using the pseudo-data $m_\sigma, \rho_0, \varepsilon_0, M, K, J, L, \zeta$ in the optimization procedure [26]. The newly developed parameter sets G3 and IOPB-I, which are used in the Lagrangian density, are determined by the Simulated Annealing Method (SAM). The SAM procedure is useful when the number of parameters is more than the observables [27−29] and provides better results in global minimum within several local minima. In this technique, the system from a highly unstable temperature state cools down slowly to a stable system by reaching a frozen temperature. The χ^2 values are considered as least square fit

TABLE 3.1

Values of New Parameter Sets G3 [8] and IOPB-I [9] along with that of Earlier Sets NL3 [20], FSUGold2 [26], FSUGarnet [30], and G2 [10,11] and Their Nuclear Matter Properties

	NL3	FSUGold2	FSUGarnet	IOPB-I	G2	G3
m_σ	508.194	497.479	496.731	500.487	520.206	524.901
m_ω	782.501	782.500	782.187	782.187	782.000	781.248
m_ρ	763.000	763.000	762.468	762.468	770.000	769.980
m_δ	000.000	000.000	000.000	000.000	000.000	979.377
g_σ	010.176	010.397	010.484	010.396	010.508	009.831
g_ω	012.788	013.557	013.715	013.351	012.786	011.603
g_ρ	008.985	008.970	013.891	011.126	009.511	012.094
g_δ	000.000	000.000	000.000	000.000	000.000	002.011
k_3	001.484	001.231	001.368	001.496	003.238	002.606
k_4	−05.660	−00.205	−01.397	−02.932	000.694	001.694
η_1	000.000	000.000	000.000	000.000	000.650	000.424
η_2	000.000	00.0000	000.000	000.000	000.110	000.114
η_ρ	000.000	000.000	000.000	000.000	000.390	000.645
ς_0	000.000	004.705	004.410	003.103	002.642	001.010
Λ_ω	000.000	000.001	000.043	000.024	000.000	000.038
α_1	000.000	000.000	000.000	000.000	000.000	002.000
α_2	000.000	000.000	000.000	000.000	000.000	−01.468
f_ω	000.000	000.000	000.000	000.000	000.000	000.220
f_ρ	000.000	000.000	000.000	000.000	000.000	001.239
α_1	000.000	000.000	000.000	000.000	000.000	−00.087
α_2	000.000	000.000	000.000	000.000	000.000	−00.484
ρ_0	000.148	000.150	000.153	000.149	000.153	000.148
E/A	−16.299	−16.280	−16.230	−16.100	−16.070	−16.020
k_∞	271.760	238.000	229.500	022.650	215.000	243.960
J	037.400	037.620	030.950	033.300	036.400	031.840
L	118.200	112.800	051.040	063.580	101.2 00	049.310
M^*/M	000.600	000.593	000.578	000.593	000.664	000.699

Note: The nucleon mass M is 939.0 MeV. All the coupling constants are dimensionless, except k_3 which is in fm^{-1}. All the masses are in MeV. The lower portion of the table is the nuclear matter properties.
Source: The values are taken from Ref. [8,35].

with model parameters p_i. χ^2 can be expressed as $\chi^2 = \frac{1}{N_d - N_p} \sum_{i=1}^{N_d} \left[\frac{M_i^{exp} - M_i^{th}}{\sigma_i} \right]^2$, where N_d and N_p represent the number of experimental data points and the fitting parameters, respectively. M_i^{exp} and M_i^{th} are the observables, and σ_i is the error in the measurement. To follow the SAM procedure, one should fit the schedule and requirements acceptable parameter spacing considering χ^2 least square fit. The schedule is decided by the Cauchy annealing formula as $T(K) = \frac{T_i}{K+1}$, where T_i and $T(K)$ are the control parameters before and after the completion of Kth step, respectively. After completing each arrangement of $100N_p$ or $10N_p$ successfully, whichever is earlier, the K-value increases by unity.

3.4 Results and Discussions

Here, we discuss our results calculated for finite nuclei, infinite nuclear matter system, neutron stars. In the finite nuclei Section 3.4.1, we have discussed the results obtained by various force parameter sets. Here, binding energy, root mean square (rms) radii, and their implications to various phenomena are vividly analyzed. The isotopic shift for Pb isotope, neutron-skin thickness, and search for a new magic number in the superheavy valley are highlighted. In the second part of the results, we have discussed the recently detected gravitational waves. In this regard, the structure of neutron stars and the tidal deformability parameter λ are reported using the latest EOSs obtained from the E-RMF approximation.

3.4.1 Finite Nuclei

For finite nuclei, the bulk properties such as binding energy, charge radius, and neutron skin thickness are mainly discussed. We have taken different force parameters such as NL3 [20], FSUGold2 [26], FSUGarnet [30], IOPB-I [9], G2 [10,11], and G3 [8] to calculate these properties for some selected nuclei, i.e., $^{16}_{8}O$, $^{40}_{20}Ca$, $^{68}_{28}$, $^{90}_{40}Zr$, $^{100}_{50}Sn$, and $^{208}_{82}Pb$. The calculated results are given in Table 3.2 and compared with experimental values [31−33] wherever available. It can be seen that all the parameter sets reproduce the binding energy and charge radius with good agreements with experimental data. The presence of excess neutrons gives rise to neutron skin thickness, which is defined as $\Delta R_{np} = R_n - R_p$, where R_n and R_p are the rms radii of neutron and proton distributions, respectively. It is important to note that for finite nuclei, there is a large uncertainty in the experimental measurement of neutron distribution radius. And the current values of neutron radius and neutron skin thickness for $^{208}_{82}Pb$ is $5.78^{+0.16}_{-0.18}$ and $0.33^{+0.16}_{-0.18}$fm, respectively. It is clear from Table 3.2 that our calculation for neutron skin thickness for $^{208}_{82}Pb$ is also matched well in the case of NL3 and IOPB-I parameters, but it is underestimated for FSUGarnet and G3 parameter sets.

3.4.1.1 Binding Energy

The differences between the calculated and experimental binding energies for 70 spherical nuclei [34] are shown in Figure 3.1. The triangles, stars, squares, diamonds, and circles represent the results for NL3, FSUGold2, FSUGarnet, G2, and G3 parameters, respectively. From the figure, it is clear that G3 set reproduces the experimental data well [31]. It is also found that the rms deviations for the binding energy are 2.977, 3.062, 3.696, 3.827, and 2.308 MeV for NL3, FSUGold2, FSUGarnet, G2, and G3, respectively. The rms error on the binding energy for G3 parameter set is smaller when compared to those for other sets.

TABLE 3.2

The Calculated Binding Energy Per Particle (BE/A), Charge Radius (r_{ch}), and Neutron Skin Thickness $\Delta R_{np} = R_n - R_p$ with Available Experimental Data [31,32]

Nucleus	Obs.	NL3	FSUGold2	FSUGarnet	IOPB-I	G2	G3	Expt.
${}^{16}_{8}$O	BE/A	7.917	7.876	7.876	7.977	7.952	8.037	7.97
	r_{ch}	2.714	2.690	2.690	2.705	2.718	2.707	2.699
	ΔR_{np}	−0.026	−0.028	−0.028	−0.027	−0.028	−0.028	−
Ca	BE/A	8.540	8.528	8.528	8.577	8.529	8.561	8.551
	r_{ch}	3.466	3.438	3.438	3.458	3.453	3.459	3.478
	ΔR_{np}	−0.046	−0.051	−0.051	−0.049	−0.049	−0.049	−
${}^{68}_{28}$	BE/A	8.698	8.692	8.692	8.707	8.682	8.690	8.682
	r_{ch}	3.870	3.861	3.861	3.873	3.861	3.892	−
	ΔR_{np}	0.262	0.184	0.184	0.223	0.240	0.190	−
${}^{90}_{40}$Ca	BE/A	8.695	8.693	8.693	8.691	8.684	8.266	8.709
	r_{ch}	4.253	4.231	4.231	4.253	4.240	4.497	4.269
	ΔR_{np}	0.115	0.065	0.065	0.091	0.102	−0.079	−
${}^{100}_{50}$Sn	BE/A	8.301	8.298	8.298	8.284	8.248	8.359	8.258
	r_{ch}	4.469	4.426	4.426	4.464	4.470	4.732	−
	ΔR_{np}	−0.073	−0.078	−0.078	−0.077	−0.079	0.243	−
${}^{208}_{82}$Pb	BE/A	7.885	7.902	7.902	7.870	7.853	7.863	7.867
	r_{ch}	5.509	5.496	5.496	5.521	5.498	5.541	5.501
	ΔR_{np}	0.283	0.162	0.162	0.221	0.256	0.180	$0.33^{+0.16}_{-0.18}$

Note: Energies are in MeV, and charge radii are in fm.
Source: The results are taken from Ref. [8].

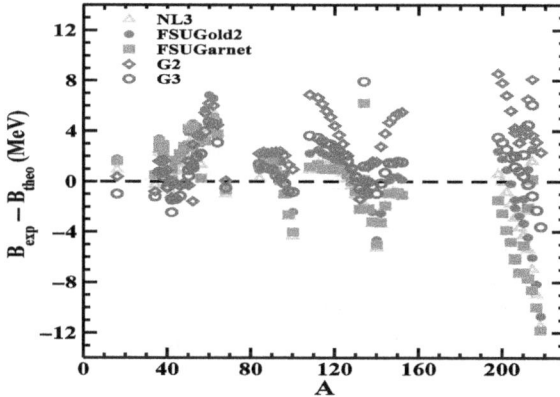

FIGURE 3.1
Difference between experimental [31] and theoretical binding energies as a function of mass numbers for NL3 [20], FSUGold2 [26], FSUGarnet [30], G2 [10,11], and G3 [8] parameter sets. (The figure is taken from Ref. [8].)

3.4.1.2 Isotopic Shift

The isotopic shift is defined as $\Delta r_c^2 = R_c^2 (A) - R_c^2 (208) \, fm^2$, where $R_c^2 (A)$ and $R_c^2 (208)$ are the mean square radii of Pb isotopes having mass number A and that of ^{208}Pb, respectively. The isotopic shift Δr_c^2 for Pb nucleus is shown in Figure 3.2. From the figure, one can observe that Δr_c^2 increases with mass number monotonously till A=208 ($\Delta r_c^2 = 0$ for ^{208}Pb) and then gives a sudden kink. It was first pointed by Sharma et al. [36] that the

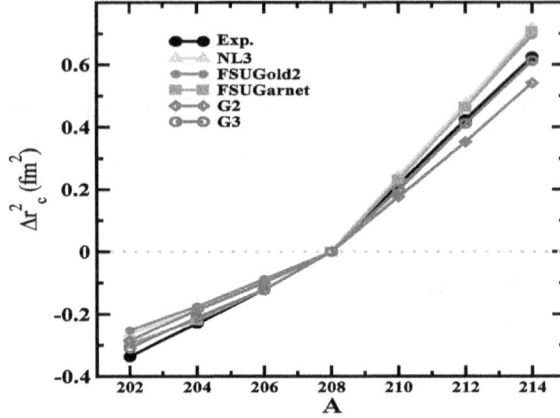

FIGURE 3.2

The isotopic shift $\Delta r_c^2 = R_c^2(208) - R_c^2(A)(fm)^2$ of Pb isotopes taking R_c of $^{208}_{82}$Pb as the standard value. Calculations of the NL3 [20], FSUGold2 [26], FSUGarnet [30], G2 [10,11], and G3 [8] parameter sets are compared with experimental data [31]. (The figure is taken from Ref. [8].)

non-relativistic parameter sets fail to show this effect. However, this effect is well explained when a relativistic set like NL-SH [36] is used. The NL3, FSUGold2, FSUGarnet, G2 and G3 sets also appropriately predict this shift in Pb isotopes. A further inspection of the results reveals that the agreement of G3 is comparatively better than the other relativistic parameterizations.

3.4.1.3   Neutron Skin Thickness (ΔR_{np})

The difference in the rms radii of neutrons and protons distribution, $\Delta R_{np} = R_n - R_p$, is called neutron skin thickness. In Figure 3.3, the variations of ΔR_{np} from ^{40}Ca to ^{238}U for the NL3, FSUGold2, FSUGarnet, G2, and G3 parameter sets as a function of proton–neutron asymmetry I = (N − Z)/A are shown. The experimental data are also shown in the figure. Trzcińska et al. extracted ΔR_{np} for 26 stable nuclei ranging from ^{40}Ca to ^{238}U in the antiprotons experiments at CERN [37,38]. The data can be fitted by $\Delta R_{np} = (0.90 \pm 0.15)I + (-0.03 \pm 0.02)$ [37,39] and are represented by the shaded regions in Figure 3.3. Most of the ΔR_{np} obtained with NL3, FSUGold2, FSUGarnet, and G2 overestimate the data and deviate from the experimentally extrapolated region (the shaded region). On the other hand, the ΔR_{np} calculated by the G3 set found to be within the observational range. It is confirmed that larger the asymmetry, more the overestimation of ΔR_{np} by the NL3, FSUGold2, and G2 parameter sets. The overestimation of ΔR_{np} for the NL3, FSUGold2, and G2 parameter sets is due to the absence (or negligible strength) of $\omega - \rho$ cross-coupling [40]. This cross-coupling plays a crucial role in the determination of neutron skin thickness ΔR_{np} without affecting much the properties of finite nuclei. It is shown in Ref. [41] that the derivative of neutron matter EOS at a sub-saturation density is strongly correlated with the ΔR_{np}. Furthermore, one can verify that the behavior of the neutron matter EOS should also depend on the incompressibility coefficient K_∞. Earlier parameters like TM1*, G1, and G2 corresponding to the Lagrangian density similar to the ones used in Ref. [8] give higher values of J. In Ref. [8], this shortcoming is attempted to improve and the force parameter G3 comprising J = 31.8 MeV and K_∞ = 243.9 MeV is constructed (see Table 3.1).

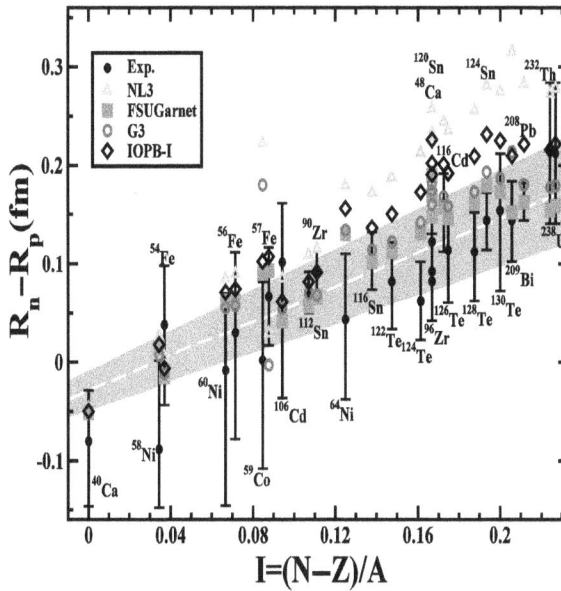

FIGURE 3.3
The differences between neutron and proton rms radii ΔR_{np} obtained for NL3 [20], FSUGold2 [26], FSUGarnet [30], G2 [10,11], and G3 [8] are plotted as a function of isospin asymmetric $I = (N - Z)/A$. The experimental data are also displayed [38,39]. The orange-shaded region represents $\Delta R_{np} = (0.90 \pm 0.15)I + (-0.03 \pm 0.02)$. (The figure is taken from Ref. [8].)

3.4.1.4   Prediction of Magic Number in Superheavy Valley

Recently, the synthesis of superheavy elements has become an interesting topic in the study of nuclear physics. A large amount of manpower has been used to know the latest element. In this regard, to identify the next proton and neutron magic number combination beyond $^{208}_{82}$Pb is the main concern. Therefore, our main task is to find out the next double closed nucleus after $Z = 82$ and $N = 126$. The two distinct but well-defined non-relativistic Skyrme-Hartee-Fock (SHF) and RMF or E-RMF formalisms are developed. For SHF, the well-known parameterizations like FITZ, SIII, SkMP, SLy4 in RMF, the NL3, G1, G2, and NL-Z2 parameter sets are used. These are very successful models for the entire mass table including the proton- and neutron-rich regions. The pairing correlations are done by using the BCS-pairing scheme in a constant strength approach. Thus, the average pairing gaps for neutrons Δ_n and protons Δ_p are obtained as the output in the calculation process [7].

A wide range of isotopes starting from the proton-rich to the neutron-rich regions ($Z = 112$ to $Z = 130$) are scanned in the superheavy valley. The following three criteria are adopted to detect the magicity of a nucleus [42]:

1. The average pairing gap for proton Δ_p and neutron Δ_n at the magic number is minimum.

2. The binding energy per particle is maximum compared to the neighboring one, i.e., there must be a sudden decrease (jump) in two-neutron (or two-proton) separation energy S_{2n} just after the magic number in an isotopic or isotonic chain.

3. At the magic number, the shell correction energy E_{shell} is more negative. In other words, a pronounced energy gap in the single-particle levels $\varepsilon_{n,p}$ appears at the magic number.

The Δ_p obtained by SHF and RMF utilizing the FITZ, SIII, SLy4, SkMP and NL3, NL-Z2, G1, G2 force are plotted in Figure 3.4. From the figure, it is clear that the value of Δ_p is almost zero for the whole $Z = 120$ isotopic chain in both the theoretical approaches. A smaller value of Δ_p is observed at $Z = 114$ and 124 isotopes in few cases. The Δ_p results for the whole chain of elements $Z = 112 - 130$ indicate that probably could be the next magic number in the SHE valley. Now, our task is to locate the corresponding neutron magic number for the $Z = 120$ magic proton. To predict the corresponding neutron shell closure, the estimated neutron pairing gap Δ_p for $Z = 112 - 130$ with increasing neutron number is shown in Figure 3.5. Some arc-like structures with vanishing Δ_n at neutron numbers $N = 182, 208$ and $N = 172, 184, 258$ for the SHF and RMF sets, respectively, are obtained. In general, the binding energy per particle (BE/A) is maximum for doubly closed nuclei than the neighboring one. For example, the BE/A with SHF (FITZ set) for $A = 300, 302, 304$ and $Z = 120$ are $7.046, 7.048$, and 7.044MeV corresponding to $N = 180, 182$, and 184, respectively. Similarly with SLy4 force, these values are 6.950, 6.952, and 6.933 MeV [42]. This is reflected in the sudden jump of the two-neutron separation energy S_{2n} from a higher to a lower value in an isotopic chain as shown in Figure 3.6. The S_{2n} can be calculated from the difference between the ground-state binding energies of two isotopes, which is defined as $S_{2n}(Z, N) = BE(Z, N) - BE(Z, N - 2)$. It provides sufficient information about the structure of a nucleus. The sudden fall in S_{2n} shows the extra stability for a particular nucleus. From Figure 3.6, we can notice such effect, i.e., jump in S_{2n} at $N182$ and 208 with SHF. For the isotopic chain of $Z = 120$ with the nuclear forces obtained through our parameter, jump in S_{2n} is clearly shown at $N = 172, 184, 258$ in RMF calculations, which can be seen in Figure 3.6. These results give the shell closure properties of the neutron numbers.

Strutinsky introduced the concept of shell correction in liquid-drop model, which determines the shell closure [43]. Positive shell correction energy E_{shell} reduces the binding energy, whereas negative E_{shell} increases the stability of the nucleus. The SHF result of E_{shell} is shown in Figure 3.6. It is clear from the figure that the extra stability is seen for

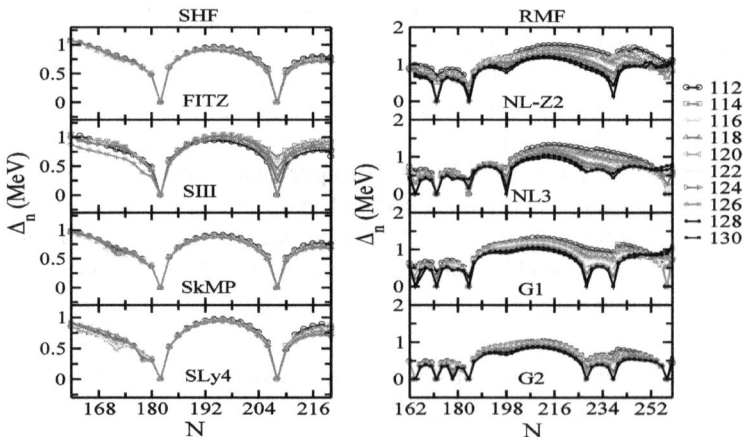

FIGURE 3.4
The proton average pairing gap Δ_p for $Z = 112-126$ with $N = 162-220$ and $Z = 112-130$ with $N = 162-260$. (The figure is taken from Ref. [42].)

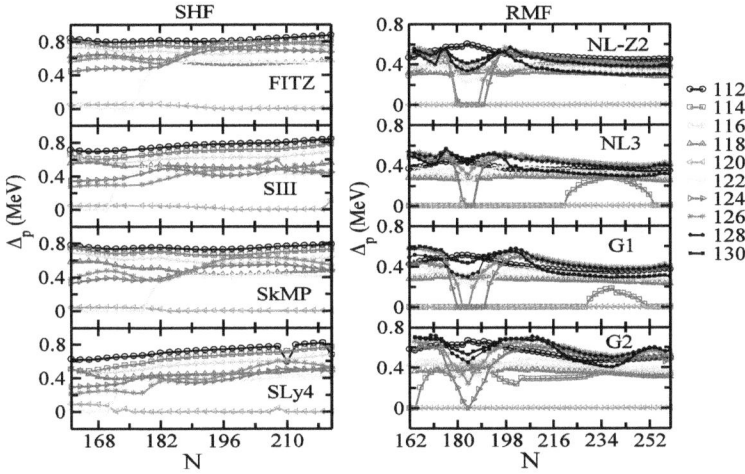

FIGURE 3.5
The neutron pairing gap Δ_n for with $162 - 220$ and with $N = 162 - 260$. (The figure is taken from Ref. [42].)

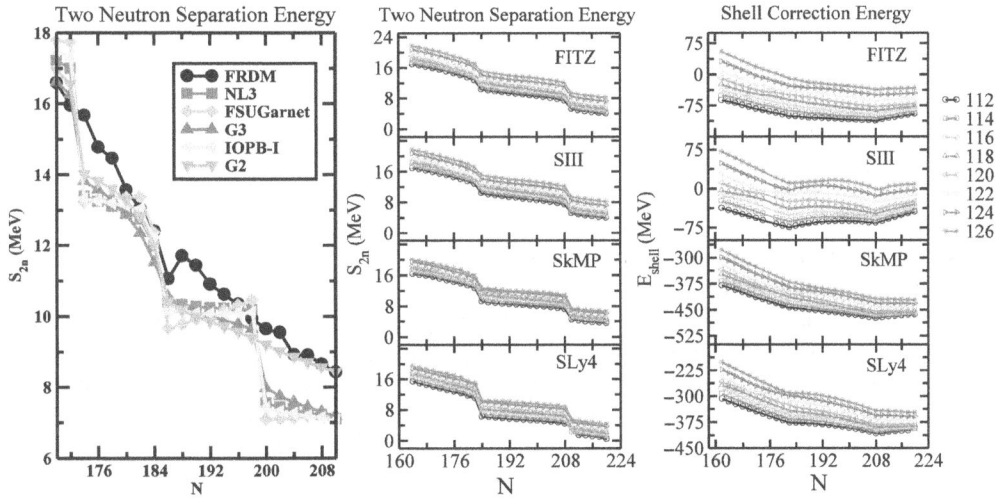

FIGURE 3.6
The two-neutron separation energy S_{2n} and the shell correction energy E_{shell} for $Z = 112 - 126$ and $N = 162 - 220$ in the SHF and RMF models.

the nuclei with $A = 302, 328$ and $Z = 120$. Similar results of large negative shell energy for RMF calculation at neutron numbers $172, 184$, and 258 are reported in Ref. [44].

Few levels of the neutron single-particle spectra near the Fermi surface are calculated by the new set G3 along with the earlier forces NL3, G1, and G2 for the superheavy nuclei $^{292}120$ and $^{304}120$ (Figure 3.7). It is observed from the figure that there is a large gap between the last occupied and first unoccupied levels in all the forces. In the case of $^{292}120$, the significant gap is at $N = 172$, and it is $N = 184$ for $^{304}120$. Thus, the large gaps of $^{292,304}120$ signify the evolution of magic numbers in the superheavy valley agreeing with the three criteria of Bhuyan and Patra [42].

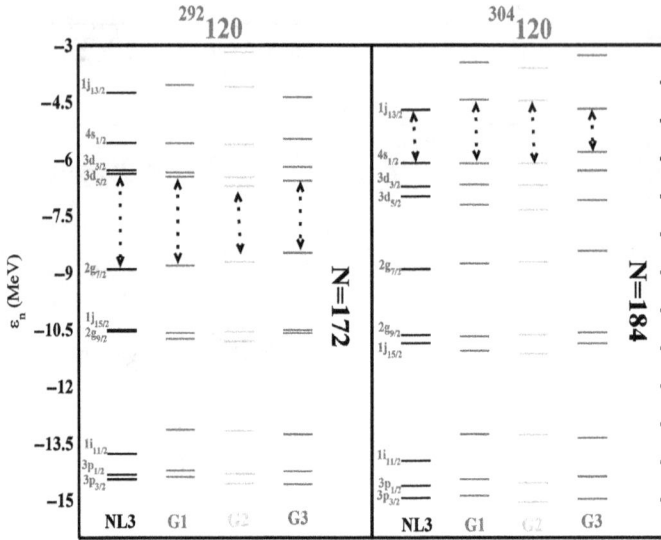

FIGURE 3.7
The neutron single particle spectrum near the Fermi surface.

3.4.2 Infinite Nuclear Matter

In Figure 3.8, various theoretical predictions for infinite nuclear matter, symmetric nuclear matter, and pure neutron matters are compared. In Figure 3.8a, the energy per neutron in pure neutron matter at sub-saturation densities is plotted. The parameter sets NL3, FSUGold2, FSUGarnet, and G2 deviate significantly from the shaded region. The non-relativistic forces labeled as Baldo-Maieron, Friedman, and AFDMC are designed for sub-saturated matter density as marked by solid square, solid triangle, and solid circle, respectively. This is important to mention that the trend for B/N in pure neutron matter at low densities obtained by G3 passes well through the shaded region. In Figure 3.8b and (c), the EOSs for symmetric matter and neutron matter at higher densities are shown for various parameter sets. Except NL3, all other parameter sets along with G3 pass through the shaded region. Arumugam et al. [45] reported that even in high-density region, the EOS overestimates the experimental data as long as the self-coupling of the ω-meson field and some cross-couplings are not included in the interaction Lagrangian. The self-coupling inclusions of the ω-meson field soften the EOS [22−25]. It is also known that δ meson softens the symmetry energy at low densities and makes it stiffer at high densities [14].

3.4.2.1 Equation of State (EOS) for Nuclear Matter

Earlier, we tested the validity of the EOSs obtained with various parameter sets of nuclear and neutron matters. In this section, NL3 [20], FSUGold [48], FSUGold2 [26], and G2 [10,11] forces are used to evaluate the mass and radius of the hybrid star (with hyperon). The G2 force parameter has a large number of couplings, and each coupling has its own effect on various properties of hyper nuclear system. Figure 3.9 shows the EOSs for the G2, FSUGold2, FSUGold, and NL3 parameter sets. From Figure 3.9a, it is clear that all EOSs follow similar trends. Among these sets, NL3 having a large compressibility K_∞ gives the stiffer EOS, whereas FSUGold predicts the softer EOS. This is due to the large and positive k_4 along with the inclusion of isoscalar-isovector coupling (Λ) in the FSUGold set [48].

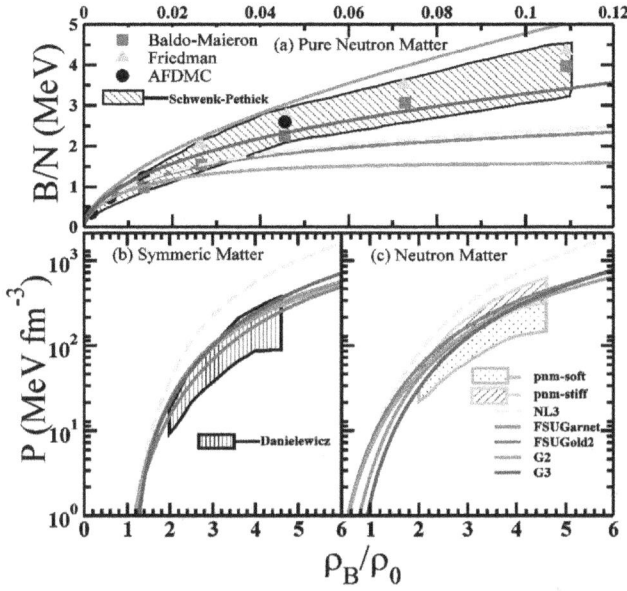

FIGURE 3.8

(a) Binding energy per neutron versus neutron density. (b and c) EOSs for symmetric matter and neutron matter at higher densities for various parameter sets. The experimental data for higher-density region are taken from Ref. [46,47]. (The figure is taken from Ref. [8].)

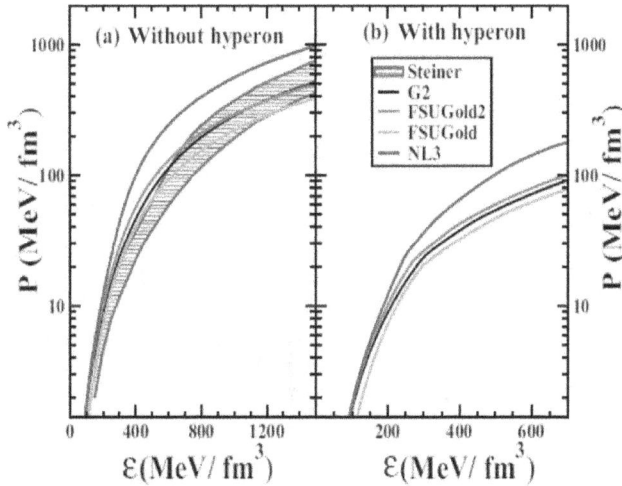

FIGURE 3.9

The EOSs for nuclear and hyperon star matters are compared with the empirical data [49]. (The figure is taken from Ref. [35].)

For understanding the softer and stiffer EOSs given by various models, one can compare their couplings and other parameters of the sets (see Table 3.1). We noted a large variation in the M, K_∞, E/A, the symmetry energy J, and its slope L at saturation density. The symmetry energy coefficient J comes from liquid-drop model of the nucleus and accounts to the contribution made by the isospin asymmetry to the total energy of the nucleus. It is defined as $J = \left(\frac{\partial^2}{\partial t^2} \frac{\varepsilon}{\rho} \right) \vee_{t=0}$, where $t = \frac{\rho_n - \rho_p}{\rho}$, ε is the energy density, ρ is the total number density, and ρ_n and ρ_p are the number densities of protons and neutrons, respectively. $\frac{\varepsilon}{\rho}$ is the energy per nucleon. For higher $\varepsilon \sim 500$–1400 MeV fm^{-3}, all sets are found in the empirical region, except the NL3 set which has the lowest effective mass M.

Figure 3.9b shows a kink on the nucleon-hyperons EOS at $\varepsilon \sim 400$–500 eV fm^{-3}. This ~ 200–300 MeV) shows the appearance of hyperons in the dense matter. As compare to the attractive part of the scalar interaction generated by the σ meson, the repulsive component of the vector potential becomes more important. Therefore, the coupling of the hyperon-nucleon strength becomes weak. At a given baryon density, the inclusion of hyperons significantly lowers the pressure compared to the EOS without hyperons. This is due to the higher energy of the hyperons, as the neutrons are replaced by the low-energy hyperons.

3.4.2.2 Neutron Star

The G2, FSUGold2, FSUGold, and NL3 parameter sets are also used to estimate the mass and radius of the static neutron star composed of neutrons, protons, electrons, and muons. For a given EOS, the TOV equations [50] are integrated with the boundary conditions $P(0) = P_c$ and $M(0) = 0$. Here, P_c and $M(0)$ are the pressure and mass of the star at the origin $r = 0$, respectively. The value of r $(= R)$, where the pressure gets vanish completely, gives the idea about the surface of the star. Thus, the mass M and radius R of the static neutron and hyperon stars can be determined uniquely at each central. Comparing with the precisely measured properties of two massive $(2M_\odot)$ neutron stars and the extraction of stellar radii from X-ray observation [51,52], the calculated results for the maximum mass as a function of radius is shown in Figure 3.10a and b. According to the recent data, the

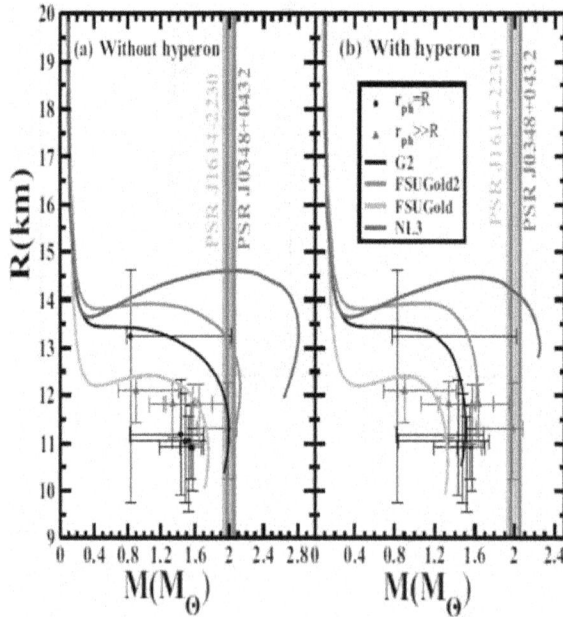

FIGURE 3.10
The mass-radius profile for the G2, FSUGold2, FSUGold and NL3 parameter sets. The solid circles ($r_{ph}=R$) and triangles ($r>>R$) are the observational constraints [49], where $r_{ph} =$ photospheric radius. The vertical shaded region is the recent observation [51,52]. (The figure is taken from Ref. [35].)

maximum mass predicted by any theoretical model [49] should reach the limit of 2.0 $M_\odot$. The G2 EOS of a nucleonic matter compact star gives a mass of 1.99 $M_\odot$ and a radius of 11.25 km.

Steiner et al. [49] estimate the most probable neutron star radius in the range of $11-12$ km and the mass of 1.4 $M_\odot$ from the X-ray observation data. Their predicted EOS is also relatively softer in the density range of $1-3$ M_0, where M_0 is the nuclear saturation density. The stiff EOS like NL3 predicts a larger stellar radius of 13.23 km and a maximum mass of 2.81 $M_\odot$. FSUGold2 EOS suggests a larger and heavier neutron star with a mass of 2.12 $M_\odot$ and a radius of 12.12 km, whereas EOS of FSUGold gives the mass of 1.75 $M_\odot$ and a radius of 10.76 km. It is important to see here that although FSUGold and FSUGold2 are from the same RMF model with similar terms in the Lagrangian, their results for a neutron star are quite different. This is due to the impact of M_0 and also the large values of the slope parameter $L = 112.8 \pm 16.1$ MeV (see Table 3.1) in FSUGold2 EOS at high densities [53]. One can observe from the figure that the stiffer NL3 EOS gives the maximum neutron star mass 2.81 $M_\odot$ and the presence of hyperon matter reduces the mass to 2.25 $M_\odot$. The reason behind this is that the presence of hyperon matter under β-equilibrium condition softens the EOS as they are more massive than nucleons. Most of the RMF models, in the absence of δ mesons, which satisfy the observational constraint of 2 $M_\odot$, yield $R_{1.4}>13$ km [54]. The model DDHδ [55], which includes the δ-meson contributions, yields $R_{1.4}$ similar to the results obtained in the present case. The value of M_{max} is consistent with maximum masses so far observed for neutron stars like PSR J1614-2230, which has $M = 1.9-0.04$ $M_\odot$ [51], and PSR J0348+0432, which has $M = 2.01-0.04$ $M_\odot$ [52]. The value of $R_{1.4}= 12.69$ km is also in good agreement with the empirical value of $R_{1.4}= 10.7-13.1$ km, which is consistent with the observational analysis [56].

3.4.2.3 Tidal Deformability

After getting the EOSs for various parameter sets, the tidal deformabilities of a non-rotating neutron star are evaluated. When a spherical star is placed in an external static quadrupole tidal field ε_{ij}, the star gets deformed (quadrupole deformation). This deformation is measured by [57]

$$\lambda = \frac{-Q_{ij}}{\varepsilon_{ij}} = \frac{2}{3}k_2 R^5$$

and

$$\Lambda = \frac{2k_2}{3C^5},$$

where Q_{ij} is the induced quadrupole moment of the star in binary and ε_{ij} is the static external quadrupole tidal field of the companion star. λ is the tidal deformability parameter depending on the EOS via both the neutron star radius and a dimensionless quantity k_2. This k_2 is called the second Love number [57,58]. In general theory of relativity, we have to distinguish k_2 between electric type (gravitational field generated by mass) and magnetic type (gravitational field generated by mass current, i.e. motion of mass) [35,59]. Λ is the dimensionless version of λ, and C is the compactness parameter ($C = M/R$). The expression for electric tidal Love number is as follows [54]:

$$k_2 = \frac{8}{5}\left(1 - 2C^2\right)\left[2C\left(y - 1\right) - y + 2\right]\{2C\}. \tag{3.19}$$

The value of $y \equiv y(r)$ can be calculated by solving the following first-order differential equation [59,60]:

$$r\frac{dy(r)}{dr} + y(r)^2 + y(r)F(r) + r^2 Q(r) = 0$$

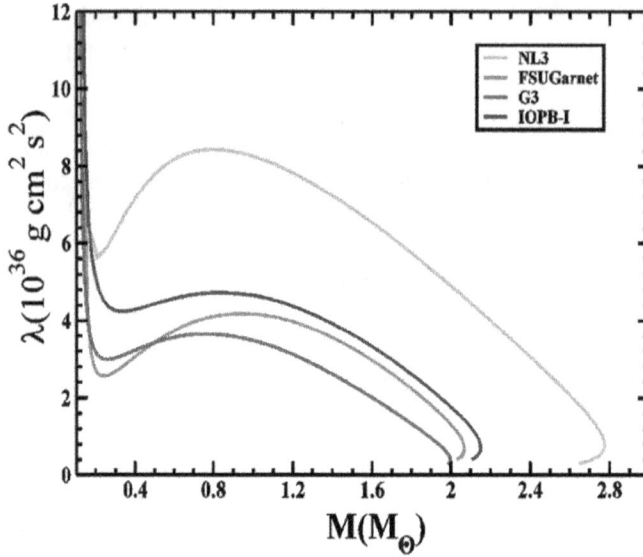

FIGURE 3.11
Tidal deformability λ as a function of neutron star mass with different EoSs. (The figure is taken from Ref. [61].)

with

$$F(r) = \frac{r - 4\pi r^3\left[\varepsilon(r) - P(r)\right]}{r - 2M(r)} Q(r) = \frac{4\pi r\left(5\varepsilon(r) + 9P(r) + \frac{\varepsilon(r)+P(r)}{\partial P(r)/\partial \varepsilon(r)} - \frac{6}{4\pi r^2}\right)}{r - 2M(r)}$$

$$-4\left[\frac{M(r) + 4\pi r^3 P(r)}{r^2\left(1 - \frac{2M(r)}{r}\right)}\right]^2. \tag{3.20}$$

One of the important calculations regarding the tidal deformability of the single neutron star (NS) as well as binary neutron stars (BNS), which has been recently discussed in GW170817 [62], is also studied. It indicates that λ strongly depends on the radius of the NS and the value of k_2. As the radius of the NS increases, the deformation by the external field becomes larger, since an increase in gravitational gradient takes place with an increase in radius. Therefore, a stiff EOS gives large deformation and a soft EOS gives small deformation in the BNS system. Figure 3.11 shows the tidal deformability as a function of NS mass. In particular, λ takes a wide range of values $\lambda = (1 - 8) \times 10^{36} \mathrm{g\ cm^2\ s^2}$ as shown in Figure 3.11. For the G3 parameter set, the tidal deformability λ is very low in the mass region of 0.5–2.0 $M_\odot$ when compared to other sets. This is because the star exerts high central pressure and energy density, which results in the formation of a compact star. However, for the NL3 EOS, it turns out that because of the stiffness of the EOS, the λ value is higher. The tidal deformability of the canonical NS (1.4 $M_\odot$) of IOPB-I along with FSUGarnet and G3 EOSs are found to be 3.191, 3.552, and 2.613 g cm^2 s^2, respectively, as shown in Table 3.3, which are consistent with the results obtained by Steiner et al. [63].

We have also discussed the weighted dimensionless tidal deformability of the BNS of masses m_1 and m_2, and is defined as follows [61,64,65]:

$\tilde{\Lambda} = \frac{8}{13}$ with tidal correction given as

TABLE 3.3

The Binary Neutron Star Masses (m_1 ($M_\odot$), m_2 ($M_\odot$)) with Corresponding Radii (R_1 (km), R_2 (km)), Tidal Love Number Tidal Deformabilities (λ_1, λ_2), and Dimensionless Tidal Deformabilities (Λ_1, Λ_2)

EOS	m_1	m_2	R_1	R_2	$(k|2)_1$	$(k|2)_2$	λ_1	λ_2	Λ_1	Λ_2	$\tilde{\Lambda}$	$\delta\tilde{\Lambda}$	M_c	R_c
NL3	1.20	1.20	14.70	14.70	0.11	0.11	7.83	7.83	2983.15	2983.15	2983.15	0.000	1.04	10.35
	1.40	1.20	14.73	14.70	0.10	0.11	7.23	7.83	1267.07	2983.15	1950.08	183.66	1.13	10.26
	1.54	1.26	14.74	14.70	0.10	0.11	6.74	7.67	729.95	729.95	2303.95	1308.91	1.21	10.18
FSU	1.20	1.20	12.94	12.94	0.11	0.11	3.96	3.96	1469.32	1469.32	1469.32	0.000	1.04	8.98
Garnet	1.40	1.20	12.98	12.94	0.10	0.11	3.55	3.96	622.06	1469.32	959.22	90.97	1.13	8.90
	1.54	1.26	12.96	12.96	0.09	0.10	3.16	3.86	343.73	1146.73	638.35	88.89	1.21	8.82
IOPB-I	1.20	1.20	13.22	13.22	0.11	0.11	4.37	4.37	1654.23	1654.23	1654.23	0.000	1.04	9.20
	1.40	1.20	13.24	13.22	0.10	0.11	3.91	4.37	680.79	1654.23	1067.64	107.34	1.13	9.10
	1.54	1.26	13.23	13.23	0.09	0.10	3.52	4.25	384.65	1253.00	703.58	94.73	1.21	8.99
G3	1.20	1.20	12.47	12.47	0.10	0.10	3.11	3.11	1776.65	1776.65	1776.65	0.000	1.04	9.33
	1.40	1.20	12.42	12.47	0.09	0.10	2.61	3.11	461.03	1776.65	976.80	183.27	1.13	8.94
	1.54	1.26	12.33	12.46	0.08	0.10	2.19	2.98	239.49	883.46	474.83	75.17	1.21	8.31

Note: $\tilde{\Lambda}$, $\delta\tilde{\Lambda}$, M_c ($M_\odot$), and R_c (km) are the dimensionless tidal deformability, tidal correction, chirp mass, and radius of the binary neutron star, respectively.

$$\delta\tilde{\Lambda} = \frac{1}{2}\left[\sqrt{1-4\eta}\left(1 - \frac{13272}{1319}\eta + \frac{8944}{1319}\eta^2\right)(\Lambda_1 + \Lambda_2)\right.$$
$$\left. + \left(1 - \frac{15910}{1319}\eta + \frac{32850}{1319}\eta^2 + \frac{3380}{1319}\eta^2\right)(\Lambda_1 - \Lambda_2)\right], \tag{3.21}$$

where $\eta = m_1 m_2/M^2$ is the symmetric mass ratio, m_1 and m_2 are the binary masses, $M = m_1 + m_2$ is the total mass, and Λ_1 and Λ_2 are the dimensionless tidal deformability of the BNS for the case $m_1 \geq m_2$ [66,67]. The calculated results for the Λ_1 and Λ_2 weighted tidal deformability $\tilde{\Lambda}$ of the recently developed EOSs are displayed in Table 3.3. In Figure 3.12, the different dimensionless tidal deformability corresponding to progenitor masses of the NS is displayed. It can be seen that the IOPB-I along with FSUGarnet and G3 is in good agreement with the 90% and 50% probability contour of GW170817 [62] for a low-spin rotating neutron star $|\chi| \leq 0.05$. Recently, the LIGO/VIRGO detectors have measured the value of $\tilde{\Lambda}$ whose results are more precise than the results found by considering the individual values of Λ_1 and Λ_2 of the BNS [66,67]. It is noticed that the value of $\tilde{\Lambda} \leq 800$ in the low-spin case and $\tilde{\Lambda} \leq 700$ in the high-spin case within the 90% credible intervals are consistent with the 680.79, 622.06, and 461.03 of the 1.4 $M_\odot$ binary for the IOPB-I, FSUGarnet, and G3 parameter sets, respectively (see Table 3.3). The chirp mass M_c and chirp radius R_c of the BNS system are also displayed in Table 3.3 which is defined as follows:

$$M_c = (m_1 + m_2)^{3/5}(m_1 + m_2)^{-1/5} \text{ and } R_c = 2M_c\tilde{\Lambda}^{(-1/5)}. \tag{3.22}$$

The precise mass measurements of the NSs have been reported in Refs. [51,52]. However, so far no observation has been confirmed regarding the radius of the most massive NS. Recently, a LIGO/VIRGO has measured chirp mass $1.188^{+0.004}_{-0.002}$ $M_\odot$ with a very good precision. With the help of this, one can easily calculate the chirp radius R_c of the BNS system and can find that the chirp radius is in the range of $7.867 \leq R_c \leq 10.350$ km for equal and unequal-mass BNS system as shown in Table 3.3.

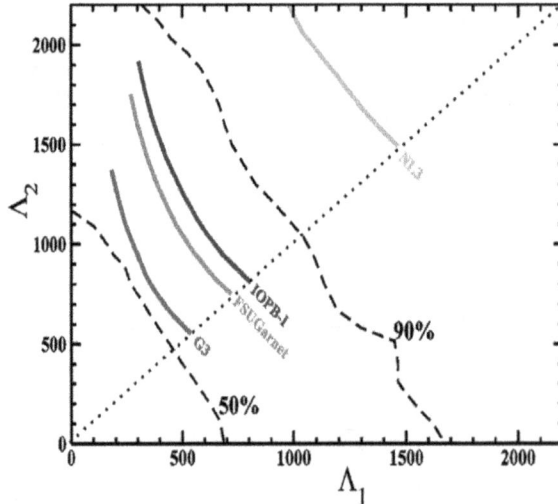

FIGURE 3.12
Different values Λ generated by using IOPB-I along with NL3, FSUGarnet, and G3 EOSs are compared with the 90% and 50% probability contours in the case of low-spin neutron star $|\chi| \leq 0.05$ as given in Figure 3.5 of GW170817 [62]. (The figure is taken from Ref. [61].)

3.5 Summary and Conclusions

In summary, the present chapter reviews the RMF formalism starting from the original Walecka model to the presently available E-RMF formalism. Based on the development of the model, the evolution of various forces is presented. The successes and limitations of the models are highlighted, so that one can think about the extension or modification of the formalisms. After describing briefly about the development of the model and optimization of the parameter sets, the applications of the forces are discussed for various nuclear phenomena. In this chapter, we show the applicability of the recently developed E-RMF parameter sets G3 and IOPB-I by calculating the binding energy, rms radius, and neutron skin thickness for some selected known nuclei. The reported results are found to match well with the experimental data. Then, the discussions are elaborated for the calculation of binding energy, neutron skin thickness, and isotopic shift of Pb isotopes, which approves the validity of the nuclear models. The discussion is then shifted to the superheavy region to have a glimpse on the current status of the stability valley. From both the relativistic and non-relativistic calculations, it is concluded that the possible *island of stability* could be in the region of $Z = 120$ and $N = 172 - 184$ around the proton magic number 120. The pairing gaps for proton and neutron, the two-neutron separation, the shell correction and single-particle energies for a wide range of nuclei are analyzed. The general set of magic numbers beyond $^{208}_{82}\text{Pb}$ are found to be $Z = 120$ and $N = 172$, 182, 184, 208, and 258. It is important to note that the widely discussed proton magic number $Z = 114$ in the past four decades is found to be feebly magic in nature, and $Z = 120$ could be the next magic number after $Z = 82$ situated in the superheavy island.

The E-RMF model, which includes all types of self- and cross-couplings, is used in the calculation of nuclear matter and neutron stars. The symmetry energy coefficient J and incompressibility k_∞ associated with earlier forces of E-RMF model, which are larger than the empirical values, are cured in the new parameter set G3. The rms errors on the total binding energy for the new sets G3 and IOPB-I are noticeably smaller than those for the commonly used forces NL3, FSUGold2, FSUGarnet, and G2. The neutron skin thickness with G3 calculated over a wide range of masses is in good agreement with the available experimental data. The neutron matter EOS at sub-saturation densities corresponding to G3 shows reasonable improvement over other sets. The maximum mass for the neutron star is compatible with the measurements, and the radius of the neutron star with the canonical mass agrees quite well with the empirical data. The G3 force parameter yields a small canonical radius $R_{1.4}$ when compared to other predictions of the RMF models, which are compatible with the observational constraint of 2 $M_\odot$. The prediction of neutron star mass $\approx$ 2 $M_\odot$ with the G3 set is an interesting feature of the recently developed forces.

Acknowledgments

We are thankful to B. K. Agrawal for fruitful discussions and useful suggestions. Two of the authors (KCN and RNP) acknowledge Institute of Physics (IOP), Bhubaneswar for providing facilities. This work is supported by the Department of Science and Technology (DST), Govt. of India, Project No. EMR/2015/002517.

References

[1] Walecka J. D, A theory of highly condensed matter, *Ann. Phys. (New York)* 1974; 83:491.

[2] Duerr H. P. and Teller E, Interaction of Antiprotons with Nuclear Fields, *Phys. Rev.* 1956; 101:494.

[3] Duerr H. P, Relativistic Effects in Nuclear Forces, *Phys. Rev.* 1956; 103:469.

[4] Miller L. D, and Green A. E. S, Relativistic Self-Consistent Meson Field Theory of Spherical Nuclei, *Phys. Rev. C* 1972; 5:241.

[5] Chin S. A. and Walecka J. D., An equation of state for nuclear and higher-density matter based on relativistic mean-field theory, *Phys. Lett. B* 1974; 52:24.

[6] Boguta J. and Bodmer A. R., Relativistic calculation of nuclear matter and the nuclear surface, *Nucl. Phys. A* 1977; 292:413.

[7] Del Estal M., Centelles M., Vnas.X and Patra S. K., Effects of new nonlinear couplings in relativistic effective field theory, *Phys. Rev. C* 2001; 63:024314.

[8] Kumar Bharat, Singh S. K., Agrawal B. K. and Patra S. K, New parameterization of the effective field theory motivated relativistic mean field model, *Nucl. Phys. A* 2017; 966:197.

[9] Kumar Bharat, Agrawal B. K. and Patra S. K., New relativistic effective interaction for finite nuclei, infinite nuclear matter, and neutron stars, *Phys. Rev. C* 2018 ; 97, 045806.

[10] Furnstahl. R. J., Serot B. D. and Tang H. B., Analysis of chiral mean-field models for nuclei, *Nucl. Phys. A* 1996; 598:539.

[11] Furnstahl. R. J., Serot B. D. and Tang H. B., A chiral effective Lagrangian for nuclei, *Nucl. Phys. A* 1997; 615:441.

[12] Uller H. M. and Serot B. D., Relativistic mean-field theory and the high-density nuclear equation of state, Nucl. *Phys. A* 1996; 606:508.

[13] Serot B. D. and Walecka J. D., Recent Progress in Quantum Hadrodynamics, *Int. J. Mod. Phys. E* 1997; 6:515.

[14] Singh S. K., Biswal S. K., Bhuyan M. and Patra S. K., Effects of δ mesons in relativistic mean field theory, *Phys. Rev. C* 2014; 89:044001.

[15] Negele, J.W., Structure of finite nuclei in the local-density approximation. *Phys. Rev. C* 1970; 1:1260

[16] Horowitz C. J. and Serot Brian D., Self-consistent hartree description of finite nuclei in a relativistic quantum field theory, *Nucl. Phys. A* 1981;368:503.

[17] Brockmann R. and Machleidt R., Relativistic nuclear structure. I. Nuclear matter, *Phys. Rev. C* 1990; 42:1965.

[18] Reinhard P. G., Rufa M., Maruhn J., Greiner W., and Friedrich J., Nuclear ground-state properties in a relativistic Meson-Field theory, *Z. Phys. A* 1986; 323:13.

[19] Moeller P. and Nix J., R, Los Alamos National Laboratory Report No. LA-UR-86-3983, 1986; 29; Abrahamyan S., et al., Measurement of the neutron radius of ^{208}Pb through parity violation in electron scattering, *Phys. Rev. Lett.* 2012; 108:112502.

[20] Lalazissis G.A., Köning J. and Ring P., Ground-state properties of even–even nuclei in the relativistic mean-field theory, *Phys. Rev. C* 1997; 55:540.

[21] Danielewicz P., Lacey R and Lynch W. G., Determination of the equation of state of dense matter, *Science* 2002; 298:1592.

[22] Bodmer A. R., Relativistic mean field theory of nuclei with a vector meson self-interaction, *Nucl. Phys. A* 1991; 526:703.

[23] Gmuca S., Finite-nuclei calculations based on relativistic mean-field effective interactions, *Nucl. Phys. A* 1992; 547:447.

[24] Bunta J. K. and Gmuca S., Asymmetric nuclear matter in the relativistic mean-field approach with vector cross interaction, *Phys. Rev. C* 2003; 68:054318.

[25] Sugahara Y. and Toki H., Relativistic mean-field theory for unstable nuclei with non-linear σ and ω terms, *Nucl. Phys. A* 1994; 579:557.

[26] Wei-Chai Chen and Piekarewicz J., Building relativistic mean field models for finite nuclei and neutron stars, *Phys. Rev. C* 2014; 90:044305.

[27] Kirkpatrick, S., Gelatt, C. and Vecchi, M., Simulated annealing methods. *J. Stat. Phys.* 1984; 34:975–986.

[28] Ingber, L., Very fast simulated re-annealing. *Math. Comput. Model.* 1989; *12*(8): 967–973.

[29] Cohen, B., *Training Synaptic Delays in a Recurrent Neural Network*. Tel-Aviv, Israel: Tel-Aviv University 1994.

[30] Chen, W.-C., and Piekarewicz. J, Searching for isovector signatures in the neutron-rich oxygen and calcium isotopes, *Phys. Lett. B* 2015; 748:284.

[31] Wang, M., Audi, G., Wapstra, A.H., Kondev, F.G., MacCormick, M., Xu, X. and Pfeiffer, B., The Ame2012 atomic mass evaluation, *Chin. Phys. C* 2012; 36:1603.

[32] Angeli and Marinova K. P, Table of experimental nuclear ground state charge radii: An update, *At. Data Nucl. Data Tables* 2013; 99:69.

[33] Sonzogni, A., NNDC chart of nuclides. In *International Conference on Nuclear Data for Science and Technology* 2007, pp. 105–106. EDP Sciences.

[34] Klúpfel, P., Reinhard, P.-G., Búrvenich, T.J. and Maruhn, J.A., Variations on a theme by Skyrme: A systematic study of adjustments of model parameters *Phys. Rev. C* 2009; 79:034310.

[35] Kumar Bharat, Biswal S. K. and Patra S. K., Tidal deformability of neutron and hyperon stars within relativistic mean field equations of state, *Phys. Rev. C* 2017; 95:015801.

[36] Sharma, M.M., Lalazissis, G.A. and Ring, P., Rho meson coupling in the relativistic mean field theory and description of exotic nuclei. *Phys. Lett. B* 1993; 317:9.

[37] Trzcińska, A., Jastrzębski, J., Lubiński, P., Hartmann F.J., Schmidt, R., von Egidy, T., and Klos, B., Neutron density distributions deduced from antiprotonic atoms, *Phys. Rev. Lett.* 2001; 87:082501.

[38] Jastrzębski, J., Trzcińska, A., Lubiński, P., Hartmann F.J., von Egidy, T., and Wycech, S., Neutron density distributions from antiprotonic atoms compared with hadron scattering data, *Int. J. Mod. Phys. E* 2004; 13:343.

[39] Viñas, X., Centelles, M., Roca-Maza, X., and Warda, M., Density dependence of the symmetry energy from neutron skin thickness in finite nuclei, *Eur. Phys. J. A* 2014; 50:27.

[40] Horowitz, C. J., and Piekarewicz, J., Neutron star structure and the neutron radius of ^{208}Pb,, *Phys. Rev. Lett.* 2001; 86:5647; Neutron radii of ^{208}Pb and neutron stars, *Phys. Rev. C* 2001; 64:062802(R).

[41] Brown, B.A., Neutron radii in nuclei and the neutron equation of state, *Phys. Rev. Lett.* 2000; 85:5296.

[42] Bhuyan, M., and Patra, S.K., Magic nuclei in Superheavy valley, Mod. *Phys. Lett. A* 2012; 27:1250173.

[43] Strutinsky, V.M., Shell effects in nuclear masses and deformation energies. *Nucl. Phys. A* 1967; *95*(2), pp. 420−442.

[44] Sil, T., Patra, S.K., Sharma, B.K., Centelles, M., and Vias, X., Superheavy nuclei in a relativistic effective Lagrangian model, *Phys. Rev. C* 2004; 69:044315.

[45] Arumugam, P., Sharma, B.K., Sahu, P.K., Patra, S.K., Sil, T., Centelles, M., and Viñas, X., Versatility of field theory motivated nuclear effective Lagrangian approach, *Phys. Lett. B* 2004; 601, p. 51.

[46] Gezerlis, A., and Carlson, J., Low-density neutron matter, *Phys. Rev. C* 2010; 81:025803.

[47] Dutra, M., Lourenço, O., Martins, J.S., Delfino, A., Stone, J.R., and Stevenson, P.D., Skyrme interaction and nuclear matter constraints. *Phys. Rev. C*, 2012; 85(3):035201.

[48] Todd-Rutel, B.G., and Piekarewicz, J., Neutron-rich nuclei and neutron stars: A new accurately calibrated interaction for the study of neutron-rich matter, *Phys. Rev. Lett.* 2005; 95(12):122501.

[49] Steiner, A.W., Lattimer, J.M., and Brown, E.F., The equation of state from observed masses and radii of neutron stars, *Astrophys, J.*, 2010; 722:33.

[50] Oppenheimer, J.R., and Volkoff, G.M., On massive neutron cores, *Phys. Rev.* 55(4), p. 374.; R. C. Tolman, ibid. 1939; 055:364.

[51] Demorest, P.B., Pennucci, T., Ransom S.M., Roberts, M.S.E., and Hessels, J.W.T., A two-solar-mass neutron star measured using Shapiro delay, *Nature (London)* 2010; 467:1081.

[52] Antoniadis, J., Freire, P.C., Wex, N., Tauris, T.M., Lynch, R.S., van Kerkwijk, M.H., Kramer, M., Bassa, C., Dhillon, V.S., Driebe, T., and Hessels, J.W., A massive pulsar in a compact relativistic binary, *Science* 2013; 340(6131):1233232.

[53] Horowitz, C.J., and Piekarewicz, J., Neutron radii of 208 Pb and neutron stars. *Phys. Rev. C* 2001; 64(6):062802.

[54] Alam, N., Agrawal, B.K., Fortin, M., Pais, H., Providência, C., Raduta, A.R. and Sulaksono, A., Strong correlations of neutron star radii with the slopes of nuclear matter incompressibility and symmetry energy at saturation. *Phys. Rev. C* 2016; 94(5):052801.

[55] Gaitanos, T., Di Toro, M., Typel, S., Baran, V., Fuchs, C., Greco, V. and Wolter, H.H., On the Lorentz structure of the symmetry energy. *Nucl. Phys. A*, 2004; 732:24−48.

[56] Lattimer, J.M. and Lim, Y., Constraining the symmetry parameters of the nuclear interaction. *Astrophys. J.* 2013; 771(1):51.

[57] Hinderer, T., Lackey, B.D., Lang, R.N. and Read, J.S., Tidal deformability of neutron stars with realistic equations of state and their gravitational wave signatures in binary inspiral. *Phys. Rev. D*, 2010; 81(12):123016.

[58] Flanagan, É.É. and Hinderer, T., Constraining neutron-star tidal Love numbers with gravitational-wave detectors. *Phys. Rev. D*, 2008; 77(2):021502.

[59] Landry, P., and Poisson, E., Relativistic theory of surficial Love numbers. *Phys. Rev. D*, 2014; 89(12):124011.

[60] Hinderer, T., Tidal Love numbers of neutron stars. *Astrophys J.* 2008; 677(2):1216.

[61] Kumar, B., Patra, S.K., and Agrawal, B.K., New relativistic effective interaction for finite nuclei, infinite nuclear matter, and neutron stars. *Phys. Rev. C* 2018; 97(4):045806.

[62] Abbott, B.P., Abbott, R., Abbott, T.D., Acernese, F., Ackley, K., Adams, C., Adams, T., Addesso, P., Adhikari, R.X., Adya, V.B. and Affeldt, C., GW170817: Observation of gravitational waves from a binary neutron star inspiral. *Phys. Rev. Lett.* 2017, 119(16):161101.

[63] Steiner, A.W., Gandolfi, S., Fattoyev, F.J., and Newton, W.G., Using neutron star observations to determine crust thicknesses, moments of inertia, and tidal deformabilities, *Phys. Rev. C*, 2015; 91(1):015804.

[64] Favata, M., Systematic parameter errors in inspiraling neutron star binaries, *Phys. Rev. Lett.* 2014; 112(10):101101.

[65] Wade, L., Creighton, J. D. E., Ochsner, E., Lackey, B. D., Farr, B. F., Littenberg, T. B., and Raymond, V., *Phys. Rev. D*, 2014; 89:103012.

[66] Abbott, B.P., Abbott, R., Abbott, T.D., Acernese, F., Ackley, K., Adams, C., Adams, T., Addesso, P., Adhikari, R.X., Adya, V.B. et. al., Search for post-merger gravitational waves from the remnant of the binary neutron star merger GW170817, *Astrophys. J. Lett.* 2017; 851:L16 (13pp).

[67] Radice, D., Perego, A., Zappa, F. and Bernuzzi, S., GW170817: joint constraint on the neutron star equation of state from multimessenger observations. *Astrophys. J. Lett.*, 2018; 852(2):L29.

4

Nuclear Symmetry Energy in Finite Nuclei

M.K. Gaidarov, A.N. Antonov, and D.N. Kadrev

Institute for Nuclear Research and Nuclear Energy

P. Sarriguren

IEM-CSIC

E. Moya de Guerra

Universidad Complutense de Madrid

CONTENTS

4.1   Introduction

The study of the nuclear matter symmetry energy that essentially characterizes the isospin-dependent part of the equation of state (EOS) of asymmetric nuclear matter (ANM) is currently an exciting topic of research in nuclear physics [1–6]. In fact, applications of ANM are broad, ranging from the structure of rare isotopes [7,8] to the properties of neutron stars [9,10] and the dynamical process of nuclear reactions [11]. The transition from ANM to finite nuclei is a natural and important way to learn more about the nuclear symmetry energy (NSE) which is poorly constrained by experimental data on ground-state nuclear properties.

Measurements of nuclear masses, densities, and collective excitations have allowed to resolve some of the basic features of the EOS of nuclear matter. However, the asymmetrical properties of the EOS due to different neutron and proton numbers remain more elusive to date, and the study of the isospin-dependent properties of asymmetric nuclear matter [12–16] and the density dependence of the NSE remains a prime objective. As can be seen, e.g., in Refs. [17–21], an increasing wide range of theoretical conclusions are being proposed on that density dependence, as well as on some associated nuclear characteristics. The new radioactive ion beam facilities at CSR (China), FAIR (Germany), RIKEN (Japan), SPIRAL2/GANIL (France), and the upcoming FRIB (USA) will provide the possibility of exploring the properties of nuclear matter and nuclei under the extreme condition of large isospin asymmetry.

Nowadays, the experimental information about the symmetry energy is fairly limited. The need to have information for this quantity in finite nuclei, even theoretically obtained, is a major issue because it allows one to constrain the bulk and surface properties of the nuclear energy-density functionals (EDFs) quite effectively. The symmetry energy is not a directly measurable quantity and has to be extracted indirectly from observables that are related to it (see, e.g., the review in Ref. [22]). The neutron skin thickness of nuclei is a sensitive probe of the NSE, although its precise measurement is difficult to obtain. At present, neutron skin thicknesses are derived from pygmy dipole resonance measurements [23], data from antiprotonic atoms [4], and other methods for its extraction like reactions and giant resonances. This allows one to constrain the parameters describing the NSE. Therefore, in the present chapter, a particular attention is paid to investigation of the relation between the neutron skin thickness and some nuclear matter properties in finite nuclei, such as the symmetry energy at the saturation point, symmetry pressure (proportional to the slope of the bulk symmetry energy), and asymmetric compressibility, considering nuclei in given isotopic chains and within a certain theoretical approach [24–26]. In addition to various linear relations between several quantities in bulk matter and for a given nucleus that have been observed and tested within different theoretical methods (e.g., nonrelativistic calculations with different Skyrme parameter sets and relativistic models [4,7,8,27–29]), we are looking forward to establish a possible correlation between the skin thickness and these quantities and to clarify to what extent this correlation is appropriate for a given isotopic chain. The evolution of the symmetry energy as we increase the number of neutrons is also studied.

The NSE, as a fundamental quantity in nuclear physics and astrophysics, represents a measure of the energy gain in converting isospin asymmetric nuclear matter to a symmetric system. Its value depends not only on the density ρ but also on the temperature T. The thermal behavior of the symmetry energy has a role in changing the location of the nuclear drip lines as nuclei warm up. Also, it is of fundamental importance for the liquid-gas phase transition of ANM, the dynamical evolution mechanisms of massive stars, its crustal composition or its thickness [30], and the supernova explosion [31]. Although the temperature dependence of single-particle properties in nuclear and neutron matter was also broadly investigated (e.g., Refs. [32,33]), the problem of accurate treatment of the thermodynamical properties of hot finite nuclei is still challenging. In this respect, our main goal is to study the NSE in finite nuclei and its volume and surface components at zero [34] and finite temperatures [35,36], as well as other properties such as the T-dependent nucleon densities and related root-mean-square (rms) radii and the possibility of formation of neutron skins.

The symmetry energy of finite nuclei at saturation density is often extracted by fitting ground-state masses with various versions of the liquid-drop mass (LDM) formula within liquid-drop models [37–39]. It has also been studied in the random phase approximation based on the Hartree-Fock (HF) approach [40] or effective relativistic Lagrangians with density-dependent meson-nucleon vertex functions [41], energy density functionals (EDF) of Skyrme force [27,42,43], and relativistic nucleon-nucleon interaction [44,45]. In the present chapter, the symmetry energy is studied *in a wide range of finite nuclei* on the basis of the Brueckner [46,47] and Skyrme EDFs for nuclear matter and using the coherent density fluctuation model (CDFM) (e.g., Refs. [48,49]). The latter is a natural extension of the Fermi gas model based on the generator coordinate method [49,50] and includes long-range correlations of collective type. The numerous applications of the CDFM show its capability to be employed as an alternative way to make a transition from the properties of nuclear matter to the properties of finite nuclei.

In our works, we have chosen medium-heavy and heavy spherical nuclei from Ni ($A = 74$–84), Sn ($A = 124$–152), and Pb ($A = 202$–214) isotopic chains [24] and deformed Kr ($A = 82$–120) and Sm ($A = 140$–156) isotopes [25]. Most of these nuclei are far from the stability line and are of interest for future measurements with radioactive exotic beams. Moreover, there is a significant interest to study the neutron distribution and rms radius in ^{208}Pb, aiming at precise determinations of the neutron skin in this nucleus [51–55]. In the light of the new precise spectroscopic measurements of the neutron-rich ^{32}Mg nucleus, which lies in the much explored "island of inversion" at $N = 20$ in Ref. [26], we have performed a systematic study of the nuclear ground-state properties of neutron-rich and neutron-deficient Mg isotopes (including characteristics related to the ρ-dependence of the NSE) with $A = 20$–36. The densities of these nuclei were calculated within a deformed HF+BCS approach with Skyrme-type density-dependent effective interactions [56,57]. When studying the temperature dependence of the symmetry energy and its components [35,36] to calculate the T-dependent proton and neutron densities and kinetic energy densities, we applied the self-consistent Skyrme Hartree-Fock-Bogoliubov (HFB) method using the cylindrical transformed deformed harmonic-oscillator basis through the HFBTHO code [58,59].

The structure of this chapter is as follows. In Section 4.2, we present a brief description of the theoretical formalism (definitions of ANM properties, basic expressions for the model density distributions and nuclear radii obtained in the deformed HF+BCS and HFB methods with Skyrme forces, CDFM formalism, Brueckner and Skyrme energy-density functionals, NSE and its volume and surface components of hot nuclei) used to calculate the intrinsic quantities in finite nuclei. The numerical results and discussions of the NSE for the isotopic chains of spherical (Ni, Sn, Pb) and deformed (Kr, Sm) nuclei, neutron-rich and neutron-deficient Mg isotopes, as well as its temperature dependence for the same spherical nuclei, are presented in Section 4.3. Concluding remarks are given in Section 4.4.

4.2 Theoretical Formalism

4.2.1 The Key EOS Parameters in Nuclear Matter

The symmetry energy $S(\rho)$ is defined by a Taylor series expansion of the energy per particle for ANM in terms of the isospin asymmetry $\delta = (\rho_n - \rho_p)/\rho$:

$$E(\rho, \delta) = E(\rho, 0) + S(\rho)\delta^2 + O(\delta^4) + \cdots , \qquad (4.1)$$

where $\rho = \rho_n + \rho_p$ is the baryon density with ρ_n and ρ_p denoting the neutron and proton densities, respectively (see, e.g., [28]). Odd powers of δ are forbidden by the isospin symmetry, and the terms proportional to δ^4 and higher orders are found to be negligible.

Near the saturation density ρ_0, the energy of isospin-symmetric matter, $E(\rho, 0)$, and the symmetry energy, $S(\rho)$, can be expanded as

$$E(\rho, 0) = E_0 + \frac{K}{18\rho_0^2}(\rho - \rho_0)^2 + \cdots , \qquad (4.2)$$

and

$$\begin{aligned} S(\rho) &= \frac{1}{2} \left. \frac{\partial^2 E(\rho, \delta)}{\partial \delta^2} \right|_{\delta=0} \\ &= a_4 + \frac{p_0}{\rho_0^2}(\rho - \rho_0) + \frac{\Delta K}{18\rho_0^2}(\rho - \rho_0)^2 + \cdots . \end{aligned} \qquad (4.3)$$

The parameter a_4 is the symmetry energy at equilibrium ($\rho = \rho_0$). The pressure p_0^{ANM}

$$p_0^{ANM} = \rho_0^2 \left. \frac{\partial S}{\partial \rho} \right|_{\rho=\rho_0} \qquad (4.4)$$

and the curvature ΔK^{ANM}

$$\Delta K^{ANM} = 9\rho_0^2 \left. \frac{\partial^2 S}{\partial \rho^2} \right|_{\rho=\rho_0} \qquad (4.5)$$

of the NSE at ρ_0 govern its density dependence and thus provide important information on the properties of the NSE at both high and low densities. The widely used "slope" parameter L^{ANM} is related to the pressure p_0^{ANM} [Eq. (6.4)] by

$$L^{ANM} = \frac{3p_0^{ANM}}{\rho_0}. \qquad (4.6)$$

We remark that our present knowledge of these basic properties of the symmetry term around saturation is still very poor (see the analysis in Ref. [29] and references therein). In particular, we note the uncertainty of the symmetry pressure at ρ_0 (sometimes it can vary by a factor of three) which is important for structure calculations. In practice, predictions for the symmetry energy vary substantially: *e.g.*, $a_4 \equiv S(\rho_0) = 28 - 38$ MeV.

4.2.2 Deformed HF+BCS Formalism and HFB Method with Skyrme Forces

The results that are shown and discussed in Refs. [24–26,34] have been obtained from self-consistent deformed HF calculations with density-dependent Skyrme interactions [56] and accounting for pairing correlations. Pairing between like nucleons has been included by solving the BCS equations at each iteration with a fixed pairing strength that reproduces the odd-even experimental mass differences.

The spin-independent proton and neutron densities are given by [57,60]

$$\rho(\vec{R}) = \rho(r, z) = \sum_i 2v_i^2 \rho_i(r, z), \qquad (4.7)$$

in terms of the occupation probabilities v_i^2 resulting from the BCS equations and the single-particle densities ρ_i

$$\rho_i(\vec{R}) = \rho_i(r, z) = |\Phi_i^+(r, z)|^2 + |\Phi_i^-(r, z)|^2 , \qquad (4.8)$$

with

$$\Phi_i^{\pm}(r,z) = \frac{1}{\sqrt{2\pi}}$$
$$\times \sum_{\alpha} \delta_{\Sigma,\pm 1/2}\, \delta_{\Lambda,\Lambda\mp}\, C_{\alpha}^i\, \psi_{n_r}^{\Lambda}(r)\, \psi_{n_z}(z) \tag{4.9}$$

and $\alpha = \{n_r, n_z, \Lambda, \Sigma\}$. The functions $\psi_{n_r}^{\Lambda}(r)$ and $\psi_{n_z}(z)$ entering Eq. (4.9) are defined in terms of Laguerre and Hermite polynomials:

$$\psi_{n_r}^{\Lambda}(r) = \sqrt{\frac{n_r}{(n_r+\Lambda)!}}\, \beta_{\perp}\, \sqrt{2}\, \eta^{\Lambda/2}\, e^{-\eta/2}\, L_{n_r}^{\Lambda}(\eta), \tag{4.10}$$

$$\psi_{n_z}(z) = \sqrt{\frac{1}{\sqrt{\pi}2^{n_z}n_z!}}\, \beta_z^{1/2}\, e^{-\xi^2/2}\, H_{n_z}(\xi), \tag{4.11}$$

with

$$\beta_z = (m\omega_z/\hbar)^{1/2}, \quad \beta_{\perp} = (m\omega_{\perp}/\hbar)^{1/2},$$
$$\xi = z\beta_z, \quad \eta = r^2\beta_{\perp}^2. \tag{4.12}$$

The normalization of the densities is given by

$$\int \rho(\vec{R})d\vec{R} = X, \tag{4.13}$$

with $X = Z, N$ for protons and neutrons, respectively.

The multipole decomposition of the density can be written in terms of even λ multipole components as [56,57]

$$\rho(r,z) = \sum_{\lambda} \rho_{\lambda}(R)P_{\lambda}(\cos\theta). \tag{4.14}$$

In the calculations, for the density distribution $\rho(r)$, we use the monopole term $\rho_0(R)$ in the expansion (4.14).

The mean square radii for protons and neutrons are defined as

$$<r_{\mathrm{p,n}}^2> = \frac{\int R^2 \rho_{\mathrm{p,n}}(\vec{R})d\vec{R}}{\int \rho_{\mathrm{p,n}}(\vec{R})d\vec{R}}, \tag{4.15}$$

and the rms radii are given by

$$r_{\mathrm{p,n}} = <r_{\mathrm{p,n}}^2>^{1/2}. \tag{4.16}$$

Then, the neutron skin thickness is usually characterized by the difference of neutron and proton rms radii:

$$\Delta R = <r_{\mathrm{n}}^2>^{1/2} - <r_{\mathrm{p}}^2>^{1/2}. \tag{4.17}$$

A self-consistent approach based on the simultaneous treatment of temperature-dependent density distributions and kinetic energy density is related to the finite temperature formalism for the HFB method. In it, the nuclear Skyrme HFB problem is solved by using the transformed harmonic-oscillator basis [58]. The HFBTHO code solves the finite-temperature HFB equations assuming axial and time-reversal symmetry. These equations are formally equivalent to the HFB equations at $T = 0$ MeV if the expressions of the density matrix ρ and pairing tensor κ are redefined as

$$\rho = UfU^{\dagger} + V^*(1-f)V^T,$$
$$\kappa = UfV^{\dagger} + V^*(1-f)U^T, \tag{4.18}$$

where U and V are the matrices of the Bogoliubov transformation (here T means transpose) and f is the temperature-dependent Fermi-Dirac factor given by

$$f_i = \left(1 + e^{E_i/k_B T}\right)^{-1}. \tag{4.19}$$

In this expression, E_i is the quasiparticle energy of the state i and k_B is the Boltzmann constant. In HFBTHO, the Fermi level λ is determined at each iteration from the conservation of particle number in the BCS approach [58]:

$$N(\lambda) = \sum_i \left[v_i(\lambda)^2 + f_i(\lambda)\left(u_i(\lambda)^2 - v_i(\lambda)^2\right)\right], \tag{4.20}$$

where the BCS occupations are given by

$$v_i^2 = \frac{1}{2}\left[1 - \frac{e_i - \lambda}{E_i^{BCS}}\right], \quad u_i^2 = 1 - v_i^2, \tag{4.21}$$

and $E_i^{BCS} = \left[(e_i - \lambda)^2 + \Delta_i^2\right]^{1/2}$. Note that at $T = 0$ MeV, the Fermi-Dirac factors are zero, and one recovers the usual expressions for ρ and κ in Eq. (4.18) and for the number of particles in Eq. (4.20).

In our calculations, the following Skyrme force parametrizations are used: SLy4 [61], Sk3 [62], SGII [63], LNS [64], and SkM* [65]. These are among the most extensively used Skyrme forces that work successfully for describing finite nuclei properties.

4.2.3   The Coherent Density Fluctuation Model (CDFM)

The CDFM was suggested and developed in Refs. [48,49]. The model is related to the delta-function limit of the generator coordinate method [49,50]. It is shown in the model that the one-body density matrix of the nucleus $\rho(\mathbf{r}, \mathbf{r}')$ can be written as a coherent superposition of the one-body density matrices (OBDM) for spherical "pieces" of nuclear matter (so-called "fluctons") with densities

$$\rho_x(\mathbf{r}) = \rho_0(x)\Theta(x - |\mathbf{r}|), \tag{4.22}$$

where

$$\rho_0(x) = \frac{3A}{4\pi x^3}. \tag{4.23}$$

The generator coordinate x is the radius of a sphere containing Fermi gas of all A nucleons uniformly distributed in it. It is appropriate to use for such a system OBDM of the form:

$$\rho_x(\mathbf{r}, \mathbf{r}') = 3\rho_0(x)\frac{j_1(k_F(x)|\mathbf{r} - \mathbf{r}'|)}{(k_F(x)|\mathbf{r} - \mathbf{r}'|)}$$
$$\times \Theta\left(x - \frac{|\mathbf{r} + \mathbf{r}'|}{2}\right), \tag{4.24}$$

where j_1 is the first-order spherical Bessel function and

$$k_F(x) = \left(\frac{3\pi^2}{2}\rho_0(x)\right)^{1/3} \equiv \frac{\alpha}{x} \tag{4.25}$$

with

$$\alpha = \left(\frac{9\pi A}{8}\right)^{1/3} \simeq 1.52 A^{1/3} \tag{4.26}$$

is the Fermi momentum of such a formation. Then, the OBDM for the finite nuclear system can be written as a superposition of the OBDM's from Eq. (4.24):

$$\rho(\mathbf{r}, \mathbf{r}') = \int_0^\infty dx |\mathcal{F}(x)|^2 \rho_x(\mathbf{r}, \mathbf{r}'). \tag{4.27}$$

The density $\rho(\mathbf{r})$ in the CDFM is expressed by means of the same weight function $|\mathcal{F}(x)|^2$:

$$\rho(\mathbf{r}) = \int_0^\infty dx |\mathcal{F}(x)|^2 \frac{3A}{4\pi x^3} \Theta(x - |\mathbf{r}|) \tag{4.28}$$

normalized to the mass number:

$$\int \rho(\mathbf{r}) d\mathbf{r} = A. \tag{4.29}$$

If one takes the delta-function approximation to the Hill-Wheeler integral equation in the generator coordinate method, they get a differential equation for the weight function $\mathcal{F}(x)$ [49,50]. Instead of solving this differential equation, we adopt a convenient approach to the weight function $|\mathcal{F}(x)|^2$ proposed in Refs. [48,49]. In the case of monotonically decreasing local densities (*i.e.*, for $d\rho(r)/dr \leq 0$), the latter can be obtained by means of a known density distribution $\rho(r)$ for a given nucleus (from Eq. (4.28)):

$$|\mathcal{F}(x)|^2 = -\frac{1}{\rho_0(x)} \frac{d\rho(r)}{dr}\bigg|_{r=x}. \tag{4.30}$$

The normalization of the weight function is:

$$\int_0^\infty dx |\mathcal{F}(x)|^2 = 1. \tag{4.31}$$

4.2.4 Energy-Density Functionals for Infinite Nuclear Matter

Considering the pieces of nuclear matter with density $\rho_0(x)$ (Eq. 4.23), one can use for the matrix element $V(x)$ of the nuclear Hamiltonian the corresponding nuclear matter energy from the method of Brueckner *et al.* [46,47]. In this energy-density method, the expression for $V(x)$ reads

$$V(x) = AV_0(x) + V_C - V_{CO}, \tag{4.32}$$

where

$$\begin{aligned}
V_0(x) = {}& 37.53[(1 + \delta)^{5/3} + (1 - \delta)^{5/3}]\rho_0^{2/3}(x) \\
& + b_1\rho_0(x) + b_2\rho_0^{4/3}(x) + b_3\rho_0^{5/3}(x) \\
& + \delta^2[b_4\rho_0(x) + b_5\rho_0^{4/3}(x) + b_6\rho_0^{5/3}(x)]
\end{aligned} \tag{4.33}$$

with

$$\begin{aligned}
b_1 &= -741.28, \quad b_2 = 1179.89, \quad b_3 = -467.54, \\
b_4 &= 148.26, \quad\ \ b_5 = 372.84, \quad\ \ b_6 = -769.57.
\end{aligned} \tag{4.34}$$

$V_0(x)$ in Eq. (4.32) corresponds to the energy per nucleon in nuclear matter (in MeV) with the account for the neutron-proton asymmetry. V_C is the Coulomb energy of protons in a "flucton":

$$V_C = \frac{3}{5} \frac{Z^2 e^2}{x} \tag{4.35}$$

and the Coulomb exchange energy is:

$$V_{CO} = 0.7386 Z e^2 (3Z/4\pi x^3)^{1/3}. \tag{4.36}$$

Thus, in the Brueckner EOS (Eq. 4.33), the potential symmetry energy turns out to be proportional to δ^2. Only in the kinetic energy, the dependence on δ is more complicated. Substituting $V_0(x)$ in Eq. (4.3) and taking the second derivative, the symmetry energy $S^{ANM}(x)$ of the nuclear matter with density $\rho_0(x)$ (the coefficient a_4 in Eq. (4.3)) can be obtained:

$$S^{ANM}(x) = 41.7 \rho_0^{2/3}(x) + b_4 \rho_0(x)$$
$$+ b_5 \rho_0^{4/3}(x) + b_6 \rho_0^{5/3}(x). \tag{4.37}$$

The corresponding analytical expressions for the pressure $p_0^{ANM}(x)$ and asymmetric compressibility $\Delta K^{ANM}(x)$ of such a system in the Brueckner theory have the form:

$$p_0^{ANM}(x) = 27.8 \rho_0^{5/3}(x) + b_4 \rho_0^2(x)$$
$$+ \frac{4}{3} b_5 \rho_0^{7/3}(x) + \frac{5}{3} b_6 \rho_0^{8/3}(x) \tag{4.38}$$

and

$$\Delta K^{ANM}(x) = -83.4 \rho_0^{2/3}(x) + 4 b_5 \rho_0^{4/3}(x)$$
$$+ 10 b_6 \rho_0^{5/3}(x). \tag{4.39}$$

Our basic assumption within the CDFM is that the symmetry energy, the slope, and the curvature for finite nuclei can be defined weighting these quantities for nuclear matter (with a given density $\rho_0(x)$ (Eq. 4.23)) by means of the weight function $|\mathcal{F}(x)|^2$. Thus, in the CDFM, they will be an infinite superposition of the corresponding nuclear matter quantities. Following the CDFM scheme, the symmetry energy, the slope, and the curvature have the following forms:

$$s = \int_0^\infty dx |\mathcal{F}(x)|^2 S^{ANM}(x), \tag{4.40}$$

$$p_0 = \int_0^\infty dx |\mathcal{F}(x)|^2 p_0^{ANM}(x), \tag{4.41}$$

$$\Delta K = \int_0^\infty dx |\mathcal{F}(x)|^2 \Delta K^{ANM}(x). \tag{4.42}$$

We note that in the limit case when $\rho(r) = \rho_0 \Theta(R - r)$ (ρ_0 being the saturation density of symmetric nuclear matter) and $|\mathcal{F}(x)|^2$ becomes a delta-function (see Eq. (4.30)), Eq. (4.40) reduces to $S^{ANM}(\rho_0) = a_4$.

For the Skyrme EDF that we use in our works [34–36], it has the form:

$$\mathcal{E}(r, T) = \frac{\hbar^2}{2m_{n,k}} \tau_n + \frac{\hbar^2}{2m_{p,k}} \tau_p$$
$$+ \frac{1}{2} t_0 \left[\left(1 + \frac{1}{2} x_0\right) \rho^2 - \left(x_0 + \frac{1}{2}\right) (\rho_n^2 + \rho_p^2) \right]$$
$$+ \frac{1}{12} t_3 \rho^\alpha \left[\left(1 + \frac{x_3}{2}\right) \rho^2 - \left(x_3 + \frac{1}{2}\right) (\rho_n^2 + \rho_p^2) \right]$$
$$+ \frac{1}{16} \left[3 t_1 \left(1 + \frac{1}{2} x_1\right) - t_2 \left(1 + \frac{1}{2} x_2\right) \right] (\nabla \rho)^2$$

$$-\frac{1}{16}\left[3t_1\left(x_1+\frac{1}{2}\right)+t_2\left(x_2+\frac{1}{2}\right)\right]$$
$$\times[(\nabla\rho_n)^2+(\nabla\rho_p)^2]+\mathcal{E}_c(r), \tag{4.43}$$

where for infinite homogeneous nuclear matter, only the first three lines of Eq. (4.43) contribute. The derivative terms vanish, and the Coulomb term $\mathcal{E}_c$ is neglected. In Eq. (4.43), t_0, t_1, t_2, t_3, x_0, x_1, x_2, x_3, and α are the Skyrme parameters. The corresponding analytical expression for the symmetry energy of ANM with density $\rho_0(x)$, as well as the parameters of SGII, Sk3, and SLy4 Skyrme forces in the Skyrme EDF, can be found in Ref. [34].

4.2.5 Temperature Dependence of the Symmetry Energy and Relationships Concerning Its Volume and Surface Contributions

For finite systems, we develop an approach [35] to calculate the T-dependent symmetry energy coefficient (or NSE) for a specific nucleus starting with the local-density approximation (LDA) expression given in [66,67]

$$e_{sym}(A,T)=\frac{1}{I^2A}\int\rho(r)e_{sym}[\rho(r),T]\delta^2(r)d^3r. \tag{4.44}$$

In Eq. (4.44), $I=(N-Z)/A$, $e_{sym}[\rho(r),T]$ is the symmetry energy coefficient at temperature T of infinite nuclear matter at the value of the total local density $\rho(r)=\rho_n(r)+\rho_p(r)$, and $\delta(r)=[\rho_n(r)-\rho_p(r)]/\rho(r)$ is the ratio between the isovector and the isoscalar parts of $\rho(r)$, with $\rho_n(r)$ and $\rho_p(r)$ being the neutron and proton local densities, respectively. The symmetry energy coefficient $e_{sym}(\rho,T)$ can be evaluated in different ways. Following Refs. [66,68], we adopt here the definition

$$e_{sym}(\rho,T)=\frac{e(\rho,\delta,T)-e(\rho,\delta=0,T)}{\delta^2}, \tag{4.45}$$

where $e(\rho,\delta,T)$ is the energy per nucleon in an asymmetric infinite matter, while $e(\rho,\delta=0,T)$ is that one of symmetric nuclear matter. These quantities are expressed by $e=\mathcal{E}(r,T)/\rho$, where $\mathcal{E}(r,T)$ is the total energy density of the system. In our work, we use the Skyrme EDF (Eq. 4.43). The nucleon effective mass $m_{q,k}$ is defined through

$$\frac{m}{m_{q,k}(r)}=1+\frac{m}{2\hbar^2}\left\{\left[t_1\left(1+\frac{x_1}{2}\right)+t_2\left(1+\frac{x_2}{2}\right)\right]\rho\right.$$
$$\left.+\left[t_2\left(x_2+\frac{1}{2}\right)-t_1\left(x_1+\frac{1}{2}\right)\right]\rho_q\right\}, \tag{4.46}$$

with $q=(n,p)$ referring to neutrons or protons. The dependence on temperature of $\mathcal{E}(r,T)$ (Eq. 4.43) and $m/m_{q,k}(r)$ (Eq. 4.46) comes from the T-dependence of the densities and kinetic energy densities.

There exist various methods to obtain the kinetic energy density $\tau_q(r,T)$ entering the expression for $\mathcal{E}(r,T)$ (Eq. 4.43). One of them is, as mentioned above, to use the HFBTHO code. Another way is to use the Thomas-Fermi (TF) approximation adopted in Ref. [66], or an extension of the TF expression up to T^2 terms [69]:

$$\tau_q(r,T)=\frac{2m}{\hbar^2}\varepsilon_{K_q}=\frac{3}{5}(3\pi^2)^{2/3}$$
$$\times\left[\rho_q^{5/3}+\frac{5\pi^2m_q^2}{3\hbar^4}\frac{1}{(3\pi^2)^{4/3}}\rho_q^{1/3}T^2\right]. \tag{4.47}$$

In Eq. (4.47), the first term in square brackets is the degenerate limit at zero temperature and the T^2 term is the finite-temperature correction. We calculate the kinetic energy density using the self-consistent Skyrme HFB method and the HFBTHO code, but for a comparison, we also present the results when using $\tau_q(r, T)$ from Eq. (4.47).

The symmetry-component term in the expression for the nuclear energy given in the extended Bethe-Weizsäcker LDM can be written in the following form [36]:

$$S(T)\frac{(N - Z)^2}{A},\tag{4.48}$$

where

$$S(T) = \frac{S^V(T)}{1 + \dfrac{S^S(T)}{S^V(T)}A^{-1/3}} = \frac{S^V(T)}{1 + A^{-1/3}/\kappa(T)}\tag{4.49}$$

with

$$\kappa(T) \equiv \frac{S^V(T)}{S^S(T)}.\tag{4.50}$$

S^V is the volume symmetry energy parameter, and S^S is the modified surface symmetry energy one in the liquid model (see Ref. [70], where it is defined by S^{S*}). In the case of nuclear matter, where $A \longrightarrow \infty$ and $S^S/S^V \longrightarrow 0$, we have $S(T) = S^V(T)$. From Eq. (4.49) follow the relations of $S^V(T)$ and $S^S(T)$ with $S(T)$:

$$S^V(T) = S(T)\left(1 + \frac{1}{\kappa(T)A^{1/3}}\right),\tag{4.51}$$

$$S^S(T) = \frac{S(T)}{\kappa(T)}\left(1 + \frac{1}{\kappa(T)A^{1/3}}\right).\tag{4.52}$$

Following Refs. [34,71–73], an approximate expression for the ratio $\kappa(T)$ can be written within the CDFM:

$$\kappa(T) = \frac{3}{R\rho_0}\int_0^\infty dx|\mathcal{F}(x, T)|^2 x\rho_0(x)\left\{\frac{S(\rho_0)}{S[\rho(x, T)]} - 1\right\},\tag{4.53}$$

where $R = r_0 A^{1/3}$ [73] and $S(\rho_0)$ is the NSE at equilibrium nuclear matter density ρ_0 and $T = 0$ MeV. The weight function $|\mathcal{F}(x, T)|^2$, which depends on the temperature through the temperature-dependent total density distribution $\rho_{total}(r, T)$, is determined by

$$|\mathcal{F}(x, T)|^2 = -\frac{1}{\rho_0(x)}\left.\frac{d\rho_{total}(r, T)}{dr}\right|_{r=x},\tag{4.54}$$

where

$$\rho_{total}(r, T) = \rho_p(r, T) + \rho_n(r, T).\tag{4.55}$$

The T-dependent proton $\rho_p(r, T)$ and neutron $\rho_n(r, T)$ densities were calculated using the HFB method with transformed harmonic-oscillator basis and the HFBTHO code [58]. For the density dependence of the symmetry energy $S[\rho(x, T)]$, we use the commonly employed power parametrization:

$$S[\rho(x, T)] = S^V(T)\left[\frac{\rho(x, T)}{\rho_0}\right]^\gamma,\tag{4.56}$$

as well as some other alternative parametrizations [36].

4.3 Results and Discussion

4.3.1 Spherical Nuclei: Ni ($A = 74-84$), Sn ($A = 124-152$), and Pb ($A = 202-214$)

The analysis of the correlation between the neutron skin thickness and some macroscopic nuclear matter properties in finite nuclei is shown in Figure 4.1 for a chain of Ni isotopes. The calculations have been carried out on the basis of the Brueckner EDF for infinite nuclear matter and within the CDFM according to Eqs. (4.40)–(4.42) by using the weight functions (4.30) obtained from the self-consistent densities in Eq. (4.7). It is seen from Figure 4.1a that there exists an approximate linear correlation between ΔR and s for the even-even Ni isotopes with $A = 74-84$. We observe a smooth growth of the symmetry energy till the double-magic nucleus ^{78}Ni ($N = 50$) and then a linear decrease of s while the neutron skin thickness of the isotopes increases. This behavior is valid for all Skyrme parametrizations used in the calculations, and in particular, the average slope of ΔR for various forces is almost the same. The LNS force yields larger values of s comparing to the other three Skyrme interactions. In this case, the small deviation can be attributed to the fact that the LNS force has not been fitted to finite nuclei, and therefore, one cannot expect a good quantitative description at the same level as purely phenomenological Skyrme forces [64]. As a consequence, the neutron skin thickness calculated with LNS force has larger size with respect to the other three forces whose results for ΔR are comparable with each other. Nevertheless, it is worth to compare its predictions not only for ground-state nuclear properties but also for EOS characteristics of finite systems with those of other commonly used Skyrme forces.

We also find a similar approximate linear correlation for Ni isotopes between ΔR and p_0 in Figure 4.1b and less strong correlation between ΔR and ΔK in Figure 4.1c. As in the symmetry energy case, the behavior of the curves drawn in these plots shows the same tendency, namely the inflexion point transition at the double-magic ^{78}Ni nucleus. We would like to note that the predictions for the difference ΔR between the rms radii of neutrons and protons with Skyrme forces obtained in Ref. [60] exhibited a steep change at the same place in which the number of neutrons starts to increase in the chain of nickel isotopes. In addition, the calculated results for the two-neutron separation energies and neutron and matter rms

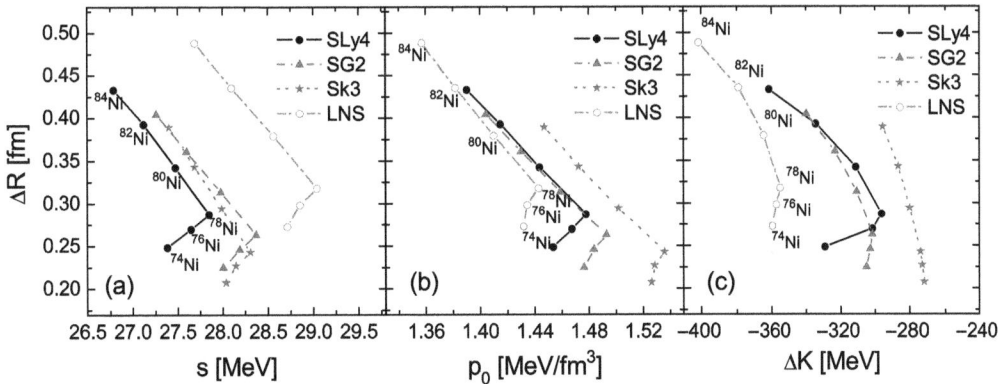

FIGURE 4.1
HF+BCS neutron skin thicknesses ΔR for Ni isotopes as a function of the symmetry energy s (a), pressure p_0 (b), and asymmetric compressibility ΔK (c) calculated with SLy4, SG2, Sk3, and LNS forces.

radii of the even Ni isotopes obtained in the relativistic Hartree-Bogoliubov framework [74] showed a quite strong kink at $N = 50$. Also one can see from Figures 4.1b and c that the calculated values for p_0 and ΔK are lower in the case of LNS force than for the other three Skyrme parameter sets.

Figure 4.2 illustrates the evolution of the symmetry energy and the change of the neutron skin for Sn isotopes with SLy4 force. It is seen from Figure 4.2 that the symmetry energy varies in the interval of 27–29 MeV. Similarly to the case of Ni isotopes with transition at specific shell closure, we observe a smooth growth of the symmetry energy till the double-magic nucleus ^{132}Sn ($N = 82$) and then an almost linear decrease of s while the neutron skin thickness of the isotopes increases. As can be seen from Figure 4.2, the values of ΔR for ^{130}Sn and ^{132}Sn obtained from our HF+BCS calculations with SLy4 force fit very well the corresponding neutron skin thicknesses of 0.23 ± 0.04 and 0.24 ± 0.04 fm of these nuclei derived from pygmy dipole resonances [23]. Moreover, both experimental and theoretical values follow a trend established by a measurement of the stable Sn isotopes [23,60].

The theoretical neutron skin thickness ΔR of Pb nuclei ($A = 202 - 214$) against the parameters of interest, s, p_0, and ΔK, is illustrated in Figure 4.3. Similarly to the three panels for Ni isotopes presented in Figure 4.1, the predicted correlations manifest the same linear dependence. However, in this case the kink observed at double-magic ^{208}Pb nucleus is much less pronounced that it can be seen at ^{78}Ni isotope and it does not change its direction. Similarly to Ni and Sn isotopes, the LNS force produces larger symmetry energies s than the other three forces also for Pb nuclei with values exceeding 30 MeV. Another peculiarity of the results obtained with LNS is the almost constant ΔK observed in Figure 4.3a.

The kinks displayed in Figure 4.1 by the Ni isotopes and in Figure 4.2 by the Sn isotopes can be attributed to the shell structure of these exotic nuclei. Indeed, the isotopic chains of Ni and Sn are of particular interest for nuclear structure calculations because of their proton shell closures at $Z = 28$ and $Z = 50$, respectively. They also extend from the proton drip line that is found nearby the double-magic ^{48}Ni (see, for instance, the discussion in Ref. [60] for a possible proton skin formation) and ^{100}Sn nuclei to the already β unstable neutron-rich double-magic ^{78}Ni and ^{132}Sn isotopes. In the case of ^{78}Ni ($N = 50$ shell closure), the filling of the $1g_{9/2}$ orbit is completed. Beyond this isotope, the $2d_{5/2}$ subshell is being filled and a thick neutron skin is built up. Similar picture is present for the Sn isotopes. Here, one finds a sudden jump beyond $N = 82$ where the $1h_{11/2}$ shell is filled and the $2f_{7/2}$ subshell

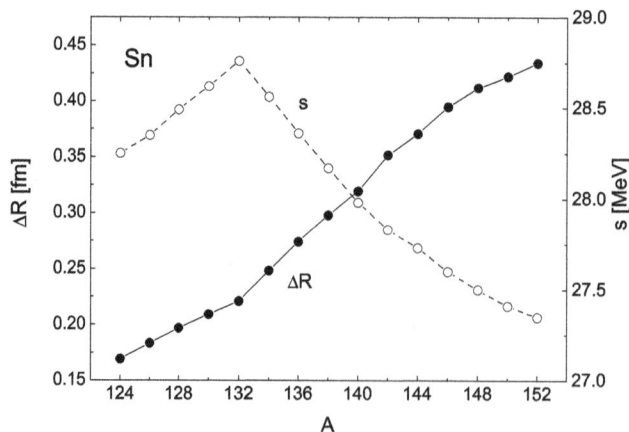

FIGURE 4.2

HF+BCS neutron skin thicknesses ΔR (solid line) and symmetry energies s (dashed line) for Sn isotopes calculated with SLy4 force.

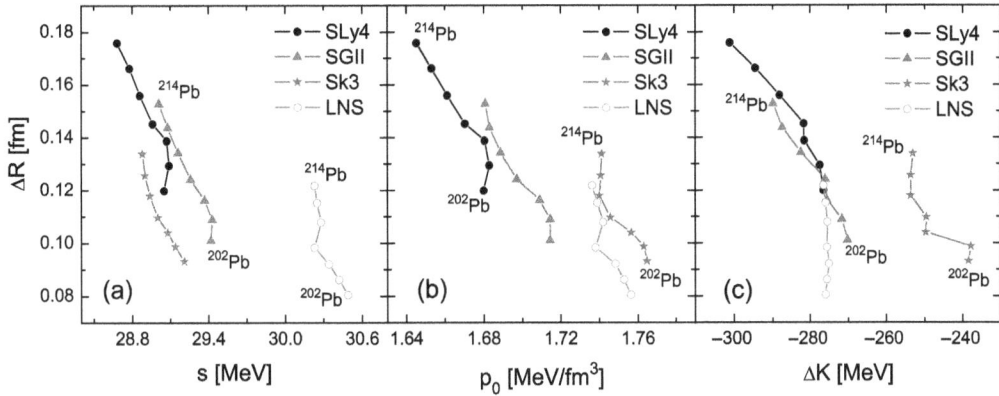

FIGURE 4.3

HF+BCS neutron skin thicknesses ΔR for Pb isotopes as a function of the symmetry energy s (a), pressure p_0 (b), and asymmetric compressibility ΔK (c) calculated with SLy4, SGII, Sk3, and LNS force.

becomes populated. The situation in Pb isotopes shown in Figure 4.3 is different from those of Ni and Sn isotopes-no kinks appear in the Pb chain considered. In the case of ^{208}Pb ($N = 126$), the $3p_{1/2}$ orbit is filled, and above this nucleus, the first occupied level is $2g_{9/2}$. As a result, the bulk and surface contributions to the neutron skin thickness of neutron-rich Sn and Pb isotopes reveal an opposite effect, i.e., the surface part dominates the bulk one in tins, while for Pb isotopes, the bulk part is larger [75]. A detailed study of the weight function $|\mathcal{F}(x)|^2$ as the key ingredient of the CDFM to understand the existence of kinks on the example of Ni and Sn isotopic chains and the lack of such a kink for the Pb isotopic chain is performed in Ref. [25].

4.3.2 Deformed Nuclei: Kr ($A = 82 - 120$) and Sm ($A = 140 - 156$)

The ground states of atomic nuclei are characterized by different equilibrium configurations related to corresponding geometrical shapes. The position of the neutron drip line is closely related to the neutron excess and the deformation in nuclei. Therefore, it is interesting to explore how the NSE changes in the presence of deformation and correlates with the neutron skin thickness within a given isotopic chain. In Ref. [60], the effects of deformation on the skin formation were studied in Kr isotopes that are well-deformed nuclei. It has been shown from the analysis on ^{98,100}Kr nuclei that although the profiles of the proton and neutron densities, as well as the spatial extensions, change with the direction in both oblate and prolate shapes, the neutron skin thickness remains almost equal along the different directions perpendicular to the surface. Thus, a very weak dependence of the neutron skin formation on the character of the deformation was found [60].

The results for the symmetry energy s as a function of the mass number A for the Kr isotopic chain ($A = 82 - 120$) are presented in Figure 4.4. We observe peaks of the symmetry energy at specific Kr isotopes, namely at semi-magic ^{86}Kr ($N = 50$) and ^{118}Kr ($N = 82$) nuclei. In addition, a flat area is found surrounded by transitional regions $A = 88 - 96$ and $A = 110 - 116$. Also, the SGII and Sk3 forces yield values of s comparable with each other that lie between the corresponding symmetry energy values when using SLy4 and LNS sets. The specific nature of the LNS force [64] (not being fitted to finite nuclei) leads to larger values of s (and to a larger size of the neutron skin thickness, as is seen from Figure 2 of Ref. [25]) with respect to the results with other three forces.

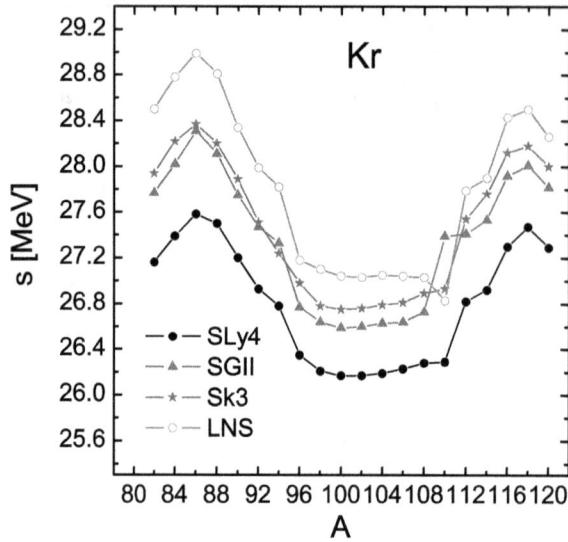

FIGURE 4.4
The symmetry energies s for Kr isotopes ($A = 82 - 120$) calculated with SLy4, SGII, Sk3, and LNS forces.

The results shown in Figure 4.4 are closely related to the evolution of the quadrupole parameter $\beta = \sqrt{\pi/5} Q/(A\langle r^2 \rangle^{1/2})$ (Q being the mass quadrupole moment and $\langle r^2 \rangle^{1/2}$ the nucleus rms radius) as a function of the mass number A that is presented in Figure 4.5. First, one can see from Figure 4.5 that the semi-magic $A = 86$ and $A = 118$ Kr isotopes are spherical, while the open-shell Kr isotopes within this chain possess two equilibrium shapes, oblate and prolate. In the case of open-shell isotopes, the oblate and prolate minima are very close in energy, and the energy difference is always less than 1 MeV. In this region of even-even Kr isotopes with very large N/Z ratio (≥ 1.7), the competition between the prolate and oblate shapes has also been studied with HFB calculations and the Gogny force in Ref. [76]. Shape coexistence in lighter Kr isotopes has also been examined [77,78]. Nevertheless, we specify in Figure 4.5 which shape corresponds to the ground state of each isotope by encircling them. Thus, the trend that the evolution of the symmetry energy shown in Figure 4.4 follows can be clearly understood. The peaks of the symmetry energy

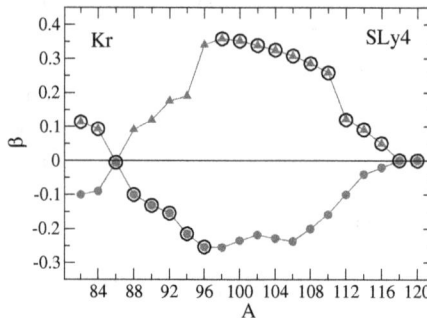

FIGURE 4.5
The quadrupole parameter β as a function of the mass number A for the even-even Kr isotopes ($A = 82 - 120$) in the case of SLy4 force.

correspond to the closed-shell nuclei that are spherical. Mid-shell nuclei ($A = 96 - 110$) are well deformed and exhibit a stabilized behavior with small values of s. The transitional regions from spherical to well-deformed shapes correspond to transitions from the peaks to the valley in the symmetry energy.

In Figures 4.6 and 4.7, we give results for Sm isotopes ($A = 140 - 156$) as a well-established example of deformed nuclei. In the calculations, all Sm isotopes are found to have a prolate shape, except for the even-even ^{144}Sm and ^{146}Sm nuclei that are spherical. Such an evolution of shape from the spherical to the axially deformed shapes in the same Sm isotopic region is in accordance with the results obtained from microscopic calculations in the relativistic mean-field theory [79]. In Ref. [79], the ground state of the semi-magic ^{144}Sm ($N = 82$) is found to be spherical (having about 12 MeV stiff barrier against deformation) and the deformation in ^{146}Sm to be still small. With the increase of the neutron number, the ground state gradually moves toward the deformed one till the well-deformed $^{154-158}$Sm [79].

The results for the correlation between the neutron skin thickness and the nuclear matter properties in finite nuclei with SLy4 Skyrme force for a chain of Sm isotopes are shown in

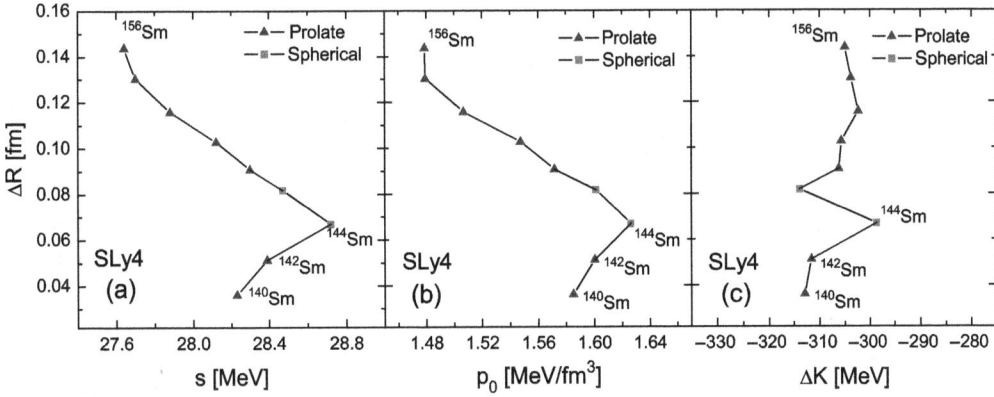

FIGURE 4.6
HF+BCS neutron skin thicknesses ΔR for Sm isotopes as a function of the symmetry energy s (a), pressure p_0 (b), and asymmetric compressibility ΔK (c) calculated with SLy4 force.

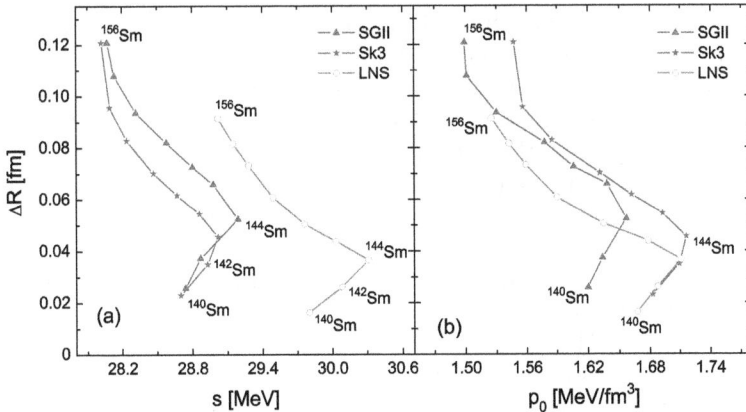

FIGURE 4.7
HF+BCS neutron skin thicknesses ΔR for Sm isotopes as a function of the symmetry energy s (a) and the pressure p_0 (b) calculated with SGII, Sk3, and LNS forces.

Figure 4.6, while those with SGII, Sk3, and LNS forces are presented in Figure 4.7. Similar to the case of Kr isotopes with transition at specific shell closure, we observe a smooth growth of the symmetry energy until the semi-magic nucleus ^{132}Sm ($N = 82$) and then an almost linear decrease of s while the neutron skin thickness of the isotopes increases [25]. An approximate linear correlation between ΔR and p_0 is also shown in Figures 4.6b and 4.7b, while Figure 4.6c exhibits a very irregular behavior of ΔR as a function of the asymmetric compressibility ΔK. Nevertheless, the values of ΔK deduced from our calculations are in the interval between −295 and −315 MeV that compare fairly well with the neutron-asymmetry compressibility ($K'_\Sigma = -320 \pm 180$ MeV) deduced from the data [80] on the breathing mode giant monopole resonances in the isotopic chains of Sm and Sn nuclei.

4.3.3 Neutron-Deficient and Neutron-Rich Mg Isotopes with $A = 20$–36

Low-lying states of neutron-rich nuclei around the neutron number $N = 20$ attract a great interest, as the spherical configurations associated with the magic number disappear in the ground states. For ^{32}Mg, from the observed population of the excited 0^+_2 state (found at 1.058 MeV) in the (t,p) reaction on ^{30}Mg, it is suggested [81] that the 0^+_2 state is a spherical one coexisting with the deformed ground state and that their relative energies are inverted at $N = 20$. Very recently, a new signature of an existence of "island of inversion" [82] has been experimentally tested by measuring the charge radii of all magnesium isotopes in the sd shell at ISOLDE-CERN [83] showing that the borderline of this island lies between ^{30}Mg and ^{31}Mg.

Within the same theoretical scheme applied for spherical and deformed nuclei, we studied the emergence of an "island of inversion" at the neutron-rich ^{32}Mg nucleus and the symmetry-energy evolution in the Mg isotopic chain ($A = 20 - 36$), thus searching for possible similarities or differences in these lighter isotopes with respect to the heavy ones [26]. Apart from the standard calculations, we make additional estimations for the ^{32}Mg nucleus by increasing the spin-orbit strength of the SLy4 effective interaction by 20%. Increasing the spin-orbit strength will bring near the neutron $f_{7/2}$ and $d_{3/2}$ orbitals facilitating the promotion of neutrons to the former. According to the Federman and Pittel mechanism [84], protons in the open $d_{5/2}$ orbital overlap substantially with $f_{7/2}$ neutrons ($\ell_p = \ell_n - 1$) generating nuclear deformation by the effect of the isoscalar part of the n-p interaction. The corresponding potential energy curve is illustrated in Figure 4.8 together with the curve from the original SLy4 interaction leading to a spherical equilibrium shape in ^{32}Mg. As a result, we find strong prolate deformation for the intruder configuration ($\beta = 0.38$). This value of the quadrupole deformation is close to the value $\beta = 0.32$ found for the generator coordinate in Ref. [83], where a slight modification of the spin-orbit strength of the effective interaction for a better description of the "island of inversion" was also applied. Thus, the "dual" nature of the ground state of ^{32}Mg that reflects the shape coexistence in ^{32}Mg is in favor to understand the structure of ^{32}Mg.

The impact of these new modified calculations on the evolution of the charge radii, especially in the region of the Mg isotopic chain where an "island of inversion" is expected, is illustrated in Figure 4.9. In addition to ^{32}Mg, we apply the same procedure also to ^{31}Mg nucleus in order to better establish the border of the island. A further increase in the charge radii of these isotopes is found. For ^{31}Mg, the charge rms radius increases from 3.117 to 3.154 fm and for ^{32}Mg from 3.137 to 3.179 fm toward the experimentally extracted values for both nuclei indicated in Figure 4.9. In general, it can be seen from Figure 4.9 that the comparison between the new values that are very close to the experimental data [83] and the previously obtained values of the charge radii of ^{31,32}Mg isotopes can define a region

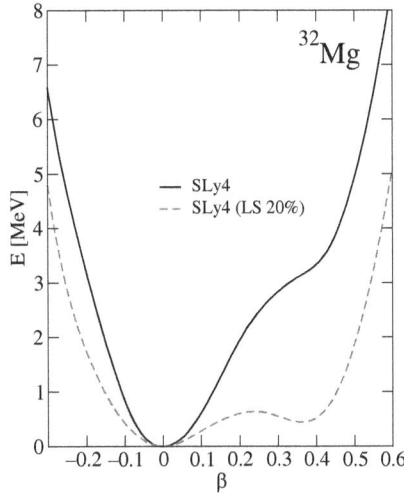

FIGURE 4.8
Potential energy curves of ^{32}Mg obtained from HF+BCS calculations with SLy4 force for the spherical case (solid line) and in the case when the spin-orbit strength of the effective SLy4 interaction is increased by 20% (dashed line).

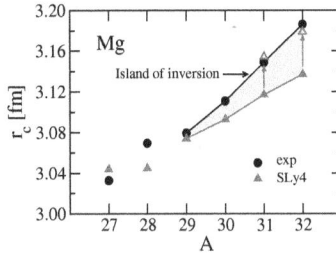

FIGURE 4.9
Theoretical (with the SLy4 Skyrme force) and experimental [83] rms charge radii r_c of Mg isotopes in the range $A = 27$–32. The open triangles represent the calculated values of r_c when the spin-orbit strength of the effective interaction is increased by 20%.

associated with the "island of inversion" which is not seen in the HF+BCS theoretical method by using the original Skyrme force fitted to stable nuclei.

The results for the symmetry energy s (Eq. 4.40) as a function of the mass number A for the whole Mg isotopic chain ($A = 20$–36) are presented in Figure 4.10. It is useful to search for possible indications of an "island of inversion" around $N = 20$ revealed also by the symmetry energy. Therefore, it is interesting to see how the trend of the symmetry energy will be changed when a prolate deformed ground state of ^{32}Mg is obtained for the magic number $N = 20$ (see Figure 4.8). We would like to note that in this case, the modification of the spin-orbit strength of the SLy4 effective interaction by increasing it by 20% leads to a smaller value of $s = 23.67$ MeV compared with the one for the spherical case $s = 24.75$ MeV. Thus, the role of deformation on the nuclear charge radii is also confirmed on the NSE. Indeed, the results shown in Figure 4.10 are related to the evolution of the quadrupole parameter as a function of the mass number A, as well as to the evolution of the charge radii [26]. We find strong deformations in the range of $A = 22$–24 that produce larger charge

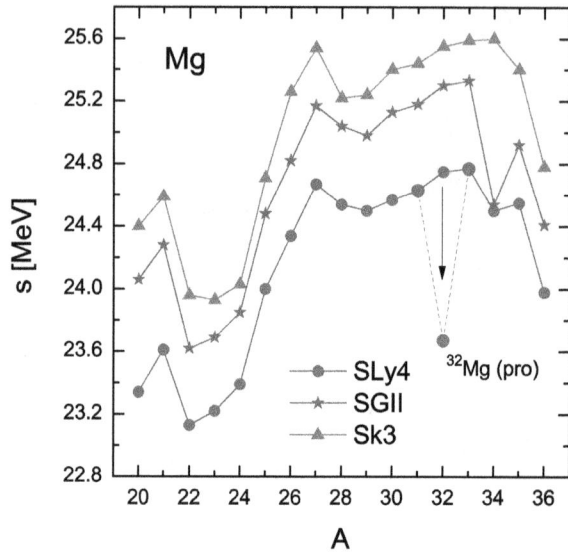

FIGURE 4.10

The symmetry energies s for Mg isotopes ($A = 20$–36) calculated with SLy4, SGII, and Sk3 forces.

radii in relation to their neighbors, and local wells in the symmetry energy. Next, there is a region of flat energy profiles that correspond to small charge radii and increasing values of the symmetry energy. Above this region, we find first spherical shapes that produce a plateau in the symmetry energy from $A = 28$ to $A = 34$ and finally prolate deformations in $A = 36$ that produce very large radii and a sharp decrease in the symmetry energy. This confirms the physical interpretation given in Ref. [85], where this fact is shown to be a result from the moving of the extra neutrons to the surface, thus increasing the surface tension but reducing the symmetry energy. Although the considered Mg chain does not contain a double-magic isotope, it is worth mentioning that we find maximum values within a plateau around the semi-magic isotope $A = 32$ that resembles the sharp peak observed for double-magic nuclei [24,25].

4.3.4 Temperature-Dependent Symmetry Energy Coefficient, Densities, Nuclear Radii, and Neutron Skins

In understanding the symmetry energy coefficient e_{sym} for finite nuclear systems and their thermal evolution, some ambiguities about their proper definition could be noted. Aiming to study the T-dependence of e_{sym} within a given isotopic chain, we introduced two new definitions of $e_{sym}(A, T)$ within the LDA, as an attempt to analyze in a more appropriate way the symmetry energy coefficient of finite nuclei within a given chain [35]. They concern the problem of calculating the term $e(\rho, \delta = 0, T)$ of Eq. (4.45) that is responsible for the contribution of the energy per particle of symmetric nuclear matter. In our LDA approach, the latter is simulated by considering the $N = Z = A/2$ nucleus analyzing two possibilities. It has been shown in Ref. [35] that for the cases when there is no such a bound nucleus HFB solution, none of the recipes used seems to be totaly justified or free from ambiguities, so that more work along this line is required.

In addition, we perform a comparative analysis of e_{sym} for several isotopes from the Ni, Sn, and Pb chains applying the LDA in a version based on Eqs. (4.43)–(4.46). The symmetric nuclear matter part of Eq. (4.45) $e(\rho, \delta = 0, T)$ is obtained approximately with densities $\rho_n = \rho_p = \rho/2$, where ρ is the total density calculated with the HFBTHO code. The kinetic energy density is from the TF method with T^2 term [69] in Eq. (4.47) calculated with the above densities. So, in this case, $\tau_n \approx \tau_p$. The results are presented in Figure 4.11, which illustrates the isotopic evolution of the symmetry energy coefficient on the examples of Ni ($A = 64$–82), Sn ($A = 124$–152), and Pb ($A = 202$–214) chains in the case of both SLy4 and SkM* Skyrme interactions used in the calculations. A smooth decrease of e_{sym} is observed with the increase of the mass number. Unfortunately, it is difficult to compare our results with other theoretical calculations of e_{sym} of nuclei from the mass range covered here except from the results for ^{208}Pb shown in Refs. [66,86] and for the mass number $A = 120$ presented in Figure 5 of Ref. [68].

In Figure 4.12, we display as examples the density distributions of protons and neutrons for double-magic ^{78}Ni and ^{132}Sn nuclei at $T = 0$, 2, and 4 MeV obtained by using the SLy4 and SkM* parametrizations within the HFB method. The tendency in the behavior of proton and neutron densities of ^{78}Ni and ^{132}Sn obtained with a given Skyrme force (SLy4 or SkM*)

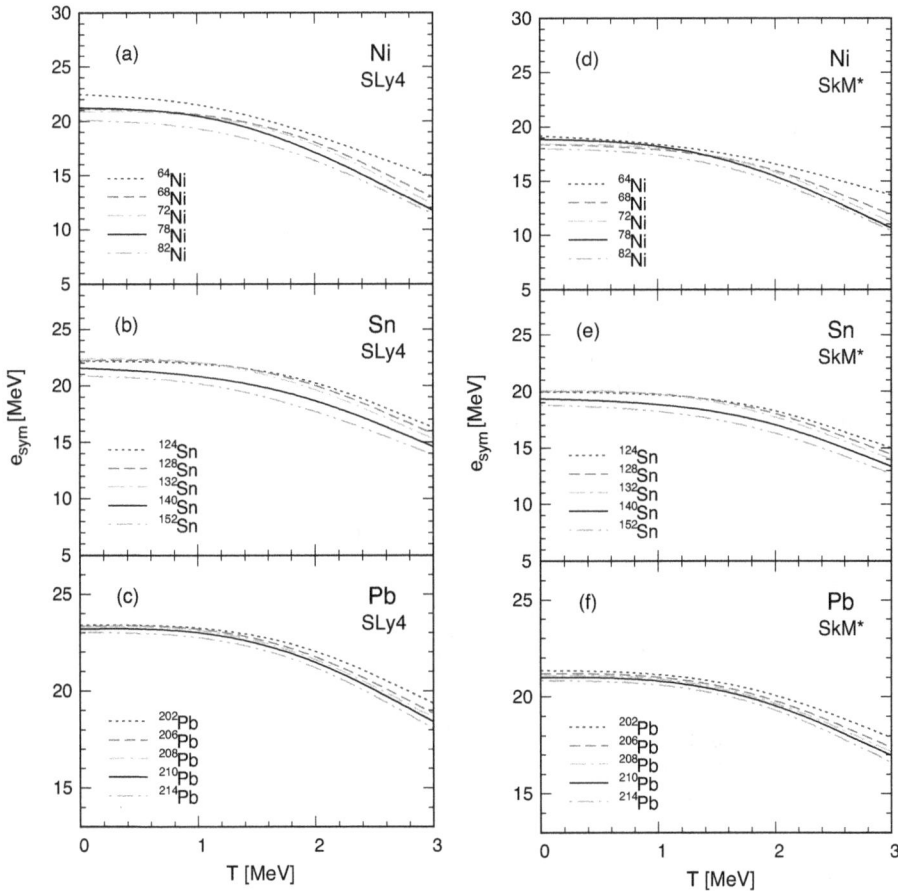

FIGURE 4.11

Temperature dependence of the symmetry energy coefficient e_{sym} obtained for several nuclei from Ni ($A = 64$–82) [(a) and (d)], Sn ($A = 124$–152) [(b) and (e)], and Pb ($A = 202$–214) [(c) and (f)] isotopic chains in HFB method with SLy4 (left panel) and SkM* (right panel) forces.

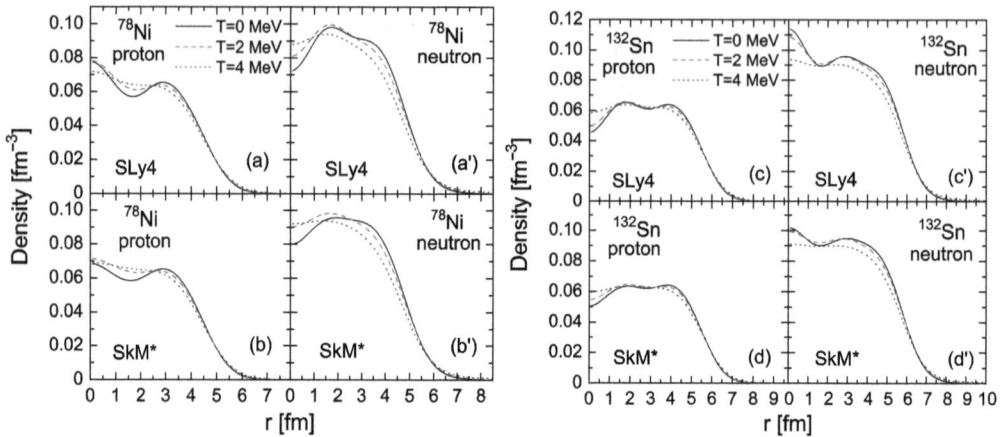

FIGURE 4.12
HFBTHO density distributions of protons and neutrons for ^{78}Ni [(a), (a′), (b), (b′)] and ^{132}Sn [(c), (c′), (d), (d′)] at $T = 0$ MeV (solid line), $T = 2$ MeV (dashed line), and $T = 4$ MeV (dotted line) obtained using the SLy4 and SkM* parametrizations.

is similar. For example, the use of both parametrizations leads to a depression of the proton densities in the interior of ^{132}Sn (Figures 4.12c and d) being larger at zero temperature and to a growth in the same region for the neutron densities, but in an opposite direction relative to T (Figures 4.12c′ and d′). As a consequence, a spatial extension of both densities at the surface region is observed with the increase of T. Namely this region is responsible for the emergence of a neutron skin (e.g., Ref. [60]).

In Figure 4.13, we show the neutron and proton rms radii and the neutron skin thickness as a function of the mass number A for the Sn ($A = 124$–152) isotopic chain calculated by using SLy4 force. First, it can be seen that the proton rms radii increase more slowly than the neutron ones, which is valid for all the isotopic chains and temperatures. This is naturally

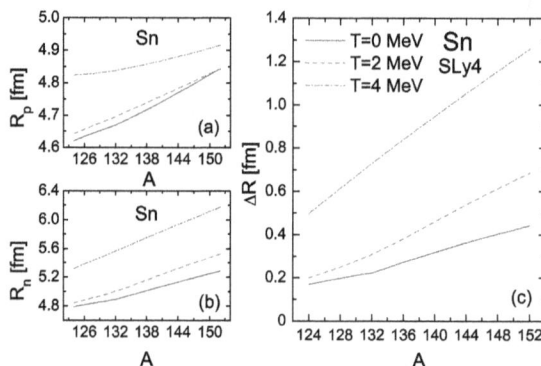

FIGURE 4.13
Mass dependence of the proton R_p (a) and neutron R_n (b) radius of the Sn isotopes ($A = 124$–152) calculated with SLy4 interaction at $T = 0$ MeV (solid line), $T = 2$ MeV (dashed line), and $T = 4$ MeV (dash-dotted line). Neutron skin thickness ΔR as a function of A (c) for the Sn isotopes.

expected in isotopic chains where the number of protons remains fixed. In addition, while the results of both radii at $T = 0$ and $T = 2$ MeV are close to each other with increasing A, one can see a steep increase of their values when the nucleus becomes very hot ($T = 4$ MeV). As can be seen from Figure 4.13c, the neutron skin thickness exhibits the same trend as the rms radii. It grows significantly with the increase of T being much larger at $T = 4$ MeV than at lower temperatures $T = 0$, 2 MeV. The mechanism of formation of neutron skin in tin isotopes has been studied in Ref. [87], where the changes in the neutron skin were attributed mainly to the effect of temperature on the occupation probabilities of the single-particle states around the Fermi level. In Ref. [87], a more limited Sn isotopic chain up to ^{120}Sn was considered. Our results for larger A in this chain (from $A = 124$ to $A = 152$) also show a slow increase of the neutron skin size. The enhancement of the proton and neutron radii at high temperatures leads to a rapid increase in the neutron skin size. We would like to note that at zero temperature, the use of HFBTHO temperature-dependent densities in the present approach confirms the observation in Ref. [60] (where the densities were calculated within a deformed Skyrme HF+BCS approach), namely that a pronounced neutron skin can be expected at $A > 132$ in Sn isotopes.

4.3.5 Temperature Dependence of the Volume and Surface Components of the Nuclear Symmetry Energy

Studying the T-dependence of the NSE $S(T)$, its volume $S^V(T)$ and surface $S^S(T)$ components, as well as their ratio $\kappa(T) = S^V(T)/S^S(T)$, we observed a certain sensitivity of the results to the value of the parameter γ in Eq. (4.56). In order to make a choice of its value, we imposed the following physical conditions: i) the obtained results for the considered quantities at $T = 0$ MeV to be equal or close to those obtained for the same quantities in our previous works for the NSE, its components and their ratio (Ref. [34,35]), and ii) their values for $T = 0$ MeV to be compatible with the available experimental data (see, e.g., the corresponding references in [34]).

In Figures 4.14 and 4.15, the results for the mentioned quantities are given as functions of the mass number A for the isotopic chains of Ni and Sn nuclei for temperatures $T = 0$–3 MeV calculated using the SkM* Skyrme force. The results are presented for two values of the parameter $\gamma = 0.3$ and 0.4. The reason for this choice is related to the physical criterion mentioned above. It can be seen that at $T = 0$ MeV and $\gamma = 0.4$, the value of κ is around 2.6. This result is in agreement with our previous result obtained in the case of the Brueckner EDF in Ref. [34], namely $2.10 \leq \kappa \leq 2.90$. The latter is compatible with the published values of κ extracted from nuclear properties presented in Ref. [73] from the IAS and skins [88] ($2.6 \leq \kappa \leq 3.0$) and from masses and skins [71] ($2.0 \leq \kappa \leq 2.8$). In the case of $\gamma = 0.3$, our result for $T = 0$ MeV is $\kappa = 1.65$ which is in agreement with the analyses of data in Ref. [73] ($1.6 \leq \kappa \leq 2.0$), as well as with the results in Ref. [34] in the case of Skyrme EDF, namely, for the Ni isotopic chain $1.5 \leq \kappa \leq 1.7$ and for the Sn isotopic chain $1.52 \leq \kappa \leq 2.1$, all obtained with SLy4 and SGII forces. The comparison of the results for S at $T = 0$ MeV with those illustrated in Figure 4.11 shows that they agree with our present values of S within the range of $\gamma = 0.3$–0.4.

It can be seen from Figures 4.14 and 4.15 that the quantities $S(T)$, $S^V(T)$, and $S^S(T)$ decrease with increasing temperatures ($T = 0$–3 MeV), while $\kappa(T)$ slowly increases when T increases. This is true for both isotopic chains of Ni and Sn nuclei. Here, we would like to note that the values of γ between 0.3 and 0.4 that give an agreement of the studied quantities with data, as well as with our previous results for $T = 0$ MeV, are in the lower part of the estimated limits of the values of γ (e.g., in the case of $\gamma = 0.5 \pm 0.1$ [73]). It can be seen also from Figures 4.14 and 4.15 that there are "kinks" in the curves of $S(T)$,

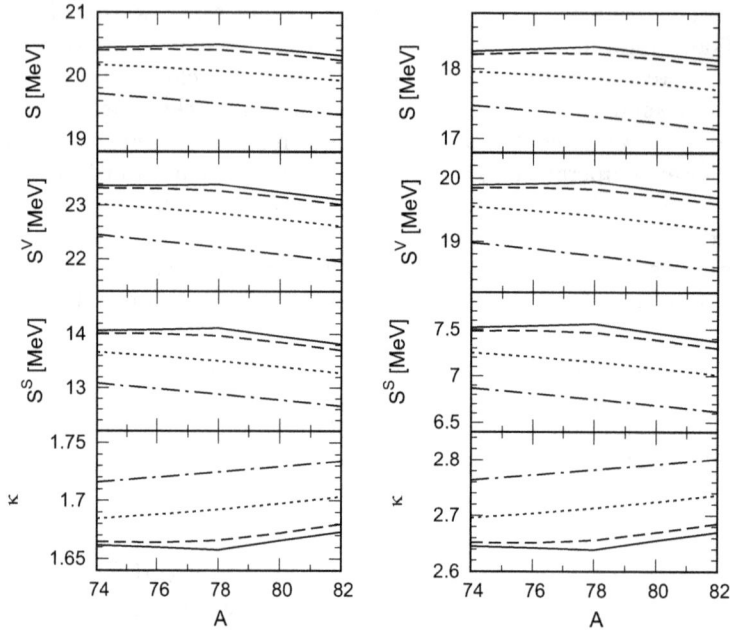

FIGURE 4.14
Mass dependence of the NSE $S(T)$, its volume $S^V(T)$ and surface $S^S(T)$ components and their ratio $\kappa(T)$ for nuclei from the Ni isotopic chain at temperatures $T = 0$ MeV (solid line), $T = 1$ MeV (dashed line), $T = 2$ MeV (dotted line), and $T = 3$ MeV (dash-dotted line) calculated with SkM* Skyrme interaction for values of the parameter $\gamma = 0.3$ (left panel) and $\gamma = 0.4$ (right panel).

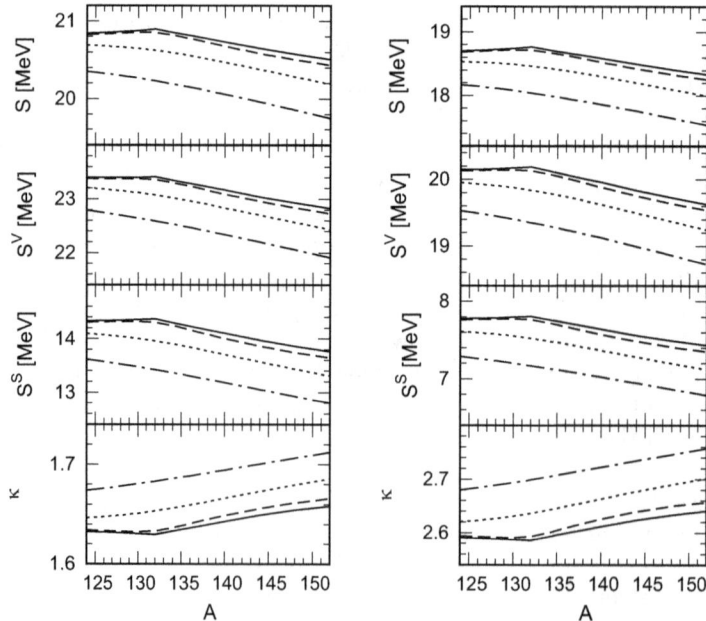

FIGURE 4.15
Same as in Figure 4.14, but for nuclei from the Sn isotopic chain.

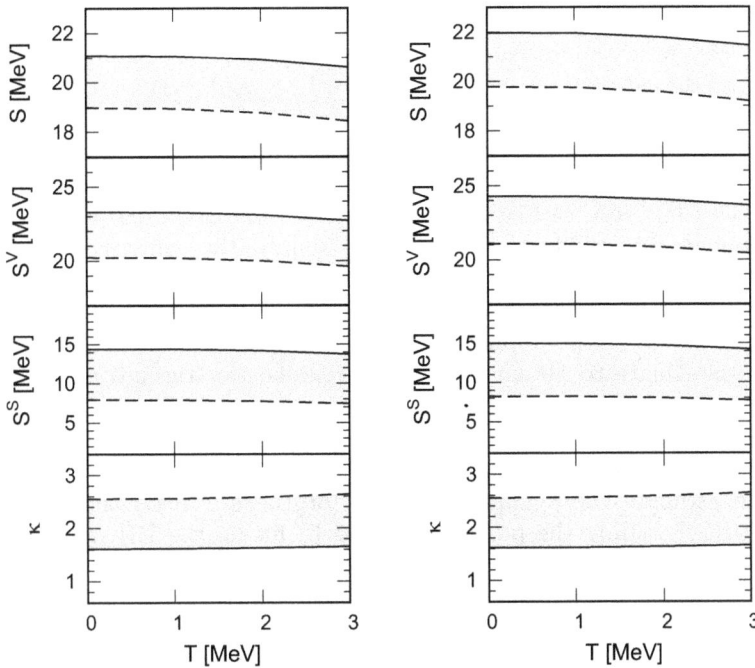

FIGURE 4.16
Temperature dependence of the NSE $S(T)$, its volume $S^V(T)$ and surface $S^S(T)$ components, and their ratio $\kappa(T)$ obtained for values of the parameter $\gamma = 0.3$ (solid line) and $\gamma = 0.4$ (dashed line) with SkM* (left panel) and SLy4 (right panel) forces for ^{208}Pb nucleus.

$S^V(T)$, $S^S(T)$, and $\kappa(T)$ for $T = 0$ MeV in the cases of double closed-shell nuclei ^{78}Ni and ^{132}Sn. This had been observed also in our previous work for $S(T)$ [35], as well as for its volume and surface components and their ratio κ at $T = 0$ MeV in Ref. [34].

In Figure 4.16, the results are given for the T-dependence of $S(T)$, $S^V(T)$, $S^S(T)$, and $\kappa(T)$ for the double-magic ^{208}Pb nucleus obtained using both SkM* and SLy4 Skyrme forces. The results are presented by grey areas between the curves for the values of the parameter $\gamma = 0.3$ and $\gamma = 0.4$. It can be seen that $S(T)$, $S^V(T)$, and $S^S(T)$ decrease, while $\kappa(T)$ slowly increases with the increase in the temperature for both Skyrme forces.

4.4 Conclusions

In this chapter, a theoretical approach to the nuclear many-body problem combining the deformed HF+BCS formalism with Skyrme-type density-dependent effective interactions [56] or HFB method [58] and the coherent density fluctuation model [48,49] has been used to study nuclear symmetry energy and other properties of finite nuclei. For this purpose, we examined three chains of spherical neutron-rich Ni, Sn, and Pb isotopes and two chains of deformed Kr and Sm isotopes. Most of them are located far from the stability line and represent an interest for future measurements with radioactive exotic beams. In addition, we investigated the nuclear properties of Mg isotopes that go from the proton drip-line nucleus ^{20}Mg up to ^{36}Mg, which approaches the neutron drip line. Five Skyrme parametrizations were involved in the calculations: SG2, Sk3, SLy4, LNS, and SkM*.

The main emphasis of this study was to demonstrate the capability of CDFM to be applied as an alternative way to make a transition *from the properties of nuclear matter to the properties of finite nuclei* investigating the NSE and the related characteristics in finite nuclei. This has been carried out on the base of the Brueckner and Skyrme energy-density functionals for infinite nuclear matter. Mainly, the symmetry energy has been determined from analyzing nuclear masses within liquid-drop models, while much more effort has been recently devoted to extracting the value of its slope parameter. Instead of this, in the present work, we applied the CDFM scheme. One of the advantages of the CDFM is the possibility to obtain transparent relations for the intrinsic EOS quantities analytically by means of a convenient approach to the weight function. The key element of this model is the choice of density distributions which were taken from self-consistent deformed HF+BCS or HFB calculations. We would like to note that our new method allows one to estimate the symmetry energy (or the symmetry-energy coefficient) in finite nuclei, thus avoiding the problems related to fitting the HF energies to an LDM parametrization.

We have found that there exists an approximate linear correlation between the neutron skin thickness of spherical even-even nuclei from the Ni, Sn, and Pb isotopic chains and deformed nuclei from Kr and Sm chains and their symmetry energies. A similar linear correlation between ΔR and p_0 is also found to exist, while the relation between ΔR and ΔK is less pronounced. For all chains considered except the Pb isotopes, an inflection point transition at specific shell closure, in particular at double-magic ^{78}Ni and ^{132}Sn and semi-magic ^{86}Kr and ^{144}Sm nuclei, appears for these correlations of the neutron skins with s and p_0. For the Mg isotopes in the considered chain, the correlations between ΔR and s and ΔR and p_0 show up a peculiarity at $A = 27$ just reflecting the transition regions between different nuclear shapes.

In general, the results of the T-dependent symmetry coefficients $e_{sym}(A, T)$ calculated for various isotopes are in good agreement with theoretical predictions for some specific nuclei reported in the literature. At the same time, however, the difference between the results obtained by exploring the suggested two new expressions in our work and by other procedures points out the ambiguities arising from the different definitions for symmetry coefficients of finite nuclear systems. Following the trend of the corresponding proton and neutron rms radii, the neutron skin thickness grows significantly with the increase of T within a given isotopic chain. With increasing T, the NSE S, its volume S^V, and surface S^S components decrease, while their ratio κ slightly increases for all the isotopes in the three chains for both SkM* and SLy4 Skyrme forces and for all used density dependencies of the symmetry energy. The same conclusion can be drawn for the thermal evolution of the mentioned quantities for the three considered double-magic ^{78}Ni, ^{132}Sn, and ^{208}Pb nuclei. In the cases of the double-magic ^{78}Ni and ^{132}Sn nuclei, we observe "kinks" for $T = 0$ MeV in the curves of $S(T)$, $S^V(T)$, $S^S(T)$, and $\kappa(T)$, but not in the case of Pb isotopes. It is also worth mentioning how the kinks are blurred and eventually disappear as T increases, demonstrating its close relationship with the shell structure.

In summary, we reviewed our microscopic theoretical approach to study important macroscopic nuclear matter quantities in finite nuclei and their relation to surface properties of neutron-rich exotic nuclei. The obtained results concerning several aspects of the density and temperature dependence of the symmetry energy could provide a possibility to test the properties of the nuclear EDFs and characteristics related to NSE, e.g., the neutron skin thickness of finite nuclei.

References

[1] P. Danielewicz, R. Lacey, and W. G. Lynch, *Science* **298**, 1592 (2002).

[2] M. A. Famiano *et al.*, *Phys. Rev. Lett.* **97**, 052701 (2006).

[3] D. V. Shetty, S. J. Yennello, and G. A. Souliotis, *Phys. Rev. C* **76**, 024606 (2007).

[4] M. Centelles, X. Roca-Maza, X. Viñas, and M. Warda, *Phys. Rev. Lett.* **102**, 122502 (2009).

[5] B.-A. Li, L.-W. Chen, and C. M. Ko, *Phys. Rep.* **464**, 113 (2008).

[6] *Topical issue on Nuclear Symmetry Energy.* Guest editors: B.-A. Li, A. Ramos, G. Verde, I. Vidaña. *Eur. Phys. J.* A **50** 2 (2014).

[7] B. Alex Brown, *Phys. Rev. Lett.* **85**, 5296 (2000).

[8] S. Typel and B. Alex Brown, *Phys. Rev. C* **64**, 027302 (2001).

[9] C. J. Horowitz and J. Piekarewicz, *Phys. Rev. Lett.* **86**, 5647 (2001).

[10] J. M. Lattimer and M. Prakash, *Phys. Rep.* **442**, 109 (2007).

[11] V. Baran, M. Colonna, V. Greco, and M. Di Toro, *Phys. Rep.* **410**, 335 (2005).

[12] K. Oyamatsu, I. Tanihata, Y. Sugahara, K. Sumiyoshi, and H. Toki, *Nucl. Phys. A* **634**, 3 (1998).

[13] W. Zuo, I. Bombaci, and U. Lombardo, *Phys. Rev. C* **60**, 024605 (1999).

[14] P. Roy Chowdhury, D. N. Basu, and C. Samanta, *Phys. Rev. C* **80**, 011305(R) (2009).

[15] L.-W. Chen, B.-J. Cai, C. M. Ko, B.-A. Li, C. Shen, and J. Xu, *Phys. Rev. C* **80**, 014322 (2009).

[16] P. Gögelein, E. N. E. van Dalen, Kh. Gad, S. A. Hassaneen, and H. Müther, *Phys. Rev. C* **79**, 024308 (2009).

[17] Z. H. Li, U. Lombardo, H.-J. Schulze, W. Zuo, L. W. Chen, and H. R. Ma, *Phys. Rev. C* **74**, 047304 (2006).

[18] J. Piekarewicz and M. Centelles, *Phys. Rev. C* **79**, 054311 (2009).

[19] I. Vidaña, C. Providência, A. Polls, and A. Rios, *Phys. Rev. C* **80**, 045806 (2009).

[20] F. Sammarruca and P. Liu, *Phys. Rev. C* **79**, 057301 (2009).

[21] P. Danielewicz, arXiv:1003.4011 [nucl-th] (2010).

[22] D. V. Shetty and S. J. Yennello, Pramana **75**, 259 (2010); arXiv:1002.0313 [nucl-ex] (2010).

[23] A. Klimkiewicz *et al.*, *Phys. Rev. C* **76**, 051603(R) (2007).

[24] M. K. Gaidarov, A. N. Antonov, P. Sarriguren, and E. Moya de Guerra, *Phys. Rev. C* **84**, 034316 (2011).

[25] M. K. Gaidarov, A. N. Antonov, P. Sarriguren, and E. Moya de Guerra, *Phys. Rev. C* **85**, 064319 (2012).

[26] M. K. Gaidarov, P. Sarriguren, A. N. Antonov, and E. Moya de Guerra, *Phys. Rev. C* **89**, 064301 (2014).

[27] L.-W. Chen, C. M. Ko, and B.-A. Li, *Phys. Rev. C* **72**, 064309 (2005).

[28] A. E. L. Dieperink, Y. Dewulf, D. Van Neck, M. Waroquier, and V. Rodin, *Phys. Rev. C* **68**, 064307 (2003).

[29] R. J. Furnstahl, *Nucl. Phys.* **A706**, 85 (2002).

[30] A. W. Steiner, *Phys. Rev. C* **77**, 035805 (2008).

[31] E. Baron, J. Cooperstein, and S. Kahana, *Phys. Rev. Lett.* **55**, 126 (1985).

[32] Ch. C. Moustakidis, *Phys. Rev. C* **76**, 025805 (2007).

[33] F. Sammarruca, arXiv:0908.1958 [nucl-th] (2009).

[34] A. N. Antonov, M. K. Gaidarov, P. Sarriguren, and E. Moya de Guerra, *Phys. Rev. C* **94**, 014319 (2016).

[35] A. N. Antonov, D. N. Kadrev, M. K. Gaidarov, P. Sarriguren, and E. Moya de Guerra, *Phys. Rev. C* **95**, 024314 (2017).

[36] A. N. Antonov, D. N. Kadrev, M. K. Gaidarov, P. Sarriguren, and E. Moya de Guerra, *Phys. Rev. C* **98**, 054315 (2018).

[37] W. D. Myers and W. J. Swiatecki, *Nucl. Phys. A* **81**, 1 (1966).

[38] P. Möller *et al.*, *At. Data Nucl. Data Tables* **59**, 185 (1995).

[39] K. Pomorski and J. Dudek, *Phys. Rev. C* **67**, 044316 (2003).

[40] A. Carbone, G. Colò, A. Bracco, Li-Gang Cao, P. F. Bortignon, F. Camera, and O. Wieland, *Phys. Rev. C* **81**, 041301(R) (2010).

[41] D. Vretenar, T. Nikšić, and P. Ring, *Phys. Rev. C* **68**, 024310 (2003).

[42] S. Yoshida, and H. Sagawa, *Phys. Rev. C* **73**, 044320 (2006).

[43] L.-W. Chen, C. M. Ko, B.-A. Li, and J. Xu, *Phys. Rev. C* **82**, 024321 (2010).

[44] C.-H. Lee, T. T. S. Kuo, G. Q. Li, and G. E. Brown, *Phys. Rev. C* **57**, 3488 (1998).

[45] B. K. Agrawal, *Phys. Rev. C* **81**, 034323 (2010).

[46] K. A. Brueckner, J. R. Buchler, S. Jorna, and R. J. Lombard, *Phys. Rev.* **171**, 1188 (1968).

[47] K. A. Brueckner, J. R. Buchler, R. C. Clark, and R. J. Lombard, *Phys. Rev.* **181**, 1543 (1969).

[48] A. N. Antonov, V. A. Nikolaev, and I. Zh. Petkov, *Bulg. J. Phys.* **6**, 151 (1979); *Z. Phys. A* **297**, 257 (1980); *ibid* **304**, 239 (1982); *Nuovo Cimento A* **86**, 23 (1985); A. N. Antonov, E. N. Nikolov, I. Zh. Petkov, Chr V. Christov, and P. E. Hodgson, *ibid* **102**, 1701 (1989); A. N. Antonov, D. N. Kadrev, and P. E. Hodgson, *Phys. Rev. C* **50**, 164 (1994).

[49] A. N. Antonov, P. E. Hodgson, and I. Zh. Petkov, *Nucleon Momentum and Density Distributions in Nuclei* (Clarendon Press, Oxford, 1988); *Nucleon Correlations in Nuclei* (Springer-Verlag, Berlin-Heidelberg-New York, 1993).

[50] J. J. Griffin and J. A. Wheeler, *Phys. Rev.* **108**, 311 (1957).

[51] http://hallaweb.jlab.org/parity/prex.

[52] S. Abrahamyan *et al.*, *Phys. Rev. Lett.* **108**, 112502 (2012).

[53] A. Tamii *et al.*, *Phys. Rev. Lett.* **107**, 062502 (2011).

[54] C. J. Horowitz *et al.*, *Phys. Rev. C* **85**, 032501(R) (2012).

[55] F. J. Fattoyev, J. Piekarewicz, and C. J. Horowitz, *Phys. Rev. Lett.* **120**, 172702 (2018).

[56] D. Vautherin, *Phys. Rev. C* **7**, 296 (1973).

[57] E. Moya de Guerra, P. Sarriguren, J. A. Caballero, M. Casas, and D. W. L. Sprung, *Nucl. Phys. A* **529**, 68 (1991).

[58] M. V. Stoitsov, N. Schunck, M. Kortelainen, N. Michel, H. Nam, E. Olsen, J. Sarich, and S. Wild, *Comp. Phys. Comm.* **184**, 1592 (2013).

[59] M. V. Stoitsov, J. Dobaczewski, W. Nazarewicz, and P. Ring, *Comput. Phys. Comm.* **167**, 43 (2005).

[60] P. Sarriguren, M. K. Gaidarov, E. Moya de Guerra, and A. N. Antonov, *Phys. Rev. C* **76**, 044322 (2007).

[61] E. Chabanat, P. Bonche, P. Haensel, J. Meyer, and R. Schaeffer, *Nucl. Phys. A* **635**, 231 (1998).

[62] M. Beiner, H. Flocard, N. Van Giai, and P. Quentin, *Nucl. Phys. A* **238**, 29 (1975).

[63] N. Van Giai and H. Sagawa, *Phys. Lett. B* **106**, 379 (1981).

[64] L. G. Cao, U. Lombardo, C. W. Shen, and N. Van Giai, *Phys. Rev. C* **73**, 014313 (2006).

[65] J. Bartel, P. Quentin, M. Brack, C. Guet, and H.-B. Håkansson, *Nucl. Phys. A* **386**, 79 (1982); H. Krivine, J. Treiner, and O. Bohigas, *Nucl. Phys. A* **336**, 155 (1980).

[66] B. K. Agrawal, J. N. De, S. K. Samaddar, M. Centelles, and X. Viñas, *Eur. Phys. J. A* **50**, 19 (2014).

[67] S. K. Samaddar, J. N. De, X. Viñas, and M. Centelles, *Phys. Rev. C* **76**, 041602(R) (2007).

[68] J. N. De and S. K. Samaddar, *Phys. Rev. C* **85**, 024310 (2012).

[69] S. J. Lee and A. Z. Mekjian, *Phys. Rev. C* **82**, 064319 (2010).

[70] A. W. Steiner, M. Prakash, J. M. Lattimer, and P. J. Ellis, *Phys. Rep.* **411**, 325 (2005).

[71] P. Danielewicz, *Nucl. Phys. A* **727**, 233 (2003).

[72] P. Danielewicz, arXiv: 0607030 [nucl-th] (2006).

[73] A. E. L. Dieperink and P. Van Isacker, *Eur. Phys. J. A* **32**, 11 (2007).

[74] J. Meng, *Phys. Rev. C* **57**, 1229 (1998).

[75] M. Warda, X. Viñas, X. Roca-Maza, and M. Centelles, *Phys. Rev. C* **81**, 054309 (2010).

[76] S. Hilaire and M. Girod, *Eur. Phys. J. A* **33**, 237 (2007).

[77] P. Sarriguren, E. Moya de Guerra, and A. Escuderos, *Nucl. Phys. A* **658**, 13 (1999); *Phys. Rev. C* **64**, 064306 (2001).

[78] H. Flocard, P. Quentin, A.K. Kerman, and D. Vautherin, *Nucl. Phys. A* **203**, 433 (1973).

[79] J. Meng, W. Zhang, S.-G. Zhou, H. Toki, and L.S. Geng, *Eur. Phys. J. A* **25**, 23 (2005).

[80] M. M. Sharma, W. T. A. Borghols, S. Brandenburg, S. Crona, A. van der Woude, and M. N. Harakeh, *Phys. Rev. C* **38**, 2562 (1988).

[81] K. Wimmer *et al.*, *Phys. Rev. Lett.* **105**, 252501 (2010).

[82] E. K. Warburton, J. A. Becker, and B. A. Brown, *Phys. Rev. C* **41**, 1147 (1990).

[83] D. T. Yordanov *et al.*, *Phys. Rev. Lett.* **108**, 042504 (2012).

[84] P. Federman and S. Pittel, *Phys. Rev. C* **20**, 820 (1979).

[85] C. J. Horowitz, E. F. Brown, Y. Kim, W. G. Lynch, R. Michaels, A. Ono, J. Piekarewicz, M. B. Tsang, and H. H. Wolter, *J. Phys. G* **41**, 093001 (2014).

[86] Z. W. Zhang, S. S. Bao, J. N. Hu, and H. Shen, *Phys. Rev. C* **90**, 054302 (2014).

[87] E. Yüksel, E. Khan, K. Bozkurt, and G. Colò, *Eur. Phys. J. A* **50**, 160 (2014).

[88] P. Danielewicz, arXiv: 0411115 [nucl-th] (2004).

5

Theoretical Description of Low-Energy Nuclear Fusion

Raj Kumar

Thapar Institute of Engineering and Technology

M. Bhuyan

University of Malaya
Duy Tan University

D. Jain

Mata Gujri College

B. V. Carlson

Instituto Tecnológico de Aeronáutica

CONTENTS

5.1   Introduction

Nuclear fusion is one of the future sources of clean energy. It can revolutionize the world. The basic understanding of the dynamics involved in the fusion is a must. The process in which two nuclei, or else a nucleus of an atom and a subatomic particle, i.e., proton and neutron, collide to produce one or more nuclides is known as the nuclear reaction. The nuclear reaction dynamics is usually classified into three categories by the energy of incident projectile as low energy ($E \leq 15$ MeV/nucleon), intermediate energy ($15 \leq E \leq 500$ MeV/nucleon), and high energy ($E \geq 500$ MeV/nucleon). In low-energy reactions, the mean nuclear force field acting between the two nuclei dominates in comparison to the high-energy reactions where direct nucleon-nucleon interactions take place. In the intermediate-energy reactions, both the aspects play their role. The heavy-ion (HI)-induced, low-energy nuclear reactions are a topic of interest for the nuclear physics community as they lead to a better understanding of nuclear forces with the formation of heavy nuclei away from the valley of stability providing the much-needed nuclear structure information. In a complete fusion reaction, a composite system is expected to be formed after an intimate contact and transient amalgamation of the projectile and target nucleus leading to the formation of a fully equilibrated compound nucleus. During the interaction, the relative motion energy is distributed among the intrinsic degree of freedom through a series of nucleon-nucleon interactions. Finally, the system reaches a thermodynamic equilibrium before the decay. The energy of the system is the energy of the relative motion in the center-of-mass reference frame plus the reaction Q value; the angular momentum is the vector sum of the orbital angular momentum and the spins of the two nuclei. More details regarding the basic concept of a fusion reaction can be found in Ref. [1].

In nuclear fusion, the nucleon-nucleon interaction is the key to the proper understanding of any nuclear phenomena. The emphasis is laid on the interaction between the incident particle and the target nucleus since the nuclear fusion process is more sensitive to the total potential in the nuclear surface region [2]. The total potential as a function of distance consists of the sum of the long-range Coulomb repulsive force, centrifugal interaction, and short-range nuclear potential terms. The centrifugal and Coulomb parts of the interaction potential are well known. However, the nuclear interaction part is not fully understood yet. Furthermore, the Coulomb potential alone cannot define the barrier, so an appropriate choice of nuclear potential is also necessary to understand reaction dynamics. At present, there are no transparent experimental means to extract information about fusion barriers directly. All experiments measure fusion cross sections [2], and then with the help of theoretical models, one can extract the fusion barriers. Theoretical models are helpful in understanding the nuclear interactions at both macroscopic and microscopic levels. Thus, for a better understanding of the world around us, the more accurate microscopic methods should be exploited for calculating the ion-ion interaction, between the colliding nuclei.

The simplest analytical expression for calculating the nucleus-nucleus interaction is known as the proximity potential within the phenomenological models. These proximity potentials are the benchmark for various macroscopic/microscopic fusion models. All these potentials are based on the proximity force theorem, according to which the nuclear part of the interaction potential can be taken as the product of a factor depending on the mean curvature of the interaction surface and a universal function (depending on the separation distance), and is independent of the masses of colliding nuclei. This concept did introduce a great amount of simplification in nuclear potential studies [3,4]. The role of different proximity potentials is the same as that of using various Skyrme forces in the Skyrme Energy Density Formalism (SEDF) approach. Here, we use the energy density formalism given by Vautherin and Brink [5] which uses the density-dependent Skyrme interactions consisting of spin-dependent and spin-independent parts within a semi-classical extended Thomas Fermi (ETF) approach [6−10].

At low energies, the interaction potential between a pair of nucleons is instantaneous so that a theory of nuclear forces can be applied to nuclear structure calculations [11,12]. The analytical derivation of potential through particle exchange is important to understand the nuclear force as well as structural properties via the nucleus-nucleus optical potential [13−15]. A more fundamental approach to NN-interactions at low energies has been formulated in Refs. [11,16] regarding an effective theory for non-relativistic nucleons. It involves a few basic coupling constants that have been determined from nucleon scattering data at low energies. Furthermore, the new effective NN-interaction-entitled R3Y potential [13,15], analogous to the M3Y form [17], can be derived from the relativistic mean-field (RMF) Lagrangian. This interaction depends on the relativistic force parameters, the coupling constant of the interacting mesons and their masses [13,15]. One can find various potentials in more details and use them in future studies for general and up-to-date views on the subject, by referring, for instance, to Refs. [16,18−20]. Furthermore, the nuclear potential is quite important in the studies of elastic scattering of light and heavy-ion systems, in particular for the simple one-dimensional barrier penetration model (BPM) of fusion reaction, barrier energy, radius, curvature, and Coulomb potential [21,22]. The widely used method to obtain the nuclear potential by integrating an NN-interaction over the matter distributions of the two colliding nuclei is called the double-folding model [21,22]. It produces a nucleus-nucleus optical potential for further use in various studies including radioactive decays [13,15,23]. Furthermore, another model called the São Paulo potential [24−26] attempts to partly overcome the difficulties of some specific parameter adjustments to the experimental data. The potential is theoretically founded on the Pauli non-locality, which arises from quantum exchange effects and has been experimentally tested in a wide variety of mass and energy regions in combination with different theoretical approaches. It does not contain any adjustable parameter and, combined with coupled channel (CC) calculations, provides a powerful tool for predicting cross sections for quite different systems and energies around the Coulomb barrier [27].

The nuclear potential is influenced by various factors, for example, the nuclear shape, orientation and internal arrangement of the nucleons within a nucleus. Nuclei may have extreme shapes, higher multipole deformations (β_λ), or extreme neutron-to-proton ratios, where the term β_λ determines the deviation or change from the spherical shape of the fusing nucleus. The contribution of these higher multipole deformations along with the choice of appropriate orientations leads to exotic shapes, which in turn are immensely useful to understand various aspects of nuclear structure and dynamics at extreme conditions, thereby providing important information for future experiments. Experiments can also create such nuclei, using artificially induced fusion or nucleon transfer reactions, employing ion beams from an accelerator. With the advent of new generation of accelerators, which are capable of accelerating not only heavy ions but also radioactive ion beams (RIB) to produce exotic nuclei, a vast amount of experimental data are being accumulated, from lighter nuclei to some superheavy ones. The collisions between the deformed and oriented nuclei have been investigated in the early 1980s, and Greiner [28] suggested that oriented ^{238}U + ^{238}U collisions could lead to a very long-lived (life time of $\sim 10^{-20}$ seconds) giant molecule. Also, if the projectile is loosely bound, then its breakup during the fusion process can change the dynamics involved. So, instead of a one-dimensional barrier, there can be multiple barriers depending upon the channels. The proper description of a fusion process, therefore, is essentially demanding to single out the relevant CCs involved [29].

The chapter is organized as follows. Section 5.2 gives a brief description of the theoretical formalisms for the total nucleus-nucleus interaction potential along with the fusion cross section within the Wong formula. The method of calculations and the results for the nuclear interaction potential are discussed in Section 5.3. The factors affecting the sub-barrier fusion regarding nuclear interaction potential and the fusion cross sections for various systems are discussed in Sections 5.4 and 5.5, respectively. Section 5.6 includes a summary along with a few concluding remarks.

5.2   Theoretical Formalisms

The total interaction potential for deformed and oriented nuclei [30] is given by the sum of nuclear, Coulomb, and centrifugal interactions:

$$V_{Total} = V_N\left(R, A_i, \beta_{\lambda i}, T, \theta_i\right) + V_C\left(R, Z_i, \beta_{\lambda i}, T, \theta_i\right) + V_\ell\left(R, A_i, \beta_{\lambda i}, T, \theta_i\right), \qquad (5.1)$$

with $V_\ell = \frac{\hbar^2 \ell(\ell+1)}{2 I_{NS}}$. Here, $I_{NS}(= \mu R^2)$ is a non-sticking moment-of-inertia. The Coulomb potential for a multipole-multipole interaction between two separated nuclei is given by [31,32]

$$V_C = \frac{Z_1 Z_2 e^2}{R} + 3 Z_1 Z_2 e^2 \sum_{\lambda, i=1,2} \frac{R_i^\lambda\left(\alpha_i, T\right)}{(2\lambda+1)\, R^{\lambda+1}} Y_\lambda^{(0)}(\theta_i) \left[\beta_{\lambda i} + \frac{4}{7} \beta_{\lambda i}^2 Y_\lambda^{(0)}(\theta_i)\right]. \qquad (5.2)$$

From Eq. (5.1), one can get the ℓ-dependent barrier height V_B^ℓ, position R_B^ℓ, and the curvature $\hbar \omega_\ell$ for each value of ℓ. The temperature T is related to the incoming center-of-mass energy $E_{c.m.}$ or the compound nucleus (CN) excitation energy E_{CN}^* via the entrance channel Q_{in} value, as

$$E_{CN}^* = E_{c.m.} + Q_{in} = \frac{1}{a} A T^2 - T \quad (\text{T in MeV}), \qquad (5.3)$$

with $a = 9$ or 10, respectively, for intermediate mass or superheavy systems. $Q_{in} = B_1 + B_2 - B_{CN}$, with binding energies Bs taken from [33].

The temperature effects in both V_C and V_N are introduced via the radius vectors of two nuclei, as follows:

$$R_i\left(\alpha_i, T\right) = R_{0i}(T)\left[1 + \sum_\lambda \beta_{\lambda i} Y_\lambda^{(0)}(\alpha_i)\right] \qquad (5.4)$$

with the temperature dependence of R_{0i}, as in Ref. [34],

$$R_{0i}(T) = \left[1.28 A_i^{\frac{1}{3}} - 0.76 + 0.8 A_i^{-\frac{1}{3}}\right]\left(1 + 0.0007 T^2\right). \qquad (5.5)$$

Here, θ_i is the angle between the nuclear symmetry axis and the collision Z-axis, called the orientation angle and measured in the counterclockwise direction, and angle α_i is the angle between the symmetry axis and the radius vector of the colliding nucleus, measured in the clockwise direction from the symmetry axis (see, e.g., Figure 1 of Ref. [35]). One of the main contributions to the interaction potential is from the nuclear potential (V_N) and is calculated from various theoretical approaches. The details of various nuclear potentials are given below.

5.2.1 Semi-classical Extended Thomas Fermi (ETF) Model

The nucleus-nucleus interaction potential in SEDF based on a semi-classical ETF model [36−43] is

$$V_N(R) = E(R) - E(\infty) = \int H(\vec{r})\,d\vec{r} - \left[\int H_1(\vec{r})d\vec{r} + \int H_2(\vec{r})d\vec{r}\right] \tag{5.6}$$

where the Skyrme Hamiltonian density is

$$
\begin{aligned}
H\left(\rho, \tau, \vec{J}\right) &= \frac{\hbar^2}{2m}\tau + \frac{1}{2}t_0\left[\left(1 + \frac{1}{2}x_0\right)\rho^2 - \left(x_0 + \frac{1}{2}\right)\left(\rho_n^2 + \rho_p^2\right)\right] \\
&+ \frac{1}{12}t_3\rho^{\alpha_0}\left[\left(1 + \frac{1}{2}x_3\right)\rho^2 - \left(x_3 + \frac{1}{2}\right)\left(\rho_n^2 + \rho_p^2\right)\right] \\
&+ \frac{1}{4}\left[t_1\left(1 + \frac{1}{2}x_1\right) + t_2\left(1 + \frac{1}{2}x_2\right)\right]\rho\tau - \frac{1}{4}\left[t_1\left(x_1 + \frac{1}{2}\right) - t_2\left(x_2 + \frac{1}{2}\right)\right] \\
&\times \left(\rho_n\tau_n + \rho_p\tau_p\right) + \frac{1}{16}\left[3t_1\left(1 + \frac{1}{2}x_1\right) - t_2\left(1 + \frac{1}{2}x_2\right)\right]\left(\vec{\nabla}\rho\right)^2 \\
&- \frac{1}{16}\left[3t_1\left(x_1 + \frac{1}{2}\right) - t_2\left(x_2 + \frac{1}{2}\right)\right]\left[\left(\vec{\nabla}\rho_n\right)^2 + \left(\vec{\nabla}\rho_p\right)^2\right] \\
&- \frac{1}{2}W_0\left(\rho\vec{\nabla}\cdot\vec{J} + \rho_n\vec{\nabla}\cdot\vec{J}_n + \rho_n\vec{\nabla}\cdot\vec{J}_p\right) \\
&- A\left[\frac{1}{16}\left(t_1x_1 + t_2x_2\right)\vec{J}^2 - \frac{1}{16}\left(t_1 - t_2\right)\left(\vec{J}_p^2 + \vec{J}_n^2\right)\right].
\end{aligned}
\tag{5.7}
$$

Here, $\rho = \rho_n + \rho_p$ is the nuclear density, $\tau = \tau_n + \tau_p$ is the kinetic energy density, and $\vec{J} = \vec{J}_n + \vec{J}_p$ is the spin-orbit density for nucleon mass m. The quantities x_j, t_{jj} (j=0,1,2), x_{3i}, t_{3i}, α_i (i=1,2,3), W_0, and A are recent Skyrme interaction parameters [36,37], for example, GSkI, GSkII, SSk, KDE0, and KDE forces. More details of these force parameter and energy density functionals can be found in Refs. [36−39].

The kinetic energy density, taken up to the second-order terms for reasons of numerical convergence in ETF [40], is ($q = n$ or p)

$$
\begin{aligned}
\tau_q(\vec{r}) &= \frac{3}{5}(3\pi^2)^{2/3}\rho_q^{5/3} + \frac{1}{36}\frac{(\vec{\nabla}\rho_q)^2}{\rho_q} + \frac{1}{3}\Delta\rho_q + \frac{1}{6}\frac{\vec{\nabla}\rho_q \cdot \vec{\nabla}f_q + \rho_q\Delta f_q}{f_q} \\
&- \frac{1}{12}\rho_q\left(\frac{\vec{\nabla}f_q}{f_q}\right)^2 + \frac{1}{2}\rho_q\left(\frac{2m}{\hbar^2}\right)^2\left(\frac{W_0}{2}\frac{\vec{\nabla}(\rho + \rho_q)}{f_q}\right)^2.
\end{aligned}
\tag{5.8}
$$

with f_q as the effective mass form factor,

$$
\begin{aligned}
f_q(\vec{r}) &= 1 + \frac{2m}{\hbar^2}\frac{1}{4}\left\{t_1\left(1 + \frac{x_1}{2}\right) + t_2\left(1 + \frac{x_2}{2}\right)\right\}\rho(\vec{r}) \\
&- \frac{2m}{\hbar^2}\frac{1}{4}\left\{t_1\left(x_1 + \frac{1}{2}\right) - t_2\left(x_2 + \frac{1}{2}\right)\right\}\rho_q(\vec{r}).
\end{aligned}
\tag{5.9}
$$

Note that both τ_q and f_q are each function of $\rho(r)$ only. The spin-orbit density J is a purely quantal property, and hence, it has no contribution at the lowest (TF) order. However, the second-order contribution gives

$$\vec{j}_q(\vec{r}) = -\frac{2m}{\hbar^2}\frac{1}{2}W_0\frac{1}{f_q}\rho_q\vec{\nabla}(\rho + \rho_q), \tag{5.10}$$

which is also a function of the nuclear density. The densities are calculated using a frozen approximation. Now, the total Hamiltonian density can be written as

$$V_N(R) = \int \left\{ H\left(\rho, \tau, \vec{J}\right) - \left[H_1\left(\rho_1, \tau_1, \vec{J_1}\right) + H_1(\rho_2, \tau_2, \vec{J_2}) \right] \right\} d\vec{r} = V_P(R) + V_J(R),$$

(5.11)

where $V_P(R)$ and $V_J(R)$ are the spin-orbit density-independent and -dependent parts of the interaction potential, respectively. The temperature dependence is introduced in the formalism, by using for the nuclear density ρ_i of each nucleus, the T-dependent, two-parameter Fermi density (FD) distribution [41], which for the slab approximation is given by

$$\rho_i(z_i) = \rho_{0i}(T) \left[1 + exp\left(\frac{z_i - R_i(T)}{a_i(T)} \right) \right]^{-1}, -\infty \le z \le \infty$$

(5.12)

with $Z_2 = R - Z_1 = [R_1(\alpha_1) + R_2(\alpha_2) + s] - Z_1$, and central density,

$$\rho_{0i}(T) = \frac{3A_i}{4\pi R_i^3(T)} \left[1 + \frac{\pi^2 a_i^2(T)}{R_i^2(T)} \right]^{-1},$$

(5.13)

of nuclear radii R_i and surface thickness parameters a_i obtained by fitting the experimental data to the polynomials in nuclear mass A [41–43]. The temperature dependencies in R_{0i} and a_i are taken from Ref. [44]. Furthermore, the nuclear potential (V_N) can be calculated from a phenomenological approach as well as the energy density formalism. The details of various phenomenological nuclear potentials, which are based on proximity theorem and are denoted by Prox 1977, Prox 1988, Prox 2000, mod-Prox 1988, Bass 1980, CW 1976, BW 1991, Denisov 2002, Akyuz-Winther, etc., can be found in Refs. [45,46]. Also more details of the semi-classical approach of SEDF in the ETF approach for calculating nuclear potentials within various Skyrme forces can be found in Ref. [43].

5.2.2 Relativistic Mean-Field Approach

In the case of microscopic self-consistent mean-field calculations, the RMF theory is a standard tool for investigation of nuclear structure and reaction phenomena. It starts with the nucleons as Dirac spinors interacting through different meson fields. The original Lagrangian of Walecka has undergone several modifications to take care of various limitations, and a recent successful relativistic Lagrangian density for nucleon-meson many-body systems [47–55] is expressed as follows:

$$\begin{aligned}
\mathfrak{L} = & \overline{\psi}_i \left\{ i\gamma^\mu \partial_\mu - M \right\} \psi_i + \frac{1}{2} \partial^\mu \sigma \partial_\mu \sigma - \frac{1}{2} m_\sigma^2 \sigma^2 \\
& - \frac{1}{3} g_2 \sigma^3 - \frac{1}{4} g_3 \sigma^4 - g_s \overline{\psi}_i \psi_i \sigma - \frac{1}{4} \Omega^{\mu\vartheta} \Omega_{\mu\vartheta} \\
& + \frac{1}{2} m_\omega^2 V^\mu V_\mu - g_\omega \overline{\psi}_i \gamma^\mu \psi_i V_\mu - \frac{1}{4} \mathbf{B}^{\mu\vartheta} \cdot \mathbf{B}_{\mu\vartheta} \\
& - \frac{1}{2} m_\rho^2 \mathbf{R}^\mu \cdot \mathbf{R}_\mu - g_\rho \overline{\psi}_i \gamma^\mu \boldsymbol{\tau} \psi_i \cdot \mathbf{R}_\mu \\
& - \frac{1}{4} F^{\mu\vartheta} F_{\mu\vartheta} - e\overline{\psi}_i \gamma^\mu \frac{(1 - \tau_{3i})}{2} \psi_i A_\mu.
\end{aligned}$$

(5.14)

The ψ_i are the Dirac spinors for the nucleons. The iso-spin and the third component of the iso-spin are denoted by τ and τ_3, respectively. Here, g_σ, g_ω, g_ρ, and $e^2/4\pi$ are the coupling constants for the σ-, ω-, and ρ-mesons and photon, respectively. The constant g_2 and g_3 multiply the cubic and quadric terms of the self-interacting non-linear σ-meson field.

The masses of the σ-, ω-, and σ-mesons and nucleons are m_σ, m_ω, m_ρ, and M, respectively. The quantity A_μ stands for the electromagnetic field. The vector field tensors for the ω_μ, ρ_μ, and photon are given by

$$F^{\mu\vartheta} = \partial_\mu A_\vartheta - \partial_\vartheta A_\mu,$$
$$\omega^{\mu\vartheta} = \partial_\mu \omega_\vartheta - \partial_\vartheta \omega_\mu,$$

and

$$\vec{B}^{\mu\vartheta} = \partial_\mu \vec{\rho}_\vartheta - \partial_\vartheta \vec{\rho}_\mu, \tag{5.15}$$

respectively. From the above Lagrangian density, we obtain the field equations for the Dirac nucleons and the mesons:

$$(-i\alpha \cdot \nabla + \beta (M + g_\sigma) + g_\omega\omega + g_\rho\tau_3\rho_3)\,\psi = \epsilon\psi,$$
$$(-\nabla^2 + m_\sigma^2)\,\sigma(r) = -g_\sigma\rho_s(r) - g_2\sigma^2(r) - g_3\sigma^3(r),$$
$$(-\nabla^2 + m_\omega^2)\,V(r) = -g_\omega\rho_\omega(r),$$
$$(-\nabla^2 + m_\rho^2)\,\rho(r) = -g_\rho\rho_3(r). \tag{5.16}$$

In the limit of one-meson exchange, for a static baryonic medium, the solutions of the nucleon-nucleon potential for scalar and vector fields are given by

$$V_\sigma = -\frac{g_\sigma^2}{4\pi}\frac{e^{-m_\sigma r}}{r} + \frac{g_2^2}{4\pi}re^{-2m_\sigma r} + \frac{g_3^2}{4\pi}\frac{e^{-3m_\sigma r}}{r}$$

and

$$V_\omega = +\frac{g_\omega^2}{4\pi}\frac{e^{-m_\omega r}}{r}, V_\rho = +\frac{g_\rho^2}{4\pi}\frac{e^{-m_\rho r}}{r}. \tag{5.17}$$

The total effective NN-interaction is obtained from the scalar and vector parts of the meson fields. We call the recently developed relativistic NN-interaction potential analogous to the M3Y form [48] the R3Y potential. Here, the R3Y potential is derived for the NL3* force parameter, which is able to predict the nuclear matter as well as the properties of the finite nuclei at very high isospin asymmetries [47−51]. The relativistic effective nucleon-nucleon interaction (V_{eff}^{R3Y}) for NL3* force along with the single-nucleon exchange effects is as follows [13,14,47−55]:

$$V_{eff}^{R3Y}(r) = \frac{g_\omega^2}{4\pi}\frac{e^{-m_\omega r}}{r} + \frac{g_\rho^2}{4\pi}\frac{e^{-m_\rho r}}{r} - \frac{g_\sigma^2}{4\pi}\frac{e^{-m_\sigma r}}{r}$$
$$+ \frac{g_2^2}{4\pi}re^{-2m_\sigma r} + \frac{g_3^2}{4\pi}\frac{e^{-3m_\sigma r}}{r} + J_{00}(E)\,\delta(s) \tag{5.18}$$

On the other hand, the M3Y effective interaction, obtained from a fit of the G-matrix elements based on the Reid-Elliott soft-core NN-interaction [17], in an oscillator basis, is the sum of three Yukawa's (M3Y) with ranges 0.25 fm for a medium-range attractive part, 0.4 fm for a short-range repulsive part, and 1.414 fm to ensure a long-range tail of the one-pion-exchange potential (OPEP). The widely used M3Y effective interaction (V_{eff}^{M3Y}) is given by

$$V_{eff}^{M3Y}(r) = 7999\frac{e^{-4r}}{4r} - 2134\frac{e^{-2.5r}}{2.5r}, \tag{5.19}$$

where the ranges are in fm and the strengths are in MeV. Note that Eq. (5.19) represents the spin- and isospin-independent parts of the central component of the effective NN-interaction, and that the OPEP contribution is absent here. One can find more details in Refs. [13,14]. The nuclear interaction potential, $V_N(R)$, between the projectile (p) and the target (t) nuclei, with the respective RMF (NL3*) calculated nuclear densities ρ_p and ρ_t, is obtained by using the well-known double-folding procedure [17]:

$$V_N\left(\vec{R}\right) = \int \rho_p(\vec{r}_p)\rho_t(\vec{r}_t)V_{eff}\left(\left|\vec{r}_p - \vec{r}_t + \vec{R} \equiv r\right|\right) d^3r_p d^3r_t, \qquad (5.20)$$

The M3Y and the recently developed M3Y interaction potential (V_{eff}), proposed in Refs. [13,14], along with the zero-range pseudo-potential representing the single-nucleon exchange effects (EX), is used.

5.2.3   São Paulo Potential

Another widely used model for the heavy-ion interaction is the Sao Paulo potential. The model is based on the effects of the Pauli non-locality that arises from the exchange of nucleons between the target and the projectile. The Sao Paulo potential is a product of two terms: a velocity-dependent function multiplied by the folding potential. The folding potential depends on the densities of the partners in the collision. Thus, to obtain a parameter-free model, a systematic description of the nuclear densities is obtained.

Within this context, the model has been successfully used in the elastic scattering data analysis and also the descriptions of the heavy-ion fusion process. A normalized version of the São Paulo optical potential [24−27] reads as follows:

$$V_{SP}\left(R, E\right) = V_F\left(R\right)e^{-4v^2/c^2}, \qquad (5.21)$$

where $V_F\left(R\right) \approx V_n\left(R\right)$ is the double-folding potential given in Eq. (5.20), and c is the speed of light. The relative velocity squared of the colliding ions is given by

$$v^2\left(R, E\right) = \frac{2}{\mu}\left[E - V_C\left(R\right) - V_{SP}(R, E)\right]. \qquad (5.22)$$

Therefore, in this context, the effect of the Pauli non-locality is equivalent to a velocity-dependent nuclear interaction in Eq. (5.21). Another possible interpretation is that the local-equivalent potential may be associated directly with the folding potential in Eq. (20), with an effective nucleon-nucleon interaction, which depends on the relative speed (v) between the nucleons. More details of the model can be found in Ref. [26].

5.2.4   Coupled Channel Approach

Under ingoing-wave boundary conditions (IWBC), one can calculate the fusion probabilities by solving the corresponding coupled-channel equations. The coupled-channel formalism for direct reaction processes given by Austern [56] expands the total wave function for the internal state of the projectile ϕ_β and the radial wave functions χ_β that account for the relative motion between projectile and target as

$$\psi^{(+)} = \sum_\beta \frac{\chi_\beta(R)}{R}\phi_\beta. \qquad (5.23)$$

This leads to a set of coupled equations for the radial wave functions:

$$\frac{d^2\chi_\beta}{dR^2} + \frac{2\mu_\beta}{\hbar^2}\left[E_\beta - V_\beta(R)\right]\chi_\beta = \frac{2\mu_\beta}{\hbar^2}\sum_{\alpha \neq \beta} V_{\beta\alpha}^{coup}(R)\chi_\alpha \qquad (5.24)$$

In this expression, V_β is the interaction potential, while μ_β is the reduced mass and E_β is the relative energy for a given channel β. Here, we consider only two channels, the incoming channel and breakup channel following the fusion, without the ejected particle for the nuclear proximity potential of the Broglia and Winther [57] parameterization. The coupling potential V^{coup} is given as a derivative Woods-Saxon form with the same radius and diffuseness of the proximity potential for the incoming channel. The strength is set to 10% of the strength of the same proximity potential. The total transmission probability is then given by

$$T = \sum_\beta \left| T_\beta^2 \right| = |t_1|^2 + \frac{v_2}{v_1} |t_2|^2 , \tag{5.25}$$

where v_1 and v_2 are the velocities corresponding to channels 1 and 2, respectively. The fusion cross section, in terms of partial waves, is given by

$$\sigma = \sum_{\ell=0}^{\ell_{max}} \sigma_\ell = \frac{\pi \hbar^2}{2\mu_1 E} \sum_{\ell=0}^{\ell_{max}} (2\ell + 1)\, T_\ell\,(E) . \tag{5.26}$$

The probability of transmission for the partial wave can also be approximated by a shift of energy:

$$T_\ell \cong T_0 \left[E - \frac{\ell(\ell+1)\hbar^2}{2\mu_1 r_0^2} \right] . \tag{5.27}$$

where r_0 is the position of the barrier for the s wave [29]. For further details of the model, one can refer to Refs. [29,58].

5.2.5 The ℓ-Summed Extended Wong Model and Wong Formula

The fusion cross section is calculated using the Wong formula and its extended version. According to Wong [31], regarding ℓ-partial waves, the fusion cross section for two deformed and oriented nuclei, colliding with $E_{c.m.}$, is

$$\sigma\,(E_{c.m.}, \theta_i) = \frac{\pi}{k^2} \sum_{\ell=0}^{\ell_{max}} (2\ell + 1) P_\ell\,(E_{c.m.}, \theta_i) , \tag{5.28}$$

with $k = \sqrt{\frac{2\mu E_{c.m.}}{\hbar^2}}$ and μ as the reduced mass. P_ℓ is the transmission coefficient for each ℓ which describes the penetration of the barrier $V_{Total}(R_\ell, E_{c.m.}, \theta_i)$, given by Eq. (5.1). Using the Hill-Wheeler [59] approximation, the penetrability P_ℓ, in terms of its barrier height $V_B^\ell(E_{c.m.}, \theta_i)$, and curvature $\hbar\omega_\ell(E_{c.m.}, \theta_i)$, is

$$P_\ell = \left[1 + \exp\left(\frac{2\pi \left\{ V_B^\ell\,(E_{c.m.}, \theta_i) - E_{c.m.} \right\}}{\hbar\omega_\ell(E_{c.m.}, \theta_i)} \right) \right]^{-1} \tag{5.29}$$

with $\hbar\omega_\ell$ evaluated at the barrier position $R = R_B^\ell$ corresponding to the barrier height V_B^ℓ, given as

$$\hbar\omega_\ell\,(E_{c.m.}, \theta_i) = \hbar \left[d^2 V_T^\ell(R)/dR^2 |_{R=R_B^\ell} /\mu \right]^{1/2} \tag{5.30}$$

and R_B^ℓ obtained from the condition,

$$\left| dV_T^\ell(R)/dR \right|_{R=R_B^\ell} = 0 \tag{5.31}$$

Noting that the ℓ-summed expression Eq. (5.28) uses the ℓ-dependent potentials (Eq. (5.1)), its ℓ-summation is carried out for the ℓ_{max} determined empirically for a best fit to the measured cross section, and the angles θ_i are integrated to give the fusion cross section:

$$\sigma\left(E_{c.m.}\right) = \int_{\theta_i}^{\pi/2} \sigma\left(E_{c.m.}, \theta_i\right) sin\theta_1 d\theta_1 sin\theta_2 d\theta_2. \qquad (5.32)$$

It is to be noted that these extracted ℓ_{max} values are in agreement with the sharp cut-off model at above barrier energies. Instead of solving Eq. (5.28) explicitly, which requires the complete ℓ-dependent potentials $V_T^\ell(R, E_{c.m.}, \theta_i)$, Wong [31] carried out the ℓ-summation in this equation *approximately* under specific conditions:

$$
\left.
\begin{aligned}
&(i)\ \hbar\omega_\ell \approx \hbar\omega_0 \\[2mm]
&(ii)\ V_B^\ell \approx V_B^0 + \tfrac{\hbar^2\ell(\ell+1)}{2\mu R_B^0}
\end{aligned}
\right\}
\qquad (5.33)
$$

Using these approximations, and replacing the ℓ-summation in Eq. (5.28) by an integral, gives, on integration, the $\ell = 0$ barrier-based Wong formula:

$$\sigma\left(E_{c.m.}, \theta_i\right) = \frac{R_B^0 \hbar\omega_0}{2E_{c.m.}} \ln\left\{ 1 + \exp\left[\frac{2\pi}{\hbar\omega_0}(E_{c.m.} - V_B^0) \right] \right\}, \qquad (5.34)$$

which, by using Eq. (5.32), would also give $\sigma(E_{c.m.})$.

5.3   Results and Discussions

In this section, our interest is to determine the performance of various phenomenological models using effective non-relativistic and relativistic interactions on the fusion characteristics. Before turning to the fusion cross section, it is essential and also crucial to know the nucleus-nucleus interaction (i.e., nuclear potential V_N of Eq. (5.1)) from these interactions. Furthermore, we have also discussed the recently developed relativistic R3Y nucleon-nucleon interaction potential for selective NL3* force parameter along with the widely used M3Y potential. The calculations corresponding to the relativistic mean field are limited to the spherical symmetry to generate the nucleus-nucleus interaction potential. One may consider the coupling between fusion and other degrees of freedom to generate a multidimensional potential barrier, which enhanced the fusion probabilities.

5.3.1   Nuclear Potential from Skyrme Energy Density Formalism

The semi-classical approach of the SEDF [60] covers a huge set of interaction forces that are represented by the Skyrme Hamiltonian density which contains different sets of parameters to explain the finite nuclear properties. The use of different Skyrme forces for calculating the interaction barriers leads to different barrier characteristics. As a representative case, we show the total interaction potential V_T for ^{64}Ni $+ ^{100}$Mo reaction for various Skyrme forces in Figure 5.1. The results illustrate for the system at single $E_{c.m.}$ and fixed θ_i values, the difference between various Skyrme forces in the frozen density approximation at $\ell = 0$. From the figure, it is clear that the barrier characteristics (height, position, and curvature) are force dependent, and hence could be used to account for the fusion excitation function. Moreover, once can observe that the Skyrme parameter sets SIII and SkM* give identical barriers and hence provide identical fits to fusion cross sections.

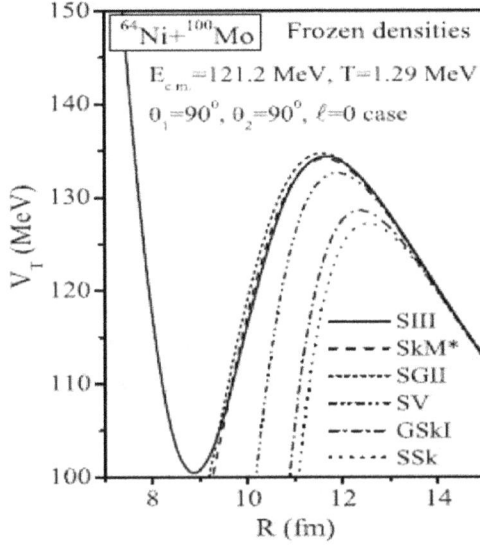

FIGURE 5.1
Total interaction potentials calculated for ^{64}Ni + ^{100}Mo reaction at a fixed $E_{c.m.}$ and fixed orientations θ_i at $\ell = 0$, using SIII, SkM*, SGII, SV, GSkI, and SSk in the frozen density approximation.

5.3.2 Nuclear Potential from Relativistic Mean-Field Theory

The nuclear interaction potential V_n (R) between the projectile (p) and target (t) nuclei can be calculated using the well-known double-folding procedure as discussed in Eq. (5.20) for RMF matter densities ρ_p and ρ_t for M3Y and recently developed relativistic R3Y potential. The R3Y interaction is obtained for the NL3* force parameter in the present analysis [55], in which an effective Lagrangian is taken to describe the nucleons interaction through effective mesons and electromagnetic fields. It is worth mentioning that the applicability of our newly introduced R3Y interaction potentials has been used for the radioactivity studies of several highly unstable proton- and neutron-rich nuclei using preformed cluster decay model (PCM) of Gupta and co-workers [13]. Furthermore, the non-linear terms in the σ-field play an important role in the nuclear matter study and the detailed nuclear structure inherited by the density when calculating the proton and cluster decay properties (mostly a surface phenomenon) [13,14].

As a representative case, the results obtained for the total interaction potentials along with the Coulomb potential V_C(R) and the nucleus-nucleus interactions without Coulomb for M3Y+EX and R3Y+EX interactions for ^{58}Ni + ^{58}Ni system are displayed in Figure 5.2. From the figure, we note that the natures of the total $V_T(R)$ and the nuclear $V_n(R)$ potentials are similar for both the R3Y+EX and M3Y+EX NN-interactions. Quantitatively, both the nuclear potentials obtained from M3Y and R3Y differ significantly particularly in the central region, and this difference is reduced as a function of the radial distance. Furthermore, the height of the barrier for M3Y NN-interaction is a bit higher as compared to the R3Y case (shown more clearly in the inset of Figure 5.2). For example, the R3Y+EX is being more attractive with a barrier height lower by a few keV, compared to the M3Y+EX NN-interaction, as is illustrated in the inset of Figure 5.2. More details of the studies can be found in Ref. [23].

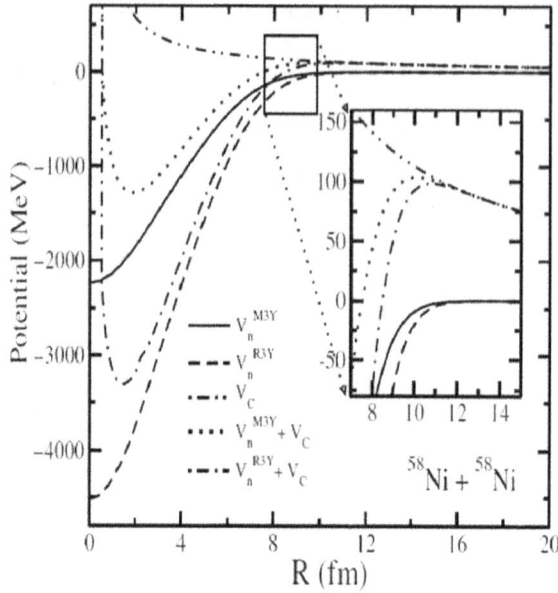

FIGURE 5.2

Total interaction potentials $V_T(R) = V_n(R) + V_C(R)$ for the Ni-based reaction such as ^{58}Ni + ^{58}Ni are obtained for the M3Y and the R3Y interactions with the NL3* densities.

5.3.3 Nuclear Potentials for São Paulo Potential

The results obtained for the total interaction potentials from the double-folding approach and ℓ-dependent local-equivalent São Paulo (V_{SP}) potentials for the $\alpha + {}^{58}$Ni system at $E_{lab} = 139$ MeV are shown Figure 5.3. From the figure, we note that the nature of the total potentials is similar for both the double-folding and ℓ-dependent local-equivalent São Paulo (V_{SP}) potentials. Quantitatively, the nuclear potentials obtained from the double-folding approach differ significantly from the São Paulo potential particularly in the central region,

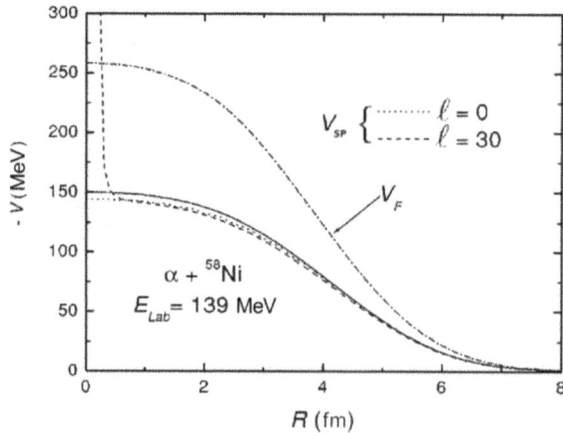

FIGURE 5.3

Double-folding (V_F) and ℓ-dependent local-equivalent São Paulo (V_{SP}) potentials for the $\alpha + {}^{58}$Ni system at $E_{lab} = 139$ MeV. More details can be found in Ref. [26].

and this difference diminishes as a function of the radial distance. Furthermore, the height of the barrier for São Paulo potential is a bit higher as compared to the double-folding potential (see Figure 5.3). More details of the studies can be found in Ref. [26].

5.3.4 Various Phenomenological Nuclear Potentials

In this section, we study the effect of various proximity potentials of different isospin- and asymmetry-dependent parameters on the total interaction potential for the fusion of ^{24}Ne + ^{208}Pb, which leads to the formation of ^{232}U at optimum orientations [61] along with their deformations up to β_2 [33] using various versions of nuclear potentials, i.e., Prox 1977, Prox 1988, Prox 2000, Bass 1980, CW 1976, BW 1991, Denisov 2002, and mod-Prox 1988. For more details on these potentials, see Ref. [45]. Since the interaction potential is the main recipe of all the theoretical models, we have covered a wide range of barrier characteristics, as shown clearly in Figure 5.4. For example, Prox 2000 gives the highest barrier, and mod-Prox 1988 gives the lowest barrier, while the barriers corresponding to other potentials lie in between these two interaction limits. It is clear from this figure that the potentials whose barrier characteristics are closer to Prox 1977 and Prox 1988 are expected to behave similarly to these potentials. Also, the barrier is much lower for mod-Prox 1988. The barrier characteristics of CW 1976, BW 1991, and Bass 1980 behave similarly to Prox 1977 and Prox 1988 and contrarily to Prox 2000 and Denisov 2002, and hence the first five of these are expected to behave similarly. For more details on this work, we refer the reader to Refs. [45,62].

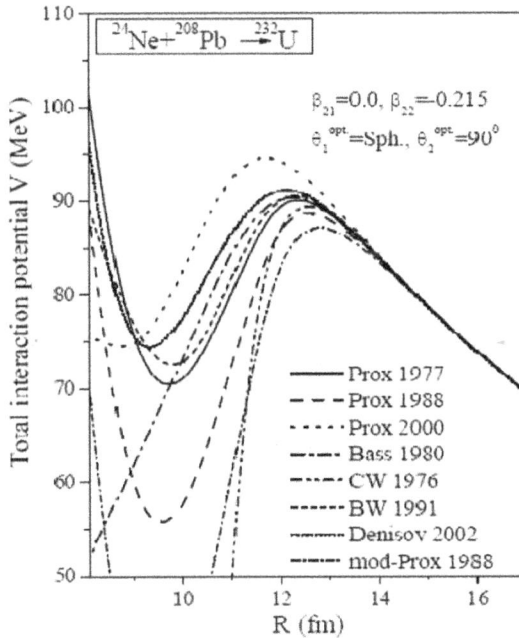

FIGURE 5.4
Total interaction potentials calculated for the fusion of ^{24}Ne + ^{208}Pb → ^{232}U at fixed optimum orientation $\theta_1^{opt} = $ Sph. and $\theta_2^{opt} = 90°$, using various nuclear proximity potentials [45].

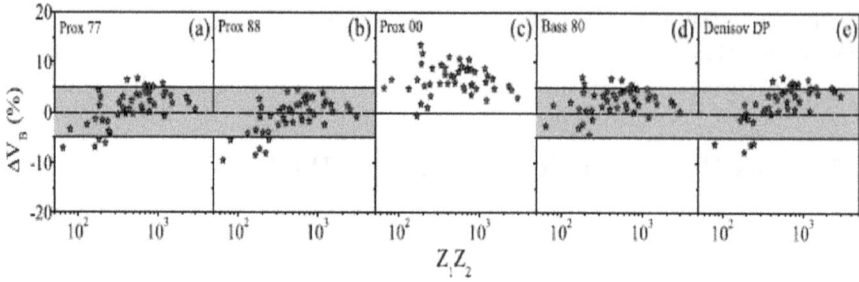

FIGURE 5.5
The percentage (%) of change in barrier height as a function of the product of charges $Z_1 Z_2$ for different versions of the proximity potential, with deformation β_2 at "optimum" orientations [62].

Furthermore, we have also carried out an extensive comparative study of fusion barrier characteristics for 52 colliding nuclei with mass asymmetry in the range of 0–0.96 for different proximity potentials [63]. We have mainly confined the study to those reactions that involve either target or projectile or both as deformed nuclei reported in Ref. [63]. Figure 5.5 shows the percentage change of the fusion barrier heights ΔV_B (%) concerning the empirical barrier heights obtained from experimental data, as a function of the product of charges $Z_1 Z_2$. This percentage change is defined as

$$\Delta V_B\,(\%) = \frac{V_B^{theor} - V_B^{expt}}{V_B^{expt}} \times 100.$$

The deformations are taken up to β_2 along with "optimum" orientation [62] in the calculations. It may be noted that the position and height of the Coulomb barrier could not be measured directly in any experiment. Therefore, the average fusion excitation functions are obtained by fitting the experimental data with the one-dimensional BPM, and the relevant characteristics of the barrier, i.e., its radius, width, and height, are extracted.

There is an improvement in the fusion barrier heights as compared to the spherical approximation to the nuclei [63]. Prox77, Prox88, Bass80, and Denisov DP reproduce the empirical barrier heights within ±5% approximately, whereas Prox00 comes in the approximate range of ±10%. This shows that barriers calculated using Prox77, Prox88, Bass80, and Denisov DP have accuracy up to ~95%. The effect of the deformation on barrier characteristics as incorporated through the use of different proximity potentials provides an interesting input to understanding the formation process in a variety of nuclear reactions. This information regarding modification of barrier characteristics could be of extreme importance for the overall understanding of the heavy-ion reaction mechanism.

5.4   Factors Affecting Sub-barrier Fusion

For a high yield of fusion below the Coulomb barrier energies, i.e., at comparatively low energies, where the structural effects of nuclei dominate, one needs to incorporate different degrees of freedom. In the following sub-sections, we will show how these degrees of freedom influence the fusion by modifying the interaction potential.

5.4.1 Deformation and Orientation

The shapes of nuclei play an important role in nuclear fusion studies. In Ref. [65], it has been shown that the sub-barrier fusion of nuclei is enhanced with deformation. Furthermore, the presence of deformation affects the barrier characteristics, i.e., the Coulomb barrier, position, and corresponding width as well. In Figure 5.6, we show how the orientation degree of freedom θ modifies the barrier characteristics of interacting deformed nuclei. Figure 5.6(a) and (b) shows the variation of the nuclear proximity potential and the corresponding total interaction potential for the ^{48}Ca + ^{96}Zr reaction at a different orientation (θ_2) while keeping θ_1 fixed at 0°, as ^{48}Ca is spherical, at $T = 0$ MeV using the proximity potential of Blocki et al. [66] denoted by Prox77. It is clear from this figure that with the change of orientation angle, the depth of nuclear proximity potential, as well as its barrier characteristics, changes significantly. Thus, deformation and an orientation degree of freedom introduce barrier modification effect [43], which can play a significant role at near- and sub-barrier energies. For more details on this work, see Ref. [64].

5.4.2 Surface Energy Constant (γ) and Angular Momentum (ℓ)

A large N/Z ratio must be taken into account when considering reactions with drip-line or heavy and superheavy nuclei. In such case, the strong nuclear interaction must include the isospin asymmetry effect of the colliding nuclei, which can be accounted by for a slight adjustment of the coefficient of the nuclear surface energy γ_0 in the proximity potential. Here, we have considered the case of the Prox 1988 potential to account for the fusion excitation function, where the barrier height is modified to fit the experimental data and is named mod-Prox 1988. The change in γ leads to change in the barrier characteristics, i.e., barrier height (V_B^ℓ), position (R_B^ℓ), and oscillator frequency ($\hbar\omega_\ell$). The modification of

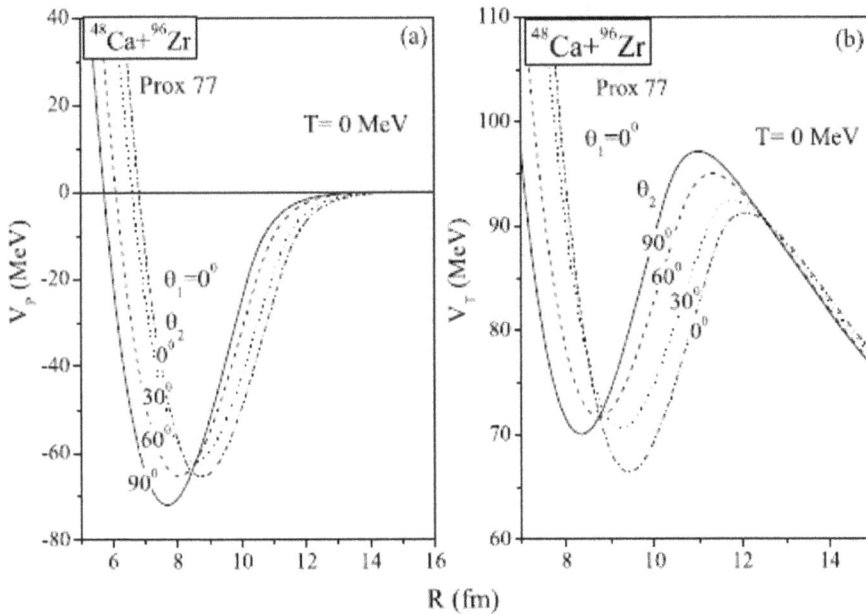

FIGURE 5.6

(a) Nuclear interaction potential and (b) total interaction potential for the ^{48}Ca + ^{96}Zr system at different θ_2 values keeping $\theta_0 = 0°$ fixed at $T = 0$ MeV.

the surface energy constant γ can change the nuclear potential keeping the universal function $\phi(s)$ the same. The value of the coefficient γ_0 used for Prox 1988 is 1.2496 MeV fm^{-2} which was slightly adjusted to 1.65 MeV fm^{-2} for the case of mod-Prox88. A comparison of proximity potentials obtained using Prox 1977, Prox 1988, and Prox 2000 is shown with that of mod-Prox 1988 in Figure 5.7a, and a comparison of the corresponding total interaction potentials calculated using these proximity potentials is shown in Figure 5.7b for the ^{64}Ni + ^{100}Mo reaction at $E_{c.m.}$ = 121.2 MeV, and T = 1.29 MeV, for $\theta_1 = \theta_2$ = 90°. It is clear from the figure that a slight modification of γ_0 leads to more attractive and stronger proximity potentials and leads to lowering of the barrier, which in the density functional approach, comes with the use of different Skyrme forces. For more details of this work, see Ref. [46].

As the ℓ-value increases, the barrier characteristics (the barrier height, its position, and the oscillator frequency) change appreciably. The barrier thickness increases, and the pocket gets shallower. This observation is clear from Figure 5.8, which shows the interaction potential for ^{48}Ca+^{238}U reaction at $E_{c.m.}$ = 193.57 MeV, using an illustrative angle of orientation θ = 90° and three ℓ-values.

The barrier modifications are also evident from Table 5.1. All barrier properties change due to the ℓ-value, and this is more so for $\hbar\omega_\ell$ and V_B^ℓ. For more details, see Ref. [30].

5.4.3 Barrier Modification through a Projectile Breakup

In the case of a loosely bound projectile, there is a possibility of breakup which might result in enhancement of the fusion cross section at energies below the Coulomb barrier. This fact can explain the enhancement found for the proton halo nucleus ^{8}B [67]. A very simplified two-channel model [29] was introduced, the first being the entrance channel and the second representing the full set of continuum breakup channels. In this channel, the

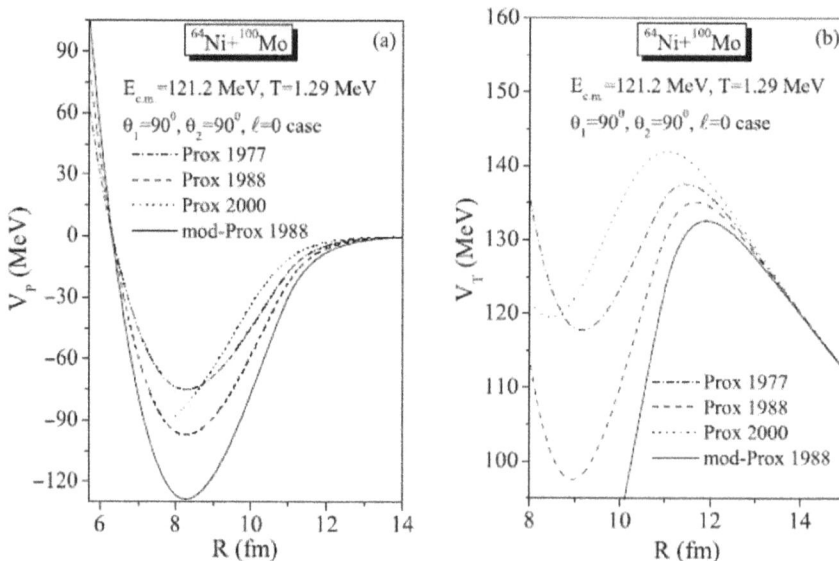

FIGURE 5.7
Comparison of (a) nuclear proximity potentials (V_P) for Prox 1977, Prox 1988, Prox 2000, and mod-Prox 1988 and corresponding (b) total interaction potentials (V_T) is made for the ^{64}Ni + ^{100}Mo reaction at $E_{c.m.}$ = 121.2 MeV (equivalently T = 1.29 MeV), by taking $\theta_1 = \theta_2$ = 90°, for the ℓ = 0.

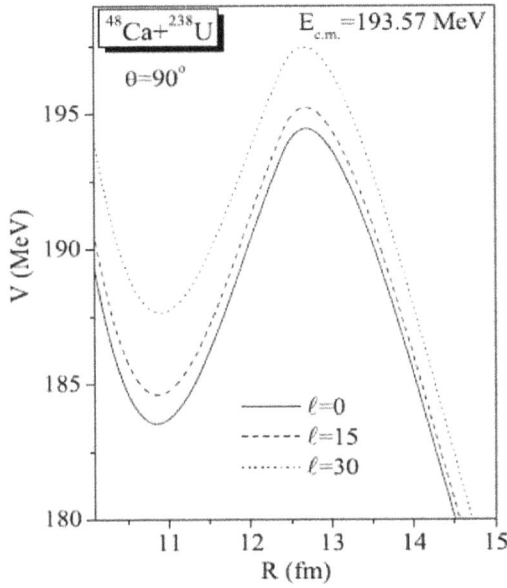

FIGURE 5.8
The interaction potential for ^{48}Ca + ^{238}U system at $E_{c.m.} = 193.57$ MeV by taking $\theta = 90°$ and $\ell = 0$, 15, and 30 $\hbar$.

TABLE 5.1
Oscillator frequencies $\hbar\omega_\ell$, barrier heights V_B^ℓ, and its positions R_B^ℓ for different ℓ-values for the interaction potential as illustrated in Figure 5.8.

ℓ ($\hbar$)	$\hbar\omega_\ell$ (MeV)	V_B^ℓ (MeV)	R_B^ℓ (MeV)
0	4.4104	194.333	12.715
15	4.4295	195.112	12.708
30	4.4780	197.355	12.689

ejected particle (neutron or proton) is neglected, and energies and the ion-ion potential are properly rescaled. This model has been applied, as representative cases of neutron or proton haloes, to the fusion with ^{58}Ni of either ^{11}Be or ^{8}B.

In Figure 5.9, the resulting ion-ion potentials for the (a) ^{8}B + ^{58}Ni and (b) ^{11}Be + ^{58}Ni reactions are shown (left panel) using the proximity potential of Broglia and Winther [57]. The corresponding ion-ion potentials of the breakup channels, i.e., for the ^{7}Be + ^{58}Ni and ^{10}Be + ^{58}Ni cases, are also shown for comparison. The energy E of the incoming channel is 20 and 17 MeV for ^{8}B and ^{11}Be, respectively, as marked by the horizontal lines in the figure. This energy can be estimated by subtracting the energy needed for the breakup and the average excitation energy $<E^*>$ in the core-nucleon relative motion, and then sharing the energy between them according to a distant breakup scenario. In this way, we consider $E_{bu} = (E - S_{1N} - <E^*>) (A-1)/A$. S_{1N} stands for the one neutron or one proton separation energy, i.e., $S_{1p} = 0.136$ MeV for ^{8}B and $S_{1n} = 0.504$ MeV for ^{11}Be.

To clarify these two different behaviors, it is useful to show the barrier distributions for both reactions. This can be done by evaluating the second-order energy derivative of the product of the cross section and the energy, or the first derivative of the transmission for $\ell = 0$. Both observables are shown in Figure 5.9b (right panel). A clear difference between the effects of proton and neutron induced on fusion is found. Both cases present

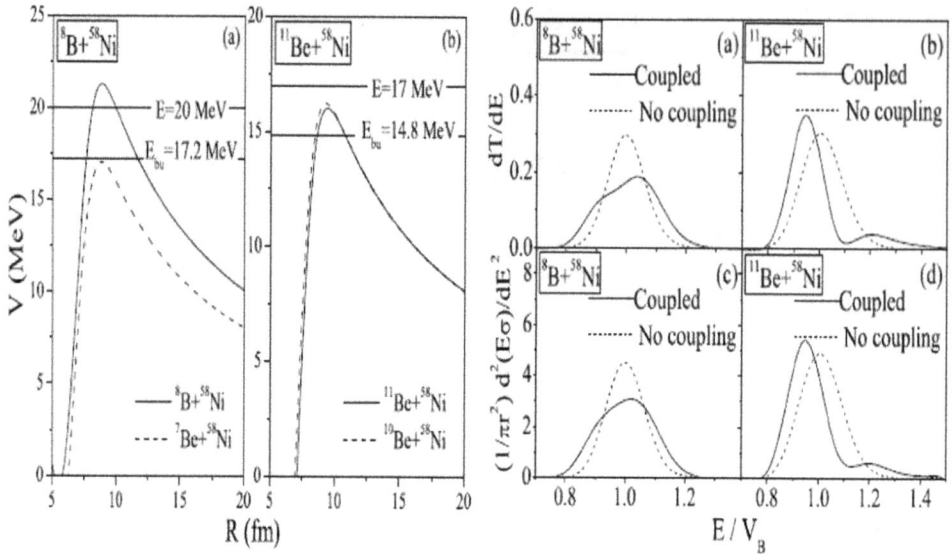

FIGURE 5.9

(a) Interaction potentials for ^{8}B + ^{58}Ni and ^{11}Be + ^{58}Ni and the respective breakup channels, ^{7}Be + ^{58}Ni and ^{10}Be + ^{58}Ni. The nuclear part of the potential is calculated using the proximity potential of Broglia and Winther [57]. (b) Barrier distributions for the ^{8}B + ^{58}Ni and ^{11}Be + ^{58}Ni fusion reactions both with and without coupling to the breakup channel. In upper panels, we show the derivative of the transmission factor for $\ell = 0$, whereas in the lower panels, we evaluate the second derivative of the product of the fusion cross section and energy.

two barriers as expected according to their interaction potential. However, in the proton case, the secondary barrier is below the barrier in the incoming channel, and so it allows a larger enhancement at low energies. Instead, in the neutron case, the secondary barrier is at higher energy. Therefore, the neutron enhancement arises from the displacement towards lower energy of the final effective Coulomb barrier. The results obtained here are similar to the effect of negative or positive Q values on barrier penetration [68,69]. The positive Q value case shows the same cross section and barrier distribution as the proton breakup case, and a parallel is found for negative Q value and neutron breakup cases.

This fact can explain the enhancement of the proton halo nucleus ^{8}B [67]. The enhancement is larger than in the neutron case, and also the energy distribution is far different. Indeed, for the neutron case, the enhancement is mainly due to a displacement in the energy of the Coulomb barrier. This can also explain why it is unclear whether or not a neutron halo produces an enhanced sub-barrier fusion. More details of this work can be found in Ref. [29].

5.5   Fusion Cross Section

The barrier characteristics of the nuclear interaction potential, i.e., barrier height, position, and frequency, from the total interaction potential are used in the Wong formula (see Eq. (5.32)) for estimating the fusion reaction cross section. In this section, we discuss some of these reactions using non-relativistic SEDF for various force parameters and the RMF formalism for the recently developed NL3* force parameter set.

5.5.1 Cross Section from Skyrme Energy Density Formalism

The fusion cross sections for the ^{64}Ni + ^{100}Mo reaction with the θ_i-integrated ℓ-summed extended Wong model, in the frozen density approximation, for the three Skyrme forces SIII, SV, and GSkI are shown in Figure 5.10. The calculated results for all the force parameters are in reasonably good agreement with the experimental data [70]. Comparing the results for different Skyrme interactions, one observes that the GSkI force fits the data well for all values of $E_{c.m.}$. It is to be noted here that the parameters of the GSkI force are obtained by fitting to several properties of isospin-rich nuclei, and hence the GSkI force is more appropriate for the ^{64}Ni +^{100}Mo reaction, an isospin-rich system. Hence, for a particular fusion reaction, the choice of a Skyrme force is crucial. For more details on this work, see Ref. [42].

5.5.2 Cross Section from Relativistic Mean-Field Theory

The barrier characteristics of the total nucleus-nucleus interaction potential, i.e., barrier height, position are used in the Wong formula (see Eq. (5.32)) for estimating the fusion reaction cross section for the system ^{64}Ni + ^{64}Ni, known for fusion hindrance phenomena. Figure 5.11 shows the comparison of the fusion cross section obtained for ^{64}Ni + ^{64}Ni near the Coulomb barrier with the experimental data [71]. The solid line shows the fusion cross section using the R3Y interaction and the dashed line using M3Y potential within the Wong formula for NL3* densities. It is observed that the R3Y interaction performs relatively better than the M3Y one in comparison with the experimental data [71] below the barrier. In other words, the R3Y interaction allows the nuclei to relax, which reduces the barrier height and hence increases the fusion cross section. It is to be noted that the fusion cross section corresponding to the R3Y interaction is always larger when compared to that of

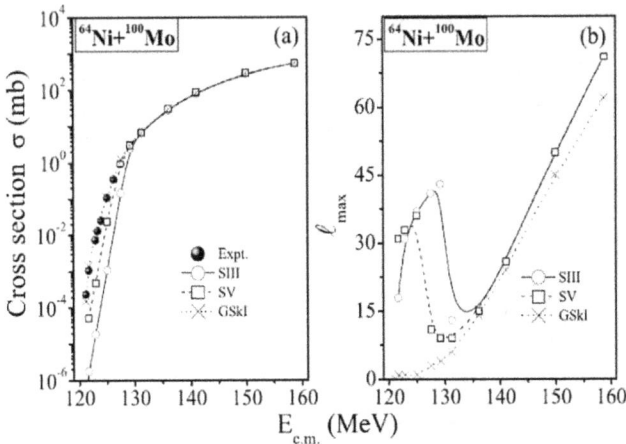

FIGURE 5.10
(a) Fusion cross sections as a function of $E_{c.m.}$, calculated by using the ℓ-summed extended Wong model, integrated over θ_i, for Skyrme forces SIII, SV, and GSkI in the frozen approximation, and compared with the experimental data for ^{64}Ni + ^{100}Mo [70]. The corresponding ℓ_{max} is shown in panel (b).

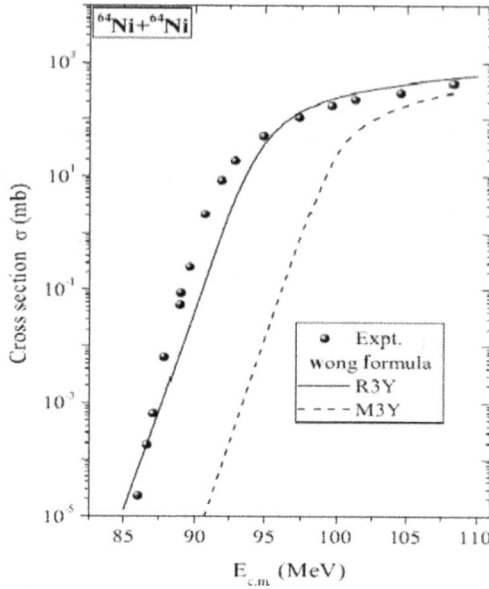

FIGURE 5.11
Fusion-evaporation cross section as a function of center-of-mass $E_{c.m.}$, calculated by using the Wong formula for R3Y (solid line) and M3Y (dashed line) NN-interactions, and compared with experimental data [71]. See the text for details.

the M3Y potential. It is clear from Figure 5.11 that the recently developed R3Y interaction with the NL3* force parameters is proven to be a relatively better choice than the M3Y for describing fusion reactions below the barrier at low energies. For a more detailed analysis, one can follow Ref. [23], for other reaction systems within the RMF formalism.

5.5.3   Effect of Orientation and Deformation on the Cross Section

Here, we show the effect of deformation on the fusion cross section at each orientation. Figure 5.12 shows the variation of the cross section at each angle for the ^{48}Ca + ^{96}Zr reaction using Prox 1977 at the highest experimental center-of-mass energy of 112.1 MeV.

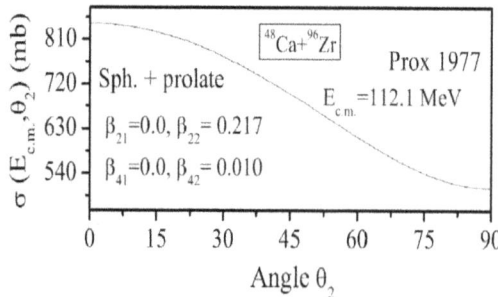

FIGURE 5.12
The fusion cross section is given at each orientation angle for the ^{48}Ca + ^{96}Zr reaction using Prox 1977 at $E_{c.m.} = 112.1$ MeV.

This reaction has a spherical projectile and a deformed target. The cross section is maximum at $\theta_2 = 0°$ for a prolate target (β_2 positive). However, for the case of an oblate target (β_2 negative), the cross section would be maximum at $\theta_2 = 90°$ (not shown here). For intermediate configurations, the fusion cross section varies smoothly as a function of the orientation angle. Therefore, one may conclude that orientation degree of freedom plays a significant role in the fusion-fission dynamics of heavy-ion reactions. For more details, see Ref. [64].

5.6 Summary and Conclusions

In this chapter, we have discussed the interrelationships between the nucleon-nucleon interaction potential and the fusion reaction cross section for a few reaction systems. Various nuclear models such as the non-relativistic Skyrme energy density functional, RMF, and various phenomenological models are included in the discussion. The double-folding procedure is used to obtain the nucleus-nucleus interaction potential using RMF densities. The well-known M3Y and recently developed relativistic nucleon-nucleon interaction entitled R3Y potential are also studied. The microscopic approaches based on an axial deformed Skyrme energy density functional for SIII, SV, GSkI, and relativistic mean field with the recently developed NL3* force have been used along with the Wong formula to provide a transparent and analytic way to calculate the fusion cross section. The total interaction potentials are calculated for the ^{64}Ni + ^{100}Mo reaction within the SEDF for a few force parameters (SIII, SkM*, SGII, SV, GSkI, and SSk), and for the ^{58}Ni + ^{58}Ni system within the RMF approach using NL3* force as well as for ^{24}Ne + ^{208}Pb using several potentials based on proximity theorem. The percentage (%) of relative change in barrier height for 52 colliding nuclei with mass asymmetry in the range of 0.00−0.96, as a function of the product of charges $Z_1 Z_2$ for various proximity potentials for deformation β_2 at optimum orientations, is also included in the present analysis. Furthermore, we have shown the effects of deformations (i.e., the shape degrees of freedom), orientations, surface energy, and angular momentum on nucleus-nucleus interaction potentials. Moreover, the interaction potentials for ^{8}B + ^{58}Ni and ^{11}Be + ^{58}Ni and the respective breakup channels for ^{7}Be + ^{58}Ni and ^{10}Be + ^{58}Ni are also studied for the proximity potential of Broglia and Winther. The fusion cross sections for ^{64}Ni + ^{64}Ni and ^{64}Ni + ^{100}Mo as a function of $E_{c.m.}$ are calculated by using the Wong formula and its extended version, respectively. The experimental data are given for comparison, wherever available. In addition, we have shown the effect of orientation on the fusion cross section of ^{48}Ca + ^{96}Zr using Prox 1977.

Acknowledgments

This work is supported in part by DAE-BRNS, Project No. 58/14/12/2019-BRNS; DST, Govt. of India, Project No. YSS/2015/000342; seed money project from TIET; FAPESP Project Nos. ((2014/26195-5 and 2017/05660-0); INCT-FNA Project No. 464898/2014-5; and the CNPq-Brasil. The authors are also thankful to S. K. Patra, Institute of Physics, and M. K. Sharma, Thapar Institute of Engineering and Technology, for discussions and Lidivania da Nobrega Silva for support throughout the work.

References

[1] G. Montagnoli, and A. M. Stefanini, *Eur. Phys. J. A* 53, 169 (2017).

[2] L. F. Canto, P. R. S. Gomes, R. Donangelo, and M. S. Hussein, *Phys. Rep.* 424, 1 (2006).

[3] N. Wang, X. Wu, Z. Li, M. Liu, and W. Scheid, *Phys. Rev. C* 74, 044604 (2006).

[4] V. Y. Denisov, *Phys. Lett. B* 526, 315 (2002).

[5] D. Vautherin, and D. M. Brink, *Phys. Rev. C* 5, 626 (1972).

[6] T. H. R. Skyrme, *Phil. Mag.* 1, 1043 (1956).

[7] B. Grammoticos, and A. Voros, *Ann. Phys.* 123, 359 (1979).

[8] B. Grammoticos, and A. Voros, *Ann. Phys.* 129, 153 (1980).

[9] M. Brack, C. Guet, and H.-B. Hakansson, *Phys. Rep.* 123, 275 (1985).

[10] J. Bartel and K. Bencheikh, *Eur. Phys. J. A.* 14, 179 (2002).

[11] S. Weinberg, *Phys. A* 96, 327 (1979).

[12] S. Weinberg, *Phys. Lett. B* 251, 288 (1990).

[13] B. B. Singh, M. Bhuyan, S. K. Patra, and R. K. Gupta, *J. Phys. G: Nucl. Part. Phys.* 39, 025101 (2012).

[14] B. B. Sahu, S. K. Singh, M. Bhuyan, S. K. Biswal, and S. K. Patra, *Phys. Rev. C* 89, 034614 (2014).

[15] M. Bhuyan, *Properties of finite nuclei using effective interaction,* LAP Lambert Academic Publishing (2014).

[16] U. van Kolck, *Prog. Part. Nucl. Phys.* 43, 337 (1999).

[17] G. R. Satchler and W. G. Love, *Phys. Rep.* 55, 183 (1979).

[18] E. Epelbaum, H. -W. Hammer and Ulf-G. Meinβer, *Rev. Mod. Phys.* 81, 1773 (2009).

[19] M. Garcon and J. W. Van Orden, *Adv. Nucl. Phys.* 26, 293 (2001).

[20] E. Epelbaum, W. Glöckle, and U.-G. Meinβer, *Nucl. Phys. A* 747, 362 (2005).

[21] D. A. Goldberg, S. M. Smith, H.G. Pugh, P.G. Roos, and N. S. Waal, *Phys. Rev. C* 7, 1938 (1973).

[22] E. Stiliarid, H. G. Bohlen, P. Frobrich, B. Gebauer, D. Kolbert, W. von Oertzen, M. Wilpert, and Th Wilpert, *Phys. Lett. B* 223, 291 (1989).

[23] M. Bhuyan and Raj Kumar, *Phys. Rev. C* 98, 054610 (2018).

[24] M. A. Cândido Ribeiro, L. C. Chamon, D. Pereira, M. S. Hussein, and D. Galetti, *Phys. Rev. Lett.* 78, 3270 (1997).

[25] L. C. Chamon, D. Pereira, M. S. Hussein, M. A. Cândido Ribeiro, and D. Galetti, *Phys. Rev. Lett.* 79, 5218 (1997).

[26] L. C. Chamon, B. V. Carlson, L. R. Gasques, D. Pereira, C. De Conti, M. A. G. Alvarez, M. S. Hussein,M. A. Candido Ribeiro, E. S. Rossi, Jr., and C. P. Silva, *Phys. Rev. C* 66, 014610 (2002).

[27] D. Pereira, C. P. Silva, J. Lubian, E. S. Rossi Jr., L. C. Chamon, G. P. A. Nobre, and T. Correa, N*ucl. Phys. A* 826, 11 (2009).

[28] W. Greiner, *International NATO Advanced Study Institute (NASI) Course on quantum Electrodynamics of Strong Fields*, Lahnstein (1981).

[29] R. Kumar, J. A. Lay, and A. Vitturi, *Phys. Rev. C.* 89, 027601 (2014).

[30] R. Kumar, M. Bansal, S. K. Arun, and R. K. Gupta, *Phys. Rev. C* 80, 034618 (2009).

[31] C. Y. Wong, *Phys. Rev. Lett.*, 31, 766 (1973).

[32] B. V. Carlson, L. C. Chamon, and L. R. Gasques, *Phys. Rev. C* 70, 057602 (2004).

[33] P. Möller, J. R. Nix, W. D. Myers, and W. J. Swiatecki, *At. Nucl. Data Tables* 59, 185, (1995).

[34] G. Royer and J. Mignen, *J. Phys. G: Nucl. Phys.* 18, 1781 (1992).

[35] R. K. Gupta, N. Singh, and M. Manhas, *Phys. Rev.* C 70, 034608 (2004).

[36] B. K. Agrawal, S. Shlomo, and V. Kim Au, *Phys. Rev. C* 72, 014310 (2005).

[37] B. K. Agrawal, S. K. Dhiman, and R. Kumar, *Phys. Rev. C.* 73, 034319 (2006).

[38] M. Brack, C. Guet, and H.-B. Hakansson, *Phys. Rep.* 123, 275 (1985).

[39] J. Friedrich and P.-G. Reinhardt, *Phys. Rev. C* 33, 335 (1986).

[40] J. Bartel and K. Bencheikh, *Eur. Phys. J. A* 14, 179 (2002).

[41] R. K. Gupta, D. Singh, and W. Greiner, *Phys. Rev. C* 75, 024603 (2007).

[42] R. K. Gupta, D. Singh, R. Kumar, and W. Greiner, J. *Phys. G: Nucl. Part. Phys.* 36, 075104 (2009).

[43] R. Kumar, M. K. Sharma, and R. K. Gupta, *Nucl. Phys. A* 870, 42 (2011).

[44] S. Shlomo, J.B. Natowitz, *Phys. Rev. C* 44, 2878 (1991).

[45] R. Kumar, M. K. Sharma, *Phys. Rev. C* 85, 054612 (2012).

[46] R. Kumar, *Phys. Rev. C* 84, 044613 (2011).

[47] M. Bhuyan, S. K. Patra, and R. K. Gupta, *Phys. Rev. C* 84, 014317 (2011).

[48] M. Bhuyan, *Phys. Rev. C* 92, 034323 (2015).

[49] M. Bhuyan, B. V. Carlson, S. K. Patra, and Shan-Gui, Zhou, *Phys. Rev.* C 97, 024322 (2018).

[50] G. A. Lalazissis, S. Raman and P. Ring, *Atm. Data. Nucl. Data. Table.* 71, 1 (1999).

[51] P.-G. Reinhard, *Rep. Prog. Phys.* 52, 439 (1989).

[52] D. Vretenar, A. V. Afanasjev, G. A. Lalazissis, and P. Ring, *Phys. Rep.* 409, 101 (2005).

[53] J. Meng, H. Toki, S. G. Zhou, S. Q. Zhang, W. H. Long, and L. S. Geng, *Prog. Part. Nucl. Phys.* 57, 470 (2006).

[54] T. Niksic, D. Vretenar, and P. Ring, *Prog. Part. Nucl. Phys.* 66, 519 (2011).

[55] G. A. Lalazissis, S. Karatzikos, R. Fossion, D. Pena, Arteaga, A. V. Afanasjev, P. Ring, *Phys. Lett. B* 671, 36 (2009).

[56] N. Austern, Y. Iseri, M. Kamimura, M. Kawai, G. Rawitscher, and M. Yahiro, *Phys. Rep.* 154, 125 (1987).

[57] R. A. Broglia and A. Winther, Heavy Ion Reactions, *Frontiers in Physics* (Addison-Wesley, Reading, MA, 1991).

[58] A. B. Balantekin and N. Takigawa, *Rev. Mod. Phys.* 70, 77 (1998).

[59] D. L. Hill and J. A. Wheeler, *Phys. Rev.* 89, 1102 (1953).

[60] M. Manhas, R. K. Gupta, *Phys. Rev. C* 72, 024606 (2005).

[61] C.H. Dasso, H. Esbensen, S. Landowne, *Phys. Rev. Lett.* 57, 1498 (1986).

[62] R. K. Gupta, M. Balasubramaniam, R. Kumar, N. Singh, M. Manhas, and W. Greiner, *J. Phys. G: Nucl. Part. Phys.* 31, 631 (2005).

[63] I. Dutt, R. K. Puri, *Phys. Rev. C* 81(2010) 064609; I. Dutt, R. K. Puri, *Phys. Rev. C* 81, 064608 (2010).

[64] D. Jain, R. Kumar, and M. K. Sharma, *Nucl. Phys. A* 915, 106 (2013).

[65] V. Zagrebaev, W. Greiner, J. *Phys. G: Nucl. Part. Phys.* 34, 1 (2007).

[66] J. Blocki, J. Randrup, W.J. Swiatecki, and C.F. Tsang, *Ann. Phys. (N.Y.)* 105, 427 (1977).

[67] E. F. Aguilera et al., *Phys. Rev. Lett.* 107, 092701 (2011).

[68] C. H. Dasso, S. Landowne, and A. Winther, *Nucl. Phys. A* 407, 221 (1983).

[69] C. H. Dasso, J. Phys. G: *Nucl. Part. Phys.* 23, 1203 (1997).

[70] C. L. Jiang, et al., *Phys. Rev. C.* 71, 044613 (2005).

[71] C. L. Jiang et al., *Phys. Rev. Lett.* 93, 012701 (2004).

6

Cluster-Decay Model for Hot and Rotating Compound Nuclei

BirBikram Singh

Sri Guru Granth Sahib World University

Manpreet Kaur

Institute of Physics

CONTENTS

6.1   Introduction

An atomic nucleus is a finite quantum system consisting of fermions—protons and neutrons held together by nuclear interaction. The exploration of arrangement of protons and neutrons or structure of nucleus giving rise to different properties of a nucleus, lies at the core of nuclear physics. Several advancements have been made on experimental as well as theoretical modeling front to understand nuclear structure and properties. Among such theoretical approaches, the shell model successfully explained several ground- and excited-state properties of nuclei. But its assumption of independent particle motion of nucleons under the mean field, fails to describe the correlation among nucleons due to clustering. In some nuclei, the protons and neutrons conglomerate together to form clusters to increase the stability or minimize the total energy of the nuclear system. The profound evidence for alpha clustering stems from high binding energy, large abundance of alpha conjugate ($N = Z$, $A = 4n$, and $Z = 2n$; n is an integer) light nuclei and observation of alpha decay in heavy nuclei. The 2n–2p correlations, due to high stability of alpha particle, play a pivotal role in the clustering of light nuclei and account for large abundance of famous Hoyle state of ^{12}C in the universe.

In order to understand this distinctive collective phenomenon, several cluster models have been developed. The alpha particle model by Hafstad and Teller [1] was quite successful in the description of alpha clustering in light nuclei. Furthermore, Ikeda showed that near the decay threshold value, alpha conjugate nuclei can be speculated to change to molecular structure [2]. The studies of exotic clustering phenomenon in neutron-rich nuclei depict that the coupling of extra neutrons with alpha cluster core leads to the stability of molecular structures in these nuclei [3]. A great deal of work has been done in this direction, which shows the persistence of alpha clustering throughout the nuclear chart [4–8]. Some other formalisms such as antisymmetrized molecular dynamics [9] and fermionic molecular dynamics [10] have also been formulated to study clustering features in nuclei.

Experimentally, the cluster states are probed via alpha knockout or capture reactions, electromagnetic transitions, and giant dipole resonance methods [11–13]. The cluster structures have been identified in ^{18}O, ^{20,21,22}Ne, ^{24}Mg, ^{28}Si, ^{32}S, and ^{40}Ca nuclei [14,15]. The heavy-ion-induced reactions (HIRs) act as an indispensable probe to study the influence of clustering aspects on the outgoing fragments. To test the survival of preexisting alpha clustering through fusion process, the collisions between light nuclei having known as alpha structure are studied [16–18]. The measurement of alpha emission over an energy range helps to inquire the extent up to which entrance channel alpha clustering is prevailing in the outgoing fragments/clusters. Moreover, the full energy-damped fragment yields observed in light mass reaction studies are mostly described in the terms of fusion fission (FF) process of compound nucleus origin. However, in some reactions involving the alpha conjugate nuclei such as ^{28}Si + ^{12}C, ^{24}Mg + ^{12}C, ^{20}Ne + ^{12}C, and ^{16}O + ^{12}C reactions, a large enhancement in the yield of few outgoing fragments is observed [19,20]. It is interpreted in terms of competitive deep inelastic orbiting (DIO) mechanism where projectile and target instead of complete amalgamation form a long-lived orbiting dinuclear configuration involving exchange of few nucleons, which decay back near the entrance channel.

To investigate the clustering effects in light mass nuclei as well as their fragment/cluster emission in excited states, here we discuss a cluster model dubbed as dynamical cluster-decay model (DCM) [21–25] based upon the collective clusterization of quantum mechanical fragmentation theory (QMFT) [26–28]. DCM is reformulation of preformed cluster-decay model [29,30] for ground-state cluster-decay analysis, with inclusion of temperature effects in different potentials and binding energies. Here, we have explored the evolution of clustering with excitation energy in light alpha conjugate ^{20}Ne, ^{28}Si, ^{40}Ca and non-alpha conjugate ^{21}Ne, ^{22}Ne, ^{39}K nuclear systems, comparatively [31,32]. The results within QMFT are well supported by microscopic relativistic mean field theory [33,34]. Furthermore, the effect of alpha clustering on the fragmentation mechanism and yield of different Z-fragments has been investigated in reference to available experimental Z-distribution data [35].

In HIRs, in addition to clustering effects, the other degrees of freedom like projectile energy, mass and charge asymmetry of colliding nuclei, shape of nuclei, and their N/Z ratio significantly affect the reaction dynamics and fusion cross section (σ_{fusion}). The study of fusion reactions and cross sections finds its roots in the pursuit to understand the nucleosynthesis mechanism in astrophysical scenario and the extension of periodic table via synthesis of superheavy elements. The investigation of the effect of N/Z ratio of nuclear systems on the yield of different mass fragments is requisite to explore the fusion process in neutron-rich stellar environment. With the availability of radioactive-ion beams, the observation of fusion enhancement in the case of neutron-rich projectile compared to stable one has been reported at sub-barrier and near-barrier energies [36–38]. The main interest of such studies is to explore the isospin effects via populating isotopic compound systems. However, we attempt to look for such effects via isobaric compound systems.

The DCM is adequate to study the same since it treats all the decaying fragments/clusters, i.e. light particles (LPs), intermediate mass fragments (IMFs), and fission fragments on the parallel footing unlike the statistical models. Within DCM, we have investigated the N/Z dependence of decay channels in the $^{80}Zr^*$, $^{80}Sr^*$, and $^{80}Kr^*$, i.e. A = 80 isobaric nuclear systems formed at the same excitation energy $E^*_{CN} \sim 47$ MeV via different mass asymmetric reactions and its influence on the cross sections of the same [39] along with a comparison with experimental data [40].

The fusion reactions provide not only the potential way of synthesizing new elements in terrestrial and non-terrestrial environments but also provide a large amount of energy for practical purpose. Due to paramount importance, the fusion reactions have been explored extensively experimentally as well as theoretically during the past decades. Here, by using DCM, the σ_{fusion} for a variety of nuclear systems with mass ranging from light, medium, and heavy nuclei have been studied, in the reactions involving the same loosely bound projectile having the same energy (E_{lab}) on different targets by optimizing the neck-length parameter (ΔR^{emp}), the only parameter of the model, in reference to the available experimental data. Here, an emphasis has been put on the significance of ΔR^{emp}, which is related to the modification/lowering of interaction barrier, which is an in-built feature of DCM. Furthermore, this optimized ΔR^{emp} has been utilized to predict the σ_{fusion} of the reactions, for which experimental data is not available [41,42]. In another work, using the ^{20}Ne projectile with the same E/A on ^{159}Tb and ^{169}Tm targets, we find that simultaneously fitted neck-length parameter ΔR for LPs, IMFs and fission fragments in one reaction works for another reaction [43]. In this work, an interesting feature of mass asymmetry (η) dependence of neck-length parameter is also observed. This chapter has been organized as follows: The theoretical framework of dynamical cluster-decay model has been discussed in detail in Section 6.2 followed by the results and discussions in Section 6.3. The summary and conclusions of the different investigations have been presented in Section 6.4.

6.2 Methodology

6.2.1 Dynamical Cluster-Decay Model (DCM)

The DCM, based on QMFT [26–28], is used to study the decay of hot and rotating compound systems formed in heavy-ion reactions and is an extended version of the PCM, as already mentioned in the Introduction. It involves the two-step process of cluster preformation followed by the penetration through the interaction barrier, analogous to the α-decay where preformation was taken to be unity. It is worked out in terms of (i) the collective coordinate of mass (and charge) asymmetry $\eta = (A_1 - A_2)/(A_1 + A_2)$ (and $\eta_Z = (Z_1 - Z_2)/(Z_1 + Z_2)$), (ii) relative separation R, (iii) multiple deformations β_{λ_i}, and (iv) orientations θ_i of two nuclei in the same plane. These coordinates η and R, respectively, characterize the nucleon division (or exchange) between outgoing fragments and the transfer of kinetic energy of incident channel ($E_{c.m.}$) to internal excitation (total excitation (TXE) or total kinetic energy (TKE)) of the outgoing channel. The TKE and TXE of fragments are related to CN excitation energy as $E^*_{CN} + Q_{out}(T) = TKE(T) + TXE(T)$.

The decay cross section of equilibrated CN, using the decoupled approximation to R and η motions, is defined in terms of ℓ partial waves, as [21–25]

$$\sigma = \frac{\pi}{k^2} \sum_{\ell=0}^{\ell_c} (2\ell + 1) P_0 P; \qquad k = \sqrt{\frac{2\mu E_{c.m.}}{\hbar^2}} \qquad (6.1)$$

where the preformation probability (P_0) and the penetrability (P) refer to η-motion and R-motion, respectively, and ℓ_c is the critical angular momentum $\ell_c = R_a\sqrt{2\mu[E_{c.m.} - V(R_a, \eta_{in}, \ell = 0)]}/\hbar$; R_a is the first turning point, defined later, where the penetration starts. The structure effects of the CN, the distinct advantage of DCM over the statistical models, enter the model via the preformation probabilities P_0 of the fragments. In case the non-compound nucleus (nCN) component, i.e. DIO, were not measured in the yield of fragments, it can be estimated empirically, $\sigma_{DIO} = \sigma^{Expt} - \sigma_{FF}^{DCM}$, where σ_{DIO}, σ^{Expt}, and σ_{FF}^{DCM} are, respectively, the DIO, experimental and DCM calculated FF cross sections.

The P_0 is given by the solution of the stationary Schrödinger equation in η, at a fixed R = R_a:

$$\left\{ -\frac{\hbar^2}{2\sqrt{B_{\eta\eta}}}\frac{\partial}{\partial\eta}\frac{1}{\sqrt{B_{\eta\eta}}}\frac{\partial}{\partial\eta} + V_R(\eta, T) \right\}\psi^\nu(\eta) = E^\nu\psi^\nu(\eta), \tag{6.2}$$

with $\nu = 0,1,2,3...$ referring to ground-state ($\nu = 0$) and excited-state solutions summed over as a Boltzmann-like function:

$$|\psi|^2 = \sum_{\nu=0}^{\infty}|\psi^\nu|^2\, exp(-E^\nu/T). \tag{6.3}$$

Then, the probability of cluster preformation is

$$P_0(A_i) = |\psi(\eta(A_i))|^2\,\frac{2}{A_{CN}^*}\sqrt{B_{\eta\eta}}, \tag{6.4}$$

where $i = 1$ or 2 and $B_{\eta\eta}$ is the smooth hydrodynamical mass parameter [44]. For clustering effects in nuclei, we look for the maxima in the $P_0(A_i)$ (as shown in Figure 6.3) or for the energetically favored potential energy minima in the fragmentation potential $V_R(\eta, T)$. The $V_R(\eta, T)$ in Eq. (6.2), for a fixed β_{λ_i}, is the potential energy for all possible mass combinations A_i, corresponding to the given charges Z_i minimized for each mass fragmentation coordinate η_A. The minimized fragmentation potential obtained is defined as

$$V_R(\eta, T) = \sum_{i=1}^{2}\left[V_{LDM}(A_i, Z_i, T)\right] + \sum_{i=1}^{2}\left[\delta U_i\right]exp\left(-\frac{T^2}{T_0^2}\right)$$
$$+ V_c(R, Z_i, \beta_{\lambda_i}, \theta_i, T) + V_P(R, A_i, \beta_{\lambda_i}, \theta_i, T) + V_\ell(R, A_i, \beta_{\lambda_i}, \theta_i, T) \tag{6.5}$$

where V_c, V_p, and V_l are temperature-dependent Coulomb, nuclear proximity, and angular momentum-dependent potentials for deformed and oriented nuclei. $B_i = V_{LDM}(A_i, Z_i, T) + \delta U_i$ (i = 1,2) are the binding energies of two nuclei. The δU are the "empirical" shell corrections, i.e., microscopic part [45], of the binding energies and V_{LDM} is the liquid-drop energy, i.e., macroscopic part. The T-dependent liquid-drop part of the binding energy $V_{LDM}(T)$ is taken from Davidson et al. [46], based on the semi-empirical mass formula of Seeger [47], as where

$$I = a_a(Z - N), a_a = 1.0 \tag{6.6}$$

and f (Z,A) = (−1,0,1), for even-even, even-odd, and odd-odd nuclei, respectively. Temperature-dependent binding energies are obtained from Ref. [46] with its constants at T=0 refitted [21,23,24] to give the ground-state (T=0) experimental binding energies [48], and where the data is not available, the theoretical binding energies are taken from Ref. [49,50]. It is important to point out here that Gupta and co-workers have shown [51] that the modified temperature dependence of the pairing energy coefficient δ(T) is to be allowed in the temperature-dependent liquid-drop energy (see Figure 3 of Ref. [51]). We have further highlighted the significance of the appropriate $\delta(T)$ in the present calculations while comparing the clustering effects in the $N = Z$, $^{20}Ne^*$, $^{28}Si^*$ (or alpha conjugate) and $N \neq Z$, $^{21,22}Ne^*$ (or non-alpha conjugate) nuclear systems (refer to Figures 6.1 and 6.2, discussed in detail in the next section).

The Coulomb potential V_c for oriented nuclei is defined as follows:

$$V_c(R, \beta_{\lambda_i}, \theta_i, T) = \frac{Z_1 Z_2 e^2}{R(T)} + 3Z_1 Z_2 e^2 \sum_{\lambda,i=1,2}^{2} \frac{R_i^\lambda(\alpha_i, T)}{(2\lambda+1)R(T)^{\lambda+1}} Y_\lambda^{(0)}(\theta_i)$$

$$* \left[\beta_{\lambda_i} + \frac{4}{7}\beta_{\lambda_i}^2 Y_\lambda^{(0)} \right] \tag{6.7}$$

The deformation parameters β_{λ_i} of the nuclei are taken from the tables of Möller et al. [49], and the orientations θ_i are the optimum [52] or compact orientations [53] of the "hot" process. The proximity potential [54]

$$V_p(T) = 4\pi\bar{R}(T)\gamma b(T)\Phi(s(T)), \tag{6.8}$$

where γ, the nuclear surface energy constant, is given by

$$\gamma = 0.9517 \left[1 - 1.7826 \left(\frac{N-Z}{A} \right)^2 \right] MeV fm^{-2}, \tag{6.9}$$

and $b(T) = 0.99(1 + 0.009T^2)\bar{R}(T)$ is the nuclear surface thickness, $\bar{R}(T)$ is the root mean square radius of the Gaussian curvature, and $\Phi(s(T))$ is the universal function, independent of the geometry of the system but dependent on the minimum separation distance $s_0(T)$, as

$$\Phi(s(T)) = \begin{cases} -\frac{1}{2}(s-2.54)^2 - 0.0852(s-2.54)^3; s \leq 1.2511 \\ -3.437exp(-\frac{s}{0.75}); s \geq 1.2511 \end{cases} \tag{6.10}$$

Then, for a fixed R, the minimum distance s_0 is defined as $s_0 = R - X_1 - X_2 = R - R_1(\alpha_1)$ $\cos(\theta_1 - \alpha_1) - R_2(\alpha_2) \cos(180 + \theta_2 - \alpha_2)$, where, for s_0 to be minimum, the conditions on s_0 are [47]

$$\partial s_0/\partial \alpha_1 = \partial s_0/\partial \alpha_2 \tag{6.11}$$

resulting in

$$\tan(\theta_1 - \alpha_1) = -R_1'(\alpha_1)/R_1(\alpha_1) \tag{6.12}$$

$$\tan(180^0 + \theta_2 - \alpha_2) = -R_2'(\alpha_2)/R_2(\alpha_2). \tag{6.13}$$

Here, $R_i'(a_i)$ is the first-order derivative of $R_i(a_i)$ with respect to (a_i). Note that the above conditions refer to perpendiculars (normal vectors) at the points P_1 and P_2. Thus, $s_0(T)$ gives the minimum separation distance along the colliding Z-axis between any two deformed, co-planar nuclei, denoted as the neck-length parameter $\Delta R(\eta, T)$ in the following (refer to Eq. (6.19)).

The angular momentum-dependent potential is given by

$$V_\ell(T) = \frac{\hbar^2 \ell(\ell+1)}{2I(T)}, \tag{6.14}$$

where

$$I(T) = I_S(T) = \mu R^2 + \frac{2}{5}A_1 m R_1^2 + \frac{2}{5}A_2 m R_2^2 \tag{6.15}$$

is the moment of inertia for sticking limit. This limit is defined for the separation distance ΔR to be within the range of nuclear proximity (~ 2 fm). The penetration probability P in Eq. (6.1) is calculated by using the WKB integral as

$$P = exp\left[-\frac{2}{\hbar}\int_{R_a}^{R_b}\left\{2\mu\left[V(R) - Q_{eff}\right]\right\}^{1/2}dR\right], \qquad (6.16)$$

where V(R) is the scattering potential calculated as the sum of Coulomb, proximity, and angular momentum-dependent potential and with R_a and R_b as the first and second turning points satisfying

$$V(R_a, \ell) = V(R_b, \ell) = Q_{eff}(T, \ell). \qquad (6.17)$$

The ℓ- dependence of R_a is defined by

$$V(R_a, \ell) = Q_{eff}(T, \ell = \ell_{min}), \qquad (6.18)$$

which means that R_a, given by the above equation, is the same for all ℓ-values and that V (R_a, ℓ) acts like an effective Q value (Q_{eff}(T, ℓ)) for the decay of hot compound system. The ℓ_{min} value refers to the minimum value that starts contributing to WKB integral. As the ℓ-value increases, the Q_{eff}((T) value increases, and hence, V(R_a, ℓ) increases. Equation (6.16) is solved analytically [55] as shown in Figure 6.8. The first turning point of the penetration path R_a is given as

$$R_a = R_1(\alpha_1, T) + R_2(\alpha_2, T) + \Delta R(\eta, T) \qquad (6.19)$$

with the radius vector $R_i(\alpha_i, T)$ defined as

$$R_i(\alpha_i, T) = R_{0i}(T)\left[1 + \sum_\lambda \beta_{\lambda_i} Y_\lambda^{(0)}(\alpha_i)\right], \qquad (6.20)$$

where

$$R_{0i}(T) = \left[1.28A_i^{1/3} - 0.76 + 0.8A_i^{-1/3}\right](1 + 0.0007T^2) \qquad (6.21)$$

with T calculated by using $E_{CN}^* = aT^2 - T$, where a is the level density parameter. The value of a = ($\frac{A}{8}$) or ($\frac{A}{9}$) is chosen for light- and medium-mass regions, respectively. The choice of parameter R_a, for a best fit to the data, allows to relate in a simple way the V(R_a) to the top of the barrier V_B for each ℓ, by defining their difference ΔV_B as the effective "barrier lowering or barrier modification":

$$\Delta V_B = V(R_a) - V_B. \qquad (6.22)$$

Note that ΔV_B is defined as a negative quantity (shown in Figure 6.8) because the actually used barrier is effectively lowered, which is an in-built property of the DCM.

6.3   Results and Discussion

We have divided this section into three parts. Firstly, we present the results within collective clusterization approach of DCM for clustering effects and fragmentation analysis of light-mass nuclear systems. Next, we discuss the significance of N/Z ratio in the decay channels and σ_{fusion} of nuclear systems formed in HIRs. Then, we explore the dynamics of reactions induced by loosely bound projectile and study σ_{fusion} of a variety of nuclear systems in mass regimes extending from lighter to heavier one, and discuss the significance of neck-length parameter within DCM to predict σ_{fusion} of the reactions for which experimental data is still not available.

6.3.1 Clustering Effects and Fragmentation in Light-Mass Nuclear Systems

Here, we discuss the temperature evolution of clustering effects in light-mass alpha conjugate ^{20}Ne, ^{28}Si, ^{40}Ca and non-alpha conjugate ^{21}Ne, ^{22}Ne, ^{39}K nuclear systems, comparatively. In the next step, the probable impact of alpha clustering on the emission of different fragments in these nuclear systems is analyzed [31]. Figure 6.1 shows the collective potential energy surface (upper panel) and preformation profile (lower panel) of different clusters in alpha conjugate ^{20}Ne system, as a representative case, at three temperatures (T) by taking into account the modified pairing strength $\delta(T)$ in the liquid-drop energies, as discussed in the Formalism section.

In ground state, i.e., T = 0 MeV with $\delta(T)$ = 32.02 MeV, α+^{16}O is the most probable cluster configuration (Figure 6.1a), while at T = 1.59 MeV corresponding to Ikeda's threshold energy value for ^{16}O cluster [2] with $\delta(T)$ = 32.73 MeV, in addition to α+^{16}O configuration, another xα-type (x is an integer) cluster configuration—^{8}Be+^{12}C—also appears (Figure 6.1b). Furthermore, with an increase in temperature at T = 4.35 MeV

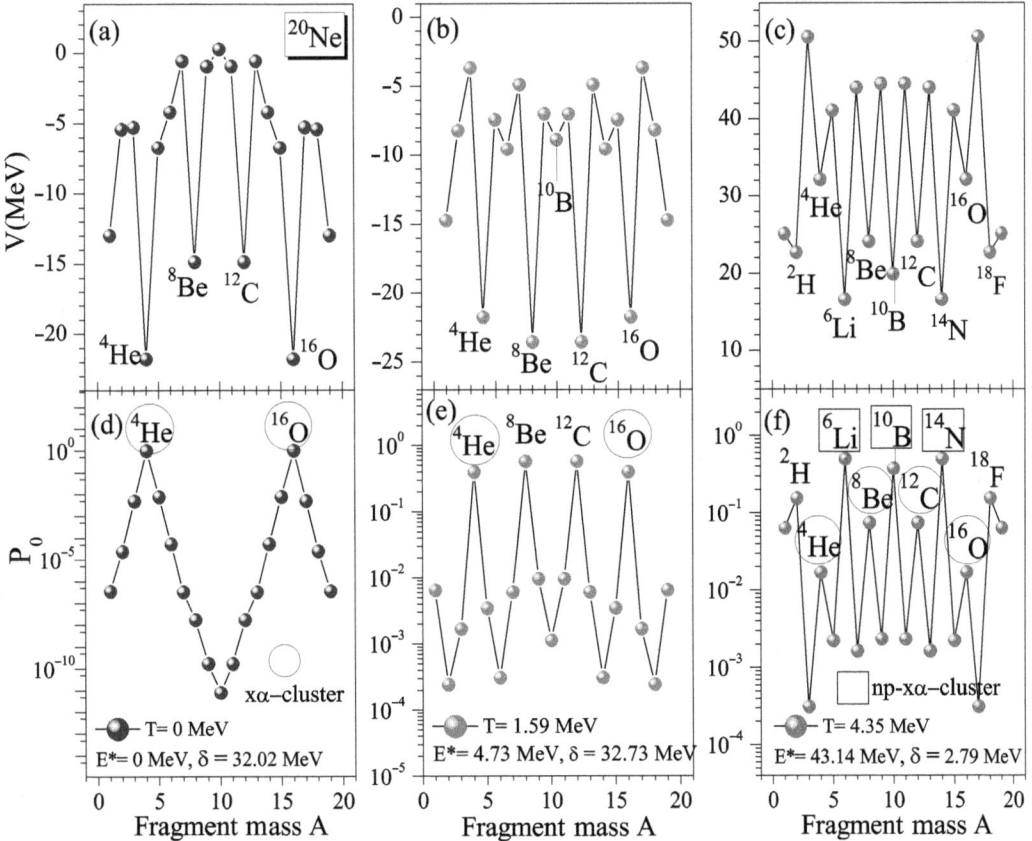

FIGURE 6.1
The fragmentation potential V (a, b, c) and preformation probability P_0 (d, e, f) of different clusters in alpha conjugate ^{20}Ne nuclear system at (a, d) T= 0 MeV (b, e) T corresponding to threshold energy [2] (c, f) T corresponding to experimental available energy [35]. This figure is based on Figure 1 of Ref. [32].

($\delta(T) = 2.79$ MeV) corresponding to available experimental energy [35], in addition to nα-type clusters, the np-xα-type clusters (n is neutron and p is proton) like ^{6}Li, ^{10}B, and ^{14}N are also energetically favorable (Figure 6.1c). These results are further reinforced in preformation probability P_0 leading to maxima corresponding to most probable clusters. This change in clustering scenario with rising temperature is associated with decreased $\delta(T)$ in macroscopic liquid-drop energies, which is an important ingredient of fragmentation potential. Figure 6.2 is similar to Figure 6.1 but for non-alpha conjugate ^{21}Ne nuclear system. Here, we note that at T = 0 MeV ($\delta(T) = 32.02$ MeV), xn-xα-type $\alpha+^{17}$O configuration is probable (Figure 6.2a and d) and at T = 2.29 MeV ($\delta(T) = 31.30$ MeV), another xn-xα-type ^{8}Be+^{13}C configuration also comes into picture (Figure 6.2b and e). At higher experimental T = 4.67 MeV ($\delta(T) = 1.53$ MeV), along with xn-xα-type clusters, np-xα-type clusters—^{10}B and ^{11}B—are also preformed (Figure 6.2c and f). These results are well supported by microscopic relativistic mean field theory [33,34].

Furthermore, these preformed clusters with a finite value of P_0 undergo tunneling through nucleus-nucleus interaction barrier with penetrability P calculated using the WKB approximation and contribute to the cross section (refer to Eq. (6.1)). The cross section has

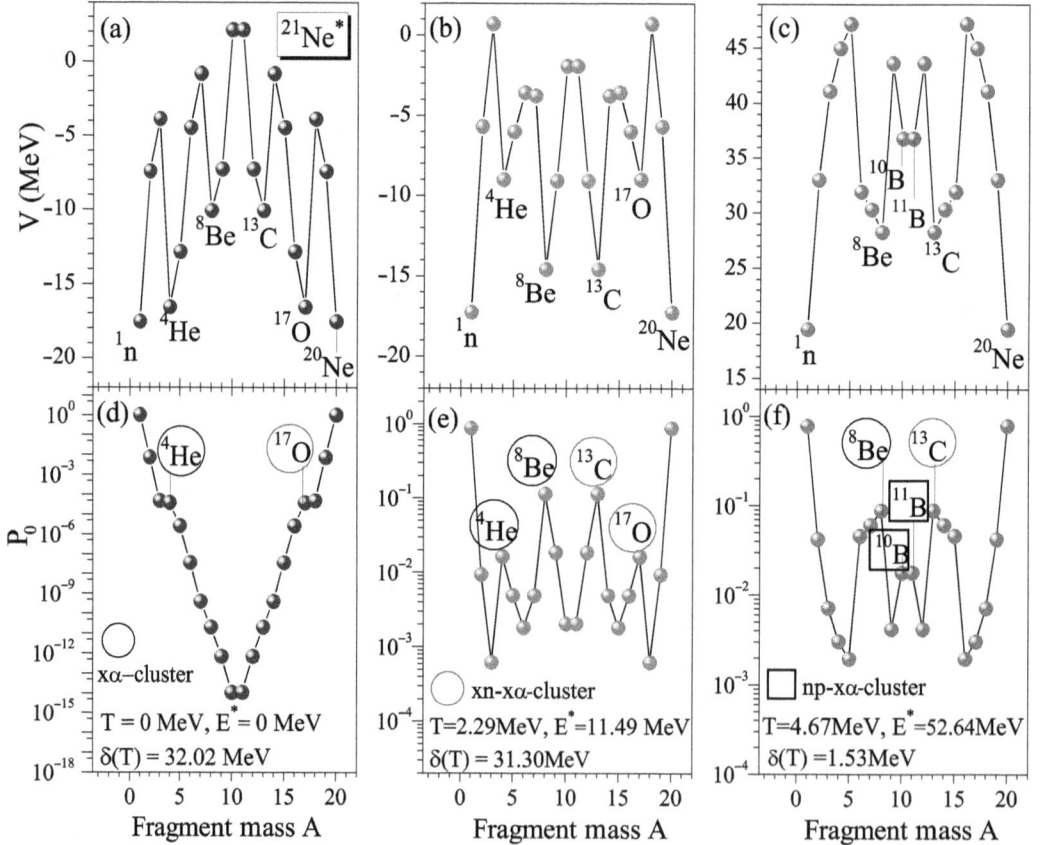

FIGURE 6.2
The fragmentation potential V (a, b, c) and preformation probability P_0 (d, e, f) of different clusters in non-alpha conjugate ^{21}Ne nuclear system at (a, d) T = 0 MeV (b, e) T corresponding to threshold energy [2] (c, f) T corresponding to experimental available energy [35]. This figure is based on Figure 2 of Ref. [32].

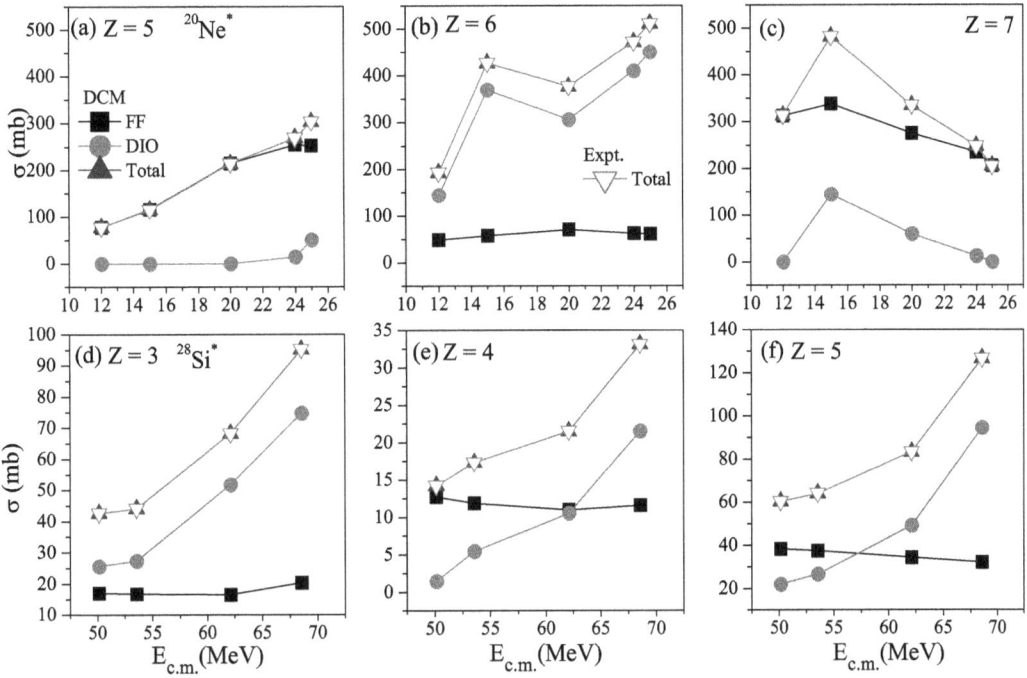

FIGURE 6.3
The excitation function for different Z-fragments in the decay of alpha conjugate systems
^{20}Ne* (a, b, c) and ^{28}Si* (d, e, f). This figure is based on Figure 3 of Ref. [32].

been calculated for different fragments/clusters in reference to availability of Z-distribution data [35] by adjusting the neck-length parameter, the only parameter of DCM, within the proximity range of ∼2.2 fm. The results present the evidence of the existence of competing reaction mechanism of FF and DIO in the yield of different Z-fragments. The DIO contribution is first empirically evaluated using fitted FF cross section within DCM (σ_{FF}^{DCM}) as $\sigma_{DIO} = \sigma^{Expt} - \sigma_{FF}^{DCM}$, and DIO contribution is fitted subsequently. Figures 6.3 and 6.4 show the fusion excitation function for different Z-fragments of alpha conjugate ^{20}Ne*, ^{28}Si* systems and alpha conjugate ^{21}Ne*, ^{22}Ne* systems, respectively. For $^{20}Ne^*$, the FF is the only decay mechanism except at the highest energy for Z = 5 (Figure 6.3a), while for Z = 7, FF and DIO are competing (Figure 6.3c). In the case of Z = 6, the DIO is the primary decay mode (Figure 6.3b). In the case of $^{28}Si^*$, the DIO contribution is high compared to FF for all Z = 3, 4, 5 fragments, and its contribution rises with increasing energy (Figure 6.3d–f). For $^{21}Ne^*$ and $^{22}Ne^*$, the DIO contribution is absent for B fragments (Figure 6.3a and d) and it competes with FF for N fragments (Figures 6.3c and f). In contrast, the DIO contributes significantly for C fragments (Figure 6.3b and e). In C yield, the large values of σ^{Expt} and σ_{DIO} in both alpha conjugate $^{20}Ne^*$ and non-alpha conjugate $^{21}Ne^*$, and $^{22}Ne^*$ nuclear systems, provide an evidence for the persistence of alpha clustering in fragmentation process, which calls for the experimental verification of the same.

6.3.2 Effect of N/Z Ratio of Nuclear Systems on the Decay Channels

Here, we present our results in the decay of isobaric compound systems ^{80}Zr*, ^{80}Sr*, and ^{80}Kr* formed in ^{40}Ca+^{40}Ca, ^{16}O+^{64}Zn, and ^{32}S+^{48}Ca reactions, respectively, at the same excitation energy $E_{CN}^* \sim 47$ MeV [39]. We calculate the contribution of different decay

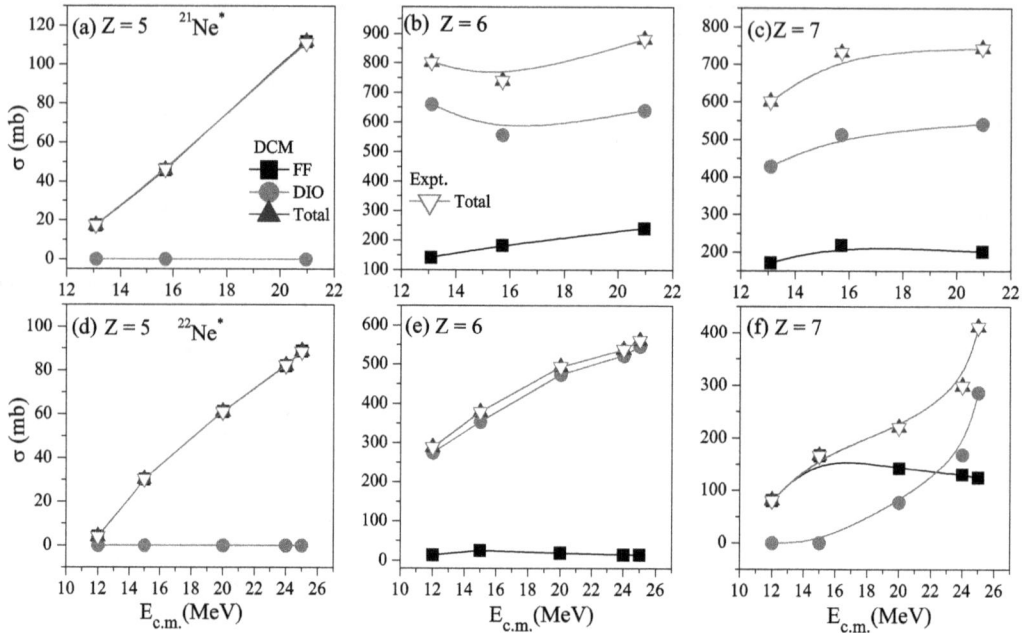

FIGURE 6.4

Same as Figure 6.3 but for ^{21}Ne* (a, b, c) and ^{22}Ne* (d, e, f). This figure is based on Figure 4 of Ref. [32].

channels, i.e. LPs, IMFs, and symmetric mass fragments (SMFs) in the total fusion cross section (σ_{fusion}) along with a comparison with experimental data [40]. Firstly, we probe the nuclear structure effects via preformation probability P_0 of different fragments in ^{80}Zr* (N/Z = 1.0), ^{80}Sr* (N/Z = 1.1), and ^{80}Kr* (N/Z = 1.2) as shown in Figure 6.5. At $\ell = 0\hbar$, the LPs are most probable (Figure 6.5a–c), while at respective $\ell = \ell_{max}$ values for all three compound systems, the LPs become less probable (Figure 6.5d–f). However, in the case of ^{80}Kr*, the LPs - 1n, ^{4}H have relatively significant P_0 at both $\ell = 0\hbar$ and $\ell = \ell_{max}$. The mass distribution evolves from U-shape to bell shape while going from $\ell = 0\hbar$ to $\ell = \ell_{max}$ with maxima near SMFs in all three compound systems. We note that maxima near SMFs as seen in ^{80}Zr* disappear in the case of n-rich ^{80}Kr* system in which IMFs start competing with SMFs. In other words, the mass distribution changes from symmetric to asymmetric one with increasing N/Z ratio of compound system. This spectroscopic factor P_0 is further used as one of the important factors in the calculation of σ_{fusion} cross section.

Table 6.1 shows the calculated σ_{fusion} for LPs, IMFs, and SMFs and a comparison with experimental data. These σ_{fusion} are further shown in Figure 6.6a depicting the percentage cross section σ_x/σ_{fusion} (x stands for LPs, IMFs, and SMFs) as a function of N/Z ratio of compound system. It is clear from the figure that LPs have the maximum contribution in σ_{fusion}, and with increasing N/Z ratio, their percentage contribution increases. There is a competition among IMFs and SMFs contribution. But as N/Z ratio increases, the SMFs contribution decreases accompanied by an increase in IMFs contribution. The calculated σ_{fusion} (= $\sigma_{LPs}+\sigma_{IMFs}+\sigma_{SMFs}$) are in good agreement with experimental data [40]. Figure 6.6b depicts the variation of σ_A with fragment mass A for extreme values of N/Z ratio, i.e., for ^{80}Zr* and ^{80}Kr*. For $5 \leq A \leq 14$, the even-odd staggering of σ_A is clearly seen in the inset of Figure 6.6b. This even-odd staggering decreases with increasing N/Z

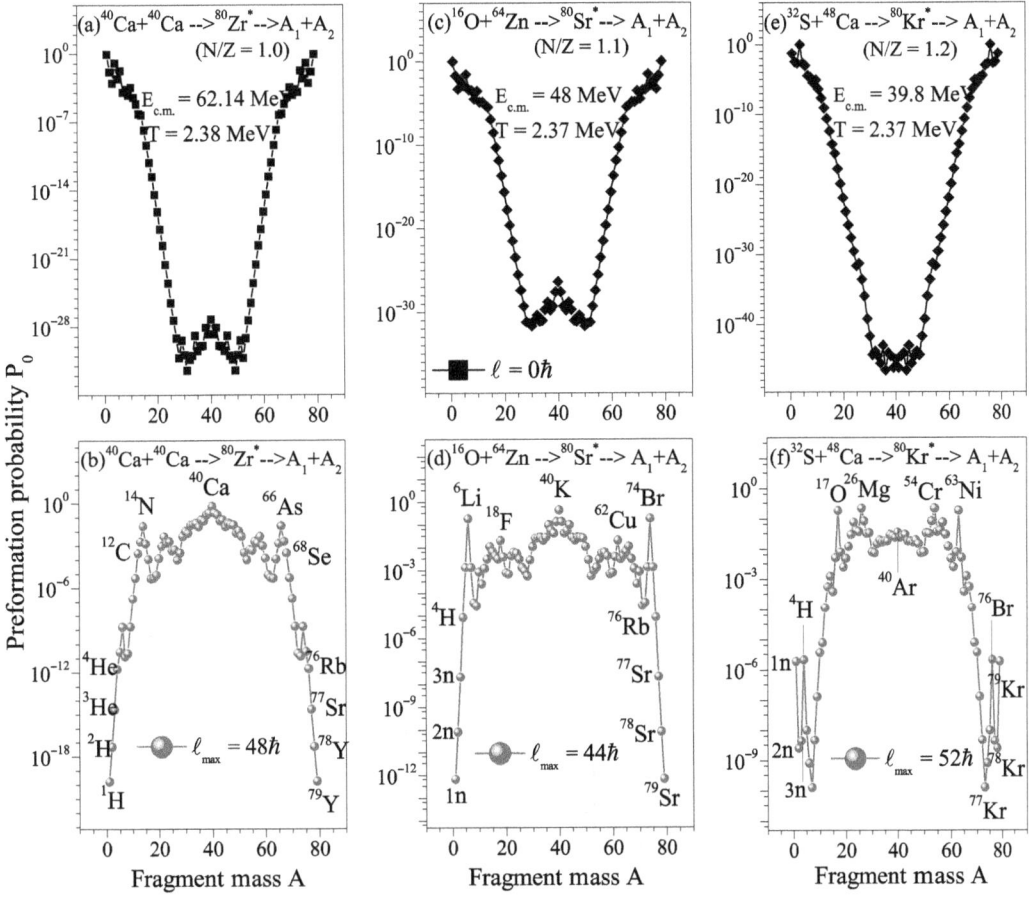

FIGURE 6.5
Variation of the preformation probability P_0 with the fragment mass in the decay of ^{80}Zr* (N/Z = 1.0), ^{80}Sr* (N/Z = 1.1), and ^{80}Kr* (N/Z = 1.2), respectively, having $E^*_{CN} \sim 47$ MeV at $\ell = 0\hbar$ (upper panel) and $\ell = \ell_{max}$ (lower panel). This figure is based on Figure 3 of Ref. [39].

ratio of compound system. It is important to note that the structure of Figure 6.6b follows that of Figure 6.5 of P_0, which shows that staggering effect arises due to nuclear structure effects in the fragment cross section, probed through P_0.

6.3.3 Fusion Cross Sections, Neck-Length Parameter and Predictability of DCM

6.3.3.1 ^{7}Li-, ^{7}Be-, and ^{9}Be-Induced Reactions Leading to A $\approx$30–200

Here, we discuss the dynamics of reactions involving the loosely bound projectiles ^{7}Li, ^{7}Be, and ^{9}Be near the Coulomb barrier [41]. The σ_{fusion} of the number of reactions is addressed by analyzing the peculiar behavior of neck-length parameter ΔR and barrier modification in order to make predictions of σ_{fusion} for some reactions not yet studied experimentally. In practice, three different sets of reactions with loosely bound projectiles ^{7}Li, ^{7}Be, and ^{9}Be having $E_{lab} = 10, 17$, and 28 MeV, respectively, on different targets leading to formation

TABLE 6.1

The DCM Calculated Fusion Cross Sections for the Emissions of LPs, IMFs, and SMFs in the Decay of ^{80}Zr*, ^{80}Kr*, and ^{80}Sr* Formed at the Same $E^*_{CN} \sim 47$ MeV, Compared with the Experimental Data [40]

Nuclear System	N/Z ratio	$E_{c.m.}$ (MeV)	ℓ_{max} ($\hbar$)	ΔR (fm)		σ (mb)			σ^{DCM}_{fus} (mb)	σ^{Expt}_{fus} (mb)
				LPs	IMFs/SMFs	LPs	IMFs	SMFs		
^{80}Zr*	1.0	62.14	48	1.62	1.35	373.38	20.74	50.65	444.77	438±69.44
^{80}Sr*	1.1	48.0	44	1.97	1.55	820.48	190.78	69.89	1081.15	1095±110
^{80}Kr*	1.2	39.80	52	0.72	0.72	1.35	0.013	0.0003	1.364	1.26±0.26

This table is based on Table 1 of Ref. [39].

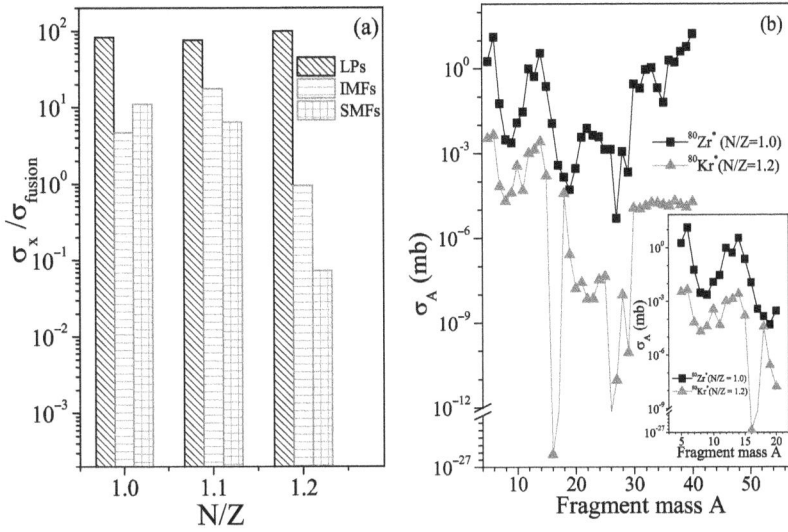

FIGURE 6.6
(a) Variation of the percentage cross section σ_x/σ_{fusion} (where x symbolizes LPs, IMFs, and SMFs) with the N/Z ratio in the decay of ^{80}Zr*, ^{80}Sr*, and ^{80}Kr* having N/Z ratios of 1.0, 1.1, and 1.2, respectively. (b) Mass dependence of σ_A in the case of ^{80}Zr* (N/Z = 1.0), and ^{80}Kr* (N/Z = 1.2) compound systems. This figure is based on Figures 6 and 7 of Ref. [39].

of compound systems with A $\sim$ 30–200 have been studied. By using the same projectile having the same energy, the σ_{fusion} has been fitted for a number of reactions employing the same value of ΔR called as empirical neck-length parameter ΔR^{emp}. The same value of ΔR^{emp} is further utilized to predict σ_{fusion} of reactions with the same projectile and the same energy on some stable targets, for which the experimental data is unavailable.

In Figure 6.7, the effect of increasing target mass on the collective potential energy surface in ^{7}Be-induced reactions at $E_{lab} \sim 17$ MeV is shown at two extreme ℓ-values. The magnitude of fragmentation potential is lowered as the mass of target increases at both ℓ-values. It is due to fall in temperature with an increase in target mass, since the projectile energy remains the same. Therefore, the heavier compound system with low fragmentation potential has less decay probability compared to lighter compound systems, and hence, the less σ_{fusion} is reported for heavy compound system. The energetically favored different mass fragments, i.e., LPs, IMFs, and SMFs, are preformed having some definite preformation probability P_0 and then will undergo penetration through barrier formed by Coulomb, proximity, and angular momentum-dependent potentials and contribute toward the $\sigma_{fusion}(= \sigma_{LPs} + \sigma_{IMFs} + \sigma_{SMFs})$.

The barrier characteristics for compound systems formed in ^{7}Be-induced reactions are shown in Figure 6.8. It is clear that the barrier height increases while moving from lighter ^{34}Cl* compound system to heavier compound system ^{72}As* (Figure 6.8a). It indicates that σ_{fusion} decreases with an increase in mass of compound system. The decays of ^{34}Cl* and ^{72}As* into ^{14}N + ^{20}Ne and ^{14}N+ ^{58}Fe exit channels, respectively, are shown in Figure 6.8b. It is noted that for ^{34}Cl*, the area under curve is less in comparison to ^{72}As*, and as a result the penetrability and hence the σ_{fusion} are high for the former case (Figure 6.9).

Table 6.2 presents the barrier lowering/modification factor ΔV_B (see Eq. (6.22)) in the case of ^{7}Be-induced reactions. It is noted that the magnitude of barrier modification is nearly the same at ℓ_{min}-value and almost constant at $\ell = \ell_{max}$. It brings forth an interesting

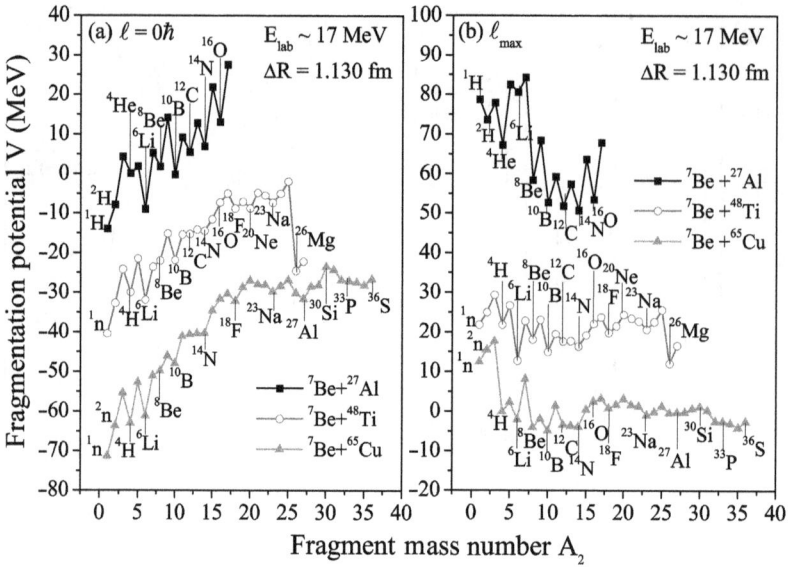

FIGURE 6.7

The fragmentation potentials V (A_2) for the compound systems having mass A $\approx$30–70 formed in ^{7}Be-induced reactions at incident energy $E_{lab} \approx$ 17 MeV for (a) $\ell = 0\hbar$ and (b) $\ell = \ell_{max}$. This figure is based on Figure 5 of Ref. [41].

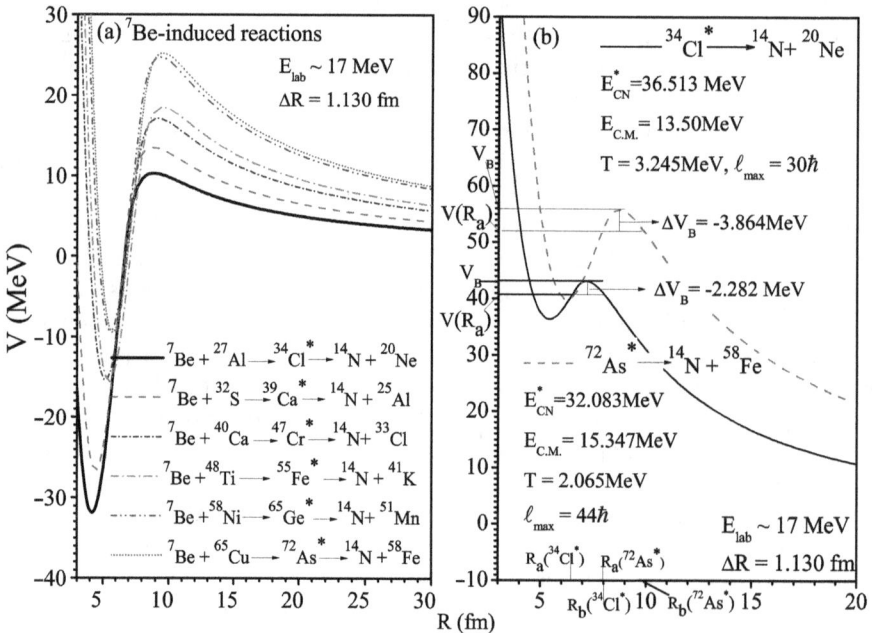

FIGURE 6.8

(a) The barrier height V_B (MeV) for the nuclei induced by ^{7}Be projectiles at $\ell = 0\hbar$. (b) The first and second turning points of ^{34}Cl* and ^{72}As* formed in ^{7}Be + ^{27}Al and ^{7}Be + ^{65}Cu reactions at the respective ℓ_{max} values. This figure is taken from Ref. [42].

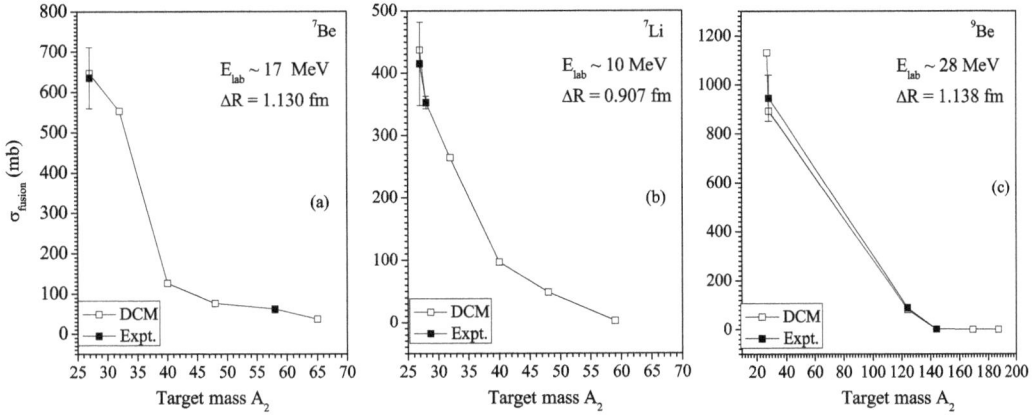

FIGURE 6.9
The calculated fusion cross section σ_{fusion} for different reactions using different loosely bound projectile (a) ^{7}Li, (b) ^{7}Be, and (c) ^{9}Be, compared with the experimental data. This figure is based on Figure 8 in Ref. [41].

TABLE 6.2
The Barrier Modification Factor $\Delta V_B = V(R_a) - V_B$ at Different ℓ-Values for the Interaction Potentials Calculated for ^{7}Be-Induced Reactions at $E_{lab} \sim 17$ MeV

Reaction	ℓ_{min} ($\hbar$)	ℓ_{max} ($\hbar$)	ΔV_B (MeV) at		
			(ℓ_{min})	($\ell = 20\ \hbar$)	(ℓ_{max})
^{7}Be+^{27}Al→^{34}Cl*→^{14}N+^{20}Ne	8	30	-12.000	-6.906	-2.282
^{7}Be+^{32}S→^{39}Ca*→^{14}N+^{25}Al	0	29	-12.131	-7.144	-3.577
^{7}Be+^{40}Ca→^{47}Cr*→^{14}N+^{33}Cl	0	30	-9.449	-7.475	-3.923
^{7}Be+^{48}Ti→^{55}Fe*→^{14}N+^{41}K	0	36	-10.013	-7.690	-4.098
^{7}Be+^{58}Ni→^{65}Ge*→^{14}N+^{51}Mn	0	38	-9.304	-7.450	-4.030
^{7}Be+^{65}Cu→^{72}As*→^{14}N+^{58}Fe	0	44	-9.454	-7.881	-3.864

This table is taken from Ref. [42].

observation that in the reactions having the same projectile, the same energy, and the same ΔR, the amount of ΔV_B is nearly the same. Exploiting this key observation from the systematic analysis of a set of reactions, the σ_{fusion} can been predicted for other reactions using the same projectile, the same energy, and the same ΔR, but on different targets forming different compound systems. The cross sections for ^{7}Be-induced reactions at $E_{lab} \sim$ 17 MeV are shown in Table 6.3. The empirically fitted $\Delta R^{emp} = 1.130$ fm, in reference to available experimental data [56,57], is further used to predict the σ_{fusion} for other reactions presented in Table 6.3. Following the same approach, the σ_{fusion} has been studied in the case of other loosely bound projectiles ^{7}Li and ^{9}Be using the $\Delta R^{emp} = 0.907$ fm and 1.138 fm, with reference to experimental data [56,58–64]. The comparative trend of variation of σ_{fusion} with target mass for three different loosely bound projectiles ^{7}Li, ^{7}Be, and ^{9}Be is analyzed in Figure 6.10. As discussed earlier, there is a decrease in σ_{fusion} for heavier compound system, and here, also the pattern is the same for heavy targets. For a particular projectile, the fitted σ_{fusion} is in nice agreement with experimental data. The fitted value of neck-length parameter ΔR^{emp} for the particular choice of projectile and energy has been

TABLE 6.3

The DCM Calculated Fusion Cross Sections σ_{fus} for ^{7}Be-Induced Reactions on Different Target Nuclei at $E_{lab} \sim 17$ MeV, Both at $\Delta R = 0$ and $\Delta R^{emp.} = 1.130$ fm, Together with Their Comparison with the Available Experimental Data

Reaction	$E_{c.m.}$ (MeV)	E^*_{CN} (MeV)	T (MeV)	ℓ_{max} ($\hbar$)	$\sigma_{fus.}$ (mb) DCM calculated		Expt.
					$\Delta R = 0$ (fm)	$\Delta R^{emp.} = 1.130$ (fm)	
^{7}Be+^{27}Al→^{34}Cl*	13.50	36.51	3.24	30	8.003	646.74	635 ± 76
^{7}Be+^{32}S→^{39}Ca*	13.99	30.82	2.79	29	6.33×10^{-2}	553.00	–
^{7}Be+^{40}Ca→^{47}Cr*	14.47	29.95	2.49	30	4.59×10^{-5}	125.90	–
^{7}Be+^{48}Ti→^{55}Fe*	14.84	45.26	2.80	36	1.18×10^{-5}	76.05	–
^{7}Be+^{58}Ni→^{65}Ge*	14.99	27.01	2.00	38	3.34×10^{-6}	62.84	61.1 ± 6.9
^{7}Be+^{65}Cu→^{72}As*	15.35	32.08	2.06	44	3.03×10^{-5}	37.00	–

Apparently, σ_{fus} is predicted for some of the reactions. This table is taken from Ref. [42].

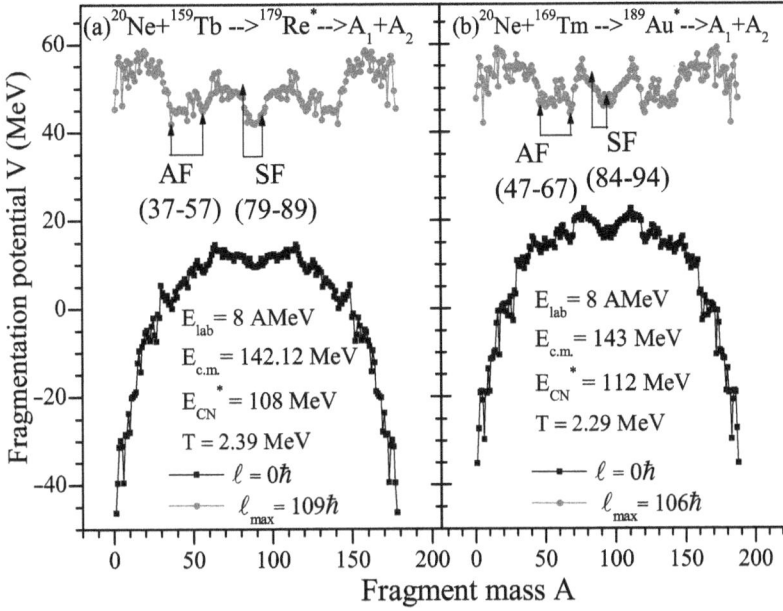

FIGURE 6.10

Mass dependence of fragmentation potential V (MeV) in the decay of (a) ^{179}Re* and (b) ^{189}Au* formed in ^{20}Ne-induced reactions with $E_{lab} = 8$ AMeV. This figure is based on Figure 1 in Ref. [43].

subsequently employed to predict σ_{fusion} for other reactions having the same projectile and energy but on some other stable targets, for which the experimental data is not available.

6.3.3.2 ^{20}Ne-Induced Reactions in Medium-Mass Region

In the context of the above results, it is quite interesting to explore the dynamics of reactions having the same projectile and energy for which the experimental data is also available. Here, we explore such reactions induced by ^{20}Ne projectile having E/A = 8 and 10 MeV/nucleon on ^{159}Tb and ^{169}Tm targets leading to medium mass composite nuclei [43], for which the cross sections of LPs and fission fragments are available experimentally [65]. Figure 6.10 depicts that n-rich LPs, i.e., $^{1,2,3}n$ and ^{4}H, are having lower fragmentation potential at $\ell = 0\hbar$, whereas at ℓ_{max} value, the asymmetric fission fragments (or heavy mass fragments (HMFs)) and SMFs are strongly minimized in comparison to LPs. For both nuclear systems, at $\ell = \ell_{max}$ value, there is strong minima at SMFs ($A_{CN}/2\pm10$), i.e., A = 79–89 and A = 84–94, plus the complementary fragments for ^{179}Re* and ^{189}Au*, respectively. In addition, another minima is seen in the neighborhood of SMFs window, which corresponds to HMFs (A = 37–57 and A = 47–67, plus the complementary fragments), respectively, for ^{179}Re* and ^{189}Au*, called as the HMFs window. The presence of HMFs window along with SMFs window advocates the presence of sub-structure in the fission of both composite systems.

The LPs and fission cross sections calculated by simultaneous fitting of ΔR of LPs and fission fragments are presented along with a comparison with experimental data [65] in Table 6.4. The calculated $\sigma_{fission}^{DCM}$ consists of asymmetric fission fragments (or HMFs) and SMFs contribution. The contribution of HMFs is more than SMFs in calculated $\sigma_{fission}^{DCM}$ for both composite systems. The large difference between the calculated $\sigma_{fission}^{DCM}$ and $\sigma_{fission}^{Expt}$

TABLE 6.4

The DCM Calculated LPs and Fission Cross Sections in the Decay of ^{179}Re* and ^{189}Au* along with a Comparison with Experimental Data [65]

E/A (MeV)	ℓ_{max} ($\hbar$)	ΔR (fm)		σ^{DCM} (mb)				σ^{Expt} (mb)		σ_{QF}^{Emp}
		LPs	HMFs/SFs	LPs	HMFs	SFs	$\sigma_{fission}$	σ_{LPs}	$\sigma_{fission}$	
^{20}Ne+^{159}Tb→^{179}Re*										
8	109	1.99	1.32	703.82	104.14	65.80	169.94	711±46	641±25	471.06
10	113	1.99	1.34	700.71	58.86	19.64	78.50	743±48	838±38	759.50
^{20}Ne+^{169}Tm→^{189}Au*										
8	106	1.99	1.32	334.64	36.26	8.46	44.72	351±59	1074±41	1025.28
10	110	1.99	1.34	300.85	15.90	2.52	18.42	300±85	1230±54	1211.58

The value of quasi fission (QF) has been evaluated empirically. This table is based on Table 1 in Ref. [43].

leading to disagreement between DCM calculated fission cross sections and data may be arising owing to the presence of non-compound nucleus quasi fission (QF) process, the contribution of which is evaluated empirically. It is important to note that here we have used different ΔR for LPs and heavy fragments showing that emission of LPs and heavy fragments occur at different time scales. The more value of ΔR for LPs shows that LPs emission occurs followed by the heavy fragments, as observed experimentally. Quite interestingly, the optimized value of ΔR of LPs and heavy fragments for one reaction, works well for the evaluation of LPs and fission cross section for another reaction having the same entrance channel mass asymmetry (η) 0.77, which is indicative of η-dependence of ΔR. The study of some more reactions with the same η will facilitate for validation of this observation.

6.4 Summary

It is seen that clustering is a generic phenomenon in light alpha and non-alpha conjugate nuclei, which is well supported by quantum mechanical fragmentation-based dynamical cluster-decay model. The temperature evolution of clustering scenario in alpha and non-alpha conjugate nuclei has been explored, comparatively, by taking into account the modified pairing strength $\delta(T)$ in liquid-drop energies. It is observed that in addition to xα- and xn-xα-type clusters in alpha conjugate and non-alpha conjugate systems, respectively, at T = 0 MeV and Ikeda's decay threshold value, the np-xα-type clusters become more probable in both alpha conjugate and non-alpha conjugate systems at higher experimental temperatures due to decreased $\delta(T)$ in liquid-drop energies. Furthermore, the presence of DIO as the primary mode of decay in C-yield as well as their large experimental cross section in ^{20,21,22}Ne*, compared to other Z-fragments gives clues about the effect of alpha clustering in the fragmentation process. The calculated cross sections are in good agreement with experimental Z-distribution data. Next, the role of N/Z ratio degree of freedom in the entrance channel and related structural as well as dynamical aspects in the fusion cross sections of A = 80 isobaric nuclear systems has been studied. With an increase in N/Z ratio of excited nuclear system, the mass distribution changes from symmetric to asymmetric one. The comparative contribution of LPs, IMFs, and SMFs in the fusion cross section is significantly influenced by the N/Z ratio of nuclear system. In another study, the significant correlation of neck-length parameter with barrier modification/lowering is highlighted in case of reactions having the same projectile and energy. Using this peculiar characteristic of neck-length parameter, the fusion cross sections predictions have been made within DCM for reactions having the same projectile and energy, but using some stable targets for which experimental data is unavailable.

Acknowledgment

This work is dedicated to the late Prof. Raj K. Gupta, Professor Emeritus, Panjab University, Chandigarh for his noble contribution in the advancement of quantum mechanical fragmentation theory (QMFT) in the form of dynamical cluster-decay model for explorations of nuclear reaction dynamics and synthesis of superheavy elements.

References

[1] L. R. Hafstad, E. Teller, *Phys. Rev.* 54, 681 (1938).

[2] K. Ikeda, N. Takigawa, and H. Horiuchi, *Prog. Theor. Phys. Suppl. E* 68, 464 (1968).

[3] W. Von Oertzen et al., *Eur. Phys. J. A* 11, 403 (2001); W. Von Oertzen , M. Freer, and Y. Kanada-Enyo, *Phys. Rep.* 432, 43 (2006); N. Itagaki, *J. Phys. Conf. Ser.* 420, 012080 (2013).

[4] R. K. Sheline and K. Wildermuth, *Nucl. Phys.* 21, 196 (1960).

[5] M. Freer, *Rep. Prog. Phys.* 70, 2149 (2007).

[6] M. Chernykh, H. Feldmeier, T. Neff, P. von Neumann-Cosel, A. Richter, *Phys. Rev. Lett.* 98, 032501 (2007).

[7] D. S. Delion, A. Sandulescu, W. Greiner, *Phys. Rev. C* 69, 044318 (2004).

[8] C. Xu, Z. Ren, G. Rpke, P. Schuck, Y. Funaki, H. Horiuchi, A.Tohsaki, T. Yamada, B. Zhou, *Phys. Rev. C* 93, 011306(R) (2016).

[9] Y. Kanada-Enyo and H. Horiuchi, *Prog. Theor. Phys. Suppl.* 142, 205 (2001).

[10] H. Feldmeier, J. Schnack, *Rev. Mod. Phys.* 72, 655 (2000).

[11] H.W. Fulbright, *Ann. Rev. Nucl. Part. Sci.* 29, 161 (1979).

[12] F. D. Becchetti et al., *Nucl. Phys. A* 339, 132 (1980); D. Jenkins, *J. Phys. Conf. Series* 436, 012016 (2013).

[13] W. B. He et al., *Phys. Rev. C* 94, 014301 (2016).

[14] G. V. Rogachevet al., *Prog. Theor. Phys. Suppl.* 196, 184 (2012).

[15] G. V. Rogachev et al., *Phys. Rev. C* 64, 051302(R) (2001); E. D. Johnson et al., *Eur. Phys. J. A* 42, 135 (2009); J. A. Scarpaci et al., *Phys. Rev. C* 82, 031301(R) (2010).

[16] M. L. Avila, G. V. Rogachev, V. Z. Goldberg et al., *Phys. Rev. C* 90, 024327 (2014).

[17] W. Von Oertzen et al., *Eur. Phys. J. A* 43, 17 (2010).

[18] J. Vadas, T. K. Steinbach, J. Schmidt et al., *Phys. Rev. C* 92, 064610 (2015).

[19] D. Shapira et al., *Phys. Lett. B* 114, 111 (1982); W. Dünnweber et al., *Phys. Rev. Lett.* 61, 927 (1988).

[20] S. Kundu et al., *Phys. Rev. C* 78, 044601 (2008) ; A. Dey et al., *Phys. Rev. C* 76, 034608 (2007) ; C. Bhattacharya et al., *Phys. Rev. C* 72, 021601(R) (2005).

[21] R. K. Gupta, R. Kumar, N. K. Dhiman, M. Balasubramaniam, W. Scheid et al., *Phys. Rev. C* 68, 014610 (2003).

[22] R. K. Gupta, S. K. Arun, R. Kumar, Niyti, *Int. Rev. Phys. (IREPHY)* 2, 369 (2008).

[23] M. Balasubramaniam, R. Kumar, R. K. Gupta, C. Beck and W. Scheid, *Phys. Rev. C* 68, 014610 (2003).

[24] B. B. Singh, M. K. Sharma, R. K. Gupta, *Int. J. Mod. Phys. E* 15, 699 (2006); ibid *Phys. Rev. C* 77, 054613 (2008).

[25] B. B. Singh, M. Kaur, V. Kaur, and R. K. Gupta, EPJ Web of Conf. 86, 00048 (2015); B. B. Singh, M. Kaur, and R. K. Gupta, *JPS Conf. Proc.* 6, 030001 (2015); M. Kaur, B. B. Singh, and S. Kaur, *AIP Conf. Proc.* 1953, 140113 (2018).

[26] J. Marhun and W. Greiner, *Phys. Rev. Lett.* 32, 548 (1974).

[27] H. J. Fink, J. Marhun, W. Scheid and W. Greiner, *Z. Phys.* 268, 321 (1974).

[28] R. K. Gupta, W. Scheid, and W. Greiner, *Phys. Rev. Lett.* 35, 353 (1975).

[29] S. K. Arun, Raj K. Gupta, B. B. Singh, S. Kanwar, and M. K. Sharma *Phys. Rev. C* 79, 064616 (2009).

[30] S. K. Arun, Raj K. Gupta, S. Kanwar, B. B. Singh, and M. K. Sharma *Phys. Rev. C* 80, 034317 (2009).

[31] M. Kaur, B. B. Singh, S. K. Patra, and Raj K. Gupta, *Phys. Rev. C* 95, 014611 (2017).

[32] M. Kaur, B. B. Singh, and S. K. Patra, *Indian J. Pure App. Phys.* 57, 584 (2019).

[33] P. Arumugam et al., *Phys. Rev. C* 71, 064308 (2005).

[34] M. Kaur, B. B. Singh, S. K. Patra, and Raj K. Gupta, *J. Nucl. Phys., Material Sci., Rad. and App.* 5, 319 (2018).

[35] M. M. Coimbra et al., *Nucl. Phys. A* 535, 161 (1991); Kundu S., Parmana 82, 727 (2014).

[36] C.J. Horowitz, H. Dussan, D.K. Berry, *Phys. Rev. C* 77, 045807 (2008).

[37] M. Alcorta, K. E. Rehm, B. B. Back, S. Bedoor et al., *Phys. Rev. Lett.* 106, 172701 (2011); J.J. Kolata, A. Roberts, A. M. Howard, D. Shapira et al., *Phys. Rev. C* 85, 054603 (2012).

[38] T. K. Steinbach, J. Vadas, J. Schmidt, C. Haycraft et al., *Phys. Rev. C* 90, 041603(R) (2014).

[39] M. Kaur, B. B. Singh, S. Kaur, and Raj K. Gupta, *Phys. Rev. C* 99, 014614 (2019).

[40] P. R. S. Gomes, M. D. Rodríguez, G. V. Martí, I. Padron et al., *Phys. Rev. C* 71, 034608 (2005); G. Montagnoli, A. M. Stefanini, C. L. Jiang, H. Esbensen et al., ibid. 85, 024607 (2012); 87, 014611 (2013).

[41] M. Kaur, B. B. Singh, M. K. Sharma, and Raj K. Gupta, *Phys. Rev. C* 92, 024623 (2015).

[42] M. Kaur, Ph.D. thesis, SGGSWU (2018).

[43] M. Kaur and B. B. Singh, *Int. J. Pure App. Phys.* 13, 162 (2017).

[44] H. Kröger and W. Scheid, *J. Phys. G* 6, L85 (1980).

[45] W. D. Myers and W. D. Swiatecki, *Nucl. Phys. A* 81, 1 (1966).

[46] N. J. Davidson, S. S. Hsiao, J. Markram, H. G. Miller and Y. Tzeng, *Nucl. Phys. A* 570, 61c (1994).

[47] P. A. Seeger, *Nucl. Phys.* 25, 1 (1961).

[48] G. Audi and A. H. Wapstra, *Nucl. Phys. A* 595, 409 (1995); G. Audi, A. H. Wapstra, and C. Thiboult, *Nucl. Phys. A* 729, 337 (2003).

[49] P. Möller, J. R. Nix, W. D. Myers and W. J. Swiatecki, *At. Data Nucl. Data Tables* 59, 185 (1995).

[50] R. K. Gupta, S. K. Arun, R. Kumar and Niyti, *Int. Rev. Phys. (IREPHY)* 2, 369 (2008).

[51] M. Bansal, R. Kumar, and R. K. Gupta, *J. Phys.: Conf. Series* 321, 012046 (2011).

[52] R. K. Gupta, M. Balasubramaniam, R. Kumar, N. Singh, M. Manhas *et al.*, *J. Phys. G* 31, 631 (2005).

[53] R. K. Gupta, M. Manhas, and W. Greiner, *Phys. Rev. C* 73, 054307 (2006).

[54] R. K. Gupta, N. Singh, M. Manhas, *Phys. Rev. C* 70, 034608 (2004).

[55] S. S. Malik and R. K. Gupta, *Phys. Rev. C* 39, 1992 (1989).

[56] K. Kalita et al., *Phys. Rev. C* 73, 024609 (2006).

[57] E. Martinez-Quiroz et al., *Phys. Rev. C* 90, 014616 (2014).

[58] C. Beck et al., *Phys. Rev. C* 67, 054602 (2003).

[59] M. Sinha, H. Majumdar, P. Basu, S. Roy, R. Bhattacharya, M. Biswas, M. K. Pradhan, and S. Kailas, *Phys. Rev. C* 78, 027601 (2008).

[60] K. Bodek et al., *Nucl. Phys. A* 339, 353 (1980).

[61] G. V. Marti et al., *Phys. Rev. C* 71, 027602 (2005).

[62] P. R. S. Gomes et al., *Phys. Rev. C* 73, 064606 (2006).

[63] V. V. Parkar et al., *Phys. Rev. C* 82, 054601 (2010).

[64] Y. D. Fang et al., *Phys. Rev. C* 91, 014608 (2015).

[65] J. Cabrera et al., *Phys. Rev. C* 68, 034613 (2003).

7

Explorations within the Preformed Cluster Decay Model

BirBikram Singh and Mandeep Kaur

Sri Guru Granth Sahib World University

CONTENTS

7.1 Introduction

This chapter takes us back to the stimulating years of the twentieth century when the study of radioactivity led to the evolution of fields such as nuclear physics, high-energy physics, and radiation physics, and to many applications in medicine and life sciences. The term radioactivity coined by Marie Curie was not identified until Henri Becquerel, trying to understand the connection of Roentgen's X-rays with fluorescence phenomena, discovered in 1895 a mysterious radiation of uranium. Curie observed that Th also emits such radiations, and the new elements Ra and Po they discovered were strong emitters. E. Rutherford deflected the charged particles named as alpha and beta rays in magnetic and electric fields, and from scattering experiments, he deduced that atomic particles consisted primarily of empty space surrounding a well-defined central core called nucleus. By bombarding nitrogen with particles, Rutherford demonstrated the production of a different element, oxygen. The experimental demonstration of artificial transmutation of nuclei in a nuclear reaction by Rutherford in 1919 further pushed the research of atomic nucleus and nuclear properties. He was the first to artificially transmute one element into another and to elucidate the concepts of the half-life and decay constant.

7.1.1   Alpha and Cluster Radioactivity

Among the numerous radioactive decay modes, α-decay of an unstable nucleus is found to be the most common decay mode for heavy and superheavy elements (SHE). During recent decades, the study of the α-cluster decay of radioactive nuclei has once again become a stimulating and popular research topic in the nuclear physics community, as it offers a valuable tool for probing atomic structure. In 1911, Geiger and Nuttal gave a semi-empirical relationship between the α-decay half-life and the range of α-particles in the air. This explanation of alpha-decay through the quantum tunneling process was first presented by Gamow, Condon, and Gurney (1928), theoretically. Here, we have explored the evolution of α preformation probability P_0 within the trans-lead parent nuclei, specifically, having shell closure N = 126 and Z value moving away from 82.

The phenomenon of cluster radioactivity (CR), which is the spontaneous emission of fragments heavier than alpha-particle, is now established. It was first explored theoretically in 1980 [1], later on confirmed experimentally in 1984 [2], and grew up reaching a stable state in the period 1990–2010. However, this was not the only inspiration of an intense experimental study that, in the course of about thirty years, allowed to measure the decay properties of some trans-lead exotic emitters of clusters with mass numbers in the range of 14–34. The discovery of CR provoked the dynamic interest of both experimenters and theorists. However, observation of CR was relatively a difficult experimental problem. In recent decades, the emissions of clusters ranging from ^{14}C to ^{34}Si were detected.

The emission of heavy cluster ^{14}C nucleus by ^{223}Ra was detected experimentally in 1984 by Rose and Jones [2]. However, this phenomenon was already observed in 1951 from the fission data of ^{232}U by Jaffey and Hirsh for the decay of ^{24}Ne. But they did not separate it from spontaneous fission process [3]. Later on, Bonetti et al. observed the cluster decay of ^{24}Ne from ^{232}U in 1951 [4,5], of ^{28}Mg in 1989 [6], of ^{20}O in 1991 [7], and of ^{23}Fe in 1992 [8]. Other most important milestones are the first (and, so far, the one and only) fine structure measurement in 1989 [9], the experiments on hindrance factors of odd-A emitters (1992–1993) [10], the attempts to extend the measurements in the trans-tin region (1994–1997) [11], and the observation of the heaviest cluster, ^{34}Si (2000) [12]. In 2001, Bonetti et al. [13] experimentally studied the cluster decay of ^{230}U isotope through the emission of ^{22}Ne and ^{24}Ne clusters. Since then, CR related to ^{14}C, ^{20}O, ^{23}Fe, ^{22,24,26}Ne, ^{28,30}Mg, and ^{32,34}Si emissions has been identified further. The common feature of these cluster emissions is that heavier nuclei emit heavier fragments in such a way that daughter nuclei are always doubly magic ^{208}Pb or nuclei in its vicinity.

Various theoretical approaches can be employed to investigate cluster emission. Shi et al. investigated the cluster decay of some unstable nuclei by considering an interacting potential comprising both Coulomb and proximity potentials for the daughter ^{208}Pb and its neighboring nuclei [14,15]. Buck et al. studied alpha and exotic decay of radioactive nuclei using the unifying model in 1991 [16,17]. Santosh et al. [18] studied the cluster decay of some experimentally observed decay modes of some radioactive nuclei in the trans-lead region using the Coulomb proximity potential model. Tavares et al. [19] studied the exotic decay of heavy nuclei within the framework of a semi-empirical, one-parameter model based on a quantum mechanical, tunneling mechanism through a potential barrier by taking into account both centrifugal and overlapping effects in half-life evaluations. In 2007, the alpha-decay half-lives of $^{218-219}$U isotopes were experimentally measured by Leppänen et al. [20]. Recently, Warda et al. [21] studied CR of actinide nuclei using the mean-field Hartree-Fock-Bogoliubov theory with the phenomenological Gogny interaction. In 2011, Sheng et al. [22] calculated half-lives for few cluster decay modes of $^{222-236}$U isotopes and also for some other nuclei in the trans-tin and trans-lead region using the suitable liquid drop description with the variable mass asymmetry shape and effective inertial coefficient.

In 2011, Mirea et al. [23] studied the heavy cluster emission of ^{24}Ne from ^{232}U and probable candidates for cluster emission from ^{234}U using the macroscopic-microscopic model.

In addition to the above-mentioned theoretical approaches, Gamow's theory of alpha cluster emission can be theorized to characterize the heavy cluster emissions [24]. Gamow's theory of decay indicates that the preformation factor is an important parameter that provides the information about the clusterization process in the nuclear structure, i.e. the cluster is preborn before penetrating through the Coulomb barrier generated by the electrostatic interaction of the α-particle with the constituent protons and neutrons of the daughter nucleus. This is important to mention two approaches here, namely, the unified fission model (UFM), which is known as analytic super asymmetric model (ASAFM) [25], and the preformed cluster model (PCM) based on collective clusterization approach [26], to understand the innovative phenomenon of CR. Gamow's theory uses the square well potential for α-decay, whereas the UFM and PCM use more realistic potential for the decay of α-cluster as well as heavier clusters [27,28]. Moreover, the UFM and PCM conflict with each other for the exclusion and inclusion of preborn probability before tunneling through the confined Coulomb potential barrier. Thus, in the UFM, the preborn probability is taken to be unity, whereas the PCM [29,30], based on quantum mechanical fragmentation theory (QMFT) [31], assumes the clusters to be preborn inside the mother nucleus with specific probabilities calculated by solving the stationary Schrödinger equation quantum mechanically. During these successive years, Gupta and collaborators studied α-decay and the cluster decay of heavy nuclei in the trans-lead region. This chapter summarizes the numerous studies done within PCM by Gupta and collaborators in the past decades, particularly those cluster radioactive decays in which daughter nucleus is always ^{208}Pb, in order to explore the deformation effects, exclusively, for the same.

7.1.2 Shell Corrections and Cluster Radioactivity

The nuclear shell structure was successfully explored by Mayres and Jansen in 1949 [32] while investigating particular characteristics or the extra stability of nuclei having either neutron or proton number or both equal to 2, 8, 20, 28, 50, 82, and 126 also called magic numbers, which are numbers of nuclear properties corresponding to magic nuclei or their neighbors and provide fingerprints of nuclear shell closure effects. CR is an innovative kind of natural radioactivity powerfully connected with shell effects. It could be considered definitely as the spontaneous emission of a double-magic ^{208}Pb nucleus or its close neighbors from a heavier mother nucleus. Several cluster decays from various radioactive nuclei in the neighborhood of ^{208}Pb have been observed experimentally, as mentioned in the previous section.

Numerous studies have been presented for the cluster emission from parent nuclei in the trans-lead region [33,34]. The entire family of the aforesaid radioactive decays consists of massive clusters (^{12}C–^{34}Si), which are emitted from various actinides (^{221}Fm–^{242}Cm) in the trans-lead region. This decay sometimes is also referred to as lead radioactivity and points out the profound effect of shell structure on this phenomenon. All the observed heavy clusters in the trans-lead region are associated with the doubly magic ^{208}Pb (Z = 82, N = 126) or its adjoining nucleus as a daughter nucleus. The phenomenological liquid drop model (LDM) was successfully used in 1939 by Bohr and Wheeler to describe numerous properties of the induced and spontaneous fission phenomena, as latter development was also shown to continue via quantum tunneling. But fission half-lives comparable with experimental data have only been obtained by the theory after Strutinsky added in 1967 to the LDM deformation energy a microscopic shell and pairing correction based on the single-particle deformed shell model [35]. The QMFT-based fragmentation potential comprises binding energies (BE) on the basis of semi-empirical mass formula of Seeger [36] (liquid drop energy (V_{LDM}) + shell corrections (δU) given by Myres and Swiatecki [37], within the Strutinsky renormalization procedure), Coulomb potential (V_c), and nuclear or proximity potential (V_P) of the cluster and daughter nuclei in the ground-state decay of radioactive parent nucleus.

In this chapter, we work out separate contributions of V_{LDM}(T=0) and δU (T=0) in order to investigate the role of fragmentation potential as well as other aspects in the CR with and without inclusion of shell corrections, i.e. to further explore its role in the decay process mainly for the calculation of P_0. The details of the methodology followed are given in the following section.

7.2   The Preformed Cluster Decay Model

Many theoretical models were advanced [38–41] to understand the process of exotic cluster decay in terms of nuclear α-decay or nuclear fission. These models fall into two main categories: i) unified fission models (UFM) and ii) preformed cluster models (PCM). Although the presumptions of the UFM and PCM are completely different, both seem to be working nicely in order to understand the spontaneous cluster/α-decay process. In the UFM, the cluster decay is dealt simply as a barrier penetration problem, whereas in the PCM, it is considered to happen in two steps: preformation of cluster with a certain probability P_0, and its penetration P across the interaction potential barrier existing between two cluster and daughter nuclei. The preformed cluster decay model [29,30,42,43] has been developed by adopting mainly Gamow's theory of α-decay. Here, instead of a square well potential, a more realistic nuclear potential, the nuclear proximity potential, is used and also a preformation probability P_0 is associated with each of the emitted clusters. In Gamow's theory of α-decay, the preformation probability for α-decay is assumed to be unity, since only α-cluster is considered to be emitted. In the PCM, preformation probability is different for different clusters. Moreover, it decreases with the increasing size of the cluster. It is relevant to mention here that in the PCM, the preformation probabilities for all the possible clusters can be obtained, whereas in another model like Blendowske et al. [44], it has been evaluated empirically. The decay constant in the PCM is defined as

$$\lambda = \nu_0 P P_0. \tag{7.1}$$

The corresponding half-life is given by

$$T_{\frac{1}{2}} = \frac{ln2}{\lambda}. \tag{7.2}$$

Here, ν_0 is the impinging frequency with which the cluster hits the barrier. P is the penetration probability that gives the probability of penetration of the barrier formed by the outgoing deformed and oriented fragment. P_0 is the probability of the formation of the cluster within the mother nucleus, calculated within QMFT.

The QMFT is worked out in terms of the following collective variables:

i.   Relative separation coordinate R between the two interacting nuclei or, particularly, two fragments (or, equivalently, the length parameter $\lambda = L/2R_0$, with L as the length of the nucleus and R_0 as the radius of an equivalent spherical nucleus).

ii. The deformation coordinates $\beta_{\lambda i}$ ($\lambda = 2,3,4...$ and i=1,2) of the incoming and outgoing nuclei.

iii. The orientation degrees of freedom $\theta_i(i = 1, 2)$ of the deformed nuclei.

iv. Azimuthal angle ϕ between the principal planes of the two interacting nuclei.

v. Neck parameter ε, defined by the ratio $\varepsilon = E_0/E'$ for the interaction region ($R < R_1 + R_2$, R_i (i = 1, 2) is the radius of the two nuclei). Here, E_0 is the actual height of the barrier, and E' is the fixed barrier of the two-center oscillator. $\epsilon = 0$ represents a broad neck formation, whereas $\epsilon = 1$ provides that the neck is fully compressed in reference to the asymptotic region ($R > R_1 + R_2$).

vi. Mass and charge fragmentation coordinates [31].

The mass and charge fragmentation for separated nuclei/fragments for the two body channels are defined by the mass and charge-asymmetry coordinates as $\eta = \frac{A_1 - A_2}{A}$; $\eta_Z = \frac{Z_1 - Z_2}{Z}$.

A_1, A_2 and Z_1, Z_2 are the mass numbers and the charge numbers of two incoming or outgoing nuclei, respectively. A and Z are the mass number and charge number of the compound nucleus, respectively. In the PCM, the two coordinates η and R refer, respectively, to the nucleon division (or exchange) between the daughter and cluster, and the transfer of positive Q-value to the total kinetic energy ($E_1 + E_2$) of two nuclei as they are produced in the ground state, as already pointed out earlier.

Solution of the stationary Schrödinger equation gives the probability P_0 of finding the mass fragmentation η at a fixed R on the decay path:

$$P_0(A_2) \propto |\psi^\nu(A_2)|^2 \tag{7.3}$$

with $\nu = 0,1,2,3...$ referring to ground-state ($\nu = 0$) and excited-state ($\nu = 1,2,3$) solutions.

Thus, the structure information of the parent nucleus is contained in P_0 via the fragmentation potential, which is calculated as

$$V(\eta, R, \ell) = -\sum_{i=1}^{2} B_i(A_i, Z_i, \beta_{\lambda i}) + V_c(R, Z_i, \beta_{\lambda i}, \theta_i, \phi)$$
$$+ V_p(R, A_i, \beta_{\lambda i}, \theta_i, \phi) + V_\ell(R, A_i, \beta_{\lambda i}, \theta_i, \phi). \tag{7.4}$$

The first part of Eq. (7.4), i.e. binding energy B of a nucleus as the sum of liquid drop energy $V_{LDM}(T)$ [45] and shell correction δU. The detailed descriptions of $V_{LDM}(T)$ and shell correction δU are given in chapter 2 of Ref. [46]. In the present work, the significant nuclear structure information carried by the shell corrections via calculated values of P_0 has been explored by these calculations (discussed in detail in Section 7.3.3), with and without δU via fragmentation potential within the collective clusterization process of the QMFT, to study the ground-state decay of radioactive nuclei. V_C, V_P, and V_ℓ are the Coulomb, nuclear proximity, and angular-momentum dependent potentials, respectively. For ground-state decays, $\ell = 0$ is a good approximation [27].

In Eq. (7.4), the proximity potential V_P for deformed and oriented nuclei (see Figure 2 of [47]) is given as

$$V_P(s_0) = 4\pi \bar{R} \gamma b \Phi(s_0). \tag{7.5}$$

$\Phi(s_0)$ is the universal function, independent of the shapes of nuclei or the geometry of nuclear system, but dependent on the minimum separation distance s_0, as

$$\Phi(s_0) = -\frac{1}{2}(s_0 - 2.54)^2 - 0.0852(s_0 - 2.54)^3 - 3.437 exp(-\frac{s_0}{0.75}). \tag{7.6}$$

Here, s_0 is defined in units of b, i.e. s_0 is s_0/b, where b is the diffuseness. For determining the shortest distance s_0 between any two colliding nuclei, we use the expression of

Gupta et al. [47], obtained for all possible orientations of two equal or unequal, axially symmetric, deformed nuclei lying in one plane (Figure 2 of [47]). For a fixed R, the minimum distance s_0 is defined as

$$s_0 = R - R_1 - R_2 \tag{7.7}$$

with the projections X_i along the Z-axis given as

$$X_1 = R_1(\alpha_1)cos(\theta_1 - \alpha_1)$$
$$X_2 = R_2(\alpha_2)cos(180 + \theta_2 - \alpha_2) \tag{7.8}$$

and b is the diffuseness of the nuclear surface given by

$$b = \left[\pi/2\sqrt{3}\ln 9 \right]_{t_{10-90}}, \tag{7.9}$$

where t_{10-90} is the thickness of the surface in which the density profile changes from 90% to 10%. The projections along the collision Z-axis of nuclei. The mean curvature radius $\bar{R}$ for deformed, oriented nuclei is given by

$$\frac{1}{\bar{R}^2} = \frac{1}{R_{11}R_{12}} + \frac{1}{R_{21}R_{22}} + \frac{1}{R_{11}R_{21}} + \frac{1}{R_{11}R_{22}} \tag{7.10}$$

with the four principal radii of curvature R_{i1} and R_{i2} of the two reaction partners given by Eq. (7.4) of [47], and the radius vectors

$$R_i(\alpha_i) = R_{0i}\left[1 + \sum_\lambda \beta_{\lambda i}Y_\lambda^{(0)}(\alpha_i)\right] \tag{7.11}$$

and

$$R_{0i} = \left[1.28A_i^{1/3} - 0.76 + 0.8A_i^{-1/3}\right]. \tag{7.12}$$

The Coulomb potential for two interacting, deformed, and oriented nuclei is given as [48,49]

$$V_c(Z_i, \beta_{\lambda i}, \theta_i) = \frac{Z_1 Z_2 e^2}{R} + 3Z_1 Z_2 e^2 \sum_{\lambda, i=1,2} \frac{R_i^\lambda(\alpha_i)}{(2\lambda + 1)R^{\lambda+1}}Y_\lambda^{(0)}(\theta_i)$$

$$\left[\beta_{\lambda i} + \frac{4}{7}\beta_{\lambda i}^2 Y_\lambda^{(0)}(\theta_i)\right] \tag{7.13}$$

with R_i from Eq. (7.11). $Y_\lambda^{(0)}(\theta_i)$ is the spherical harmonics function. The mass parameters $(B_{\eta\eta})$, entering the P_0 calculation via the kinetic energy term, are the smooth classical hydrodynamical masses [50].

The penetrability P in Eq. (7.1) is the WKB integral between R_a and R_b, the first and second turning points, respectively. In other words, the tunneling begins at $R = R_a$ and terminates at $R = R_b$, with $V(R_b) = Q$-value for ground-state decay. Thus, as per Figure 7.1, the transmission probability P consists of three contributions [42,43]:

1. the penetrability P_i from R_a to R_i,

2. the (inner) de-excitation probability W_i at R_i and

3. the penetrability P_b from R_i to R_b

giving the penetration probability as

$$P = P_i W_i P_b. \tag{7.14}$$

The shifting of the first turning point from R_a to R_0, the compound nucleus radius, gives the penetrability P similar to that of Shi and Swiatecki [51] for spherical nuclei, which is known not to fit the experimental data without the adjustment of assault frequency ν_0. Following the excitation model of Greiner and Scheid [52], we take the de-excitation probability $W_i = 1$ for a heavy cluster decays, which reduces Eq. (7.14) to

$$P = P_i P_b, \tag{7.15}$$

where P_i and P_b are calculated using the WKB approximation, as

$$P_i = \exp\left[-\frac{2}{\hbar} \int_{R_a}^{R_i} \{2\mu[V(R) - V(R_i)]\}^{1/2} dR\right] \tag{7.16}$$

and

$$P_b = \exp\left[-\frac{2}{\hbar} \int_{R_i}^{R_b} \{2\mu[V(R) - Q]\}^{1/2} dR\right]. \tag{7.17}$$

Here, R_a and R_b are the first and second turning points, respectively. This means that the tunneling begins at $R = R_a$ and terminates at $R = R_b$, with $V(R_b) = Q$-value for ground-state decay.

The first (inner) turning point R_a is chosen at $R_a = R_t + \Delta R$, where $R_t = R_1 + R_2$, and the outer turning point is taken at R_b to give the Q-value of the reaction, i.e. $V(R_b) = Q$. ΔR is the neck-length parameter that assimilates the neck formation effects of two-center shell model [53]. This method of introducing the neck-length parameter R is also used in our dynamical cluster decay model (DCM) and in the scission-point [54] and saddle-point [55,56] (statistical) fission models for decay of a hot and rotating compound nucleus.

7.3 Alpha, Cluster Radioactivity, and Preformed Cluster Decay Model

In the framework of the preformed cluster decay model (PCM), α-decay and the spontaneous decay of different radioactive nuclei, leading to daughter nuclei in the trans-Pb region, have been studied over a wide range of the mass region. In this section, we address the spontaneous decay processes of different radioactive nuclei, and the roles of deformations and orientations have been investigated. The calculations are done using a spherical choice of fragmentation and with the inclusion of quadrupole and hexadecapole deformations. The results showing the effects mentioned above were illustrated in the form of fragmentation potential and preformation probability to extract information about various spontaneous decays present in the trans-Pb mass region. The role of shell corrections is also explored within the framework of the PCM in the trans-Pb region. This section is divided into three subsections, namely, α radioactivity, CR in the trans-Pb region, and the role of shell corrections in CR.

7.3.1 α-Radioactivity and α-Cluster Preformation Probability P_0^α

The various properties of the α-decay for trans-lead nuclei $83 \leq Z \leq 92$ with $N = 126$ have been studied within the formalism PCM, which is based on collective cauterization approach of QMFT [57]. The preformation probability (P_0^α) and its penetration (P^α) through the nuclear interaction barrier give the decay constant (λ) of the ground-state decay of radioactive nuclei. The various ground-state decay properties P_0^α, P^α, λ, and half-life time $T_{1/2}$ have been studied within the PCM. The consequence of increasing atomic charge (Z) for the decay of parent nuclei on P_0^α, P^α, λ, and $T_{1/2}$ has also been observed here. The observed results for the decay $T_{1/2}$ of mother nuclei under study are very well compared with the existing experimental data [58].

The variation of fragmentation potential with fragment mass (A_2) is depicted in Figure 7.1a. We observe here that the light mass clusters $(A \leq 4)$ are more favored than the heavier fragments. Among these light mass clusters, ^{4}He or alpha cluster is in strong competition with the other light mass clusters. These results are further reflected in Figure 7.1b, i.e. in the case of preformation probability for the ground-state decay of radioactive parent nuclei ^{218}U.

Figure 7.2a presents the P_0^α and P^α as functions of atomic number Z of the trans-lead parent nuclei. We observe here that P_0^α and P^α increase with increasing Z, i.e. as moving away from the shell closure $Z = 82$. Consequently, the values of P_0^α and P^α are major contributors to calculate the value of λ for the α-decay of the parent nuclei. It is to be noted here that the value of λ increases with an increase in Z, which results in decrease in the value of $T_{1/2}$, i.e. the stability of the parent nuclei decreases with increasing Z as shown in Figure 7.2b. The results for the PCM calculated $T_{1/2}$ are also very well compared with the experimental data.

7.3.2 Cluster Radioactivity in trans-Pb Region and Effects of Deformation

The ground-state cluster decays of various nuclei in the trans-Pb region have been observed by many experimental groups around the globe, which include ^{14}C, ^{18,20}O, ^{22}Ne, ^{23}F,

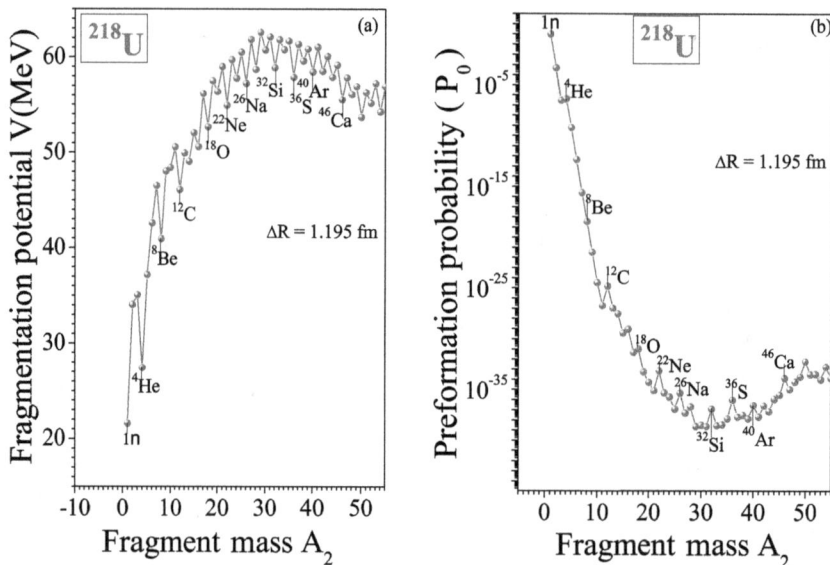

FIGURE 7.1
The variation of (a) fragmentation potential and (b) preformation probability with fragment mass A_2 for the ground-state decay of ^{218}U. This figure is based on Figure 1 of Ref. [57].

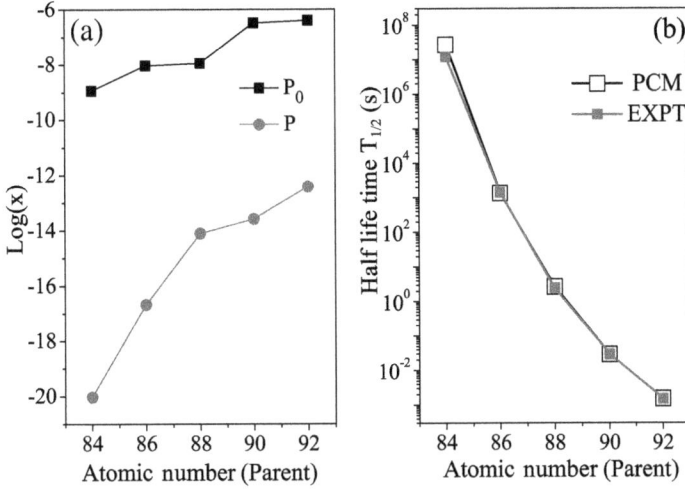

FIGURE 7.2
The variation of (a) Log(X), where X is $P_0{}^\alpha$ and P^α (b) half-life with atomic mass of trans-lead parent nuclei. This figure is based on Figure 2 of Ref. [57].

^{24,26}Ne, ^{28,30}Mg, and ^{32,34}Si. It is well established that these decays and their origin in the closed shell effects of daughter nuclei (^{208}Pb or its neighboring nuclei), understood via various theoretical studies including PCM of Gupta and collaborators [30]. These studies confined to only those heavy clusters in which a daughter formed is always a ^{208}Pb, which is spherical due to its being a double closed-shell nucleus (Z = 82, N = 126). Besides shell structure, the role of deformations and orientations has been investigated. The Q-value of these decays shows an exciting structure corresponding to measured decay half-lives, as expected, further highlighting the role of the shell structure. The calculated decay half-lives for all such cluster decays are in good agreement with measured values.

In Figure 7.3, we observe that for both the parents, there is no change in potential energy surface (PES) for the cluster mass up to $A_2 = 15$. For $A_2 \geq 15$, many new minima are seen in going from spherical to deformed. In Figure 7.3a, new minima at ^{28}Mg and ^{40}S are found in going from spherical to β_2 alone, and in the case of including higher multipole deformations (β_2, β_3, β_4), many new clusters such as ^{16,18,20}O, ^{24,26}F, and ^{40}S are energetically favored, which are not experimentally detected, and also get ruled out in calculations because of their small penetrability P. However, the relative preformation faction P_0 varies significantly, even for lighter clusters of masses $A_2 \leq 15$. Only the experimentally observed clusters with the ^{208}Pb daughter product are considered. Similar remarks apply to the panel (b), e.g. the observed ^{24}Ne cluster for the parent nucleus ^{232}U is at the minimum only in the spherical case. This is also illustrated for the instance of (β_2, β_3, β_4) via the charge dispersion potential $V(\eta_Z)$ (dashed the vertical line in panel (b)), where the minimum is found to lie at ^{24}F. Note, however, that both ^{24}F and ^{24}O are forbidden by penetrability P. Also, the ^{28}Mg cluster decay of ^{232}U, for which only the upper limit of $T_{1/2}$ is measured experimentally, becomes more and more favorable as deformation changes from spherical to β_2, and then to (β_2, β_3, β_4) case.

An analogy between measured and calculated $log_{10}T_{1/2}$-values for the choice of quadrupole deformations alone (β_2) and spherical ($\beta_2 = 0$) considerations is presented in Figure 7.4a. In Figure 7.4b, the Q-values of all observed clusters are plotted as a function of cluster mass. The structure is observed here corresponding to $log_{10}T_{1/2}$-values in Figure 7.4a, which highlights once again the well-observed role of shell effects played in the phenomenon

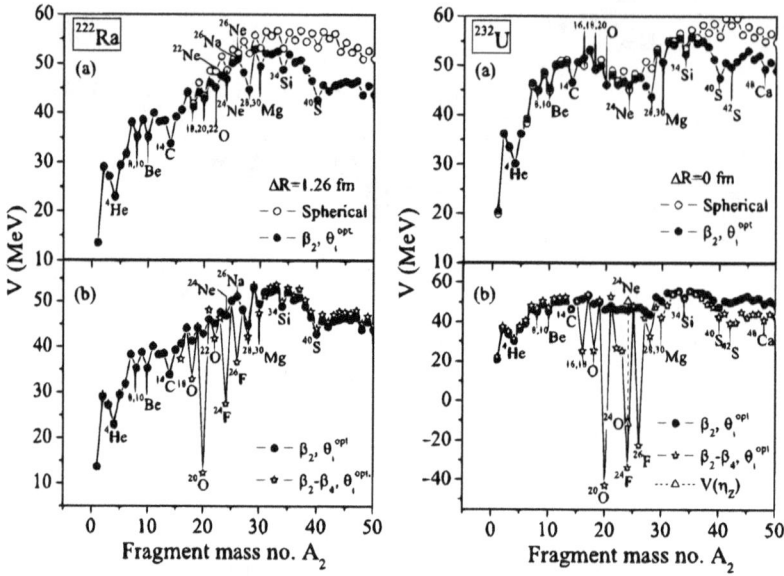

FIGURE 7.3
The fragmentation potentials for mother nuclei ^{222}Ra and ^{232}U, for cases of (a) spherical compared with quadrupole deformation β_2, and (b) quadrupole deformation β_2 alone compared with quadrupole plus octupole plus hexadecapole deformations (β_2-β_4), taken into account for all possible decaying fragments. This figure is based on Figure 2 of Ref. [29].

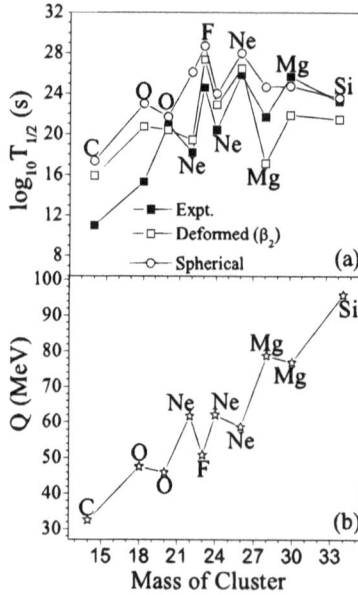

FIGURE 7.4
(a) Comparison of measured values of $log_{10}T_{1/2}$ for cluster decays ^{14}C, $^{18;20}$O, ^{22}Ne, ^{23}F, $^{24;26}$Ne, $^{28;30}$Mg, and $^{32;34}$Si of parents ^{222}Ra, ^{226,228}Th, ^{230}U, ^{231}Pa, ^{232,234}U, ^{236,238}Pu, and ^{242}Cm, respectively, having a daughter ^{208}Pb in each case, with the calculated $log_{10}T_{1/2}$ for cases of spherical (β_2), quadrupole deformations alone (β_2), and quadrupole, octupole, hexadecapole deformations (β_2, β_3, β_4). (b) Q-values of the cluster decays. This figure is based on Figure 5.3 of Ref. [59].

TABLE 7.1

Half-Life Times and Other Characteristic Quantities for Cluster Decay of Parent Nuclei Having Daughter Nuclei ^{208}Pb in Each Case, Using the PCM of Gupta and Collaborators, Extended to Include Deformation and Orientation Effects of Decaying Fragments

| | | | | **PCM** | | | | |
| | | | | **Half-Life Time ($T_{1/2}$)** | | | | |
Parent	*Cluster*	R_a	Q_{MN}	β_2	β_2, β_3, β_2	Sph	Expt Half-Life time ($T_{1/2}$)
^{222}Ra	^{14}C	$R_t+1.26$	32.47	15.89	11.20	17.36	11.01
^{226}Th	^{18}O	R_t	47.55	20.73		23.00	$\leq$15.3
^{228}Th	^{20}O	$R_t+0.50$	45.91	20.40		21.72	20.87
^{231}Pa	^{23}F	$R_t+0.25$	50.81	27.38	09.28	28.73	$\geq$24.61
^{230}U	^{22}Ne	R_t	61.69	19.41	01.27	26.15	$\geq$18.2
^{232}U	^{24}Ne	R_t	62.03	22.88	11.19	24.04	21.05
^{234}U	^{26}Ne	$R_t+0.50$	58.65	26.49	16.15	28.00	25.06
^{236}Pu	^{28}Mg	R_t	78.75	17.11		24.68	21.67
^{238}Pu	^{30}Mg	$R_t+0.50$	76.82	21.87	32.57	24.77	25.70
^{242}Cm	^{24}Si	R_t	95.78	21.47	33.96	23.59	23.24

The results are compared with the experimental data given in Refs. [60,61]. This table is based on Table 5.1 of Ref. [59].

of CR. Furthermore, the comparison of calculated and measured values of $log_{10}T_{1/2}$, for the two choices of deformed and spherical fragments of parent nuclei, points out the vital role being played by deformed and oriented nuclei, in addition to the role of well-established shell effects, which will be discussed in Section 7.3.3. We see that for almost all the cluster decays, the comparisons become good with the choice of deformed and oriented nuclei, except for few more massive cluster decays. The heavier clusters may follow the experimental data if better values of hexadecapole deformation β_2 are made available. The generalized orientation considerations may further add to the cause.

Table 7.1 summarizes the above results and characteristic quantities in our calculations for the cluster decay of parents ^{222}Ra, ^{226,228}Th, ^{230}U, ^{231}Pa, ^{232,234}U, ^{236,238}Pu, and ^{242}Cm. The outcomes for all the three choices are compiled here in this table, i.e. for spherical ($\beta_2=0$), quadrupole deformations alone ($\beta_2=0$), and the quadrupole, octupole, hexadecapole deformations ($\beta_2=0, \beta_3=0$, and $\beta_4=0$). The experimental results for $log_{10}T_{1/2}$ are given here for comparisons [60,61]. The Q-value for individual cluster decay is also presented here along with the fitted ΔR values in few cases, and in remaining cases, ΔR is zero. One can see here that β_2 alone provides decent comparisons with the experimental data among all the three choices. However, with the higher multipole deformations (β_2-β_4) included, the $log_{10}T_{1/2}$ is underestimated, possibly due to the inappropriate calculated β_4 values, except in one case of ^{14}C cluster decay of ^{222}Ra, where the comparison with measured half-life time turn into good only with the addition of higher multi-pole deformations.

7.3.3 Shell Correction and Cluster Radioactivity

In the present work, the role of shell corrections (δU) is investigated for the various radioactive parent nuclei in the trans-lead region [62], i.e. ^{222}Ra, ^{226,228}Th, ^{230,232,234}U, ^{236,238}Pu, and ^{242}Cm specifically, which lead to doubly magic ^{208}Pb daughter nucleus through emission of ^{14}C, ^{18,20}O, ^{22,24,26}Ne, ^{28,30}Mg, and ^{34}Si clusters, respectively, within the QMFT-based fragmentation potential [31]. The QMFT-based fragmentation potential comprises binding energies (BE) (liquid drop energy (V_{LDM}) + shell corrections (δU)), within the Strutinsky renormalization procedure [35]), Coulomb potential (V_c), and nuclear or proximity potential (V_P) of the cluster and daughter nuclei in the ground-state decay of radioactive parent nucleus. It is relevant to mention here that the contributions of V_{LDM} (T = 0) and δU (T = 0) in the BE have been analyzed within the Strutinsky renormalization procedure.

Figure 7.5 shows the variation of different components of fragmentation potential V (MeV) with fragment mass A for the radioactive nuclei (^{242}Cm). Figure 7.5a shows the variation of liquid drop potential energy with fragment mass A. This first decreases with an increase in the mass of the fragment, and then starts to increase. The variation of shell correction with fragment mass is shown in Figure 7.5b, which points out the minima or valleys for particular fragments. Coulomb potential is shown in Figure 7.5c, and nuclear potential or proximity potential shown in Figure 7.5d shows their respective behaviors which are complementary to each other. The above analysis suggests that the structure of the fragmentation potential is mainly due to the shell corrections, and the same is investigated in terms of P_0 in Figure 7.7. We further investigate the role of deformations and orientations for Coulomb and proximity potential shown in Figure 7.5c and d, respectively. The Coulomb potential is the same for the spherical and deformed considerations for some fragments in the decay of ^{242}Cm, but for $A \sim 51-200$ window, the Coulomb potential is more for the spherical case than the deformed one. One more observation is noticed in Coulomb potential that it shows small minima at cluster ^{34}Si with a complementary ^{208}Pb daughter. But converse is true for the proximity potential, and the proximity potential is more for the deformed case than the spherical one for $A \sim 51-200$ window.

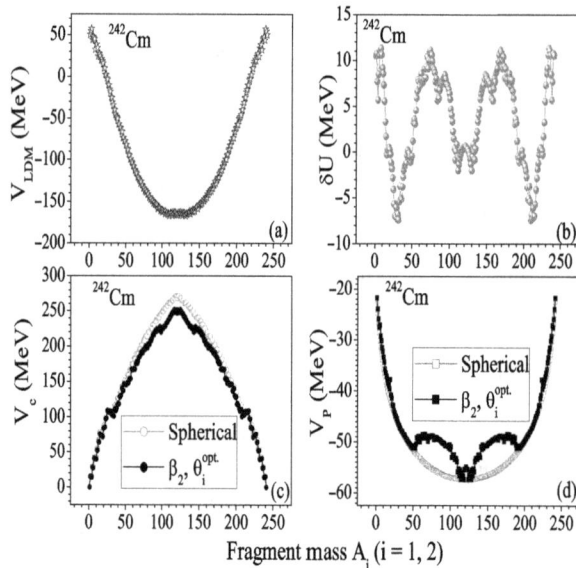

FIGURE 7.5
The variation of (a) liquid drop potential energy (V_{LDM}), (b) shell corrections (δU), (c) Coulomb potential (V_c), and (d) proximity potential (V_P) with fragment mass A. This figure is based on Figure 1 of Ref. [62].

Figure 7.6 shows the variation of fragmentation potential V (MeV) with fragment mass A for radioactive nuclei ^{242}Cm. Figure 7.6a shows that the experimentally observed clusters [61] ^{34}Si and ^{208}Pb are not strongly minimized by taking fragmentation potential as the sum of V_{LDM}, V_C, and V_P for spherical consideration. As we added δU, i.e. shell corrections in this potential, the structure of the fragmentation potential changes. The observed cluster ^{34}Si with ^{208}Pb daughter is strongly minimized. This is due to the fact that ^{208}Pb is doubly magic, and due to this nuclear property of magicity, the shell corrections shows a significant role. A similar type of observation is seen for quadrupole deformations in Figure 7.6b. Here, we also observe that there is no change in the potential energy surface for the cluster mass up to $A = 51$. For $A > 51$, many new minima are observed as we go from spherical consideration to deformed one. But we are not interested in the structure of the other fragments, except ^{34}Si; therefore, we observe that there is no significant change in the structure of decaying cluster with inculcation of deformation and orientations.

Following the fragmentation potential, the preformation probability (P_0) is also calculated for the ground-state decay of ^{242}Cm as shown in Figure 7.7a and b for spherical and deformed consideration, respectively. Again, we observe that there is no significant change in structure for the fragment mass up to $A = 51$, but for $A > 51$, many pronounced peaks are observed as we go from spherical choice of fragments to deformed one for both the cases, i.e. with and without shell corrections. Secondly, we observe here that the emitted cluster ^{34}Si with ^{208}Pb daughter is strongly preformed with the inculcation of shell correction term. On the other hand, without shell correction, its preformation probability is less as compared to its neighboring fragments.

Similarly, Figure 7.8 shows the calculations for P_0 in the decay of parents ^{222}Ra, ^{226,228}Th, ^{230,232,234}U, ^{236,238}Pu, and ^{242}Cm through emission of clusters ^{14}C, ^{18,20}O, ^{22,24,26}Ne, ^{28,30}Mg, and ^{34}Si, respectively. Here, we observe that the value of P_0 is much higher for the emitted clusters when shell corrections are included (filled circles) than the case without shell corrections (open circles). In general, we also see that the magnitude of P_0 decreases for the emitted clusters as the mass of the parent nucleus increases.

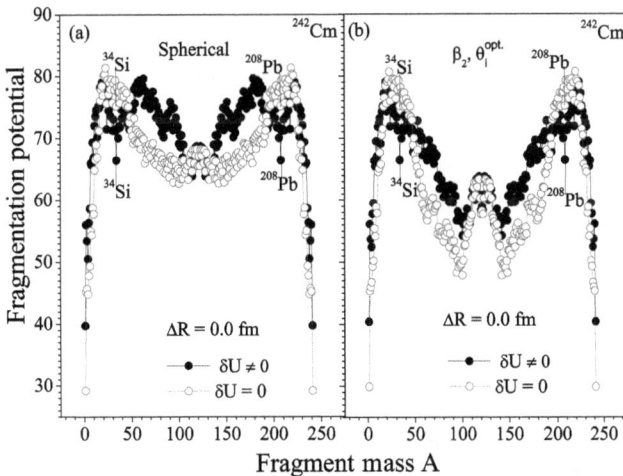

FIGURE 7.6

Fragmentation potential as a function of fragment mass A for the ground-state decay of ^{242}Cm with and without shell corrections for (a) spherical (b) deformed (β_2) choice of fragments. This figure is based on Figure 2 of Ref. [62].

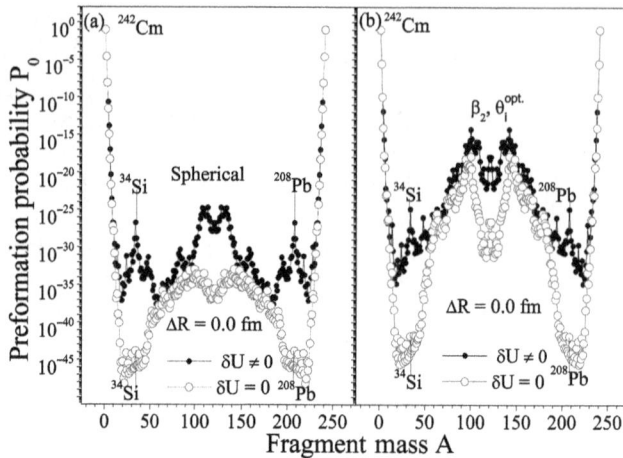

FIGURE 7.7
The same as for Figure 7.6, but for the preformation probability P_0. This figure is based on Figure 3 of Ref. [62].

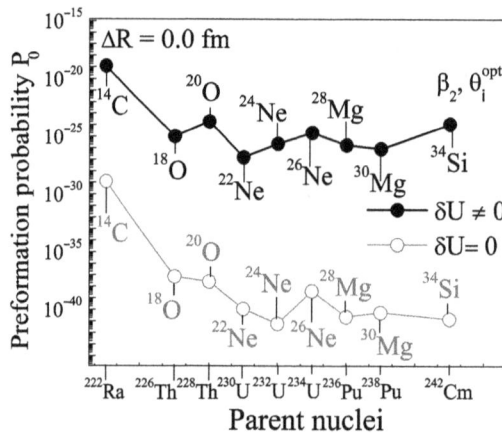

FIGURE 7.8
Preformation probability (P_0) with mass of parent nuclei for the ground-state decay with and without δU. This figure is based on Figure 4 of Ref. [62].

7.4 Conclusions

The ground-state decay of radioactive nuclei via the emission of α-cluster as well as the heavy cluster has been studied using QMFT-based PCM in the trans-lead region. It has been observed that the α-cluster is sharply minimized in the ground-state decay of ^{218}U, consequently having a higher P_0. We see that P_0^{α} and penetrability P^{α} increase with an increase in the value of Z, chosen for the parent nuclei having $N = 126$. Due to the rising value of P_0^{α} and P^{α}, there is a decrease in the value of $T_{1/2}$, i.e. the stability of the parent nuclei decreases with increasing Z. The calculated value of the $T_{1/2}$ using PCM is very well compared with the experimental data. In addition to the α-cluster decay, CR is also explored in the trans-lead region. The role of multipole deformations and orientations of nuclei are included in CR, and it is found that both deformations and orientations have vital effects on measured decay half-lives for cluster decay of parent nuclei having daughter ^{208}Pb always. The calculated values of cluster decay half-lives show better fits with calculations within PCM where the effects of deformations are included up to quadrupole deformation only. This work points out the importance of deformation and orientation effects in cluster decays of radioactive nuclei. Also, the Q-values calculated from available binding energies are shown to be very important for predicting the cluster-decay half-lives. The shell structure plays an essential role in the trans-Pb region due to the doubly magic daughter nucleus ^{208}Pb. The importance of shell corrections is explored quantitatively in the CR process, where the B with the contributions of V_{LDM} and δU reads as B $= V_{LDM} + \delta U$, within the Strutinsky renormalization procedure. Evidently, the fragmentation potential and preformation probability contain the nuclear structure information of the parent nuclear system and the decaying fragments. The calculations show that without δU, experimentally observed clusters, i.e. ^{14}C, ^{18,20}O, ^{22}Ne, ^{23}F, ^{24,26}Ne, ^{28,30}Mg, and ^{34}Si, are not strongly minimized or preformed. The results suggest that δU plays a crucial role in addressing the cluster dynamics of a variety of nuclear systems studied over here.

Acknowledgments

This Work is dedicated to Late Prof. Raj K. Gupta for his noble contribution to Nuclear Physics Community.

Bibliography

[1] A. Sandulescu, D. N. Poenaru, and W. Greiner, *Sov. J. Part. Nuclei* 11, 528 (1980).

[2] H. J. Rose and G. A. Jones, *Nature (London)* 307, 245 (1984).

[3] A. H. Jaffey and A. Hirsch unpublished data: quoted in: R. Vandenbosch and J. R. Huizenga, *Nuclear Fission* (Academic press, New York, 1973).

[4] R. Bonetti, E. Fioretto, C. Migliorino, A. Pasinetti, F. Barranco, E. Vigezzi and R. A. Broglia, *Phys. Lett.* B241, 179 (1990).

[5] S. W. Barwick, P. B. Price and J. D. Stevenson, *Phys. Rev. C* 31, 1984 (1985).

[6] S. P. Tretyakova et al., *Z. Phys. A* 333 (1989) p. 349;S. Wang, D. Snowden-Ifft, P.B. Price et al., *Phys. Rev C* 39, 1647 (1989).

[7] R. Bonetti et al., *Nucl. Phys. A* 556, 115 (1993).

[8] P. B. Price et al., *Phys. Rev. C* 46, 1939 (1992).

[9] L. Brillard et al., *C R Acad. Sci. Paris* 309 Ser II, 1105 (1989).

[10] R. Bonetti et al., *Nucl. Phys. A* 576, 21 (1994); R. Bonetti et al., *Nucl. Phys. A* 562, 32 (1993).

[11] A. Guglielmetti et al., *Phys. Rev. C* 56, R2912 (1997).

[12] A. A. Ogloblin et al., *Phys. Rev. C* 61, 034301 (2000).

[13] R. Bonetti et al., *Nucl. Phys. A* 686, 64 (2001).

[14] Y. J. Shi and W. J. Swiatecki, *Nucl. Phys. A* 438, 450 (1985).

[15] Y. J. Shi and W. J. Swiatecki, *Nucl. Phys. A* 464, 205 (1987).

[16] B. Buck, A. C. Merchant, and S. M. Perez, *J. Phys. G, Nucl. Part. Phys.* 17, 91 (1991).

[17] B. Buck, A.C.Merchant, S.M. Perez, and P. Tripe, *J. Phys. G, Nucl. Part. Phys.* 20, 351 (1994).

[18] K. P. Santhosh and A. Joseph, *Pramana J. Phys.* 59, 679 (2002).

[19] O. A. P. Tavares, L. A. M. Roberto, and E. L. Medeiros, *Phys. Scr.* 76, 375 (2007).

[20] A. P. Leppänen et al., *Phys. Rev. C* 75, 054307 (2007).

[21] M. Warda and L. M. Robledo, *Phys. Rev. C* 84, 044608 (2011).

[22] Z. Sheng, D. Ni, and Z. Ren, *J. Phys. G, Nucl. Part. Phys.* 38, 055103 (2011).

[23] M. Mirea, A. Sandulescu, and D. S. Delion, *Nucl. Phys. A* 870–871, 23 (2011).

[24] G. Gamow, *Eur. Phys. J. A* 51, 204 (1928).

[25] D. N. Poenaru, W. Greiner, M. Ivascu, and A. Sandulescu, *Phys. Rev. C* 32, 2198 (1985).

[26] S. S. Malik and R. K. Gupta, *Phys. Rev. C* 39, 1992 (1989).

[27] R. K. Gupta and W. Greiner, *Int. J. Mod. Phys. E* 3, 335 (1994).

[28] S. Kumar and R. K. Gupta, *Phys. Rev. C* 55, 218 (1997).

[29] S. K. Arun, R. K. Gupta, B. B. Singh, S. Kanwar, and M. K. Sharma, *Phys. Rev. C* 79, 064616 (2009).

[30] S. K. Arun, R. K. Gupta, S. Kanwar, B. B. Singh, and M. K. Sharma, *Phys. Rev. C* 80, 034317 (2009).

[31] H. J. Fink, W. Greiner, R. K. Gupta, S. Liran, J. H. Maruhn, W. Scheid, and O. Zohni, Proc. Int. Conf. on Reactions between Complex Nuclei, Nashville, USA, June 10–14, 2, 21 (1974); R. K. Gupta, et al., Proc. Int. School on Nucl. Phys., Predeal, Romania 75, (1974); J. Maruhn and W. Greiner, *Phys. Rev. Lett.* 32, 548 (1974).

[32] M. G. Mayer, *Phys. Rev.* 75, 1969 (1949); O. Haxel, J. H. D. Jansen, H. E Suess, *Phys. Rev.* 75, 1766 (1949).

[33] B. B. Singh, S. K. Patra, and R. K. Gupta, *Phys. Rev. C* 82, 014607 (2010).

[34] D. N. Poenaru, M. lvascu, A. Sandhulescu, and W. Greiner, *J. Phys. G: Nucl. Phys.* 10, L183 (1984).

[35] V. M. Strutinsky, *Nucl. Phys. A* 95, 420 (1967).

[36] P. A. Seeger, *Nucl. Phys.* 25, 1 (1961).

[37] W. Myers and W. J. Swiatecki, *Nucl. Phys.* 81, 1 (1966).

[38] G. A. Pik-Pichak, *Yad. Fiz.* **44**, 1421 (1986).

[39] R. Blendowske and H. Walliser, *Phys. Rev. Lett.* 61, 1930 (1988).

[40] B. Buck and A. C. Merchant, *J. Phys. G: Nucl. Phys.* 15, 615 (1989).

[41] A. Sandulescu, R. K. Gupta, F. Carstoiu, M. Horoi, and W. Greiner, *Int. J. Mod. Phys. E* 1, 374 (1992).

[42] R. K. Gupta, in Proceedings of the 5th International Conference on *Nuclear Reaction Mechanisms*, Varenna, edited by E. Gadioli (Ricerca Scientifica ed Educazione Permanente, Milano, 1988), 416 (1988).

[43] R. K. Gupta, in *Heavy Elements and Related New Phenomena*, edited by W. Greiner and R. K. Gupta (World Scientific, Singapore, 1999), Vol. II, 731 (1999).

[44] R. Blendowske, T. Fliessbach, and H. Walliser, *Nucl. Phys. A* 464, 75 (1987).

[45] P. Moller, J. R. Nix, W. D. Myers, and W. J. Swiatecki, *At. Data. Nucl. Data Tables* 59, 185 (1995).

[46] M. Kaur, Ph.D. Thesis, SGGSWU, Fatehgarh Sahib, India (2018).

[47] R. K. Gupta, N. Singh, and M. Mahhas, *Phys. Rev. C* 70, 034608 (2004).

[48] C. Y. Wong, *Phys. Rev. Lett.* 31, 766 (1973).

[49] R. K. Gupta, M. Balasubramaniam, R. Kumar, N. Singh, M. Mahhas, and W. Greiner, *J. Phys. G: Nucl. Part. Phys.* 31, 631 (2005).

[50] H. Kroger and W. Scheid, *J. Phys. G: Nucl. Phys.* 6, L85 (1980).

[51] Y. J. Shi and W. J. Swiatecki, *Phys. Rev. Lett.* 54, 300 (1985).

[52] M. Greiner and W. Scheid, *J. Phys. G: Nucl. Part. Phys.* 12, L229 (1986).

[53] J. Maruhn and W. Greiner, *Phys. Rev. Lett.* 32, 548 (1974).

[54] T. Matsuse, C. Beck, R. Nouicer, and D. Mahboub, *Phys. Rev. C* 55, 1380 (1997).

[55] S. J. Sanders, D. G. Kovar, B. B. Back, C. Beck, D. J. Henderson, R. V. F. Janssens, T. F. Wang, and B. D. Wilkins, *Phys. Rev. C* 40, 2091 (1989).

[56] S. J. Sanders, *Phys. Rev. C* 44, 2676 (1991).

[57] R. Kaur, B. B. Singh, M. Kaur, B. S. Sandhu and M. Kaur, AIP Conference proceedings, 1953, 140102 (2018).

[58] J. K. Tuli, Nuclear Wallet Cards (April 2005).

[59] B. B. Singh, Ph.D. Thesis, Thapar University, Patiala, India (2009).

[60] R. K. Gupta and W. Greiner, *Int. J. Mod. Phys. E* 3, 335 (Suppl., 1994).

[61] R. Bonetti and A. Guglielmetti, *Romanian Reports Phys.* 59, 301 (2007).

[62] M. Kaur, B. B. Singh and M. K. Sharma, AIP Conference proceedings, 1953, 140066 (2018).

8

Studies on Synthesis and Decay of Superheavy Elements with $Z = 122$

K. P. Santhosh and V. Safoora

Kannur University

CONTENTS

8.1   Introduction

The predictions of shell closure stability around $N = 184$ ("island of stability") pave the way to the synthesis of superheavy elements (SHEs), and now it becomes a topic at the forefront in experimental and theoretical nuclear physics research. The predicted stability was situated around the nucleus $^{298}114$ ($N = 184$), constituted an island, and was separated from the usual peninsula of relatively long-lived nuclei by an 'ocean' of full instability [1]. A remarkable progress in the synthesis of SHEs has been achieved by complete fusion reactions of ^{48}Ca on actinide targets (hot fusion) [2−14] and medium heavy projectile on lead/bismuth targets (cold fusion) [15−18]. Recently, the syntheses of the heaviest elements with $Z = 114$–118 using the hot fusion reactions and with $Z = 110$–113 using the cold fusion reactions have been reported. The element ^{294}Og (Oganesson) with proton number $Z = 118$ and neutron number $N = 176$ is the highest element observed so far [7]. The ^{48}Ca-induced fusion reactions ^{48}Ca + $^{239-240,242,244}$Pu ($Z = 114$) [2], ^{48}Ca + ^{243}Am ($Z = 115$) [3,4], ^{48}Ca + ^{245,248}Cm ($Z = 116$) [5], ^{48}Ca + ^{249}Bk ($Z = 117$) [6–8], and ^{48}Ca + ^{249}Cf ($Z = 118$) [9] lead to the synthesis of elements with $Z = 114$−118. Studies are still going on to synthesize elements with $Z > 118$. Many attempts are made to synthesize elements with $Z = 119$ and 120 using the fusion evaporation reactions ^{50}Ti + ^{249}Bk→$^{299}119$, ^{58}Fe + ^{244}Pu → $^{302}120$, ^{54}Cr + ^{248}Cm → $^{302}120$, and ^{64}Ni + ^{238}U → $^{302}120$ [19,20].

Evidences for the existence of SHE with $Z = 122$ was reported by Marinov et al. [21] in 2010 using inductively coupled plasma-sector field mass spectrometry (ICPMS). But Dellinger et al. [22] could not confirm this study using accelerator mass spectrometry (AMS), and therefore, the study of Marinov et al. [21] is not authenticated. By solving Langevin equations for the shape degrees of freedom, Litnevsky et al. [23] studied the evaporation residue (ER) formation cross sections using reactions leading to $Z = 122$, namely, ^{58}Fe + ^{248}Cm, ^{64}Ni + ^{244}Pu, and ^{90}Zr + ^{208}Pb. Giardina et al. [24] studied the possibilities to synthesize elements with $Z = 122$ using the reaction ^{54}Cr + ^{249}Cf and calculated ER cross section as 140 fb. Cap et al. [25] predicted that there is no chance to synthesize elements with $Z = 122$ using the symmetric reaction ^{154}Sm+^{154}Nd.

The cross section of superheavy nuclei produced in a heavy-ion fusion evaporation reaction is divided into three stages. Initially, the heavy-ion projectile overcomes the Coulomb barrier and is trapped in the potential pocket, and a composite system is formed, then the composite system evolves into an excited compound nucleus (CN), and finally, the CN cools down by emission of light particles for surviving fission. In nuclear reaction, the concept of probability of CN formation, P_{CN}, was put forward by Bohr [26]. The probability of CN defines the probability that the system will go from the configuration of two nuclei in contact into the configuration of CN, and the value of P_{CN} depends on the projectile-target used (charge and mass asymmetry in the entrance channel), quasi-fission barrier, bombarding energy, and hence the excitation energy. The survival probability of excited CN, W_{sur}, depends on the excitation energy, fission barrier, and neutron separation energy. Each step of the SHE formation should be systematically investigated to find the formation cross section. Our group developed a phenomenological model for production cross section (PMPC) to describe the three stages of ER formation [27]. Various models like fusion by diffusion model [28], dinuclear system model [29], and macroscopic-microscopic model [30] have been put forward to describe the three stages.

One of the main components for studying SHE is to observe their decay modes; thus, decay studies have achieved greater importance in the field of SHE. The dominant decay modes of SHE are α decay and spontaneous fission (SF). To detect an SHE through alpha particle channel, the SHE should have a smaller alpha decay half-life compared to SF half-life. We have developed two models, the Coulomb and proximity potential model (CPPM) [31] and Coulomb and proximity potential model for deformed nuclei (CPPMDN) [32], to describe alpha decay half-lives. Recently, we developed a modified generalized liquid drop model (MGLDM) [33], by incorporating the proximity potential of Blocki et al. [34] to the generalized liquid drop model to study alpha decay.

Using PMPC, in the present chapter, we calculated the possible isotopic production cross section for $Z = 122$ at energies near and above the Coulomb barrier by taking Coulomb and proximity potential as the interaction barrier. The studies on the decay of SHE with $Z = 122$ have also been performed in this chapter using the MGLDM.

8.2 Phenomenological Model for Production Cross Section (PMPC)

The cross section of SHE production in a heavy-ion fusion reaction with subsequent emission of x neutrons is the product of capture cross section, the fusion probability, and survival probability:

$$\sigma_{ER}^{xn} = \frac{\pi}{k^2} \sum_{\ell=0}^{\infty} (2\ell + 1) T(E, \ell) P_{CN}(E, \ell) W_{sur}^{xn}(E^*, \ell). \tag{8.1}$$

The capture cross section at a given center-of-mass energy E can be written as the sum of the cross section for each partial wave ℓ:

$$\sigma_{capture} = \frac{\pi}{k^2} \sum_{\ell=0}^{\infty} (2\ell + 1) T(E, \ell). \tag{8.2}$$

Wong [35] approximated the various barriers for different partial waves by inverted harmonic oscillator potentials of height E_ℓ and frequency ω_ℓ, and arrived at the total cross section for the fusion of two nuclei. For energy E, the probability for the absorption of ℓth partial wave given by Hill-Wheeler formula [36] is

$$T(E, \ell) = \{1 + \exp[2\pi(E_\ell - E)/\hbar\omega_\ell]\}^{-1}. \tag{8.3}$$

Using some parameterizations in the region $\ell = 0$ and replacing the sum in Eq. (8.2) by an integral, Wong gave the total/capture cross section as

$$\sigma_{capture} = \frac{R_0^2 \hbar\omega_0}{2E} \ln\left\{1 + \exp\left[\frac{2\pi(E - E_0)}{\hbar\omega_0}\right]\right\}, \tag{8.4}$$

where R_0 is the barrier radius, E_0 is the barrier height, and $\hbar\omega_0$ is the curvature of the inverted parabola for $\ell = 0$.

The effective potential between two nuclei during interaction is

$$V = \frac{Z_1 Z_2 e^2}{r} + V_P(z) + \frac{\hbar^2 \ell(\ell + 1)}{2\mu r^2}, \tag{8.5}$$

where Z_1 and Z_2 are the atomic numbers of projectile and target, respectively; r is the distance between the centers of the colliding nuclei; z is the distance between the near surfaces of the projectile and target; ℓ is the angular momentum; and μ is the reduced mass.

The term $V_P(z)$ is the proximity potential [34] given as

$$V_P(z) = 4\pi\gamma b \frac{C_1 C_2}{C_1 + C_2} \phi\left(\frac{z}{b}\right) \tag{8.6}$$

with the nuclear surface tension coefficient:

$$\gamma = 0.9517[1 - 1.7826(N - Z)^2/A^2]. \tag{8.7}$$

The universal proximity potential ϕ is given as

$$\phi(\xi) = -4.41 \exp(-\xi/0.7176), \text{ for } \xi \geq 1.9475, \tag{8.8}$$

$$\phi(\xi) = -1.7817 + 0.9270\xi + 0.01696\xi^2 - 0.05148\xi^3, \text{ for } 0 \leq \xi \leq 1.9475, \tag{8.9}$$

$$\phi(\xi) = -1.7817 + 0.9270\xi + 0.0143\xi^2 - 0.09\xi^3, \text{ for } \xi \leq 0 \tag{8.10}$$

with $\xi = z/b$, where the width (diffuseness) of nuclear surface $b \approx 1\text{fm}$ and C_i is the Susmann central radii.

For R_i, we use the semi-empirical formula in terms of mass number A_i as

$$R_i = 1.28A_i^{1/3} - 0.76 + 0.8A_i^{-1/3}. \tag{8.11}$$

Ambruster [37] has suggested an expression for the CN formation probability as

$$P_{CN} = 0.5 \exp(-c(x_{eff} - x_{thr})). \tag{8.12}$$

We used the energy-dependent expression for fusion probability [38,39] to calculate P_{CN}, and it is given by

$$P_{CN}(E, \ell) = \frac{\exp\{-c(x_{eff} - x_{thr})\}}{1 + \exp\left\{\frac{E_B^* - E^*}{\Delta}\right\}}, \tag{8.13}$$

where E^* is the excitation energy of the CN, E_B^* denotes the excitation energy of the CN when the center-of-mass beam energy is equal to the Coulomb and proximity barriers, Δ is an adjustable parameter ($\Delta = 4 MeV$), and x_{eff} is the effective fissility defined as

$$x_{eff} = \left[\frac{(Z^2/A)}{(Z^2/A)_{crit}} \right] (1 - \alpha + \alpha f(K)) \tag{8.14}$$

with $(Z^2/A)_{crit}$, $f(K)$ and K are given by

$$(Z^2/A)_{crit} = 50.883 \left[1 - 1.7286 \left(\frac{(N-Z)}{A} \right)^2 \right], \tag{8.15}$$

$$f(K) = \frac{4}{K^2 + K + \frac{1}{K} + \frac{1}{K^2}}, \tag{8.16}$$

$$K = (A_1/A_2)^{1/3}, \tag{8.17}$$

where Z, N and A represent the atomic number, neutron number, and mass number, respectively. A_1 and A_2 are the mass numbers of projectile and target, respectively. x_{thr} and c are adjustable parameters, and $\alpha = 1/3$. The best fit to the cold fusion reaction are the c and x_{eff} values of 136.5 and 0.79, respectively. For hot fusion reaction, the best fit for $x_{eff} \leq 0.8$ is $c = 104$ and $x_{thr} = 0.69$, while for $x_{eff} \geq 0.8$, the values are $c = 82$ and $x_{thr} = 0.69$. These constants are suggested by Loveland [39].

The survival probability W_{sur} is the probability for the CN to decay to the ground state of the final residual nucleus via evaporation of light particles and gamma ray for avoiding fission process. The survival probability under the evaporation of x neutrons is

$$W_{sur} = P_{xn}(E_{CN}^*) \prod_{i=1}^{i_{max}=x} \left(\frac{\Gamma_n}{\Gamma_n + \Gamma_f} \right)_{i,E^*}, \tag{8.18}$$

where the index "i" is equal to the number of emitted neutrons; P_{xn} is the probability of emitting exactly xn neutrons [40]; E^* is the excitation energy of the CN; and Γ_n and Γ_f represent the decay width of neutron evaporation and fission, respectively. To calculate Γ_n/Γ_f, Vandenbosch and Huizenga [41] have suggested a classical formalism:

$$\frac{\Gamma_n}{\Gamma_f} = \frac{4A^{2/3}a_f(E^* - B_n)}{K_0 a_n [2a_f^{1/2}(E^* - B_f)^{1/2} - 1]} \exp[2a_n^{1/2}(E^* - B_n)^{1/2} - 2a_f^{1/2}(E^* - B_f)^{1/2}], \tag{8.19}$$

where A is the mass number of the nucleus considered, E^* is the excitation energy, and B_n is the neutron separation energy. The constant K_0 is taken as 10 MeV. $a_n = A/10$ and $a_f = 1.1a_n$ are the level density parameters of the daughter nucleus and the fissioning nucleus at the ground state and saddle configurations, respectively. B_f is the fission barrier, and this height is a decisive quantity in the competition between processes of neutron evaporation and fission.

Fission barrier height B_f can be determined as the sum of liquid drop fission barrier B_f^{LD} (macroscopic part) and shell correction terms (microscopic part):

$$B_f(E^*) = B_f^{LD} + S \exp(-E^*/E_D) \quad \text{MeV}, \tag{8.20}$$

where E_D is the shell damping energy, which describes how much the shell effects fall off as the excitation energy of the decaying nucleus increases:

$$E_D = 5.48 A^{1/3}/(1 + 1.3A^{-1/3}) \quad \text{MeV}. \tag{8.21}$$

Liquid drop fission barrier is very low or equal to zero for heavy elements with $Z > 109$. So, the fission barrier in the heaviest nuclei is mainly defined by shell corrections energy, and its value depends on excitation energy. The shell correction S is taken from mass table of Moller et al. [42].

The survivability of heavy fissile nuclei strongly depends on the fission barrier of a CN. Variation of fission barrier about 1 MeV can change the ER cross sections of about one order of magnitude. The fission barrier decreases exponentially as excitation energy increases. The charged particle evaporation is neglected because of the presence of Coulomb barrier and larger proton binding energy.

The alpha decay half-lives are calculated using the MGLDM [33] of our group. SF half-lives are calculated using the formula of Santhosh et al. [43]:

$$\log_{10}(T_{1/2}/yr) = a\frac{Z^2}{A} + b\left(\frac{Z^2}{A}\right)^2 + c\left(\frac{N-Z}{N+Z}\right) + d\left(\frac{N-Z}{N+Z}\right)^2 + eE_{shell} + f, \quad (8.22)$$

where $a = -43.25203$, $b = 0.49192$, $c = 3674.3927$, $d = -9360.6$, $e = 0.8930$, and $f = 578.56058$. E_{shell} is the shell correction energy.

8.3 Results and Discussion

A detailed theoretical study on the synthesis of SHE with $Z = 122$ via fusion evaporation channel and its decay has been performed. In order to study the possibility of synthesizing SHE with $Z = 122$, and for predicting the suitable projectile-target pair, we have used our PMPC. Also the present chapter aims at studying the behavior of the isotopes of SHE with $Z = 122$, against alpha decay, SF, and proton decay, thereby conferring a theoretical prediction on the modes of decay of these isotopes. MGLDM [33] and the shell effect-dependent semi-empirical formula of Santhosh et al. [43] are used to compute alpha half-lives and SF half-lives, respectively.

The calculations of the capture cross section, probability of CN formation, and survival probability, W_{sur}, are extremely sensitive to projectile-target pair used, Coulomb barrier, barrier position, potential parameters, center-of-mass energy, excitation energy, fission barrier, and neutron separation energy. We have proved the reliability of our model through the previous studies [44–46], where our calculated fusion excitation functions are in agreement with experimental values and are shown in Table 8.1. We have also predicted the production cross section of $Z = 119$ and $Z = 120$ using the same model [47,48]. Here, we have used the same model with the same set of parameters which are described in Section 8.2. Since we could replicate the experimentally measured cross sections for SHEs with $Z = 114 - 118$, we extend our work for predicting the cross section of SHE with $Z = 122$, which are not yet synthesized in the laboratory.

The probable hot fusion reactions identified to synthesize elements with $Z = 122$ are ^{50}Ti $+ \, ^{253,257}$Fm $\rightarrow \, ^{303,307}$122, ^{54}Cr $+ \, ^{248-254}$Cf $\rightarrow \, ^{302-308}$122, ^{58}Fe $+ \, ^{240-248,250}$Cm $\rightarrow$ $^{298-306,308}$122, ^{64}Ni $+ \, ^{238-242,244}$Pu $\rightarrow \, ^{302-306,308}$122, and ^{70}Zn $+ \, ^{232-239,238}$U $\rightarrow$ $^{302-306,308}$122, and the cold fusion reactions are ^{76}Ge $+ \, ^{228-230,232}$Th $\rightarrow \, ^{304-306,308}$122, ^{82}Se $+ \, ^{226,228}$Ra $\rightarrow \, ^{308,310}$122, and ^{87}Sr $+ \, ^{208-210}$Po $\rightarrow \, ^{295-297}$122. The plots of ER cross section using the above-mentioned fusion reaction in xn ($x = 2$–5 for hot fusion and $x = 1$–2 for cold fusion) evaporation channel are presented in Figures 8.1–8.6. The calculated maximum value of $\sum_{x=2}^{5} \sigma_{ER}(xn)$ for hot fusion reactions and $\sum_{x=1}^{2} \sigma_{ER}(xn)$ for cold fusion reactions are listed in Tables 8.2 and 8.3. The corresponding Coulomb barrier, barrier position, curvature of the inverted harmonic oscillator, neutron separation energy,

TABLE 8.1

Comparison of Calculated ER Cross Section and the Corresponding Excitation Energy with that of Experiments for SHE with $Z = 114$–118

Reaction	Expt.			Our Work				
	E^* (MeV)	σ_{ER} (pb)	Ref.	E^* (MeV)	$\sigma_{capture}$ (mb)	σ_{fusion} (mb)	σ_{ER} (pb)	Ref.
^{48}Ca + ^{242}Pu → ^{290}Fl	38–42.4	$\sigma_{3n} = 3.6^{+3.4}_{-1.7}$	[5]	40.2	41.9	4.6E–3	$\sigma_{3n} = 4.2$	[44]
	38–42.4	$\sigma_{4n} = 4.5^{+3.6}_{-1.9}$	[5]	40.2	41.9	4.6E–3	$\sigma_{4n} = 2.95$	
	48–52	$\sigma_{4n} = 0.6^{+0.9}_{-0.5}$	[49]	50	266.9	4.5E–2	$\sigma_{4n} = 0.5$	
	48–52	$\sigma_{5n} = 0.6^{+0.9}_{-0.5}$	[49]	50	266.9	4.5E–2	$\sigma_{5n} = 0.3$	
^{48}Ca + ^{244}Pu → ^{292}Fl	39–43	$\sigma_{3n} = 1.7^{+2.5}_{-1.1}$	[2]	40.5	18.2	1.9E–3	$\sigma_{3n} = 4.1$	[44]
	39–43	$\sigma_{4n} = 5.3^{+3.6}_{-2.1}$	[2]	40.5	18.2	1.9E–3	$\sigma_{4n} = 3.9$	
	39.8–43.9	$\sigma_{4n} = 9.8^{+3.9}_{-3.1}$	[50]	42	54	7.1E–3	$\sigma_{4n} = 8.6$	
^{48}Ca + ^{243}Am → ^{291}Mc	42.4–46.5	$\sigma_{4n} = 0.9^{+3.2}_{-0.8}$	[4,51]	44	333.1	2.1E–2	$\sigma_{4n} = 0.9$	-
	38–42.3	$\sigma_{3n} = 2.7^{+4.8}_{-1.6}$	[4,51]	39	212.1	1.2E–2	$\sigma_{3n} = 1.2$	
^{48}Ca + ^{248}Cm → ^{296}Lv	34–38.3	$\sigma_{3n} = 8.5^{+6.4}_{-3.7}$	[52]	35	109	5.1E–3	$\sigma_{3n} = 6.02$	
	36.8–41.1	$\sigma_{3n} = 1.1^{+1.7}_{-0.7}$	[20]	38.9	203	4.9E–3	$\sigma_{3n} = 0.9$	[45]
	36.8–41.1	$\sigma_{4n} = 3.3^{+2.5}_{-1.4}$	[20]	38.9	203	4.9E–3	$\sigma_{4n} = 4.6$	
	40.9	$\sigma_{3n} = 0.9^{+2.1}_{-0.7}$	[20]	40.9	253	6.4E–3	$\sigma_{3n} = 0.3$	
	40.9	$\sigma_{4n} = 3.4^{+2.7}_{-1.6}$	[20]	40.9	253	6.4E–3	$\sigma_{4n} = 3.4$	
^{48}Ca + ^{249}Bk → ^{297}Ts	30.4–34.7	$\sigma_{3n} = 0.7^{+1.7}_{-0.57}$	[8]	32.6	78.42	1.6E–3	$\sigma_{3n} = 2.2$	[46]
	33.2–37.4	$\sigma_{3n} = 0.5^{+1.1}_{-0.4}$	[6]					
	35	$\sigma_{3n} = 3.6^{+6.1}_{-2.5}$	[7]	35	143.4	3.4E–3	$\sigma_{3n} = 1.2$	
	32.8–37.5	$\sigma_{3n} = 1.1^{+1.2}_{-0.6}$	[8]					
	37.2–41.4	$\sigma_{4n} = 1.3^{+1.5}_{-0.6}$	[6]					
	39	$\sigma_{4n} = 2.0^{+2.2}_{-1.1}$	[7]	39	238.0	6.4E–3	$\sigma_{4n} = 1.12$	
	37–41.9	$\sigma_{4n} = 1.5^{+1.1}_{-0.5}$	[8]					
^{48}Ca + ^{249}Cf → ^{297}Og	40.3–44.8	$\sigma_{4n} = 2.4^{+3.3}_{-1.4}$	[8]	43	335.2	9.6E–3	$\sigma_{4n} = 0.5$	
	32.1–36.6	$\sigma_{3n} = 0.5^{+1.6}_{-0.3}$	[9]	34	164.9	1.01E–3	$\sigma_{3n} = 0.6$	[27]

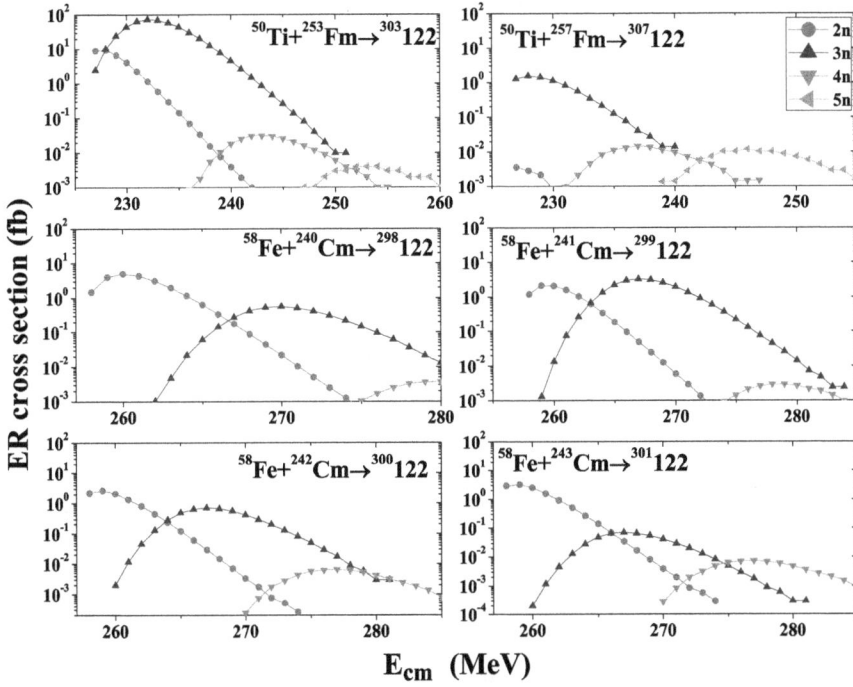

FIGURE 8.1
Evaporation residue cross section for the hot fusion reactions ^{50}Ti $+ ^{253,257}$Fm $\rightarrow ^{303,307}$122 and ^{58}Fe $+ ^{240-243}$Cm $\rightarrow ^{298-301}$122.

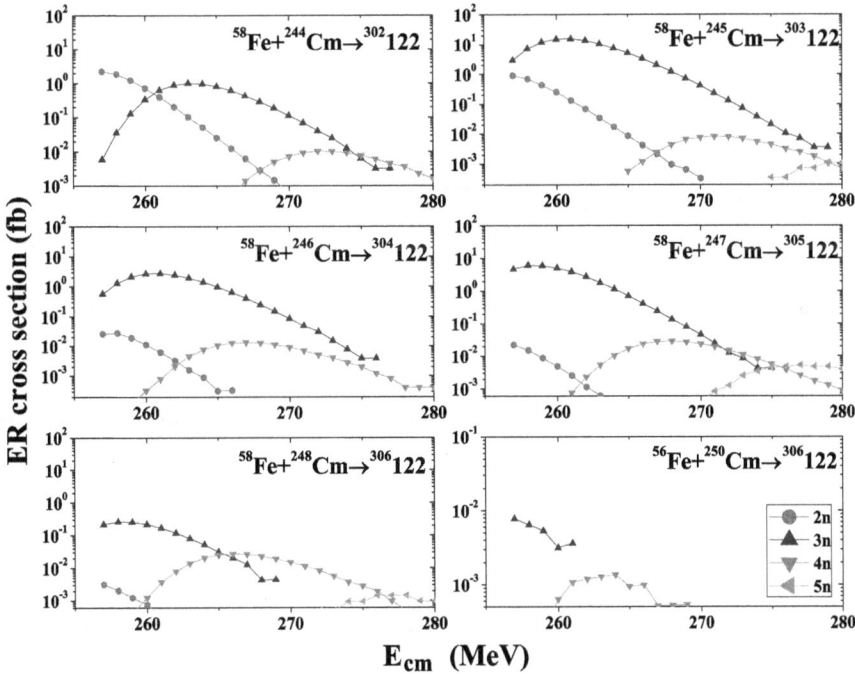

FIGURE 8.2
Evaporation residue cross section for the hot fusion reaction ^{58}Fe$+ ^{243-248,250}$Cm $\rightarrow ^{301-306,308}$122.

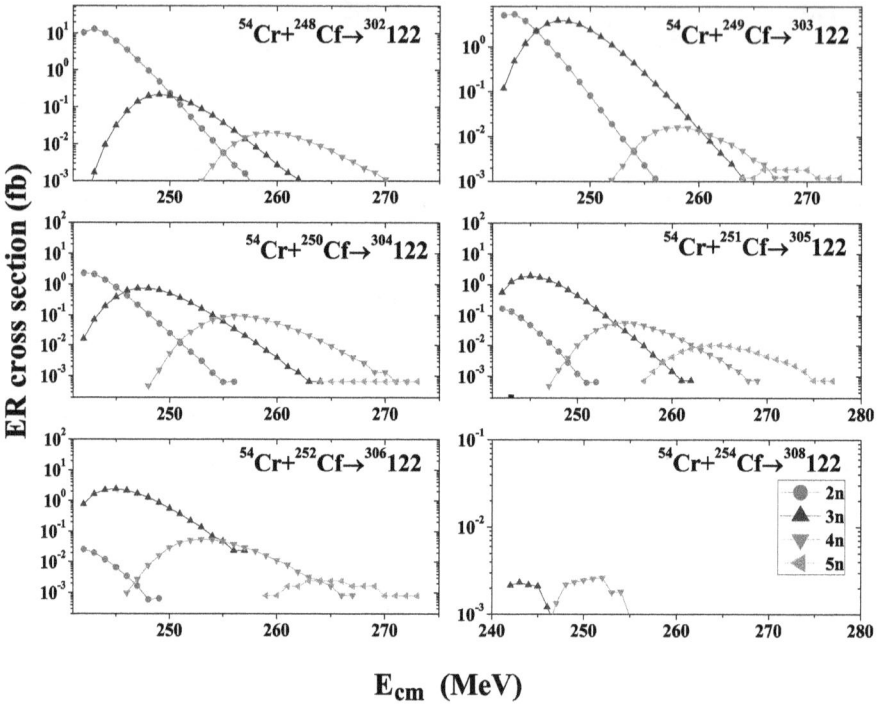

FIGURE 8.3
Evaporation residue cross section for the hot fusion reaction $^{54}\text{Cr} + {}^{248-252,254}\text{Cf} \rightarrow {}^{302-306,308}122$.

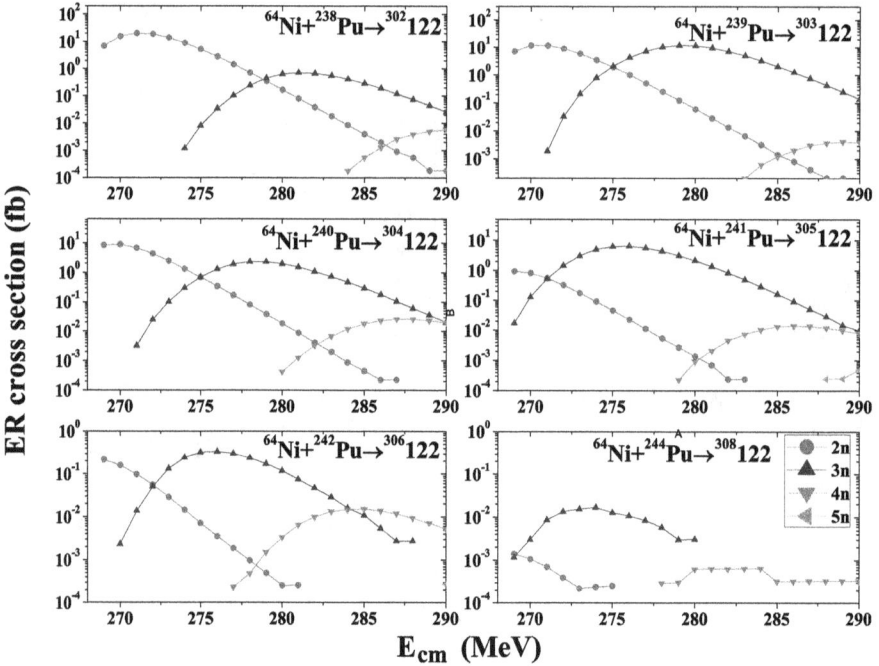

FIGURE 8.4
Evaporation residue cross section for the hot fusion reaction $^{64}\text{Ni} + {}^{238-242,244}\text{Pu} \rightarrow {}^{302-306,308}122$.

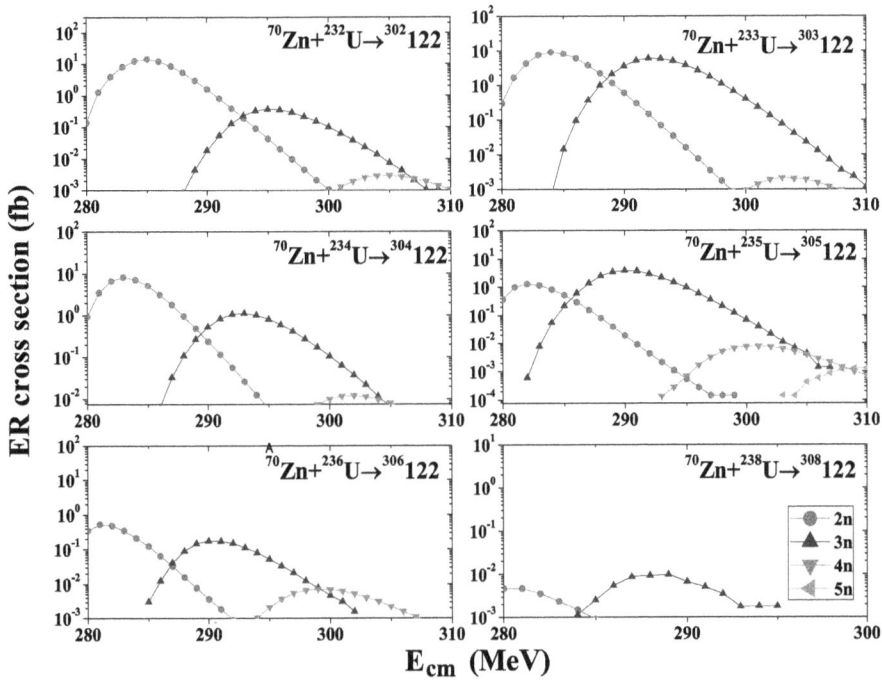

FIGURE 8.5
Evaporation residue cross section for the hot fusion reaction ^{70}Zn + $^{232-236,238}$U → $^{302-306,308}$122.

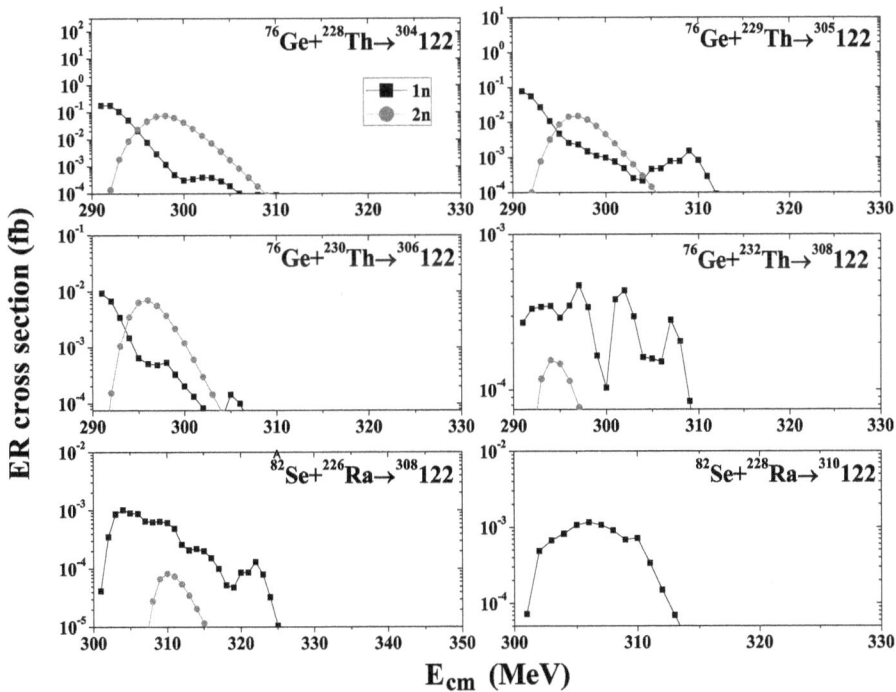

FIGURE 8.6
Evaporation residue cross section for the cold fusion reactions ^{76}Ge + $^{228-230,232}$Th → $^{304-306,308}$122 and ^{82}Se + ^{226,228}Ra → ^{308,310}122.

TABLE 8.2

Probable Hot Fusion Reactions to Synthesize Elements with $Z = 122$

Reaction	CN	Neutron no. of CN	V_B (MeV)	R_B (fm)	B_{1n} (MeV)	$\hbar\omega$ (MeV)	B_f ($E^* = 0$)	Highest $\sum_{x=2}^{5} \sigma\, ER(xn)$ (fb)
58Fe + 240Cm	298122	176	257.574	12.910	8.211	1.760	5.53	5.071
58Fe + 241Cm	299122	177	257.390	12.934	7.411	1.754	6.26	2.223
58Fe + 242Cm	300122	178	257.157	12.951	7.791	1.750	5.73	2.627
58Fe + 243Cm	301122	179	256.985	12.943	6.721	1.751	5.97	3.224
54Cr + 248Cf	302122	180	241.351	13.078	7.691	1.720	5.57	13.192
58Fe + 244Cm	302122	180	256.783	12.973	7.691	1.744	5.57	2.254
64Ni + 238Pu	302122	180	268.024	13.096	7.691	1.702	5.57	20.29
70Zn + 232U	302122	180	279.801	13.130	7.691	1.682	5.57	13.982
50Ti + 253Fm	303122	181	226.188	13.079	6.371	1.714	5.66	74.734
54Cr + 249Cf	303122	181	241.164	13.102	6.371	1.715	5.66	39.525
58Fe + 245Cm	303122	181	256.591	12.999	6.371	1.739	5.66	15.463
64Ni + 239Pu	303122	181	267.815	13.112	6.371	1.698	5.66	12.06
70Zn + 233U	303122	181	279.568	13.150	6.371	1.677	5.66	9.101
54Cr + 250Cf	304122	182	241.014	13.110	7.381	1.712	5.17	7.564
58Fe + 246Cm	304122	182	256.376	12.989	7.381	1.740	5.17	2.785
64Ni + 240Pu	304122	182	267.612	13.120	7.381	1.696	5.17	8.68
70Zn + 234U	304122	182	279.342	13.176	7.381	1.671	5.17	8.102
54Cr + 251Cf	305122	183	240.801	13.108	6.001	1.712	5.10	19.152
58Fe + 247Cm	305122	183	256.186	13.007	6.001	1.736	5.10	6.088
64Ni + 241Pu	305122	183	267.399	13.136	6.001	1.692	5.10	6.439
70Zn + 235U	305122	183	279.127	13.184	6.001	1.669	5.10	3.910
54Cr + 252Cf	306122	184	240.611	13.124	7.011	1.708	4.46	2.352

(Continued)

TABLE 8.2 (*Continued*)
Probable Hot Fusion Reactions to Synthesize Elements with $Z = 122$

Reaction	CN	Neutron no. of CN	V_B (MeV)	R_B (fm)	B_{1n} (MeV)	$\hbar\omega$ (MeV)	B_f ($E^* = 0$)	Highest $\sum_{x=2}^{5} \sigma\, ER(xn)$ (fb)
^{58}Fe + ^{248}Cm	306122	184	255.971	13.041	7.011	1.728	4.46	0.266
^{64}Ni + ^{242}Pu	306122	184	267.190	13.156	7.011	1.688	4.46	0.338
^{70}Zn + ^{236}U	306122	184	278.912	13.187	7.011	1.668	4.46	0.535
^{50}Ti + ^{257}Fm	307122	185	225.522	13.183	5.051	1.692	3.64	1.548
^{54}Cr + ^{254}Cf	308122	186	240.263	13.148	6.681	1.703	2.88	0.023
^{58}Fe + ^{250}Cm	308122	186	255.577	13.063	6.681	1.722	2.88	0.005
^{64}Ni + ^{244}Pu	308122	186	266.776	13.174	6.681	1.683	2.88	0.017
^{70}Zn + ^{238}U	308122	186	278.469	13.230	6.681	1.658	2.88	0.001

TABLE 8.3

Probable Cold Fusion Reactions to Synthesize Elements with $Z = 122$

Reaction	CN	Neutron no. of CN	V_B (MeV)	R_B (fm)	B_{1n} (MeV)	$\hbar\omega$ (MeV)	B_f ($E^* = 0$)	Highest $\sum_{x=1}^{2} \sigma ER(xn)$ (fb)
^{76}Ge + ^{228}Th	304122	182	290.357	13.165	7.381	1.662	5.17	0.181
^{76}Ge + ^{229}Th	305122	183	290.122	13.182	6.001	1.658	5.10	0.078
^{76}Ge + ^{230}Th	306122	184	289.892	13.210	7.011	1.652	4.46	0.010
^{76}Ge + ^{232}Th	308122	186	289.430	13.231	6.681	1.646	2.88	0.005
^{82}Se + ^{226}Ra	308122	186	299.652	13.226	6.681	1.637	2.88	0.001
^{82}Se + ^{228}Ra	310122	188	299.172	13.277	5.431	1.626	3.43	0.001

TABLE 8.4

The Calculated Alpha Decay Half-Lives Compared with the Corresponding SF Half-Lives for the Isotopes $^{302-308,310}122$ and Its Decay Products

Parent Nuclei	Q_α (MeV)	T_{SF} (s)	$T^\alpha_{1/2}$ (s)	Decay Mode	Parent Nuclei	Q_α (MeV)	T_{SF} (s)	$T^\alpha_{1/2}$ (s)	Decay Mode
$^{302}122$	14.828	4.503E+14	5.594E−09	α	$^{303}122$	14.628	1.094E+14	1.173E−08	α
$^{298}120$	13.297	3.402E+11	1.070E−06	α	$^{299}120$	13.787	7.814E+10	1.213E−07	α
^{294}Og	12.425	2.153E+08	1.942E−05	α	^{295}Og	11.985	1.856E+08	1.730E−04	α
^{290}Lv	11.124	4.390E+05	5.413E−03	α	^{291}Lv	11.074	2.016E+05	6.948E−03	α
^{286}Fl	9.522	1.128E+03	3.858E+01	α	^{287}Fl	9.452	6.152E+02	6.160E+01	α
^{282}Cn	9.481	1.943E+00	1.124E+01	SF	^{283}Cn	9.181	1.205E+00	9.680E+01	SF
$^{304}122$	14.638	6.387E+12	1.087E−08	α	$^{305}122$	14.608	7.033E+11	1.180E−08	α
$^{300}120$	13.757	6.326E+09	1.328E−07	α	$^{301}120$	13.677	9.020E+08	1.802E−07	α
^{296}Og	12.335	1.845E+07	2.808E−05	α	^{297}Og	12.445	3.079E+06	1.571E−05	α
^{292}Lv	10.874	5.603E+04	2.165E−02	α	^{293}Lv	10.834	1.520E+04	2.645E−02	α
^{288}Fl	9.222	1.450E+02	3.276E+02	SF	^{289}Fl	9.112	4.698E+01	7.306E+02	SF
$^{306}122$	14.678	1.917E+10	8.644E−09	α	$^{307}122$	15.328	2.895E+08	7.164E−10	α
$^{302}120$	13.607	3.727E+07	2.348E−07	α	$^{303}120$	13.567	2.912E+06	2.693E−07	α
^{298}Og	12.545	1.478E+05	9.299E−06	α	^{299}Og	12.565	1.418E+04	8.134E−06	α
^{294}Lv	10.964	1.155E+03	1.176E−02	α	^{295}Lv	10.924	1.539E+02	1.433E−02	α
^{290}Fl	8.892	6.128E+00	3.970E+03	SF	^{291}Fl	9.322	9.821E−01	1.372E+02	SF
$^{308}122$	15.348	3.990E+06	6.428E−10	α	$^{310}122$	15.348	2.849E+04	4.595E−06	α
$^{304}120$	13.607	5.530E+04	2.183E−07	α	$^{306}120$	14.337	4.979E+00	9.807E−09	α
^{300}Og	12.565	3.793E+02	7.838E−06	α	^{302}Og	12.675	2.584E−01	4.299E−06	α
^{296}Lv	11.234	5.236E+00	2.295E−03	α	^{298}Lv	11.194	7.393E−03	2.671E−03	α
^{292}Fl	8.372	8.028E−02	2.910E+05	SF	^{294}Fl	8.492	1.654E−04	9.449E+04	SF

and fission barrier at $E^* = 0$ are also listed. From the table, it is clear that the reaction ^{50}Ti + ^{253}Fm has larger cross-section values. But the half-life for the target ^{253}Fm is only about 3 days. If the necessary amount of targets are prepared by the experimentalists, this reaction will be better one to synthesize elements with $Z = 122$. The reactions ^{54}Cr + ^{249}Cf leading to ^{303}122 and ^{64}Ni + ^{238}Pu leading to ^{302}122 have the next highest cross sections. While analyzing the figures, it is clear that 2n channel cross section is larger for ^{64}Ni + ^{238}Pu and 3n channel is larger for ^{50}Ti + ^{253}Fm. The calculated 4n and 5n channel ER cross sections for all the reactions are less than 0.1 fb.

The alpha decay half-lives and SF half-lives are calculated using the MGLDM [33] and semi-empirical formula of Santhosh et al. [43]. In Table 8.4, we highlight our predictions on the modes of decay of unknown isotopes of $Z = 122$. In addition to the alpha decay half-lives calculated using the MGLDM, the SF half-lives and Q_α values of the corresponding isotopes are listed in Table 8.4. The nuclei with α-decay half-lives shorter than SF half-lives will survive fission and hence decay through α particle emission. By comparing the alpha decay half-lives with the SF half-lives for the isotopes $^{298-308,310}$122, we have observed 4α chains from the SHE $^{302-308,310}$122, and hence, it can be detected in the laboratory. To study the behavior of $^{298-308,310}$122 SHE against proton decay, we have evaluated the proton separation energies for these isotopes and revealed that SHEs $^{298-301}$122 are probable proton emitters among these nuclei and easily decay through proton emission. The calculated cross section is less than 2 fb for CN having neutron number $N > 184$, and hence, these isotopes may not be able to synthesize in near future. The major finding of our studies on synthesis and decay of $Z = 122$ is that the isotopes $^{302-306}$122 can be synthesized in future, and the most promising reactions are ^{54}Cr + ^{249}Cf and ^{64}Ni + ^{238}Pu, and can be detected via α decay.

8.4   Conclusions

We have calculated the production cross section for the isotopes of SHE with $Z = 122$ and identified the most suitable reactions ^{54}Cr + ^{248}Cf, ^{64}Ni + ^{238}Pu, ^{70}Zn + ^{232}U, ^{50}Ti+^{253}Fm, ^{54}Cr + ^{249}Cf, ^{58}Fe + ^{245}Cm, and ^{64}Ni + ^{239}Pu for which the calculated ER excitation functions are greater than 10 fb. The reactions ^{54}Cr + ^{249}Cf and ^{64}Ni + ^{238}Pu are the most probable to synthesize elements with $Z = 122$. We identified that the production cross section of SHE with neutron number N >184 is less than 2 fb, and conclude that SHE with N>184 ($^{307-308,310}$122) may not be synthesized in near future. By estimating the proton separation energies, it is seen that isotopes $^{298-301}$122 may decay through proton emission. By comparing the α-decay half-lives with the SF half-lives, the isotopes $^{302-308,310}$122 show 4α chains. Through these predictions, we could emphasize the fact that the isotopes $^{302-306}$122 can be synthesized in neutron evaporation channel and can detected experimentally via alpha decay. We hope that the studies will be useful for future experimental investigations.

References

[1] Oganessian, Y. T., A. Sobiczewski, and G. M. Ter-Akopian. "Superheavy nuclei: from predictions to discovery." *Physica Scripta* 92, no. 2 (2017): 023003.

[2] Oganessian, Y. T., V. K. Utyonkov, Yu V. Lobanov et al. "Measurements of cross sections for the fusion-evaporation reactions ^{244}Pu(^{48}Ca, xn) $^{292-x}$114 and ^{245}Cm (^{48}Ca, xn) $^{293-x}$116." *Physical Review C* 69, no. 5 (2004): 054607.

[3] Oganessian, Y. T., F. Sh Abdullin, S. N. Dmitriev et al. "Investigation of the ^{243}Am+ ^{48}Ca reaction products previously observed in the experiments on elements 113, 115, and 117." *Physical Review C* 87, no. 1 (2013): 014302.

[4] Oganessian, Y. T., V. K. Utyonkov, S. N. Dmitriev et al. "Synthesis of elements 115 and 113 in the reaction ^{243}Am + ^{48}Ca." *Physical Review C* 72, no. 3 (2005): 034611.

[5] Oganessian, Y. T., V. K. Utyonkov, Yu V. Lobanov et al. "Measurements of cross sections and decay properties of the isotopes of elements 112, 114, and 116 produced in the fusion reactions ^{233,238}U, ^{242}Pu, and ^{248}Cm+ ^{48}Ca." *Physical Review C* 70, no. 6 (2004): 064609.

[6] Oganessian, Y. T., F. Sh Abdullin, P. D. Bailey et al. "Eleven new heaviest isotopes of elements Z = 105 to Z = 117 identified among the products of ^{249}Bk+ ^{48}Ca reactions." *Physical Review C* 83, no. 5 (2011): 054315.

[7] Oganessian, Y. T., F. Sh Abdullin, C. Alexander et al. "Production and Decay of the Heaviest Nuclei ^{293,294} 117 and ^{294}118." *Physical review letters* 109, no. 16 (2012): 162501.

[8] Oganessian, Y. T., F. Sh Abdullin, C. Alexander et al. "Experimental studies of the ^{249}Bk+ ^{48}Ca reaction including decay properties and excitation function for isotopes of element 117, and discovery of the new isotope 277 Mt." *Physical Review C* 87, no. 5 (2013): 054621.

[9] Oganessian, Y. T., V. K. Utyonkov, Yu V. Lobanov et al. "Synthesis of the isotopes of elements 118 and 116 in the ^{249}Cf and ^{245}Cm + ^{48}Ca fusion reactions." *Physical Review C* 74, no. 4 (2006): 044602.

[10] Oganessian, Yu Ts, F. Sh Abdullin, P. D. Bailey et al. "Synthesis of a new element with atomic number Z = 117." *Physical Review Letters* 104, no. 14 (2010): 142502.

[11] Oganessian, Y. T., V. K. Utyonkov, and K. J. Moody. "Synthesis of ^{292}116 in the ^{248}Cm+ ^{48}Ca reaction." *Physics of Atomic Nuclei* 64, no. 8 (2001): 1349–1355.

[12] Oganessian, Y. T., V. K. Utyonkoy, Yu V. Lobanov et al. "Experiments on the synthesis of element 115 in the reaction ^{243}Am (^{48}Ca, x n) $^{291-x}$115." *Physical Review C* 69, no. 2 (2004): 021601.

[13] Oganessian, Y. T., and V. K. Utyonkov. "Superheavy nuclei from ^{48}Ca-induced reactions." *Nuclear Physics A* 944 (2015): 62–98

[14] Morita, K., K. Morimoto, D. Kaji et al. "Experiment on the synthesis of element 113 in the reaction ^{209}Bi (^{70}Zn, n) ^{278}113." *Journal of the Physical Society of Japan* 73, no. 10 (2004): 2593–2596.

[15] Morita, K., K. Morimoto, D. Kaji et al. "Experiment on synthesis of an isotope ^{277}112 by ^{208}Pb+^{70}Zn reaction." *Journal of the Physical Society of Japan* 76, no. 4 (2007): 043201.

[16] Hamilton, J. H., S. Hofmann, and Y. T. Oganessian. "Search for superheavy nuclei." *Annual Review of Nuclear and Particle Science* 63 (2013): 383–405.

[17] Hofmann, S., F. P. Heßberger, D. Ackermann, G. Münzenberg, S. Antalic, P. Cagarda, B. Kindler et al. "New results on elements 111 and 112." *The European Physical Journal A-Hadrons and Nuclei* 14, no. 2 (2002): 147–157.

[18] Hofmann, S., and G. Münzenberg. "The discovery of the heaviest elements." *Reviews of Modern Physics* 72, no. 3 (2000): 733.

[19] Oganessian, Y. T., V. K. Utyonkov, Y. V. Lobanov et al. "Attempt to produce element 120 in the ^{244}Pu+^{58}Fe reaction." *Physical Review C* 79, no. 2 (2009): 024603.

[20] Hofmann, S., S. Heinz, R. Mann et al. "Review of even element super-heavy nuclei and search for element 120." *The European Physical Journal A* 52, no. 6 (2016): 180.

[21] Marinov, A., I. Rodushkin, D. Kolb et al. "Evidence for the possible existence of a long-lived superheavy nucleus with atomic mass number A = 292 and atomic number Z ≅ 122 in natural Th." *International Journal of Modern Physics E* 19, no. 01 (2010): 131–140.

[22] Dellinger, F., O. Forstner, R. Golser et al. "Search for a superheavy nuclide with A= 292 and neutron-deficient thorium isotopes in natural thorianite." *Nuclear Instruments and Methods in Physics Research Section B: Beam Interactions with Materials and Atoms* 268, no. 7–8 (2010): 1287–1290.

[23] Litnevsky, V. L., G. I. Kosenko, and F. A. Ivanyuk. "Description of fusion and evaporation residue formation cross sections in reactions leading to the formation of element Z = 122 within the Langevin approach." *Physical Review C* 93, no. 6 (2016): 064606.

[24] Giardina, G., G. Fazio, G. Mandaglio, M. Manganaro, A. K. Nasirov, M. V. Romaniuk, and C. Sacca. "Expectations and limits to synthesize nuclei with Z ≥ 120." *International Journal of Modern Physics E* 19, no. 5–6 (2010): 882–893.

[25] Cap, T., K. Siwek-Wilczyńska, and J. Wilczyński. "No chance for synthesis of super-heavy nuclei in fusion of symmetric systems." *Physics Letters B* 736 (2014): 478–481.

[26] Bohr, N. "Neutron capture and nuclear constitution." *Nature* 137 (1936): 344.

[27] Santhosh, K. P., and V. Safoora. "Theoretical studies on the synthesis of SHE $^{290-302}$Og (Z = 118) using ^{48}Ca, ^{45}Sc, ^{50}Ti, ^{51}V, ^{54}Cr, ^{55}Mn, ^{58}Fe, ^{59}Co and ^{64}Ni induced reactions." *The European Physical Journal A* 54, no. 5 (2018): 80.

[28] Siwek-Wilczyńska, K., T. Cap, M. Kowal, A. Sobiczewski, and J. Wilczyński. "Predictions of the fusion-by-diffusion model for the synthesis cross sections of Z = 114–120 elements based on macroscopic-microscopic fission barriers." *Physical Review C* 86, no. 1 (2012): 014611.

[29] Adamian, G. G., N. V. Antonenko, W. Scheid, and V. V. Volkov. "Fusion cross sections for superheavy nuclei in the dinuclear system concept." *Nuclear Physics A* 633, no. 3 (1998): 409–420.

[30] Sandulescu, A., and M. Mirea. "Cold fission from isomeric states of superheavy nuclei." *The European Physical Journal A* 50, no. 7 (2014): 110.

[31] Santhosh, K. P., and Antony Joseph. "Exotic decay in Ba isotopes via ^{12}C emission." *Pramana* 55, no. 3 (2000): 375–382.

[32] Santhosh, K. P., and C. Nithya. "Predictions on the modes of decay of even Z superheavy isotopes within the range $104 \leq Z \leq 136$." *Atomic Data and Nuclear Data Tables* 119 (2018): 33−98.

[33] Santhosh, K. P., C. Nithya, H. Hassanabadi, and Dashty T. Akrawy. "α-decay half-lives of superheavy nuclei from a modified generalized liquid-drop model." *Physical Review C* 98, no. 2 (2018): 024625.

[34] Błocki, J., J. Randrup, W. J. Światecki, and C. F. Tsang. "Proximity forces." *Annals of Physics* 105, no. 2 (1977): 427−462.

[35] Wong, C. Y. "Interaction barrier in charged-particle nuclear reactions." *Physical Review Letters* 31, no. 12 (1973): 766.

[36] Hill, D. L., and J. A. Wheeler. "Nuclear constitution and the interpretation of fission phenomena." *Physical Review* 89, no. 5 (1953): 1102.

[37] Armbruster, P. "Nuclear structure in cold rearrangement processes in fission and fusion." *Reports on Progress in Physics* 62, no. 4 (1999): 465.

[38] Zhang, J., C. Wang, and Z. Ren. "Calculation of evaporation residue cross sections for the synthesis of superheavy nuclei in hot fusion reactions." *Nuclear Physics A* 909 (2013): 36−49.

[39] Loveland, W. "Synthesis of transactinide nuclei using radioactive beams." *Physical Review C* 76, no. 1 (2007): 014612.

[40] Jackson, J. D. "A schematic model for (p, xn) cross sections in heavy elements." *Canadian Journal of Physics* 34, no. 8 (1956): 767−779.

[41] Vandenbosch, R., and J. R. Huizenga. "Nuclear Fission Academic Press." *New York* (1973).

[42] Möller, P., A. J. Sierk, T. Ichikawa, and H. Sagawa. "Nuclear ground-state masses and deformations: FRDM (2012)." *Atomic Data and Nuclear Data Tables* 109 (2016): 1−204.

[43] Santhosh, K. P., and C. Nithya. "α-decay chains of superheavy nuclei with Z=125." *Physical Review C* 97, no. 4 (2018): 044615.

[44] Santhosh, K. P., and V. Safoora. "^{48}Ca induced reactions for the synthesis of isotopes $^{284-292}$Fl." *Journal of Physics G: Nuclear and Particle Physics* 44, no. 12 (2017): 125105.

[45] Santhosh, K. P., and V. Safoora. "Studies on the synthesis of isotopes of superheavy element Lv (Z= 116)." *The European Physical Journal A* 53, no. 11 (2017): 229.

[46] Santhosh, K. P., and V. Safoora. "Predictions of probable projectile-target combinations for the synthesis of superheavy isotopes of Ts." *Physical Review C* 95, no. 6 (2017): 064611.

[47] Santhosh, K. P., and V. Safoora. "Synthesis of $^{292-303}$119 superheavy elements using Ca-and Ti-induced reactions." *Physical Review C* 96, no. 3 (2017): 034610.

[48] Santhosh, K. P., and V. Safoora. "Systematic study of probable projectile-target combinations for the synthesis of the superheavy nucleus ^{302}120." *Physical Review C* 94, no. 2 (2016): 024623.

[49] Ellison, P. A., Kenneth E. Gregorich, Jill S. Berryman et al. "New Superheavy Element Isotopes: ^{242}Pu (^{48}Ca, 5 n) ^{285}114." *Physical Review Letters* 105, no. 18 (2010): 182701.

[50] Düllmann, Ch E., M. Schädel, A. Yakushev et al. "Production and Decay of Element 114: High Cross Sections and the New Nucleus ^{277}Hs." *Physical Review Letters* 104, no. 25 (2010): 252701.

[51] Oganessian, Y. T. "Heaviest nuclei from ^{48}Ca-induced reactions." *Journal of Physics G: Nuclear and Particle Physics* 34, no. 4 (2007): R165.

[52] Oganessian, Y. T., F. Sh Abdullin, S. N. Dmitriev, et al. "New Insights into the ^{243}Am+^{48}Ca Reaction Products Previously Observed in the Experiments on Elements 113, 115, and 117." *Physical Review Letters* 108, no. 2 (2012): 022502.

9

Decay Dynamics of Ground- and Excited-State Nuclear Systems Using Collective Clusterization Approach

Manoj K. Sharma and Neha Grover

Thapar Institute of Engineering and Technology

CONTENTS

9.1   Introduction

The discovery of radioactivity by H. Becquerel (1896) and the subsequent introduction of radioactive elements polonium ($Z = 84$) and radium ($Z = 88$) by Marie and Pierre Curie in 1898 initiated a scientific breakthrough in the world of subatomic structure of matter. Furthermore, the invention of the atomic nucleus (Rutherford, 1911) gave birth to a new discipline called nuclear physics. Since then, several ground- and excited-state decay modes of different nuclei have been predicted, identified, and established, but still the quest to explore more information about these nuclear decay modes is in progress. Radioactivity is observed in naturally existing heavy nuclei, which are unstable and thus disintegrate without any external impingement. The α-decay, cluster radioactivity (CR), and spontaneous fission (SF) are well-known examples of such decays and are of the main interest in the first half of discussion. Besides this, there are some theoretical inputs [1] regarding the possible emission of heavy clusters in radioactive manner, and the process is termed as heavy particle radioactivity (HPR) [2,3].

The primary description of ground-state/spontaneous decays via quantum mechanical leakage through a potential barrier was initially suggested for α-decay by Gamow [4] (and independently by Condon and Gurney [5]) in 1928. Subsequently, the spontaneous emission of α-particles was studied by different researchers working in the area of nuclear physics. As a consequence, various alternative, and partly more rigorous, mathematical approaches [6–12] were developed to get better insight of this radioactive process. The α-disintegration is generally associated with heavy elements, and it has been noticed from the behavior of general slope of the mass defect curve that most isotopes with $A > 150$ are energetically unstable toward α-decay [13]. However, some elements of intermediate mass such as ^{142}Ce, ^{144}Nd, and ^{146}Sm have also been observed as α-emitters [14,15]. The quest for the half-life measurements of alpha-decay has always been an important subject for both experimentalists and theoreticians. The earliest systematics for the α-decay half-lives with Q-value was given by Geiger and Nuttal [16]; however, they did not explain the theory of α-decay. Later on, along with the quantum mechanical theory of α-decay, Gamow [4], Gurney, and Condon [5] interpreted the relation between the alpha decay energy and its half-life. Afterwards, the authors of Ref. [17–24] suggested different formalisms to estimate α-decay half-lives. However, these approaches were justifiable for a limited set of nuclei and were not able to give satisfactory predictions. Furthermore, Poenaru et al. [25–27] explained a detailed analysis for the α-decay life times on the basis of fission theory with reasonable success. This theory was able to reproduce the data well but with some correction factors for the neutron, proton, and magic numbers. Following this, Buck et al. [28] proposed the cluster model calculations for α-decay life times with even more comprehensive database. In this chapter, the collective clusterization method (CCM) [1,29–33] (in which not only α but all the clusters appearing in the binary decay are considered to be preformed in the mother nucleus with different probabilities) is used to address the ground- and excited-state decay mechanisms. The brief inputs related to methodology used will be discussed in Section 9.2.

Another process involved in ground-state decays termed as spontaneous fission (SF) was first predicted by Bohr and Wheeler in 1939 [34]. Shortly after, the experimental evidence of the same was observed for Uranium nucleus (with a half-life of 10^{16} years) by Flerov and Petrjak in 1940 [35]. Theoretically, similar to α-decay, the SF can also be studied via quantum tunneling phenomena. As a result, several attempts [36–44] were made to explain SF by the mechanisms similar to that invoked by Gamow in the theory of α-disintegration. Bohr and Wheeler [34] made calculations for the SF probability with an assumption that the outgoing fragment hits the barrier about 10^{21} times per second to penetrate it or to attain the optimum configuration of fission fragments. Later on, by following the same formalism, Flugge [36] and Turner [37] made an attempt to give an alternate method for the addressal of SF decay probability. However, this approach was also not very successful and was modified further by Koch et al. [38]. In other words, an extensive amount of experimental and theoretical efforts have been made by several investigators to determine/understand the half-lives and mass distributions of numerous radioactive parent nuclei [39–44]. As of now, the SF decay mode is adequately addressed for a large number of heavy nuclei with half-lives varying from $10^{-3}s$ to 10^{19} years. Furthermore, it has been observed that SF is energetically feasible for heavy and superheavy nuclei with proton number $Z \geq 92$. Also, SF is an important limiting factor that determines the stability of newly synthesized superheavy nuclei. The SF results of nuclei varying from ^{232}U to ^{264}Hs are discussed in Section 9.2.2.1.

Apart from α-decay and SF, another radioactive decay mode showing the spontaneous emission of fragments heavier than an α-particle but lighter than the fission fragment, was first predicted on the theoretical grounds by Săndulescu, Poenaru, and Greiner in 1980 [45]. They named the process as cluster radioactivity (CR), and the basis of this work was the fragmentation theory [46,47], which brings out the applicability of the quantum concept of probability and the role of shell effects, not only for the fusing partners but also for the outgoing products. Also, it is important to mention that the same methodology is used to obtain the results discussed in this chapter.

After the successful theoretical explanation of CR, Rose and Jones [48] provided the experimental evidence for the same in 1984 by observing the emission of ^{14}C from ^{223}Ra nucleus. Since this pioneering experiment, extensive investigations were carried out [49–55], and various clusters such as ^{14}C, ^{20}O, ^{23}F, and $^{24-26}$Ne from different radioactive isotopes (having charge number $Z = 87$–96) were identified with partial half-lives ranging from 10^{11} to 10^{29} seconds. Till now, ^{34}Si emitted from ^{238}U is the heaviest cluster observed with half-life ($T_{\frac{1}{2}} \approx 10^{29}$ seconds). Also, it has been noticed that the daughter nucleus in such emissions is generally a neutron/proton (or both) closed shell or nearly closed shell spherical nucleus, which implies that the shell structure of daughter nucleus and the Q-value play a very significant role in the CR process. This observation was the foremost point for the authors of Ref. [45] to predict a new CR, because in their prior calculations [56–59], the minima in fragmentation potential was observed for at least one spherical closed shell nucleus. This implies that shell closure effects play an equally important role in the CR process, as that in SF decay.

Both α-decay and heavy clusters emission follow the Gamow theory of α-decay and behave as similar processes, but differ from the SF. Numerous theoretical approaches [1,29–33,60–68] have been developed to get a comprehensive explanation of such (α-decay, cluster decay, and SF) emissions. Starting the discussion with "unified fission" models [60–62], such formalisms were initially able to address the decay mechanisms, but without introducing the concept of cluster preformation. Also, these approaches considered the fission as an explicit process and thus were not able to distinguish between α-decay, CR, and SF. Furthermore, in an alternative saddle point fission (SPF) [63–65] model, the mass and charge distribution of all fragments were considered at the saddle point, after the decay products penetrate the potential barrier. Consequently, the barrier penetrability and assault frequencies in SPF become independent of fragment's mass. Then, the scission point fission model [66–68] following the concept of continuously deformed shape of a decaying nucleus came into existence and estimated the barrier penetration at touching configuration, which signify that the interaction potential used in this case is similar to the one used in "Unified Fission Model (UFM)". Furthermore, to have more refined insight of spontaneous emission of fragments from parent nuclei, the concept of cluster preformation probability was introduced within UFM, and this approach was named as preformed cluster decay model (PCM) [1,29–33]. A detailed description of PCM based on CCM of QMFT is provided in Section 9.2.2, and some results based on this methodology are discussed in this chapter.

Besides ground-state emissions, we also aim to focus on excited-state decays such as evaporation residues (ERs), intermediate mass fragments (IMFs), heavy mass fragments (HMFs), and fission. The study of such emissions got impetus after the pioneering discovery of Bohr's hypothesis [69], which introduces the intermediate stage (excited compound nucleus (CN)) of heavy-ion-induced reactions formed due to the compete amalgamation of projectile and target nucleus as a consequence of strong interaction between their nuclear constituents. Generally, the decay of the excited CN proceeds via light particles emission or symmetric/asymmetric fission fragments along with some minor contributions of IMFs and HMFs. Many statistical approaches were proposed to perceive the comprehensive knowledge of different excited-state emissions of various nuclei and hence to extract the relevant information of related nuclear properties. To determine the ER cross sections, statistical codes based on the theory of Hauser and Feshbach [70], such as PACE [71], CASCADE [72], etc. are most commonly used theoretical approaches. Furthermore, the alternative forms or the extensions of the Hauser-Feshbach formalism like "BUSCO" code [73] for the emission of IMFs and the "extended Hauser-Feshbach method (EHFM)" [74] to analyze IMFs, HMFs, and fission are used in view of heavy-ion-induced fission reactions at scission point configuration. Apart from these, GEMINI Monte Carlo statistical emission code [75] for A$\geq$100 and transition-state model (TSM) using saddle-point configuration [76,77] are also available for fission analysis. Also, it has been observed that the decay of excited CN mainly depends upon its mass (A$_{CN}$). Emission of light particles is mostly dominated (with minor contribution of IMFs) in the decay of light mass compound systems (A$_{CN} \leq 80$). However, due to the increase in Coulomb repulsion ($\propto Z^2$), decay of intermediate mass compound systems ($80 < A_{CN} \leq 200$) finds the competition between ER and fission. For heavier compound systems (A$_{CN} > 200$), the Coulomb repulsion becomes more dominant than nuclear binding energy ($\propto A^2$); hence, the decay of such systems results mainly into fission products (either symmetric or asymmetric including HMFs). The statistical approaches (mentioned above) address the different decay modes of a CN independently. Here, we have applied the dynamical cluster decay model (DCM) [78–87], which handles different decay modes on equal footing in one set of calculations. This means that for ground state, α-decay, CR, and SF are addressed parallelly by comparing the mass fragmentation of a given nucleus. Similarly, for the CN channel, ER, IMF, HMF, and fission are estimated within collective clusterization approach, and hence, the PCM (for ground state) and the DCM (for excited state) carry distinct advantages over the competing models.

The main thrust in this chapter is to review both ground- and excited-state emissions explored using the CCM based on quantum mechanical fragmentation theory (QMFT). This approach has been used extensively for the addressal of dynamical behaviors of various nuclei (both in ground state and in excited state), which are reviewed in Sections 9.2.2 and 9.2.3, respectively. Finally, the summary and outlook is given in Section 9.3.

9.2 Collective Clusterization Method (CCM) for Ground- and Excited-State Decays

In view of CCM, the clusters/fragments are preformed in a radioactive nucleus or a CN before tunneling the potential barrier. Initially, this approach was used to study ground-state CR and related phenomena particularly for spherical shapes of nuclei. The formalism for ground-state decays was named as preformed cluster decay model (PCM) [1,29–33]. Subsequently, CCM was reformulated to study the excited-state emission of the compound systems formed in heavy-ion-induced reactions at low-energy regime. This approach was named as dynamical cluster decay model (DCM) [78–87]. Both these approaches for ground- and excited-state decays are based on QMFT, which uses the formalism of collective mass transfer and gives a unified description of two-body exit channel. This section is divided into three subsections, where Section 9.2.1 represents the brief discussion on QMFT followed by the formalism and applications of PCM for spontaneous emissions in Section 9.2.2. The methodology and results related to the DCM for excited-state emission processes are discussed in Section 9.2.3, and finally, the results are summarized in Section 9.3.

9.2.1 Dynamical (or Quantum Mechanical) Fragmentation Theory

The QMFT is based on "two center shell model" (TCSM), which uses the collective mass transfer and gives a unified description of two-body channels in both fusion and decay processes. In QMFT, the average two-body potential is computed using the macro-microscopic method of Strutinsky [88], which successfully describes both cold and hot fusion reaction dynamics. Here, the fragments are considered to be born prior (preborn) to the decay of compound system, thus offering the quantum mechanical concept of probability to explore the role of shell effects in decay process. The collective coordinates included in the QMFT are as follows:

i. The relative separation coordinate R between the two nuclei or fragments (or, equivalently, the length parameter $\lambda = L/2R_0$, where L is length and R_0 is radius of an equivalent spherical nucleus).

ii. The deformation coordinates $\beta_{\lambda i}$ (λ = 2, 3, 4 ... and i = 1, 2) of the colliding/decaying fragments.

iii. The orientation degrees of freedom θ_i ($i = 1, 2$) of the deformed nuclei.

iv. The azimuthal angle ϕ between the principal planes of the two nuclei.

v. The neck parameter ε, defined by the ratio $\varepsilon = E_0/E'$ for the interaction region $R < R_1 + R_2$ (R_i ($i = 1$, 2) is the radius of the two nuclei). Here, E_0 is the actual height of the barrier, and E' is the fixed barrier of the two-center oscillator. $\epsilon = 0$ corresponds to a broad neck formation; however, $\epsilon = 1$ shows that the neck is fully squeezed in, corresponding to the asymptotic region ($R > R_1 + R_2$).

vi. Mass and charge fragmentation coordinates [46,47], which for binary fragmentation are defined as:

$$\eta_A = \frac{A_1 - A_2}{A_1 + A_2}; \qquad \eta_Z = \frac{Z_1 - Z_2}{Z_1 + Z_2}; \tag{9.1}$$

Here, η_A and η_Z are the mass and charge asymmetry coordinates, respectively. The domain of η-values is $-1 \leq \eta \leq 1$, and thus, it allows QMFT to give a collective description of all phenomena ranging from few nucleon transfer to complete fusion and symmetric to super-asymmetric decay of nuclear system. Furthermore, by using the collective coordinates of QMFT, the Schrödinger wave equation (by using decoupled hamiltonian [89]) can be written in the form of two coordinates η and R as follows:

$$\left[-\frac{\hbar^2}{2\sqrt{B_{\eta\eta}}} \frac{\partial}{\partial \eta} \frac{1}{\sqrt{B_{\eta\eta}}} \frac{\partial}{\partial \eta} + V(\eta) \right] \psi^\nu(\eta) = E_\eta^\nu \psi^\nu(\eta) \tag{9.2}$$

and

$$\left[-\frac{\hbar^2}{2\sqrt{B_{RR}}} \frac{\partial}{\partial R} \frac{1}{\sqrt{B_{RR}}} \frac{\partial}{\partial R} + V(R) \right] \psi^\nu(R) = E_R^\nu \psi^\nu(R) \tag{9.3}$$

with

$$\psi(\eta, R) = \psi(\eta)\psi(R) \qquad and \qquad E = E_\eta + E_R \tag{9.4}$$

The states $\psi^\nu(\eta)$ are the vibrational states in the potential $V(\eta)$ and are labeled by the quantum numbers $\nu = 0, 1, 2, ...$ To pursue the calculations, CCM uses Eq. (9.2) in η coordinate. The mass parameters $B_{\eta\eta}$, working as an important input of the kinetic energy term in Eq. (9.2), cater to the dynamical mass distribution in QMFT. These mass parameters ($B_{\eta\eta}$) can be obtained either by using the asymmetric two-center shell model ($ATCSM$) wave functions in cranking formula or by using the classical hydrodynamical model of Kröger and Scheid [90], where the later approach is relatively simpler and also gives a reasonable agreement with the microscopic cranking calculations. Next, the CCM approach for spontaneous emissions (PCM) is discussed in the following section.

9.2.2   Preformed Cluster Decay Model (PCM) for Ground-State Emission Channels

The PCM is originated by adopting Gamow's theory [4] of α-decay, which uses a two-step mechanism: first, the formation of fragment(s), and second, the tunneling through the barrier. In view of proximity theorem [91], various versions of proximity potentials can be employed for the adequate addressal of decay dynamics. The effects of deformations and orientations of decaying fragments/clusters are duly incorporated in PCM methodology. In order to study the different modes of spontaneous emissions, PCM assumes that α or cluster/fragments are preborn in the parent nucleus with preformation probability (P_0), and then the preformed cluster hits the potential barrier with impinging frequency ν_0. The barrier penetration probability (P) is calculated using the Wentzel Kramers Brillouin (WKB) approximation. The preformation probability is calculated by solving the stationary Schrödinger wave equation Eq. (9.2), in which the fragmentation potential ($V(\eta, R)$ enters as an essential input and is defined as follows:

$$V(\eta, R) = -\sum_{i=1}^{2}[B(A_i, Z_i, \beta_{\lambda i})] + V_C(R, Z_i, \beta_{\lambda i}, \theta_i) + V_P(R, A_i, \beta_{\lambda i}, \theta_i) \qquad (9.5)$$

The angle θ_i in Eq. (9.5) symbolizes the orientation angle between the nuclear symmetry axis and the collision axis, measured in the anticlockwise direction [92]. Furthermore, the term $B(A_i, Z_i, \beta_{\lambda i})$ represents the binding energies of outgoing fragments. The experimental values of B_i's can be taken from the tables of Audi et al. [93] or Wang et al. [94], and for the fragments for which the experimental values of B_i's are not available, the theoretical estimates of Möller et al. [95,96] may be employed. The second and third terms represent the Coulomb and proximity potentials [30] respectively, where V_C plays an important role to address the charged particle repulsive interactions and V_P comes in to picture when two nuclei/fragments face each other across a small gap/crevice or when the parent nucleus is at the verge of dividing into two fragments. In this chapter, the roles of different proximity potentials such as Prox 1977, Prox 1988, Prox 2000, mod-Prox 1988, and Bass 1980 are worked out. For more details, one can refer to Refs. [30,91,97–101]. Subsequent to the calculation of $V(\eta, R)$, the preformation probability or spectroscopic factor (P_0) is estimated by solving the Schrödinger wave Eq. (9.2) using an appropriate fragmentation potential like the one given in Eq. (9.5). The expression of P_0 is given as follows:

$$P_0 = |\psi_R^{(0)}[\eta(A_i)]|^2 \frac{2}{A}\sqrt{B_{\eta\eta}(A_i)} \qquad (9.6)$$

where A_i symbolizes the fragment mass and 'A' represents the mass of parent nucleus. Now, to calculate the preformation probability, fragmentation potential is calculated at a fixed radius vector $R = R_a = R_1 + R_2 + \Delta R$, where $R_i(i = 1, 2)$ represents the radius of outgoing fragments and is given as follows:

$$R_i(\alpha_i) = R_{0i}\left[1 + \sum_{\lambda} \beta_{\lambda i} Y_\lambda^{(0)}(\alpha_i)\right] \qquad (9.7)$$

with nuclear radii R_{0i} as follows:

$$R_{0i} = 1.28A_i^{1/3} - 0.76 + 0.8A_i^{-1/3} \qquad (9.8)$$

where α_i in Eq. (9.7) represents the angle, which nuclear symmetry axis makes with the radius vector $R_i(\alpha_i)$, measured in clockwise direction. (For reference, see Figures 1 and 2 of [92]). In PCM, the "ΔR" is termed as neck-length parameter, which is optimized in reference to the available experimental data. After calculating P_0, the barrier tunneling probability (P) of fragment is estimated using the WKB approximation and is given as follows:

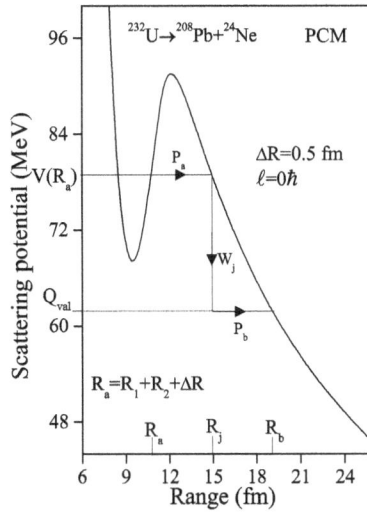

FIGURE 9.1
The scattering potential for the decay of ^{232}U into ^{208}Pb + ^{24}Ne (^{24}Ne cluster). The tunneling path is shown in terms of P_a, W_j, and P_b [102].

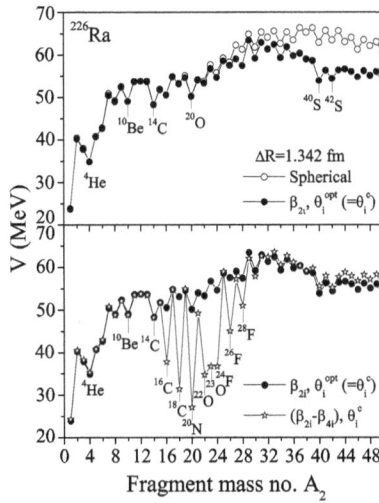

FIGURE 9.2
The fragmentation potentials for the ground-state parent nucleus ^{226}Ra considering spherical and deformed (both β_{2i} alone and β_{2i-4i}) choice of nuclei having "optimum orientations" for β_{2i} alone and "cold compact orientations" for β_{2i-4i}-deformed choice [29].

$$P = P_a W_j P_b \tag{9.9}$$

Equation (9.9) clearly indicates that the tunneling probability has three contributions, which are defined as follows:

i. The penetrability P_a from R_a to R_j

$$P_a = \exp\left[-\frac{2}{\hbar} \int_{R_a}^{R_j} \{2\mu[V(R) - V(R_j)]\}^{1/2} dR\right] \tag{9.10}$$

ii. The inner de-excitation probability W_j at R_j and is taken as $W_j = 1$ in view of the excitation model of Greiner and Scheid [103]. Thus, Eq. (9.9) becomes $P = P_a P_b$.

iii. The penetrability P_b from R_j to R_b.

$$P_b = \exp\left[-\frac{2}{\hbar}\int_{R_j}^{R_b}\{2\mu[V(R) - Q]\}^{1/2}dR\right] \qquad (9.11)$$

The potential $V(R)$ used in Eqs. 9.10 and 9.11 represents the scattering potential and is calculated as the summation of Coulomb and proximity potentials. Also, these equations clearly indicate that the tunneling begins at $R = R_a$ (first turning point) and terminates at R_b (second turning point). For more details, refer to Figure 9.1, where scattering potential for cluster decay of ^{232}U is depicted as a function of internuclear distance [102].

Finally, using P_0 and P as inputs, the decay constant (λ) and decay half-life ($T_{1/2}$) can be defined as follows:

$$\lambda = \nu_0 P_0 P, \qquad T_{1/2} = \frac{\ln 2}{\lambda} \qquad (9.12)$$

The ν_0 in Eq. (9.12) represents the impinging/assault frequency and can be calculated as follows:

$$\nu_0 = \frac{\nu}{R_0} = \frac{\sqrt{2E_2/\mu}}{R_0} \qquad (9.13)$$

The R_0 used in the above equation represents the radius of composite system, and $E_2 = \frac{1}{2}\mu v^2$ is the kinetic energy of the emitted nuclei and can also be defined in terms of Q-value [30]. Since, both outgoing cluster/fragments and daughter nuclei are produced in ground state, the entire Q-value ($Q = \sum E_i$; and $i = 1, 2$ representing the daughter and cluster/fragment) is the total kinetic energy available for the decay process, which is shared between the two nuclei and can be defined as $E = \frac{A_1}{A}Q$. In the following section, we will give an overview of different ground-state emission processes such as α-decay, CR, HPR, and SF in the framework of preformed cluster methodology.

9.2.2.1 Application of Preformed Cluster Decay Model (PCM) for Ground-State Decays

(a) **Cluster radioactivity in trans-Pb and trans-Sn region**

PCM has been used extensively to explore various spontaneous decay modes [1,29–33] such as alpha-decay, cluster emission, HPR, and SF from different radioactive nuclei. It is of extreme interest to understand the relative emergence of these spontaneous emission modes, and an effort is made here to give an overview of these ground-state decay processes in the framework of PCM. Here, the analysis of ground-state emission of nuclei having mass range $A = 105$–294 is presented. Since PCM was initially applied to study CR in the trans-Pb region, thus to begin with, the decay of ^{226}Ra parent nucleus [29] through cluster emission is discussed in Figure 9.2. This figure illustrates the fragmentation potential $V_R(\eta)$ of the above-mentioned parent nucleus at $R = R_a$ [29] for spherical, β_{2i}-deformed, and β_{2i-4i}-deformed choices of fragmentation using optimum (for β_{2i}-alone) and compact (for β_{2i-4i}) orientations [92]. Fragmentation potential represents the effort required to break a parent nucleus into possible decay channels. It is observed that the clusters with a lower magnitude of $V_R(\eta)$ have a relatively higher preformation probability. Hence, the cluster marked on the dips (see Figure 9.2) are the most probable fragments to participate in the decay process. ^{14}C is found to be the favorable cluster in the decay of ^{226}Ra nucleus for all the three choices of fragmentation (spherical, β_{2i}, and β_{2i-4i}), which is in line with the experimental result [104]. It can clearly be seen from Figure 9.2 that deformation and orientation effects do not influence the clusters of mass $A_2 \leq 15$; however, for heavier clusters, a significant change in $V_R(\eta)$ can be observed. The presence of deformation and orientation effects suggests the possibility of few more clusters ^{20}O, ^{40}S, and ^{42}S for β_{2i}-deformed case and ^{16}C, ^{18}C, ^{20}Ne, ^{22}O, ^{23}O, ^{24}F, ^{26}F, and ^{28}F for β_{2i-4i}-deformed case. The study of CR for above-mentioned parent nucleus and various other radioactive nuclei within the PCM framework is extensively discussed in [29–32,102,105].

Furthermore, the cluster decay of parent nuclei ranging from ^{222}Ra to ^{242}Cm is analyzed [30] in view of different nuclear proximity potentials such as Prox 1977, Prox 1988, and Prox 2000. The half-lives of these parent nuclei are calculated at touching configuration ($R_a = R_T$) as well as with the inclusion of neck-length parameter ($R_a = R_T + \Delta R$; $\Delta R = 0.5$ fm) using both spherical and β_2-deformed choices of nuclei and are then compared with the respective experimental data [104]. It is relevant to mention here that all the considered parent systems lead to the doubly magic ^{208}Pb fragment as daughter nuclei, which means that the shell closure effects play a very important role. Furthermore, it is clearly visible in Figure 9.3 that at $R_a = R_T$, the half-life times obtained using Prox 1977 and Prox 1988 are close to the experimental data for β_2-deformed choice of nuclei. However, Prox 2000 overestimates the half-life values for both the choices of fragmentation

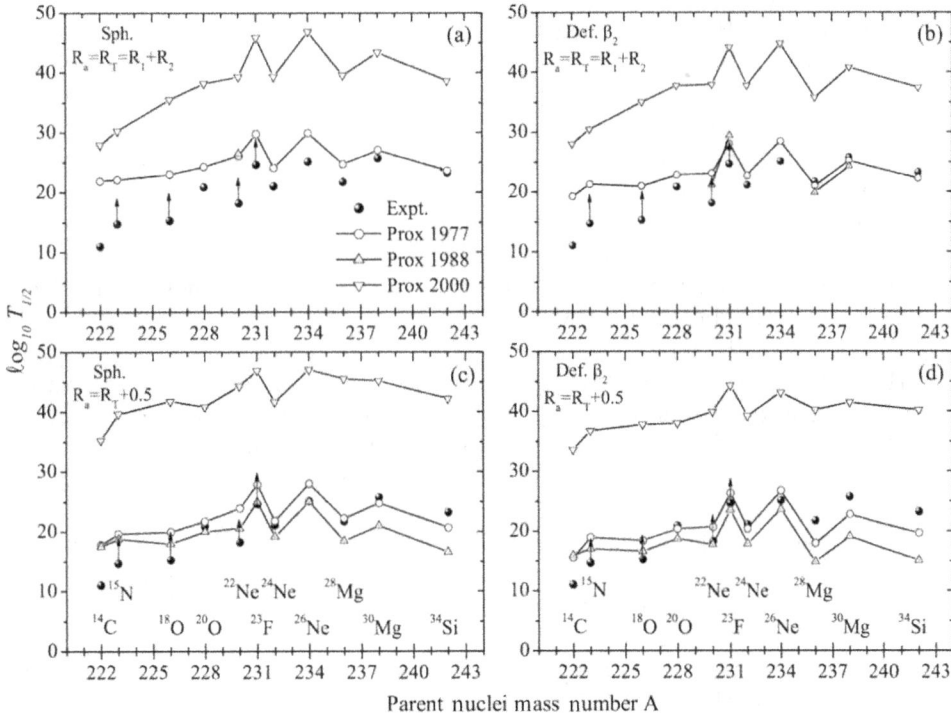

FIGURE 9.3
Comparison of the cluster decay half-lives observed experimentally [104] and the ones calculated with PCM, for various parent nuclei (with ^{208}Pb as the daughter product), using nuclear proximity potentials Prox 1977, Prox 1988, and Prox 2000 for spherical and deformed choices of nuclei at touching configuration (R_T) and $R_T + 0.5$ fm [30].

(spherical and β_2-deformed). Furthermore, the $\log_{10}T_{\frac{1}{2}}$ values calculated using modified entrance point, $R_a = R_T + 0.5$ fm are displayed in Figure 9.3c and d, respectively, for spherical and quadrupole choices of fragmentation. These panels clearly reveal that the half-lives calculated with the inclusion of neck parameter (or at modified turning point) using Prox 1977 and Prox 1988 find better comparison with the experimental data, whereas Prox 2000 is still not able to address the same. The standard rms deviation of PCM calculated half-lives from the experimental data at $R_a = R_T + 0.5$ comes out to be 3.70 (Prox 1977) and 3.72 (Prox 1988) for spherical nuclei, and for the deformed nuclei, it is equivalent to 2.96 (Prox 1977) and 4.34 (Prox 1988). Also, it is important to mention here that for ^{15}N, ^{18}O, ^{22}Ne, and ^{23}F clusters (marked with up arrow in Figure 9.3), only lower limits of half-lives are known experimentally. Thus, the calculated standard rms deviation can be improved further with more precise values of $\log_{10}T_{\frac{1}{2}}$ of these clusters. Concluding Figure 9.3, one can say that for the cluster dynamics at touching configuration for deformed nuclei, Prox 1977 is more preferable; however, with the inclusion of neck parameter (ΔR), both Prox 1977 and 1988 become comparable to each other and give better results even for the spherical choice of decaying fragments.

Furthermore, moving toward another important region of CR, i.e. CR in the trans-Sn region, the preformation factor (P_0) or equivalently mass distribution of different neuron-rich and neutron-deficient parent nuclei ^{112,146}Ba, ^{120,156}Nd, and ^{128,166}Gd [with ^{100,132}Sn (doubly magic) as daughter nuclei [105]] is analyzed for spherical and deformed case within the PCM framework and is plotted in Figure 9.4. Here, the calculations are performed at touching configuration ($R_a = R_t$) only. It is important to mention here that the P_0-factor in PCM imparts structural information of outgoing fragments/clusters, that is otherwise missing in the competing statistical models. Here, the maxima in P_0 for a particular cluster is obtained as a consequence of the minima in fragmentation potential (see Figure 1 in Ref. [105]) of that cluster. In Figure 9.4, a solid vertical line is drawn in all panels to point out the most probable and hence energetically favored cluster. Here, P_0 is a relative quantity, so its magnitude depends on the relative contribution of all the decaying fragments in the PCM and not on the decay products only. This figure shows that the preferred cluster shifts toward heavier mass region with an increase in the mass number of parent nucleus. Furthermore, the status of the favoured fragments remains intact irrespective of the inclusion of deformation effects. Also, the closed shell effects of doubly magic ^{100}Sn on ^{12}C, ^{20}Ne, and ^{28}Si and similarly ^{132}Sn on ^{14}C, ^{24}Ne, and ^{34}Si fragments are clearly evident in terms of strong maxima appearing across heavier fragments in Figure 9.4. Another important observation from Figure 9.4e exhibits that for ^{128}Gd parent nuclei, there is huge difference between the P_0 values calculated for spherical and deformed nuclei, which is possibly due to the difference in its fragmentation potential in the fission region as clearly depicted in Figure 1b of Ref. [105]. The above result holds true for all the studied [105] neutron-deficient to neutron-rich Xe-Gd parents, thereby providing the relevant information regarding the structural behavior of these exotic nuclei. Note that, here doubly magic ^{100}Sn and ^{132}Sn correspond to the magic shell closure at neutron magic $N = 50$, proton magic $Z = 50$ and at $N = 82$, $Z = 50$, respectively.

(b) **Heavy Particle Radioactivity**

Poenaru et al. [2,3] explored the CR in superheavy mass region and predicted the emission of heavier clusters having $Z > 28$. This process is termed as HPR,

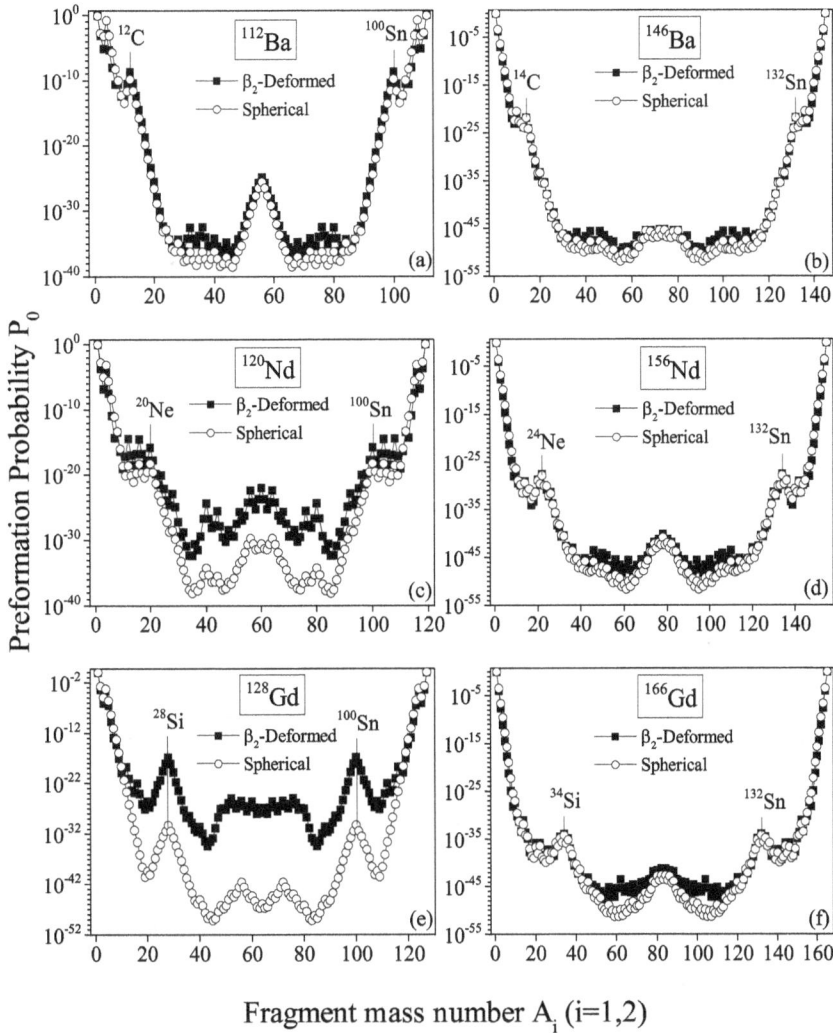

FIGURE 9.4

Fragment preformation probability P_0 plotted as a function of fragment mass A_i showing the comparison of spherical and deformed choices of fragmentation for the parent nuclei: (a) ^{112}Ba, (b) ^{146}Ba, (c) ^{120}Nd, (d) ^{156}Nd, (e) ^{128}Gd, and (f) ^{166}Gd [105].

and the same is studied in [1] within the PCM approach. As an example, the fragmentation potential for $^{294}117$ parent nucleus calculated using quadrupole deformations (β_{2i}) and "optimum" orientations θ_i^{opt} (uniquely fixed on the basis of the signs of β_{2i}) for both "hot compact" and "cold elongated" configurations is displayed in Figure 9.5. Note that the first turning point of penetration for "hot compact" is equivalent to $R_a = R_t + 1.1$ fm and that for "cold elongated" configurations is $R_a = R_t + 0.9$ fm. Figure 9.5 also depicts the influence of two different proximity potentials Prox 1977 and Prox 2000 on HPR for superheavy $^{294}117$ parent nucleus. It can be noticed from Figure 9.5 that both the nuclear potentials suggest ^{86}Br as the most probable cluster (showing a deep minimum in potential energy surface) with complementary ^{208}Pb daughter. Also, it is clear that the minima is more sharp for "hot configuration", which implies the more

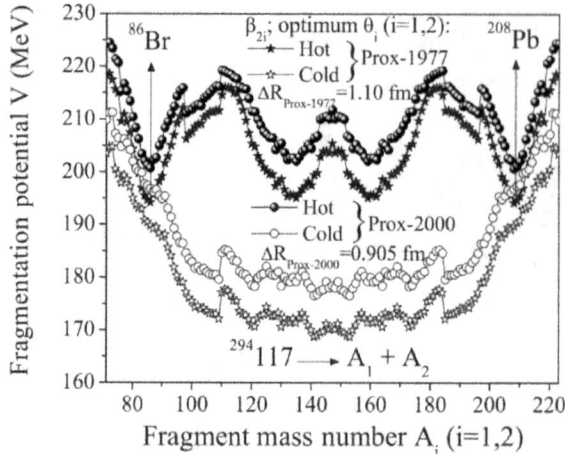

FIGURE 9.5
The variation of fragmentation potential plotted for the superheavy parent nucleus
$^{297}117$ depicting the most probable HPR candidate for hot compact (filled symbols) and
cold elongated (open symbols) configuration using Prox 1977 and Prox 2000 nuclear
potentials [1].

favorable nature of "hot configuration" in comparison to that of "cold elongated"
configuration.

(c) **Alpha-radioactivity**

After exploring CR and HPR, PCM successfully explained [102,106] other two
important modes of radioactivity, α-decay and SF very well. As an illustration,
Figure 9.6 represents the difference in the magnitude of the PCM calculated and
experimental [107] α-decay half-lives of different trans-Sn parent nuclei $^{105-111}$Te,
$^{107-112}$I, $^{109-113}$Xe, ^{114}Cs, and ^{114}Ba. Note that within PCM, α-decay half-
lives of the above-mentioned systems are worked out in view of different nuclear
potentials Prox 1977, Prox 2000, and Ngo80. The choice of the above- mentioned
nuclear potentials has been decided on the basis of tunneling range $(R_b–R_a)$
shown in Table 2 of Ref. [106]. The deviation of PCM calculated half-lives from
the experimental data is also compared with the difference obtained using the
Universal Decay Law (UDL) [108]. Figure 9.6 suggests that except for ^{113}Xe,
^{114}Cs, and ^{114}Ba parent nuclei, Prox 2000 seems to be favorable to address the
α-radioactivity of the chosen nuclei. Additionally, it has also been observed [106]
that Prox Ngo80 performs better to address the α-radioactivity for parent nuclei
having mass $A \geq 113$; however, Prox 2000 seems better only up to $A = 112$.
The deviations in half-lives attained using Prox77 and Prox Ngo80 are primarily
due to the different dependencies on isospin and asymmetry parameters of the
outgoing fragments. Moreover, the radius expressions for these nuclear potentials
are different, which results in two distinct barrier characteristics, and hence are
responsible for the variations in the estimation of half-lives. To have an explicit
description of α-decay of the above-mentioned nuclei within the PCM framework,
one may refer to Ref. [106].

(d) **Spontaneous Fission**

Besides CR, HPR, and α-decay, PCM has also proved its viability to elucidate
the mechanism of SF for different parent nuclei varying from ^{232}U to ^{264}Hs. The

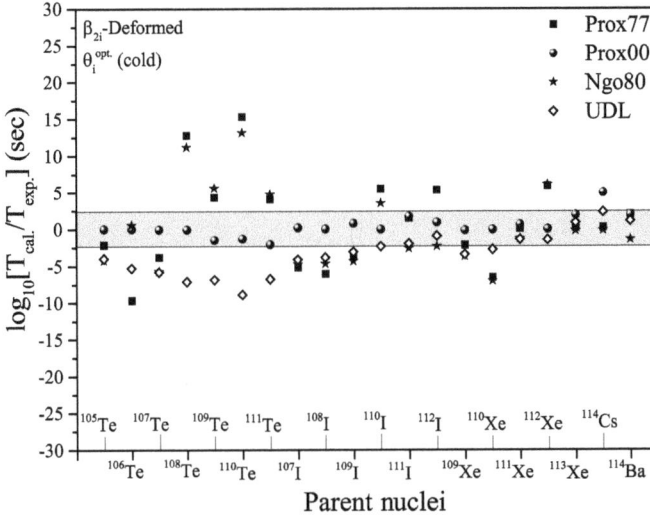

FIGURE 9.6
Deviations between the logarithms of PCM calculated half-lives and the experimental ones for different parent nuclei, using Prox77, Prox00, Prox Ngo80, and UDL [106].

decay path of various parent nuclei is demonstrated in Table 1 of Ref. [102]. PCM-based half-lives for the 45 parent nuclei calculated using two different choices of radius, namely effective sharp radius R_t and Süssmann central radius C_t, were compared with the available experimental data [109] and other model calculations [110]. The Q-values were taken from Audi and Wapstra's table [93], and wherever not available, the theoretical data of Moller and Nix [95] was used. The preference for the SF decay channel depends primarily on the minima in the fragmentation potential and hence for the cases of the largest preformation factors P_0. It is apparent from the above-mentioned table that the fragments in the range of $A_2 = 98$ to 130 (plus complementary heavy fragments) seem the prominent contributors toward SF half-lives for all the chosen fragmentation paths corresponding to ^{232}U–^{264}Hs systems. The contributing fission fragments are asymmetric for a lower mass range of chosen nuclei ($A < 249$; $Z < 98$) and become symmetric for comparatively heavy mass region of parent nuclei. Moving toward the comparison of R_t and C_t, it has been observed [102] that R_t is the suitable choice for all considered parent nuclei having $Z < 100$. However, for $Z \geq 100$, R_t seems to be preferable for neutron-deficient nuclei, and for neutron-rich nuclei, C_t is more appropriate choice.

(e) **Comparative analysis of CR, HPR, α-decay, and SF**

Finally, the comparison between the above-discussed four types of ground-state emission processes is presented in terms of their spectroscopic factor (P_0), penetrability (P), and half-lives for even $^{232-238}$U parents, and the results are displayed in Figure 9.7. It is clearly visible from Figure 9.7a that ^{4}He is always preformed with the largest preformation probability in all parents, though its corresponding barrier penetrability 'P' values are significantly small. This result is important because $T_{1/2}$ is a combined effect of both P_0 and P (ν_0 being nearly constant). Other two important observations from Figure 9.7 are (i) α-decay half-life for ^{238}U is higher than for ^{232}U, ^{234}U, and ^{236}U parents, which means

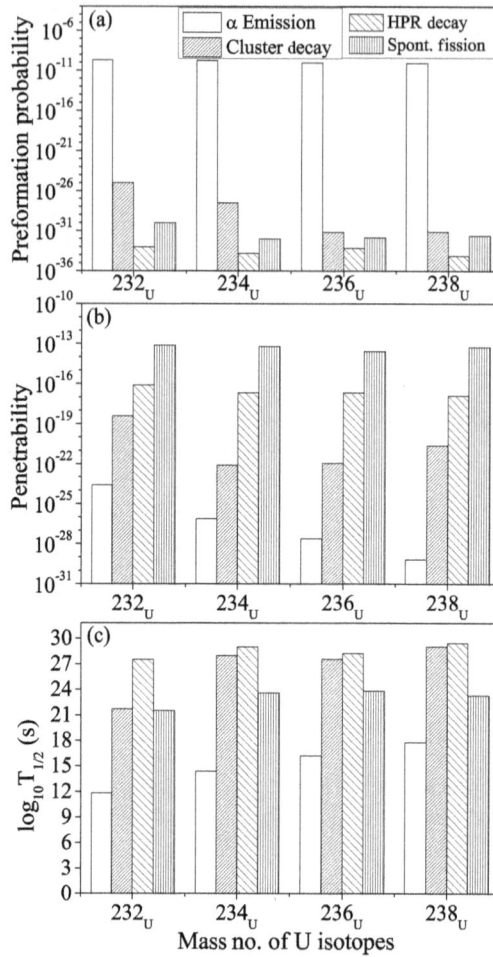

FIGURE 9.7
Histogram representation of (a) preformation probability, (b) penetrability, and (c) logarithm of decay half-lives for the four different decay modes (α-decay, cluster decay, HPR, and SF) of ground-state parent nuclei ^{232}U, ^{234}U, ^{236}U, and ^{238}U [102].

that the ^{238}U nucleus is more stable against α emission than other isotopes of Uranium. (ii) The heavy cluster ^{82}Ge seems to be preformed with a lesser P_0 value, compared to ^{24}Ne, ^{26}Ne, ^{28}Mg, and ^{34}Si, respectively, from ^{232}U, ^{234}U, ^{236}U, and ^{238}U parents. The calculated half-lives for ^{82}Ge heavy clusters lie within the present limits of experiments; hence, such studies provide interesting cases for cluster decay measurements in future. Thus, one may consider the possibility of HPR in addition to α-decay, cluster emission, and SF from ground-state nuclei. The comprehensive analysis of SF within the PCM framework can be seen in Ref. [102].

Apart from the ground-state emissions, the CCM has also been applied to explore the decay of hot and rotating compound systems (i.e. having angular momentum ($\ell \neq 0$) and temperature ($T \neq 0$)). The salient points of this methodology are discussed briefly in the next section along with relevant results prevailing to fusion evaporation and fusion fission dynamics.

9.2.3 Dynamical Cluster Decay Model (DCM) for Excited-State Emissions

The DCM for hot $(T \neq 0)$ and rotating $(\ell \neq 0)$ nuclei is originated from the QMFT and is an extended version of PCM. To study binary fragmentation process, DCM utilizes η and R coordinates of QMFT, where η-coordinate refers to the nuclear mass division, whereas "R" is used to describe the transfer of kinetic energy of incident channel $(E_{c.m.})$ to internal excitation (total excitation or total kinetic energy, TXE or TKE) of the outgoing channel. By following this decoupled approximation in η and R motion, DCM works in two steps: first, it computes the quantum mechanical preformation factor (P_0) of the fragments preborn in the CN, and second, it determines the barrier penetration probability (P) of the preformed fragments. To pursue the first step of DCM, the Schrödinger wave equation (Eq. 9.2) in η-coordinate is solved analytically for temperature $(T \neq 0)$ and angular momentum $(\ell \neq 0)$ dependent fragmentation potential $(V(\eta, R, T))$ as an essential input. The expression of $V(\eta, R, T)$ is defined as following:

$$V(\eta, R, \ell, T) = B_i(A_i, Z_i, \beta_{\lambda i}, T) + V_c(R, Z_i, \beta_{\lambda i}, \theta_i, T)$$
$$+ V_p(R, A_i, \beta_{\lambda i}, \theta_i, T) + V_\ell(R, A_i, \beta_{\lambda i}, \theta_i, T) \tag{9.14}$$

In Eq. (9.14), the angle θ_i symbolizes the orientation angle between the nuclear symmetry axis and the collision axis, measured in the anticlockwise direction [92]. $\beta_{\lambda i}$ ($\lambda = 2, 3, 4; i = 1, 2$) represents the deformations of outgoing fragments. $B_i(A_i, Z_i, \beta_{\lambda i}, T)$ in Eq. (9.14) is the binding energy of two fragments, which as per the Strutinsky renormalization procedure [88] can be expressed as the summation of the macroscopic (liquid drop part) and the microscopic (shell correction) parts. The expression for the same can be written as follows:

$$B(T) = V_{LDM}(T) + \delta U exp\left(-\frac{T^2}{T_0{}^2}\right). \tag{9.15}$$

Here, V_{LDM} (macroscopic term) is the T-dependent liquid drop energy of Davidson et al. [111], with its constants at $T = 0$ refitted in [112] to give the experimental binding energies [93] or that of Möller et al. [95], wherever not available in [93]. The microscopic shell corrections δU are the "empirical" estimates of Myers and Swiatecki [113] with the inclusion of temperature dependence from [114]. The V_C, V_P, and V_ℓ in Eq. (9.14) are, respectively, the T-dependent, Coulomb, nuclear proximity, and angular momentum dependent potentials for deformed and oriented nuclei, and the expressions for V_C, V_P, and V_ℓ are given below.

The Coulomb potential signifies the presence of repulsive force between the two interacting nuclei due to their charge and is given in the following:

$$V_C(R, Z_i, \beta_{\lambda i}, \theta_i, T) = \frac{Z_1 Z_2 e^2}{R(T)} + 3Z_1 Z_2 e^2$$
$$\times \sum_{\lambda, i=1}^{2} \frac{R_i^\lambda(\alpha_i, T)}{(2\lambda + 1)R(T)^{\lambda+1}} Y_\lambda^{(0)}(\theta_i)\left[\beta_{\lambda i} + \frac{4}{7}\beta_{\lambda i}^2 Y_\lambda^{(0)(\theta_i)}\right] \tag{9.16}$$

Here, $Y_\lambda^{(0)}(\theta_i)$ is the spherical harmonic function, the static deformations $\beta_{\lambda i}$ are obtained from the theoretical estimates of Möller and Nix [95], and the optimum orientations θ_i^{opt} for β_{2i} choice of fragments are considered as in Ref. [92].

The proximity potential plays an important role, when the distance between two interacting fragments becomes comparable to their surface thickness ($\approx 2fm$), or when a compound system is about to separate into two fragments. Based on the pocket formula of Blocki et al. [91], the expression of T-dependent V_P for deformed, oriented nuclei is given as follows:

$$V_P(s_0(T)) = 4\pi \overline{R}(T)\gamma b(T)\phi(s_0(T)) \tag{9.17}$$

where $\overline{R}(T)$ represents the mean curvature radius and γ, $\phi(s_0(T))$, and b(T) represent the surface energy constant, universal function, and nuclear surface thickness, respectively.

According to Blocki et al., the universal function, $\phi(s_0(T))$, does not depend upon the shapes of nuclei or the geometry of nuclear system, and can be written as a function of minimum separation distance s_0 as follows:

$$\Phi(s_0) = \begin{cases} -\frac{1}{2}(s_0 - 2.54)^2 - 0.0852(s_0 - 2.54)^3; & \text{for } s_0 \leq 1.2511 \\ -3.437 exp(-\frac{s_0}{0.75}); & \text{for } s_0 \geq 1.2511 \end{cases} \tag{9.18}$$

Here, negative s_0 represents the overlap region, zero signifies touching configuration, and its positive values correspond to the formation of neck between the interacting nuclei. The V_P using the above-mentioned $\phi(s_0(T))$ is termed as Prox 1977.

As an alternative to Prox 1977, one can also use various other versions of the proximity potentials [97–101] such as Prox-88, Prox-00, Bass-73, Bass-77, Bass-80, CW-76, BW-91, AW-95, Ngo-80, and Denisov DP. Besides this, the Skyrme energy density formalism (SEDF) [115] can also be employed, in which one can use different Skyrme forces such as SIII, SLy4, SSk, GSkI, and GSkII to obtain appropriate nuclear potentials. For more information of these forces, one can refer to [115].

Finally, the centrifugal potential is written as:

$$V_\ell(R, Z_i, \beta_{\lambda i}, \theta_i, T) = \frac{\hbar^2 \ell(\ell + 1)}{2I(T)} \tag{9.19}$$

where $I(T)$ indicates the moment of inertia (MOI) in sticking limit or non-sticking limit [81]. For sticking configuration, the rotation of two touching spheres is considered about their common center of mass; however, for non-sticking state, no intrinsic rotation of fragments is considered due to the small separation distance between them, and the fragment emission process is treated as punctual. Also, it is relevant to mention here that the limiting value of angular momentum in sticking configuration is greater than that in non-sticking state [81]. The expression of $I(T)$ for both configurations is as follows:

(a) **Sticking moment of inertia (I_S)**

$$I_s(T) = \mu R^2 + \frac{2}{5}A_1 m R_1^2(\alpha_1, T) + \frac{2}{5}A_2 m R_2^2(\alpha_2, T). \tag{9.20}$$

(b) **Non-sticking moment of inertia (I_{NS})**

$$I_{NS}(T) = \mu R^2 \tag{9.21}$$

where $\mu(= \frac{A_1 A_2}{A_1 + A_2}m)$ and m represent the reduced mass and nucleon mass, respectively. In experiments, the ℓ-value is computed using the non-sticking MOI [112,116]. However, in DCM calculations, it has been observed that I_S choice of MOI is more appropriate owing to the use of proximity potential (neck-length surface $\Delta R \leq 2$ fm) [116].

The fragmentation potential described in Eq. (9.14) is calculated at a fixed distance given as

$$R(T) = R_a(T) = R_1(\alpha_1, T) + R_2(\alpha_2, T) + \Delta R(T)$$
$$= R_t(\alpha, T) + \Delta R(T). \tag{9.22}$$

with radius vectors

$$R_i(\alpha_i, T) = R_{0i}(T)[1 + \sum_\lambda \beta_{\lambda i} Y_\lambda^{(0)}(\alpha_i)], \tag{9.23}$$

and temperature-dependent nuclear radii $R_{0i}(T)$

$$R_{0i}(T) = \left[1.28A_i^{1/3} - 0.76 + 0.8A_i^{-1/3}\right](1 + 0.0007T^2) \tag{9.24}$$

In Eqs. (9.22) and (9.23), α_i is an angle that the nuclear symmetry axis makes with the radius vector $R_i(\alpha_i)$, measured in the clockwise direction (for reference, see Figures 1 and 2 in Ref. [92]), and $\Delta R(\eta, T)$ is the neck-length parameter that assimilates the deformation and neck formation effects between two fragments. It is relevant to remind here that a similar neck-length parameter is used in PCM.

Next, by using $V(\eta, R, T)$ (Eq. (9.14)) as an essential input, the Schrödinger Eq. (9.2) is solved numerically, which gives the formation probability as

$$P_0 \propto |\psi^\nu(\eta)|^2 \tag{9.25}$$

where ν represents the vibrational states of nucleus. The lowest value of $\nu = 0$ refers to ground-state decays, whereas $\nu = 1, 2, 3...$ corresponds to excited-state emissions at a fixed $R = R_a$ (Eq. (9.22)). For excited-state decays or to include the temperature dependence in the preformation probability (used in PCM Eq. (9.6)), the Boltzmann-like occupation of excited states is given as

$$|\psi(\eta)|^2 = \sum_{\nu=0}^{\infty} |\psi^\nu(\eta)|^2 \exp\left(-\frac{E_\eta^\nu}{T}\right) \tag{9.26}$$

Using the above-mentioned Boltzmann expression, the P_0 for excited-state emissions can be written as follows:

$$P_0 = |\psi^\nu(\eta(A_i))|^2 \sqrt{B_{\eta\eta}} \frac{2}{A_{CN}} \tag{9.27}$$

where A_{CN} and A_i, respectively, represent the mass of CN and that of outgoing decay fragments.

After calculating P_0, the penetrability (P) in DCM is calculated through single-step penetration from R_a to R_b and can be expressed as follows:

$$P = \exp\left[-\frac{2}{\hbar}\int_{R_a}^{R_b} \{2\mu[V(R) - Q_{eff}]\}^{1/2}dR\right] \tag{9.28}$$

where R_a and R_b denote, respectively, the first and second turning points of decay path and satisfy the condition $V(R_a, T) = V(R_b, T) = Q_{eff}$, with Q_{eff} as the effective Q-value of the decay process. $V(R)$ is the scattering potential, and for each η-value in excited-state decay, it is calculated as the summation of T-dependent Coulomb, nuclear proximity, and centrifugal potentials. For more clarification, one may see Figure 9.8, where the scattering potential for ^{118}Ba$^* \rightarrow \,^{117}$Cs + ^{1}H channel at a fixed temperature $T = 2.784$ MeV (equivalently, $E_{c.m.} = 145.42$ MeV) is plotted as a function of internuclear separation distance (R) at extreme values of angular momentum (ℓ). Also, the barrier lowering parameter defined as $\Delta V_B = V(R_a) - V_B$ is demonstrated at both ℓ-values.

After obtaining P_0 and P, the decay cross sections in DCM are calculated as follows:

$$\sigma = \sum_{\ell=\ell_{min}}^{\ell_{max}} \sigma_\ell = \frac{\pi}{k^2} \sum_{\ell=\ell_{min}}^{\ell_{max}} (2\ell + 1)P_0 P; \qquad k = \sqrt{\frac{2\mu E_{c.m.}}{\hbar^2}} \tag{9.29}$$

where $\mu = [A_1 A_2/(A_1 + A_2)]m$ is the reduced mass, and ℓ_{min} and ℓ_{max} are the minimum and maximum angular momenta, respectively. Some illustrative examples pertaining to the problems related to fusion evaporation and fusion fission dynamics are discussed in the subsequent subsection.

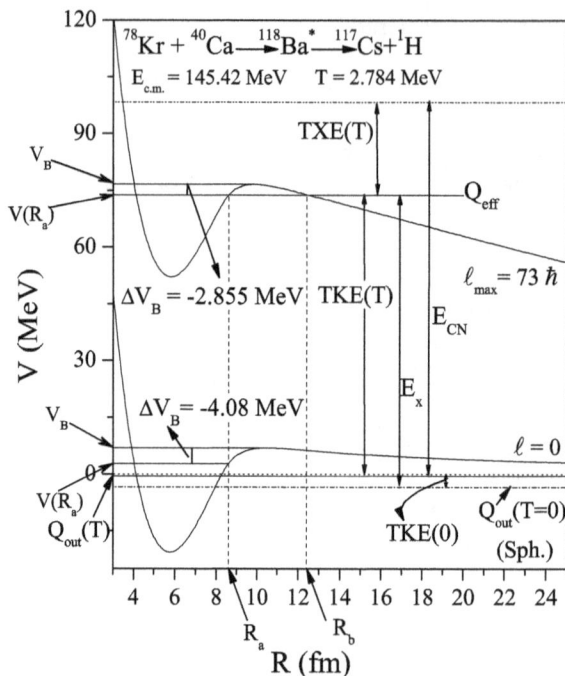

FIGURE 9.8
The scattering potential for $^{118}\text{Ba}^* \rightarrow {}^{117}\text{Cs} + {}^{1}\text{H}$ at fixed temperature $T = 2.78$ MeV (equivalently, $E_{c.m.} = 145.42$ MeV), at extreme values of angular momentum (ℓ). The barrier lowering parameter defined as $\Delta V_B = \text{V}(\text{R}_a)\text{-V}_B$ is also shown at $\ell = 0$ and $\ell = \ell_{max}$ values [87].

9.2.3.1   Dynamics of Compound Nuclei Formed in Heavy-Ion-Induced Reactions

The DCM [78–87] has been used extensively to study the decay of various compound systems pertaining to different mass regions (light, medium, heavy, and superheavy) of periodic table over a wide span of incident energies. This approach has been used to address different CN decay modes such as ER, IMFs, HMFs, and fission after proper inclusion of various effects such as excitation energy (E_{CN}), temperature (T), angular momentum (ℓ), deformations ($\beta_{\lambda i}$), and orientations (θ). This section presents few applications of DCM for compound nuclei belonging to medium and heavy mass regions. To begin with, the dynamics of intermediate-mass nucleus $^{122}\text{Xe}^*$ formed in $^{28}\text{Si} + {}^{94}\text{Zr}$ reaction [83] is discussed using (i) the spherical fragmentation, (ii) quadrupole deformations (β_{2i}) within the optimum orientation approach [79], and (iii) deformation effects up to the hexadecapole (β_{2i-4i}) with the compact orientation approach [79]. Figure 9.9 depicts the fragmentation potential plotted as a function of fragment mass (A_2) for extreme values of angular momentum at the lowest and highest energies. As mentioned in Section 9.2.2, the fragmentation potential goes as an input in the Schrodinger wave equation to calculate the preformation probability and hence the decay cross sections. From Figure 9.9a, it is observed that at a minimum ℓ-value, except for some IMFs ($5 \leq A \leq 20$) and HMFs, the structure of fragmentation potential is almost identical for all three choices (spherical, β_{2i}-deformed, and β_{2i-4i}-deformed) of fragments and their magnitude almost fold over each other. However, the variation of fragmentation potential attained at $\ell = \ell_{max}$ indicates that the deformation effects are more significant at lower $E_{c.m.}$ value. On comparing the fragmentation path attained at

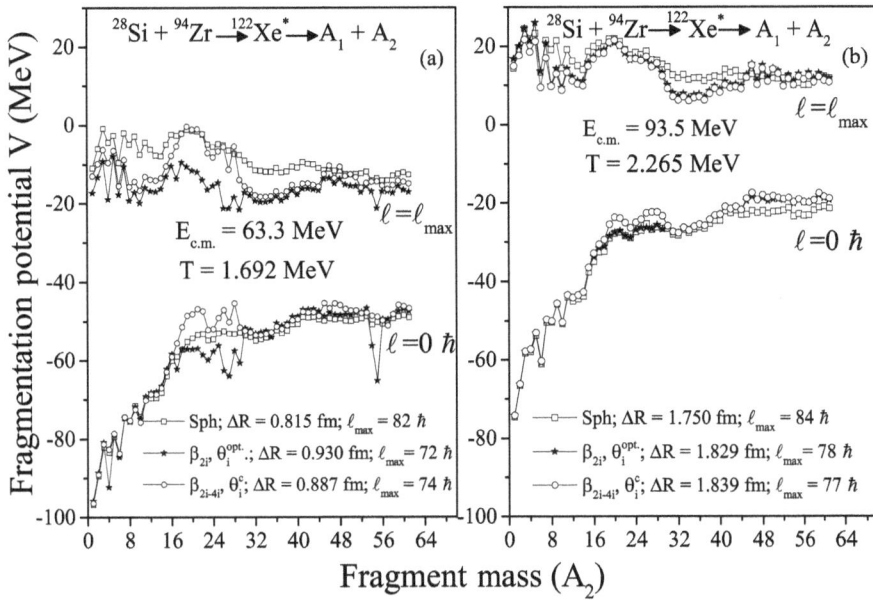

FIGURE 9.9

Fragmentation potential as a function of fragment mass (A_2) for the decay of ^{122}Xe* nucleus at (a) $E_{c.m.} = 63.3$ MeV and (b) $E_{c.m.} = 93.5$ MeV, plotted for spherical, $\beta_{2i}^{opt.}$ (static-β_{2i} with optimum orientations) and β_{2i-4i}, θ_i^c (static-β_{2i-4i} with compact orientations) deformed choices of fragments [83].

higher ℓ-values, it is observed that for the quadrupole-deformed approach, the magnitude of fragmentation potential is lower for all the fragments as compared to that for the spherical approach. Evidently, apart from few IMFs and HMFs, the fragmentation potential for hexadecapole (β_{2i-4i}) deformation overlays that of quadrupole deformation. Furthermore, analyzing the variation of fragmentation potential plotted in Figure 9.9b, it is observed that at a higher energy, i.e. $E_{c.m.} = 93.5$ MeV, the fragmentation potential is higher in magnitude as compared to a lower energy. In other words, with an increase in energy, the magnitude of fragmentation potential also increases. On the other hand, the difference in magnitude of fragmentation potential for all three approaches is very small. Also, it is observed that after incorporating the quadrupole and hexadecapole deformations, the change in structure appears at a lower energy in the form of minor fluctuations in the IMF and HMF regions, which start disappearing at higher energies. However, at both the energies, the overall structure of fragmentation potential is near symmetric. It is worth noting that in DCM, the minimum fragmentation potential signifies the maximum preformation probability (P_0) and hence represents the emergence of the most favorable decay mode. For extensive analysis of decay dynamics of the above-mentioned nucleus and its isotopic dependence, one can refer to [83].

We have taken another example of intermediate nuclei (^{150,158}Tb*) studied within the DCM framework. The calculations for ^{150,158}Tb* nuclei formed in ^{6}Li + ^{144,152}Sm reactions are made at an extensive range of energies ($E_{c.m.} = 19.2$–38.5 MeV). DCM-based calculations adequately addressed [85] the decay cross sections for both complete fusion and incomplete fusion (arises due to loosely bound character of projectile ^{6}Li). However, in this chapter, decay pattern for only complete fusion dynamics is displayed in Figure 9.10 at two extreme energies ($E_{c.m.} = 19.2$ and 38.5 MeV). The main focus is to see that how the

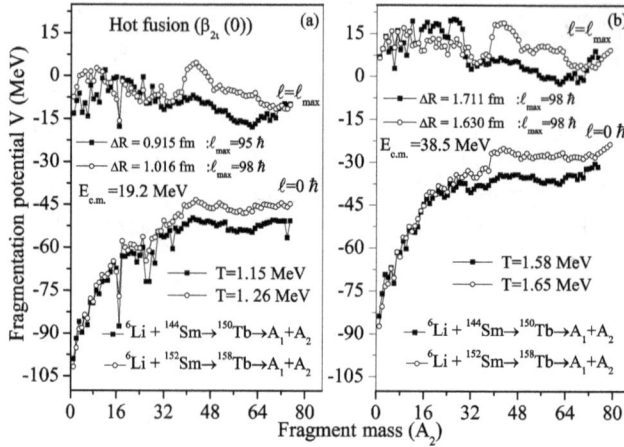

FIGURE 9.10

Fragmentation potential as a function of light fragment mass, A_2 for ^{150}Tb* and ^{158}Tb* channel at (a) $E_{c.m.} = 19.2$ MeV and (b) $E_{c.m.} = 38.5$ MeV [85].

excess neutrons in target nucleus influence the decay pattern of a composite system. Figure 9.10 clearly depicts the similarity in the behavior of fragmentation path for both ^{150,158}Tb* systems at $\ell = 0\hbar$; however, at $\ell = \ell_{max}$, significant difference can be observed in their decay patterns. For the lighter isotope ^{150}Tb*, the fission distribution seems more asymmetric as compared to that for ^{158}Tb* nucleus. Also, the magnitude of fragmentation potential of ^{158}Tb* nucleus is relatively higher for majority of decaying fragments at extreme energies. For more explication, one may refer to [85].

Moving toward heavy mass region of compound systems, DCM has been used to explain a variety of heavy mass nuclei formed via different reactions [84,86]. An illustration of few cases is presented here. First, to study the role of deformations in heavy mass region, the preformation probability (P_0) of different compound systems ^{215}Fr*, ^{223}Pa*, ^{227}Np*, and ^{233}Am* formed using different projectiles ^{18}O, ^{26}Mg, ^{30}Si, and ^{36}S, respectively on unique target ^{197}Au is displayed in Figure 9.11. This figure represents the preformation probability behavior of chosen nuclei for spherical and deformed choices of fragments within optimum orientation approach (θ_i^{opt}) for hot (equatorial) compact configuration at maximum angular momentum (ℓ_{max}) [84]. Figure 9.11 clearly indicates that deformations strongly influence the mass distribution of these nuclei. Also, the variation of P_0 advocates that for all the four nuclei, the mass distribution is symmetric for spherical approach and becomes asymmetric with the inclusion of quadrupole deformations within optimum orientation approach. Thus, it is concluded that deformations play a crucial role in the dynamics of these nuclei, being more pronounced for the heavy mass fragment and fission fragments and relatively silent for light mass fragments. This observation is in line with the pattern observed for the fragmentation potential of nuclei in Figure 1 of Ref. [84]. Another important observation from Figure 9.11 is that the preformation probability (P_0) is higher for ^{215}Fr* nucleus for IMFs and HMFs, and becomes least for the heaviest nucleus ^{233}Am*. However, the situation gets reversed in fission region for deformed choice of nuclei, whereas for spherical fragmentation, the fission region of all compound systems seems equally probable. In the framework of DCM, P_0 also helps in identifying the fragments contributing toward the decay dynamics of a CN. From the variation of P_0 plotted in Figure 9.11a and b, it is observed that the mass range of dominant fission fragments contributing toward the fission decay is higher for spherical choice as compared to that for deformed approach. It may be noted

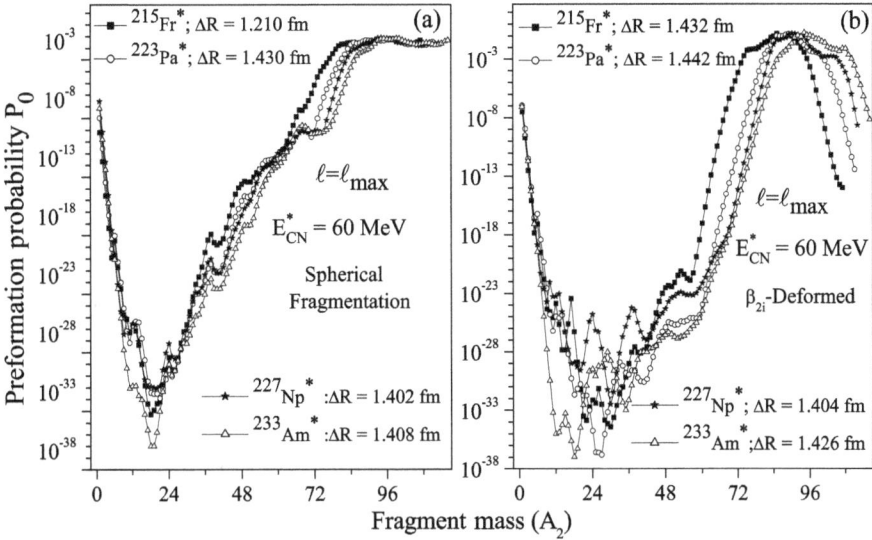

FIGURE 9.11
Preformation probability (P_0) plotted as a function of fragment mass (A_2) for the decay of ^{215}Fr*, ^{223}Pa*, ^{227}Np*, and ^{233}Am* nuclei for (a) spherical and (b) β_2-deformed choices within optimum orientation (θ_i^{opt}) approach for hot (equatorial) compact configuration at $E_{CN}^* = 60$ MeV [84].

that the comprehensive fission analysis and related emergence of shell closure effects of the above- mentioned nuclei can be seen in Ref. [84].

Finally, it is worth mentioning here that the mass distribution of heavy compound systems obtained using DCM is found in reasonable agreement with the available experimental mass distribution data. As an example, Figure 9.12 depicts the preformation probability of fragments contributing toward fission yields of ^{192,202}Pb* nuclear systems at common excitation energy $(E_{CN}^* = 49$ MeV) using β_{2i}-deformed choice of nuclei. It has been observed that mass distribution of both compound systems is nearly symmetric with the emergence of shell closure (magic shells) effect. The possible magic shells having $Z_2 = 28$, $N_2 = 50$, $Z_1 = 50$, and $N_1 = 82$ are explicitly marked in Figure 9.12. Furthermore, it has also been observed that peaks near $Z_2 = 28$ and $N_1 = 82$ are suppressed for neutron-deficient ^{192}Pb* nucleus, which is in agreement with the experimental observations reported in Refs. [117,118]. Additionally, a comparative analysis of ^{192}Pb* nuclear system at two excitation energies $(E_{CN}^* = 49$ and 38 MeV) can be observed through Figure 9.12b and c. On comparing Figure 9.12a–c with the experimental yield of Refs. [117,118], it is evident that the structural distribution obtained using DCM is found in fair agreement with experimental data. A detailed analysis of the above-mentioned isotopes of Pb is published in Ref. [86].

9.3   Summary and Outlook

Summarizing, an extensive analysis of different decay modes of various nuclei is presented using the CCM based on QMFT. Here, the PCM is employed to explain different ground-state emissions such as α-decay, cluster emission, HPR, and SF, and to explain the fragmentation analysis of different excited-state compound nuclei, DCM is applied.

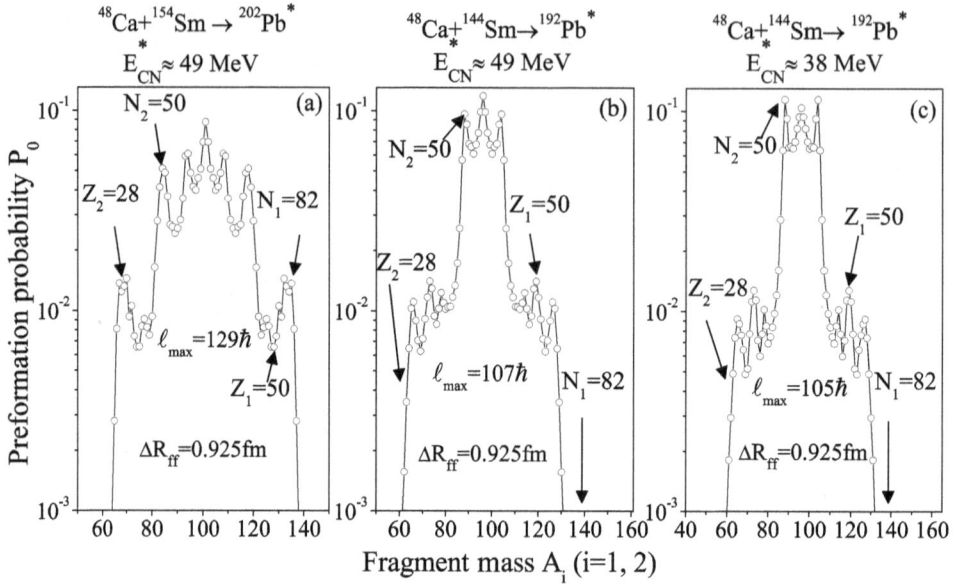

FIGURE 9.12
Preformation probability P_0 of ^{192,202}Pb* calculated using DCM plotted as a function of fragment mass A_i;(i = 1, 2) [86].

First, the CR in the trans-Pb and trans-Sn regions is demonstrated within PCM. The CR analysis of ^{226}Ra* parent nucleus states that ^{14}C is favorable cluster for both spherical and deformed (β_{2i} and β_{2i-4i}) choices of decay fragments. Furthermore, the presence of deformation and orientation effects suggested the possibility of few more clusters ^{20}O, ^{40}S, and ^{42}S for β_{2i}-deformed case and ^{16}C, ^{18}C, ^{20}Ne, ^{22}O, ^{23}O, ^{24}F, ^{26}F, and ^{28}F for β_{2i-4i}-deformed case. Besides this, the role of different proximity potentials is explored for parent nuclei varying from ^{222}Ra to ^{242}Cm, which signifies that the use of Prox 1977 and Prox 1988 potentials finds nice comparison with the experimental data for CR in the trans-Pb region.

After the trans-Pb region, the CR mechanism is explored in the trans-Sn zone for different neutron-rich and neutron-deficient parent nuclei ^{112,146}Ba, ^{120,156}Nd, and ^{128,166}Gd with ^{100,132}Sn (doubly magic) as daughter nuclei. The preformation probability (P_0) or mass distribution of the above-mentioned nuclei signifies that the outgoing clusters shift toward heavier mass region with the increase in the mass number of parent nuclei. The most probable fragments remain the same even after the inclusion of deformation effects. Besides the emission of light clusters, heavy clusters may also emerge from superheavy nuclei and the process is termed as heavy particle radioactivity. In this chapter, the HPR of ^{294}117 superheavy nucleus is explained within the PCM framework using quadrupole deformation (β_{2i}) for both "hot compact" and "cold elongated" configurations. The roles of two different proximity potentials Prox 1977 and Prox 2000 are also explored. It is observed that both nuclear potentials correspond to ^{86}Br as the most probable cluster with ^{208}Pb as daughter nucleus. The minima in fragmentation potential for ^{86}Br cluster is sharper for "hot configuration", which in turn implies that the hot configuration criteria should be preferred for the address of HPR.

Apart from the CR, two important ground-state emissions, i.e. α-decay and SF, are also discussed in this chapter. The α-decay is studied for $^{105-111}$Te, $^{107-112}$I, $^{109-113}$Xe, ^{114}Cs, and ^{114}Ba parent nuclei by using Prox 1977, Prox 2000, and Ngo80, and the PCM-based half-lives are compared with the experimental data and UDL. Calculated α-decay half-lives for nuclei ^{105}Te to ^{112}Xe are nicely addressed using Prox 2000, whereas for nuclei having mass $A \geq 113$, Prox Ngo80 seems to perform better. This switching is possibly due to the different barrier parameters in the chosen proximities. Besides this, the SF of a number of parent nuclei ranging from ^{232}U–^{264}Hs is investigated using two different choices of radii, namely effective sharp radius (R_t) and Süssmann central radius (C_t). The fragments contributing in the SF of the above-mentioned nuclei lie in the mass range ($A_2 = 98 - 130$), and the contributing fission fragments are relatively symmetric for parents having mass ($A > 249$) and charge ($Z > 98$). Additionally, the comparative analysis of the above-mentioned four different spontaneous decay modes (α-decay, CR, HPR, and SF) is also studied for different isotopes of Uranium, ^{232,234,236,238}U, in terms of their preformation probability, penetrability, and half-lives. As expected, the α-decay half-lives for chosen nuclei are lowest followed by SF, CR, and HPR.

Apart from ground-state decays, some applications of the excited-state decay are also illustrated by using an extended version of PCM termed as dynamical cluster decay model. Here, we have focused on the fragmentation analysis of intermediate and heavy nuclei in terms of their fragmentation potential ($V\eta$) and preformation probability (P_0) using both spherical and deformed choices of fragments. The fragmentation potential for intermediate ^{122}Xe* CN depicts that with the inclusion of deformations, IMF and HMF regions get influenced significantly; however, the ER and fission regions remain the same for both approaches. Next, the isotopic analysis of nuclei (^{150,158}Tb*) is discussed. It is observed that the fragmentation path for both nuclei is similar at $\ell = 0\hbar$, but it changes significantly at $\ell = \ell_{max}$. The fission distribution seems more asymmetric for lighter isotope (^{150}Tb*) in comparison to that of the heavier one (^{158}Tb*). For the decay of heavier nuclei, the mass distributions of ^{215}Fr*, ^{223}Pa*, ^{227}Np*, and ^{233}Am* compound systems are explored. It is observed that fission region becomes asymmetric for deformed choice of nuclei; however, for spherical approach, it is purely symmetric. Thus, more fragments contribute towards fission in spherical choice in comparison to that of the deformed approach. Finally, the mass distribution extracted using DCM is displayed for ^{192,202}Pb* isotopes. It is emphasized that shell effects play an important role in governing the mass distribution, which finds decent agreement with the available experimental data. The calculated results of CR, HPR, α-decay, SF, fusion evaporation, and fusion fission find nice agreement with the available experimental data.

Acknowledgments

The authors would like to thank late Prof. Raj K. Gupta, Dr. Raj Kumar, Dr. Gudveen Sawhney, Dr. Gurvinder Kaur, Dr. Kirandeep Sandhu, Dr. Rajni, and Ms. Kanishka Sharma for their contributions in the research work reviewed in this chapter. The financial support from the UGC-DAE Consortium for Scientific Research, F. No. UGC-DAE-CSR-KC/CRS/19/NP09/0920 is thankfully acknowledged.

References

[1] G. Sawhney, K. Sandhu, M. K. Sharma, and R. K. Gupta, *Eur. Phys. J. A* **50**, 175 (2014).

[2] D. N. Poenaru, R. A. Gherghescu, and W. Greiner, *Phys. Rev. Lett.* **107**, 062503 (2011).

[3] D. N. Poenaru, R. A. Gherghescu, and W. Greiner, *Phys. Rev. C* **85**, 034615 (2012).

[4] G. Gamow, *Z. Phys.* **51**, 204–212 (1928); G. Gamow, *Nature* **122**, 805-806 (1928).

[5] R. W. Gurney and E. U. Condon, *Nature* **122**, 439 (1928).

[6] G. Gamow and F. G. Houtermans, *Zeits. f. Physik* **52**, 496 (1929).

[7] M. V. Laue, *Zeits. f. Physik* **52**, 726 (1929).

[8] M. Born, *Zeits. f. Physik* **58**, 306 (1929).

[9] G. Gamow, *Physik. Zeits.* **32**, 651 (1931).

[10] T. Sexi, *Zeits. f. Physik* **81**, 163 (1933).

[11] H. A. Bethe, *Rev. Mod. Phys.; Nucl. Phys. B* **9**, 69 (1937).

[12] M. A. Preston, *Phys. Rev.* **71**, 865 (1947).

[13] J. O. Rasmussen, Jr., S. G. Thompson and A. Ghiorso, *Phys. Rev.* **89**, 33 (1953).

[14] R. D. Macearlane and T. P. Kohman, *Phys. Rev.* **121**, 1758 (1961).

[15] M. C. Gupta and R. D. Macfarlane, *J. inorg, nucl. Chem.* **32**, 3425–3432 (1970).

[16] H. Geiger and J. M. Nuttal, *Philos. Mag.* **22**, 613 (1911).

[17] H. Diamond and J. E. Gindler, *J. inorg. nucl. Chem.* **25**, 143–149 (1963).

[18] C. E. Bemis, Jr., J. Halperin and R. Eby, *J. lnorg. Nucl. Chem.* **31**, 599–604 (1969).

[19] P. O. Froman, *K. Dan. Vidensk. Selsk. Mat. -fys. Skr:* **1**, No. 3, 1–76 (1957).

[20] A. H. Wapstra, G. J. Nijgh, and R. Van Lieshout, *Nuclear Spectroscopy Tables*, North-Holland, Amsterdam, p. 37 (1959).

[21] M. Nurmia, P. Kauranen, and A. Siivola, *Phys. Rev.* **127**, 943 (1962).

[22] K. H. Keller and H. Munzel, *Z. Phys.* **255**, 419 (1972).

[23] V. E. Viola, Jr. and G. T. Seaborg, *J. inorg. nucl. Chem.* **28**, 741 (1966).

[24] P. Hornshoj, P. G. Hansen, B. Jackson, H. L. Ravn, L. Westgaad, and C. N. Nielsen, *Nucl. Phys. A* **230**, 365 (1974).

[25] D. N. Poenaru, M. lvascu and A. Sandulescu, *J. Phys. G: Nucl. Phys.* **5**, L169 (1979).

[26] D. N. Poenaru, M. Ivascu and D. Mazilu, *J. Physique - Lett.* **41**, L-589–L-590 (1980).

[27] D. N. Poenaru and M. Ivascu, *J. Physique* **44**, 791–796 (1983).

[28] B. Buck, A. C. Merchant, and S. M. Perez, *Atomic Data Nucl. Data Tables* **54**, 53–73 (1993).

[29] G. Sawhney, M. K. Sharma, and R. K. Gupta, *Phys. Rev. C* **83**, 064610 (2011).

[30] R. Kumar and M. K. Sharma, *Phys. Rev. C* **85**, 054612 (2012).

[31] G. Sawhney, K. Sharma, M. K. Sharma, and R. K. Gupta, *EPJ Web Conf.* **117**, 04013 (2016).

[32] S. K. Arun, Raj K. Gupta, S. Kanwar, B. B. Singh, and M. K. Sharma, *Phys. Rev. C* **80**, 034317 (2009).

[33] S. K. Arun and R. K. Gupta, B. B. Singh, S. Kanwar, and M. K. Sharma, *Phys. Rev. C* **79**, 064616 (2009).

[34] N. Bohr and J. A. Wheeler, *Phys. Rev.* **56**, 426 (1939).

[35] Flerov and K. A. Petrjak, *Phys. Rev.* **58**, 89 (1940).

[36] S. Flugge, *Z. Physik* **121**, 294 (1943).

[37] L. Turner, *Revs. Modern Phys.* **17**, 292 (1945).

[38] H. W. Koch, J. Mcelhinney, and E. L. Gasteiger, *Phys. Rev.* **77**, 329 (1950).

[39] G. W. Wetherill, *Phys. Rev.* **92**, 907 (1953).

[40] W. J. Swiatecki, *Phys. Rev.* **100**, 937 (1955).

[41] H. W. Schmitt, J. H. Neiler, and F. J. Walter, *Phys. Rev.* **141**, 1146 (1966).

[42] R. L. Fleischer and P. B. Price, *Phys. Rev.* **133**, B63 (1964).

[43] D. C. Hoffman and M. M. Hoffman, *Annu. Rev. Nucl. Sei.* **24**, 151 (1974).

[44] E. K. Hulet, *J. Radioanal. Nucl. Chem. Articles* **142**, 79 (1990).

[45] A. Sandulescu, D. N. Poenaru, and W. Greiner, *Fiz. Elem. Chastits At. Yadra* **11**, 1334 (1980) [Sov. *J. Part. Nucl.* **11**, 528 (1980)].

[46] J. Maruhn and W. Greiner, *Phys. Rev. Lett.* **32**, 548 (1974).

[47] R. K. Gupta, W. Scheid, and W. Greiner, *Phys. Rev. Lett.* **35**, 353 (1975).

[48] H. J. Rose and G. A. Jones, *Nature (London)* **307**, 245 (1984).

[49] D. N. Poenaru, M. Ivascu, A. Săndulescu, and W. Greiner, *J. Phys. G: Nucl. Part. Phys.* **10**, L183 (1984).

[50] S. Kumar and R. K. Gupta, *Phys. Rev. C* **49**, 1922 (1994).

[51] B. Buck, A. C. Merchant, and S. M. Perez, *Phys. Rev. Lett.* **76**, 380 (1996).

[52] Y. S. Zamyatnin, V. L. Mikheev, S.P. Tretyakova, V. I. Furman, S. G. Kadmenskii, Y. M. Chulvilskii, *Sov. J. Part. Nucl.* **21**, 231 (1990).

[53] R. K. Gupta, W. Greiner, *Int. J. Mod. Phys. E* **3**, 335 (1994).

[54] R. G. Lovas, R. J. Liotta, A. Insolia, K. Varga, and D. S. Delion, *Phys. Rep.* **294**, 265 (1998).

[55] G. Royer and R. Moustabchir, *Nucl. Phys. A* **683**, 182 (2001).

[56] A. Săndulescu, R. K. Gupta, W. Scheid and W. Greiner, *Phys. Lett. B* **60**, 225 (1976).

[57] A. Săndulescu, H. J. Lustig, J. Hahn, and W. Greiner, *J. Phys. G: Nucl. Phys.* **4**, L279 (1976).

[58] R. K. Gupta, A. Săndulescu, and W. Greiner, *Phys. Lett. B* **67**, 257 (1977); *Z. Natureforsch* **32a**, 704 (1977).

[59] R. K. Gupta, C. Pàrvulescu, A. Săndulescu, and W. Greiner, *Z. Phys. A* **283**, 217 (1977).

[60] D. N. Poenaru, M. Lvascu, A. Săndulescu, and W. Greiner, *Phys. Rev. C* **32**, 572 (1985); *J Phys. G: Nucl. Phys.* **44**, 923 (1986).

[61] D. N. Poenaru, W. Greiner, K. Depta, M. Lvascu, et al. *At. Data Tables* **34**, 423 (1986).

[62] G. Shanmugam and B. Kamalaharan, *Phys. Rev. C* **38**, 1377 (1988).

[63] S. Kumar, S. S. Malik and R. K. Gupta, *Nucl. Phys. Symp. (India)* **B31**, 61 (1988).

[64] R. K. Gupta, S. Kumar, H. Kumar, W. Scheid and W. Greiner, 7th Adriatic Int. Conf. on Nucl. Phys., Brioni, Yugoslavia, May 27–June **1**, p. 5 (1991).

[65] S. Kumar, R. K. Gupta, and W. Scheid *Int. J. Mod. Phys. E* **3**, No. 1 (1994).

[66] D. R. Saroha, R. Aroumougame, and R. K. Gupta, *Phys. Rev. C* **27**, 2720 (1983).

[67] W. Greiner and A. Săndulescu, *J. Phys. G: Nucl. Part. Phys.* **17**, S429 (1991).

[68] D. N. Poenaru, M. Mirea, W. Greiner, I. căta and Z. Mazila, *Mod. Phys. Lett.* **A5**, 2101 (1990); M. Mirea, D. N. Poenaru and W. Greiner, *II Nuovo Cimento* **105**, 571 (1992).

[69] N. Bohr, *Nature* **137**, 344–348 (1936).

[70] W. Hauser and H. Feshbach, *Phys. Rev.* **87**, 366 (1952).

[71] A. Gavron, *Phys. Rev. C* **21**, 230 (1980).

[72] F. Pühlhofer, *Nucl. Phys. A* **280**, 267–284 (1977).

[73] J. Gomez del Campo, J. L. Charvet, A. D'Onofrio, R. L. Auble, et al. *Phys. Rev. Lett.* **61**, 290 (1988).

[74] T. Matsuse, C. Beck, R. Nouicer, and D. Mahboub, *Phys. Rev. C* **55**, 1380 (1997).

[75] R. J. Charity, M. A. McMahan, G. J. Wozniak, R. J. McDonald et al., *Nucl. Phys. A* **483**, 371–405 (1988).

[76] S. J. Sanders, A. Szanto de Toledo, and C. Beck, *Phys. Rep.* **311**, 487–551 (1999).

[77] L. G. Moretto, *Nucl. Phys. A* **247**, 211–230 (1975); S. J. Sanders, D. G. Kovar, B. B. Back, C. Beck, D. J. Henderson, R. V. F. Janssens, T. F. Wang and B. D. Wilkins, *Phys. Rev. C* **40**, 2091 (1989); S. J. Sanders, *Phys. Rev. C* **44**, 2676 (1991).

[78] R. K. Gupta, edited by W. Greiner and R. K. Gupta, *Heavy Elements and Related New Phenomenon,* **Vol. II**, World Scientific, Singapore, p. 536, (Chapter 14) (1999).

[79] R. K. Gupta, M. Balasubramaniam, R. Kumar, N. Singh, M. Manhas, and W. Greiner, *J. Phys. G: Nucl. Part. Phys.* **31**, 631 (2005); R. K. Gupta, M. Manhas, and W. Greiner, *Phys. Rev. C* **73**, 054307 (2006).

[80] R. K. Gupta, in *Cluster in Nuclei*, Lecture Notes in Physics 818, **Vol. I**, edited by C. Beck (Springer-Verlag, Berlin,), p. 223 (2010).

[81] G. Kaur and M. K. Sharma, *Nucl. Phys. A* **884**, 36 (2012); G. Kaur, N. Grover, K. Sandhu, and M. K. Sharma, *Nucl. Phys. A* **927**, 232 (2014).

[82] G. Sawhney, G. Kaur, M. K. Sharma, and R. K. Gupta *Phys. Rev. C* **88**, 034603 (2013).

[83] N. Grover, I. Sharma, G. Kaur and, M. K. Sharma, *Nucl. Phys. A* **959**, 10–26 (2017).

[84] N. Grover, G. Kaur, and M. K. Sharma, *Phys. Rev. C* **93**, 014603 (2016).

[85] G. Kaur and M. K. Sharma, *Phys. Rev. C* **87**, 044601 (2013).

[86] Rajni, R. Kumar, and M. K. Sharma, *Phys. Rev. C* **90**, 044604 (2014).

[87] M. Kaur, R. Kumar, and M. K. Sharma, *Phys. Rev. C* **85**, 014609 (2012).

[88] V. M. Strutinsky, *Nucl. Phys. A* **95**, 420 (1967).

[89] R. K. Gupta, *IANCAS Bull. (India)* **6**, 2 (1990).

[90] H. Kröger and W. Scheid, *J. Phys. G* **6**, L85 (1980).

[91] J. Blocki, J. Randrup, W. J. Swiatecki, and C. F. Tsang, *Ann. Phys. (NY)* **105**, 427 (1977).

[92] R. K. Gupta, N. Singh, and M. Manhas, *Phys. Rev. C* **70**, 034608 (2004).

[93] G. Audi and A. H. Wapstra, *Nucl. Phys. A* **595**, 409–480 (1995).

[94] M. Wang, G. Audi, F. G. Kondev, W. J. Huang, S. Naimi, and X. Xu, *Chin. Phys. C* **41**, 030003 (2017).

[95] P. Moller, J. R. Nix, W. D. Myers, and W. J. Swiatecki, *At. Data Nucl. Data Tables* **59**, 185 (1995).

[96] P. Moller, A. J. Sierk, T. Ichikawa, and H. Sagawa, *At. Data Nucl. Data Tables* **109**, 1–204 (1995).

[97] R. Kumar, *Phys. Rev. C* **84**, 044613 (2011).

[98] W. Reisdorf, *J. Phys. G* **20**, 1297 (1994).

[99] W. D. Myers and W. J. Swiatecki, *Phys. Rev. C* **62**, 044610 (2000).

[100] P. R. Christensen and A. Winther, *Phys. Lett. B* **65**, 19 (1976).

[101] V. Y. Denisov, *Phys. Lett. B* **526**, 315 (2002).

[102] K. Sharma, G. Sawhney, and M. K. Sharma, *Phys. Rev. C* **96**, 054307 (2017).

[103] M. Greiner and W. Scheid, *J. Phys. G: Nucl. Part. Phys.* **12**, L229 (1986).

[104] R. Bonetti and A. Guglielmetti, *Rom. Rep. Phys.* **59**, 301 (2007).

[105] K. Sharma, G.Sawhney, M. K. Sharma and R. K. Gupta, *Eur. Phys. J. A* **55**, 30 (2019).

[106] K. Sharma, and M. K. Sharma, *Nucl. Phys. A* **986**, 1-17 (2019).

[107] L. Capponi, et al., *Phys. Rev. C* **94**, 024314 (2016).

[108] C. Qi, F. R. Xu, R. J. Liotta, R. Wyss, *Phys. Rev. Lett.* **103**, 072501 (2009).

[109] N. E. Holden and D. C. Hoffman, *Pure Appl. Chem.* **72**, 1525 (2000).

[110] X. Bao, H. Zhang, G. Royer, and J. Li, *Nucl. Phys. A* **906**, 1 (2013).

[111] N. J. Davidson, S. S. Hsiao, J. Markram, H. G. Miller, and Y. Tzeng, *Nucl. Phys. A* **570**, 61c (1994).

[112] R. K. Gupta, R. Kumar, N. K Dhiman, M. Balasubramaniam, W. Scheid, C. Beck, *Phys. Rev. C* **68**, 014610 (2003); B. B. Singh, M. K. Sharma, and R. K. Gupta *Phys. Rev. C* **77**, 054613 (2008).

[113] W. Myers and W. J. Swiatecki, *Nucl. Phys.* **81**, 1 (1966).

[114] A. S. Jensen and J. Damgaard, *Nucl. Phys. A* **203**, 578 (1973).

[115] R. K. Puri, P. Chattopadhyay, and R. K. Gupta, *Phys. Rev. C* **43**, 315 (1991); R. Kumar, M. K. Sharma, R. K. Gupta, *Nucl. Phys. A* 870–871, **42-57** (2011); D. Jain, R. Kumar, and M. K. Sharma, *Phys. Rev. C* **85**, 024615 (2012); D. Jain, M. K. Sharma, Rajni, R. Kumar, and R. K. Gupta, *Eur. Phys. J. A* **50**: 155 (2014); S. Jain, R. Kumar, and M. K. Sharma, *Eur. Phys. J. A* **54**, 203 (2018) and references therein.

[116] S. Kailas (private communication).

[117] M. G. Itkis et al., *Nucl. Phys. A* **724**, 136–147 (2004).

[118] S. Kanwar et al., *Int. J. Mod. Phys. E* **18**, 1453–1467 (2009).

10

Spectroscopic Properties of Nuclei in Generalized Seniority Scheme

Bhoomika Maheshwari

University of Malaya

Bijay Kumar Agrawal

Saha Institute of Nuclear Physics
Homi Bhabha National Institute

CONTENTS

10.1   Introduction

The nucleus is a many-body complex quantum system, which is too hard to be solved exactly. This challenging state-of-the-art is handled by the approximations called as nuclear models, which simplify the complexity to an extent without altering the nuclear properties. One of such simplifications can be achieved in the form of underlying symmetries, particularly the pairing correlations. Before entering into the details of pairing, we shortly revisit the major aspects of a general nuclear Hamiltonian, where it is necessary to solve the non-relativistic many-body Schrodinger equation:

$$H\Psi(r_1, r_2, r_3, ...r_A) = E\Psi(r_1, r_2, r_3, ...r_A) \tag{10.1}$$

where H comprises the kinetic energy and potential energy,

$$H = \sum_{i=1}^{A} T(r_i) + \sum_{i<j}^{A} V(r_{ij}) \tag{10.2}$$

where the nature of interaction acting between nucleons is assumed to be predominantly two-body in nature. Such a general Schrodinger equation would yield an infinite number of stationary states Ψ_n and their energies E_n. A practical way to solve this Schrodinger equation is to rewrite H as

$$H = \sum_{i=1}^{A}(T(r_i) + V_c(r_i)) + \left(\sum_{i<j}^{A} V(r_{ij}) - \sum_{i=1}^{A} V_c(r_i) \right) \tag{10.3}$$

where V_c is an average one-body potential which is common to all the nucleons. We can also write this Hamiltonian as

$$H = H_0 + H_{res} \tag{10.4}$$

where H_0 is the single-particle Hamiltonian, and H_{res} is termed as the residual interaction which is generally much smaller than H_0. Solutions of H_0 provide us a set of single-particle states and energies which satisfy the Schrodinger equation

$$H_0\Phi_\alpha(r_i) = E_0\Phi_\alpha(r_i) \tag{10.5}$$

where Φ_α is a product of the single-particle wave functions:

$$\Phi_\alpha(r_i) = \phi_1(r_1)\phi_2(r_2)\phi_3(r_3)\phi_4(r_4)\dots\phi_A(r_A) \tag{10.6}$$

The total energy E_0 is given by

$$E_0 = \epsilon_1 + \epsilon_2 + \epsilon_3 + \dots + \epsilon_A \tag{10.7}$$

Therefore, the total Hamiltonian H can be given by the single-particle energies and two-body matrix elements generated from residual interaction. Since nucleons are fermions, a normalized wave function of A-nucleons will be anti-symmetric in nature and has to be written as a Slater's determinant.

Although the average potential V_c takes care of the major nuclear interaction, the remaining weak residual interaction is found to be quite important in understanding nuclear spectroscopic properties. One of the simple yet elegant residual interactions is known as the pairing interaction, based on the correlation of nucleons in pairs coupled to $J = 0$. This interaction explains the extra binding of ground $J = 0$ states in even-even nuclei and results in a large energy gap between $J = 0$ and $J > 0$ states. Such pairing effects were well recognized by Racah in terms of seniority [1] in the atomic context, while classifying the electronic states having the same values of L, S, and J in LS-coupling. L, S, and J represent the orbital angular momentum, spin angular momentum, and total angular momentum, respectively. He introduced the seniority scheme to aid in the understanding of complex spectra, where v represents the number of electrons that are not paired to angular momentum $J = 0$ [1–3]. The pairing interaction was then understood as the approximation of zero-range $\delta-$interaction. The concept was similarly adopted for nuclear physics and explained the energy spectra of few particles in single-j shell, where seniority v represents the number of unpaired nucleons involved in generating a non-zero J state. This means that the state with an odd-particle in the valence orbital, $J = j$, always has the lowest seniority as $v = 1$. On the other hand, the ground states of even-even nuclei always have the lowest seniority as $v = 0$, i.e. all the particles are paired to $J = 0$ state. For the first excited states, one needs to break up at least one pair to generate non-zero angular momentum [4–9]. For example, consider the two particle in $j = 7/2$, as $(7/2)^2$ configuration. This configuration can generate spins from 0 to 6, where the $J = 0$ state has the seniority $v = 0$, while the other spins from 2 to 6 have the seniority $v = 2$. Seniority behaves as a good quantum number for $j \leq 7/2$ configurations. This description becomes cumbersome for large j values, where the possibility of many states having the same J and v starts to appear. However, partial conservation of seniority has also been shown for larger j values due to the solvability of the Hamiltonian for a few eigenstates [10–13].

The chapter is divided into six sections. Section 10.2 defines the pairing operators for a two-particle system. Section 10.3 briefly reviews the quasi-spin algebra and the role of seniority in pairing interaction for a single-j shell. Section 10.4 further discusses the extension of the concept of seniority to the generalized seniority in multi-j shell with corresponding electromagnetic selection rules. Section 10.5 shortly glances over the origin of seniority isomerism in both single-j and multi-j shells, and then provides an example of new situation of seniority isomers in Sn isotopes. At last, Section 10.6 summarizes the chapter.

10.2 Pairing Operators

A two-particle wave function (unnormalized) can be written as

$$\Phi_{j_1 j_2; JM}(r_1, r_2) = \sum_{m_1, m_2} \langle j_1 m_1 j_2 m_2 | j_1 j_2 JM \rangle \phi_{j_1 m_1}(r_1) \phi_{j_2 m_2}(r_2) \tag{10.8}$$

where $\langle j_1 m_1 j_2 m_2 | j_1 j_2 JM \rangle$ represents the Clebsch-Gordan coefficient showing the coupling of two-angular momenta. This wave function must be anti-symmetric for nucleons. So, in the second quantization, A^+ and A operators can be defined as

$$A^+(j_1 j_2; JM)|0\rangle = \mathcal{N} \sum_{m_1, m_2} \langle j_1 m_1 j_2 m_2 | j_1 j_2 JM \rangle a^+_{j_1 m_1} a^+_{j_2 m_2} |0\rangle$$

$$\langle 0|A(j_1 j_2; JM) = \mathcal{N} \langle 0| \sum_{m_1, m_2} \langle j_1 m_1 j_2 m_2 | j_1 j_2 JM \rangle a_{j_2 m_2} a_{j_1 m_1}$$

$$\langle 0|A(j_1 j_2; JM) A^+(j_1 j_2; JM)|0\rangle = \mathcal{N}^2 \sum_{m_1 m_2} \sum_{m_1' m_2'} \langle j_1 m_1 j_2 m_2 | j_1 j_2 JM \rangle$$

$$\langle j_1 m_1' j_2 m_2' | j_1 j_2 JM \rangle$$

$$\langle 0| a_{j_2 m_2'} a_{j_1 m_1'} a^+_{j_1 m_1} a^+_{j_2 m_2} |0\rangle \tag{10.9}$$

where $\mathcal{N}$ denotes the normalization constant. Using Wick's theorem,

$$\langle 0| a_{j_2 m_2'} a_{j_1 m_1'} a^+_{j_1 m_1} a^+_{j_2 m_2} |0\rangle = \delta_{m_1 m_1'} \delta_{m_2 m_2'} - \delta_{j_1 j_2} \delta_{m_1 m_2'} \delta_{m_2 m_1'}. \tag{10.10}$$

Therefore,

$$\langle 0|A(j_1 j_2; JM) A^+(j_1 j_2; JM)|0\rangle = \mathcal{N}^2 \sum_{m_1 m_2} \sum_{m_1' m_2'} \langle j_1 m_1 j_2 m_2 | j_1 j_2 JM \rangle$$

$$\times \langle j_1 m_1' j_2 m_2' | j_1 j_2 JM \rangle \left(\delta_{m_1 m_1'} \delta_{m_2 m_2'} - \delta_{j_1 j_2} \delta_{m_1 m_2'} \delta_{m_2 m_1'} \right)$$

$$= \mathcal{N}^2 (1 + \delta_{j_1 j_2} (-1)^{-J}) \sum_{m_1 m_2} \langle j_1 m_1 j_2 m_2 | j_1 j_2 JM \rangle^2$$

$$= \mathcal{N}^2 (1 + \delta_{j_1 j_2} (-1)^{-J})$$

$$= \mathcal{N}^2 \quad \text{if } j_1 \text{ not equal to } j_2$$

$$\text{OR} \quad 2\mathcal{N}^2 \quad \text{if } j_1 \text{ equal to } j_2, \ J = \text{even}$$

$$\mathcal{N} = \frac{1}{\sqrt{1 + \delta_{j_1 j_2}}} \tag{10.11}$$

For the case of pairing, $j_1 = j_2 = j$, $J = M = 0$; the pair creation and annihilation operators can be given by

$$A^+(jj; J = 0, M = 0) = \frac{1}{\sqrt{2}} \sum_{m_1, m_2} \langle jm_1 jm_2 | jj00 \rangle a^+_{jm_1} a^+_{jm_2}$$

$$= \frac{1}{\sqrt{2}} \sum_m \langle jmj, -m | jj00 \rangle a^+_{jm} a^+_{j,-m}$$

$$= \sqrt{\frac{2}{2j+1}} \sum_m \frac{(-1)^{j-m}}{2} a^+_{jm} a^+_{j,-m}$$

$$= \sqrt{\frac{2}{2j+1}} \sum_{m>0} (-1)^{j-m} a^+_{jm} a^+_{j,-m} \qquad (10.12)$$

$$A(jj; J = 0, M = 0) = \sqrt{\frac{2}{2j+1}} \sum_m \frac{(-1)^{j-m}}{2} a_{j,-m} a_{jm}$$

$$= \sqrt{\frac{2}{2j+1}} \sum_{m>0} (-1)^{j-m} a_{j,-m} a_{jm} \qquad (10.13)$$

10.3   Pairing in Quasi-spin Scheme: Seniority

Kerman [14,15] and Helmers [16] proposed the quasi-spin scheme for the description of seniority in a simple way. The quasi-spin scheme starts with the pair creation operator S^+_j for a single-j shell which can be defined as

$$S^+_j = \sqrt{\frac{2j+1}{2}} A^+(jj; J = 0, M = 0)$$

$$S^+_j = \frac{1}{2} \sum_m (-1)^{j-m} a^+_{jm} a^+_{j,-m}$$

$$= \sum_{m>0} (-1)^{j-m} a^+_{jm} a^+_{j,-m} \qquad (10.14)$$

This operator creates a paired state with $J = 0$ while acting on the vacuum state. Similarly, one can define the pair annihilation operator, which annihilates a pair and is a Hermitian conjugate of the pair creation operator. Therefore, the pair annihilation operator for a single-j shell can be defined as

$$S^-_j = \sqrt{\frac{2j+1}{2}} A(jj; J = 0, M = 0)$$

$$= \frac{1}{2} \sum_m (-1)^{j-m} a_{j,-m} a_{jm}$$

$$= \sum_{m>0} (-1)^{j-m} a_{j,-m} a_{jm} \qquad (10.15)$$

A state with seniority v in the single-j configuration j^v can, therefore, be defined as

$$S^-_j | j^v, v, J, M > = 0 \qquad (10.16)$$

If we add pair of particles coupled to $J = 0$ to this state,

$$(S_j^+)^{\frac{n-v}{2}}|j^v, v, J, M> \tag{10.17}$$

we get a state with the same seniority v in the j^n configuration, where $\frac{n-v}{2}$ is the number of pairs. This means that the states of j^n configuration with n even or odd have the seniorities v as even or odd, respectively. For example, 10^+ and $27/2^-$ isomers in Sn isotopes are usually understood as the $v = 2$ and $v = 3$ states from $h_{11/2}$ orbit [7]. The commutation relation between the pair creation operator and its Hermitian conjugate results in

$$[S_j^+, S_j^-] = \sum_m (a_{jm}^+ a_{jm}) - \frac{1}{2}(2j+1) = \hat{n}_j - \Omega = 2S_j^0 \tag{10.18}$$

where

$$S_j^0 = \frac{1}{2}\sum_m a_{jm}^+ a_{j,m} - \frac{2j+1}{4} \tag{10.19}$$

These commutation relations can be followed from the Appendix. The operator S_j^0 has a number operator n_j and a constant term. This further satisfies the relations

$$[S_j^0, S_j^+] = S_j^+, \quad [S_j^0, S_j^-] = -S_j^- \tag{10.20}$$

The proof of the above relations is given in the Appendix. Therefore, the operators S_j^+, S_j^-, and S_j^0 follow the $SU(2)$ Lie algebra, similar to the angular momentum (spin) algebra due to which this scheme was given the name as quasi-spin scheme. Note that the algebraic details can be found in Ref. [4–7].

In order to relate the scheme with pairing interaction, Racah [1] introduced the special Hermitian operator as $2S_j^+ S_j^-$. The seniority scheme is hence defined as the scheme of eigenstates of this operator

$$H_{pair} = -2GS_j^+ S_j^- = -(2j+1)GA^+(jj; J=0, M=0)A(jj; J=0, M=0) \tag{10.21}$$

where G is the pairing strength. This operator results in the same eigenvalues as $(2j+1)$ in j^2 pair-coupled states. The eigenvalues of this pairing operator can simply be obtained by

$$H_{pair} = -2GS_j^+ S_j^- = -2G(S_j^2 - S_j^0(S_j^0 - 1)) \tag{10.22}$$

$$H_{pair}(n, v) = -G\left(2s(s+1) - \frac{1}{2}(\Omega - n)(\Omega + 2 - n)\right)$$

$$= -G\frac{n-v}{2}(2\Omega + 2 - n - v) \tag{10.23}$$

where quasi-spin $s = \frac{1}{2}(\Omega - v)$ and $S_j^0 = \frac{1}{2}(n - \Omega)$, where the number operator $n = n_j$ and pair degeneracy (total number of possible pairs) $\Omega = \frac{1}{2}(2j+1)$. The limiting value of Ω can simply correspond to $\frac{n}{2}$ for a fully filled j^n configuration. The eigenvalues in Eq. (10.23) directly correspond to the eigenstates having seniority v in j^n configuration with $\frac{n-v}{2}$ number of pairs. The eigenvalue will simply become equal to $-nG$, i.e. $-(2j+1)G$ for a $v = 0$ state from Eq. (10.23), which represents the fully pair-coupled $J = 0$ state. In other words, the v cannot exceed the value Ω, which represents the middle of the j-shell. Note that the pairing and seniority play a very important role in nuclear physics, particularly in semi-magic nuclei. The semi-magic nuclei are long known to have the interesting states with maximum pairing and thus have the lowest seniorities. Furthermore, the seniority states in semi-magic nuclei and the corresponding spectroscopic properties are due to identical nucleons.

Also, we can easily get the following commutation relations between S_j^+, S_j^-, S_j^0, and a_{jm}^+, $\tilde{a}_{jm}$:

$$[S_j^0, a_{jm}^+] = \frac{1}{2} a_{jm}^+ \tag{10.24}$$

$$[S_j^+, a_{jm}^+] = 0 \tag{10.25}$$

$$[S_j^-, a_{jm}^+] = (-1)^{j-m} a_{j,-m} = -\tilde{a}_{jm} \tag{10.26}$$

and

$$[S_j^0, \tilde{a}_{jm}] = \frac{1}{2} \tilde{a}_{jm} \tag{10.27}$$

$$[S_j^+, \tilde{a}_{jm}] = 0 \tag{10.28}$$

$$[S_j^-, \tilde{a}_{jm}] = -a_{jm}^+ \tag{10.29}$$

Also, one can find the following commutation relations for a tensor operator $T_\kappa^{(s)}$:

$$[S_j^0, T_\kappa^{(s)}] = \kappa T_\kappa^{(s)} \tag{10.30}$$

$$[S_j^+, T_\kappa^{(s)}] = \sqrt{s(s+1) - \kappa(\kappa+1)} T_{\kappa+1}^{(s)} \tag{10.31}$$

$$[S_j^-, T_\kappa^{(s)}] = \sqrt{s(s+1) - \kappa(\kappa-1)} T_{\kappa-1}^{(s)} \tag{10.32}$$

One can, therefore, find that the creation operator a_{jm}^+ and the annihilation operator $(-1)^{j-m} a_{j,-m} = -\tilde{a}_{jm}$ become the $\kappa = \frac{1}{2}$ and $\kappa = -\frac{1}{2}$ components of an irreducible quasi-spin tensor of rank $s = \frac{1}{2}$. In other words, one may obtain the higher rank quasi-spin tensors by linear combinations of products of creation and annihilation operators. To achieve a κ component operator, the products may contain x annihilation operators and $x + 2\kappa$ creation operators. This results that only $\kappa = 0$ component will survive between the states having the same number of particles.

Furthermore, the single-particle operators are involved in electromagnetic moments and transition probabilities. According to the Wigner-Eckart theorem, the matrix elements of odd-tensor single-particle operator vanish in j^n configuration. Also, odd-tensor operators cannot have $s = 1$ rank with $\kappa = 0$ component. This means that the odd-tensor operators are quasi-spin scalars $s = 0$. Furthermore, one finds that the matrix elements of odd-rank tensor operator vanish between states with different values of s, i.e. seniorities (v). This can be written as follows:

$$\langle j^n v l J || T^{(k=odd)} || j^n v' l' J' \rangle = \langle s, S_j^0 | T_{\kappa=0}^{s=0} | s', S_j^0 \rangle$$

$$= (-1)^{s-S_j^0} \langle s || T^{s=0} || s \rangle \begin{pmatrix} s & 0 & s \\ -S_j^0 & 0 & S_j^0 \end{pmatrix} \tag{10.33}$$

where $s = s'$ is restricted by quasi-spin scalar. This eventually holds $v = v'$, since $s = \frac{1}{2}(\Omega - v)$, $s' = \frac{1}{2}(\Omega - v')$ and $S_j^0 = \frac{1}{2}(n - \Omega)$. The involved 3j-symbol would be equal to one. So, the similar relation remains valid in j^v configuration by simply putting $n = v$, i.e. $S_j^0 = -s$. This further results in the n-independent variation of the non-vanishing matrix elements for odd-tensor operators. This result may simply be written as follows:

$$\langle j^n v l J M | T_\kappa^{(k=odd)} | j^n v' l' J' M' \rangle = \langle j^v v l J M | T_\kappa^{(k=odd)} | j^v v' l' J' M' \rangle \, \delta_{v,v'} \tag{10.34}$$

It means that such operators do not allow the possibility of different v states, that is, the seniority v quantum number remains conserved. This can be related to the case of magnetic dipole moments in electromagnetic transitions. It means that the magnetic moments, i.e. g-factors support a particle number-independent behavior in single-j shell.

On the other hand, the even-tensor operators are the $\kappa = 0$ components of the quasi-spin vectors having rank $(s = 1)$. Again, according to the Wigner-Eckart theorem, one can have non-vanishing matrix elements between states with s and s' differing at most by 1. This further implies that the seniorities of these states, v and v', may either be equal or differ at most by 2; otherwise, the matrix elements will vanish. Moreover, the Wigner-Eckart theorem in quasi-spin space for j^n configuration says

$$\langle j^n vlJ||T^{(k=even)}||j^n v'l'J'\rangle = \langle s, S_j^0|T_{\kappa=0}^{s=1}|s', S_j^0\rangle$$

$$= (-1)^{s-S_j^0}(s||T^{s=1}||s') \begin{pmatrix} s & 1 & s' \\ -S_j^0 & 0 & S_j^0 \end{pmatrix} \quad (10.35)$$

The dependence of such matrix elements on n is due to the 3j-symbol in quasi-spin scheme and is different for $s' = s$ and $s' = s \pm 1$ as shown below:

i. When $s' = s$, i.e. $v' = v$

$$\langle s, S_j^0|T_{\kappa=0}^{s=1}|s, S_j^0\rangle = \frac{2S_j^0}{\sqrt{2s(2s+1)(2s+2)}}(s||T^{s=1}||s)$$

$$= \frac{(n-\Omega)}{\sqrt{2s(2s+1)(2s+2)}}(s||T^{s=1}||s) \quad (10.36)$$

Similar is true for j^v configuration, where $n = v$ in Eq. (10.36). Therefore, the matrix elements of involved spherical harmonics in electric transitions may be related from j^n configuration to j^v configuration as follows:

$$\langle j^n vlJM|Y_\kappa^L|j^n vl'J'M'\rangle = \left[\frac{\Omega - n}{\Omega - v}\right]\langle j^v vlJM|Y_\kappa^L|j^v vl'J'M'\rangle \quad (10.37)$$

where l and l' are the parities of initial and final states for a given electromagnetic transition, respectively. L represents the allowed value of multipolarity that is the orbital angular momentum transfer. The above result is valid for $L > 0$ and an even value. This also means that the matrix elements of even-tensor operators, $L > 0$, between the same seniority states are equal in magnitudes but having opposite signs for j^n (n particles) and j^{2j+1-n} (n holes) configurations. At the middle of the shell, $n = \frac{2j+1}{2} = \Omega$, the matrix elements vanish and change their sign to negative when n crosses middle. For a specific case of $L = 0$, the tensor $Y_0^{(0)}$ will have only diagonal elements in any scheme, which are all equal and proportional to n. The simplest way to calculate $Y_0^{(0)}$ is to recall $Y_0^{(0)} = \sum_{m,m'}(jmjm'|jj00)a_{jm}^+\tilde{a}_{jm'} = \frac{1}{\sqrt{2j+1}}\sum_m a_{jm}^+ a_{jm}$, which is simply proportional to the number operator.

ii. When $s' = s + 1$, i.e. $v' = v - 2$

$$\langle s, S_j^0|T_{\kappa=0}^{s=1}|s+1, S_j^0\rangle = (-1)^{s-S_j^0}(s||T^{s=1}||s+1) \begin{pmatrix} s & 1 & s+1 \\ -S_j^0 & 0 & S_j^0 \end{pmatrix}$$

$$= -\sqrt{\frac{4(s+S_j^0+1)(s-S_j^0+1)}{2(2s+1)(2s+2)(2s+3)}}(s||T^{s=1}||s+1)$$

$$= -\sqrt{\frac{(n-v+2)(2\Omega+2-n-v)}{2(2s+1)(2s+2)(2s+3)}}(s||T^{s=1}||s+1) \quad (10.38)$$

Therefore,

$$\langle j^n v l J M | Y_\kappa^L | j^n v - 2, l' J' M' \rangle = \left[\sqrt{\frac{(n - v + 2)(2\Omega + 2 - n - v)}{2(2\Omega + 2 - 2v)}} \right] \tag{10.39}$$

$$\langle j^v v l J M | Y_\kappa^L | j^v v - 2, l' J' M' \rangle \quad (L > 0, \text{ even})$$

iii. When $s' = s - 1$, i.e. $v' = v + 2$

$$\langle j^n v l J M | Y_\kappa^L | j^n v + 2, l' J' M' \rangle = \left[\sqrt{\frac{(n - v + 2)(2\Omega + 2 - n - v)}{2(2\Omega + 2 - 2v)}} \right] \tag{10.40}$$

$$\langle j^v v l J M | Y_\kappa^L | j^v v + 2, l' J' M' \rangle \quad (L > 0, \text{ even})$$

The matrix elements in Eqs. (10.40) and (10.41) are off-diagonal but symmetric between n and $2\Omega - n$, resulting in equal values for j^n (n particles) and j^{2j+1-n} (n holes) configurations. Note that the even rank tensors could have $\Delta v = 0$, or 2 and electric transitions, while the odd rank tensors have $\Delta v = 0$ and magnetic transitions in single-j shell. The reduced electric transition probabilities $B(EL)$ between J_i and J_f states in a single-j configuration along with the corresponding total pair degeneracy $\Omega = \frac{1}{2}(2j + 1)$ can be written as follows:

$$B(EL) = \frac{1}{2J_i + 1} |\langle j^n v l J_f || \sum_i r_i^L Y^L(\theta_i, \phi_i) || j^n v' l' J_i \rangle|^2 \tag{10.41}$$

where l, l', and L can only take even values. This further implies in the parabolic $B(EL)$ behavior. The seniority reduction formulas for the seniority conserving $\Delta v = 0$ and seniority changing $\Delta v = 2$ transitions can be obtained as follows:

$$\langle j^n v l J_f || \sum_i r_i^L Y^L(\theta_i, \phi_i) || j^n v l' J_i \rangle = \left[\frac{\Omega - n}{\Omega - v} \right]$$

$$\langle j^v v l J_f || \sum_i r_i^L Y^L(\theta_i, \phi_i) || j^v v l' J_i \rangle$$

$$(L > 0, \text{ even}) \tag{10.42}$$

$$\langle j^n v l J_f || \sum_i r_i^L Y^L(\theta_i, \phi_i) || j^n v \pm 2, l' J_i \rangle = \left[\sqrt{\frac{(n - v + 2)(2\Omega + 2 - n - v)}{4(\Omega + 1 - v)}} \right]$$

$$\langle j^v v l J_f || \sum_i r_i^L Y^L(\theta_i, \phi_i) || j^v v \pm 2, l' J_i \rangle$$

$$(L > 0, \text{ even}) \tag{10.43}$$

To sum up, the odd-tensor operators conserve the seniority for the magnetic transitions and lead to the particle number-independent behavior. This may further be related to the constant value of magnetic moments, i.e. g-factors for a given J, v state in j^n configuration. On the other hand, the even-tensor operators may conserve the seniority or may change the seniority by two for the electric transitions. As a result, the even-tensor transitions exhibit a parabolic $B(EL)$ behavior for both seniority conserving $\Delta v = 0$ and seniority changing $\Delta v = 2$ transitions. The $B(EL)$ parabola has a minimum at the middle of the shell for $\Delta v = 0$ transitions while has a maximum at the middle of the shell for $\Delta v = 2$ transitions, as shown in Figure 10.1. The dip at the middle of the shell for $B(EL)$ (L even) values leads

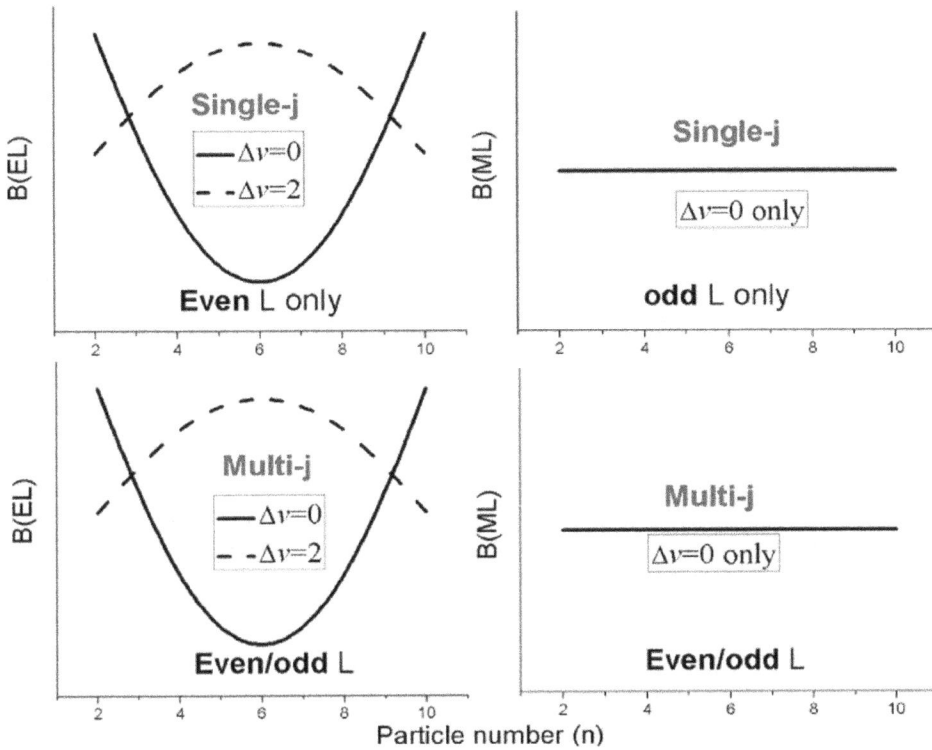

FIGURE 10.1
A schematic comparative plot of the electric and magnetic reduced transition probabilities in single-j and multi-j shells. One may observe the difference in both situations in terms of allowed nature of L.

to a large half-life than neighborhood, and results in the isomerism, i.e. long-lived excited states in nuclei. Such isomers are said to be seniority isomers. It was well expected and found that the seniority isomers occur in electric quadrupole transitions, particularly in semi-magic nuclei, where seniority behaves as a good quantum number. These results show the elegance of seniority scheme, where a problem from n particles simply reduces to v particles. For example, Eqs. (10.42) and (10.43) relate the reduced matrix elements of electric multipole operators from the states of j^n configuration to the states of j^v configuration. If one obtains the solution of matrix elements in j^v configuration, the full trend with changing nucleon n number can be followed.

10.4 Generalized Seniority

To consider the real situations of nuclear environment, one needs to consider many active orbits from a valence space. The extension of these results to the multi-j orbits is long known by using the quasi-spin scheme of Kerman [14,15] and Helmers [16]. Arima and Ichimura [17] were the first to introduce the concept of generalized seniority in multi-j shell, mainly for degenerate orbits. Talmi further extended it to the real case of several non-degenerate orbits [18,19]. Here, we present the extension of seniority scheme to generalized seniority scheme by simply extending the quasi-spin algebra in multi-j degenerate orbits [20]. A straightforward definition of multi-j pair creation operators can be obtained as [7]:

$$S^+ = \sum_j S_j^+ \tag{10.44}$$

where the summation takes care of all the orbits considered in any multi-j configuration. The commutation relation between the multi-j pair creation operator and its Hermitian conjugate may now be written as

$$[S^+, S^-] = \sum_{jm} (a_{jm}^+ a_{jm}) - \frac{1}{2} \sum_j (2j+1) = \sum_j \hat{n}_j - \Omega = 2S^0 \tag{10.45}$$

The operator S^0 further satisfies the relations

$$[S^0, S^+] = S^+, \quad [S^0, S^-] = -S^- \tag{10.46}$$

Again, the operators S^+, S^-, and S^0 form the generators of $SU(2)$ Lie algebras. The total quasi-spin of the system is, therefore, given by

$$s = s_{j_1} + s_{j_2} + s_{j_3} + s_{j_4} + \cdots \tag{10.47}$$

It, therefore, follows that the system of eigenstates of S^2, or those of $2S^+S^-$, defines a generalized seniority scheme for identical nucleons in multi-j shell. This scheme has all the features of the seniority scheme in a single-j orbital which follow from the $SU(2)$ algebra. Eigenvalues of the pairing interaction $-2GS^+S^- = -2G(S^2 - S^0(S^0 - 1))$ are given by $-G(2s(s+1) - \frac{1}{2}(\Omega - n)(\Omega + 2 - n)) = -G(\frac{n-v}{2}(2\Omega + 2 - n - v))$, since the total quasi-spin as given in Eq. (10.47) can be obtained as $s = \frac{1}{2}(\Omega - v)$ and $S^0 = \frac{1}{2}(n - \Omega)$. The seniority in single-j shell is now modified to the generalized seniority v in multi-j shell. Also, the total number of particles $n = \sum_j n_j$, and the pair degeneracy of multi-j $\Omega = \frac{1}{2}\sum_j(2j+1)$. Note that the generalized seniority v states are now akin to quasi-particle picture supporting a shared occupancy and corresponding configuration mixing. For example, the generalized seniority $v = 0$ state can be obtained by $(S^+)^n|0>$, where $n = \sum_j n_j$ and $S^+ = \sum_j S_j^+$. Hence, each state with $(S_j^+)^{n_j}|0>$ is now mixed with various configurations to generate a generalized seniority $v = 0$, $(S^+)^n|0>$ state. Further, consider a single-nucleon tensor operator with rank k and κ component in multi-j can be defined by,

$$T_\kappa^{(k)} = \frac{1}{\sqrt{(2k+1)}} \sum_{j,j'} (j||T^{(k)}||j')(a_j^+ \times \tilde{a}_{j'})_\kappa^{(k)} \tag{10.48}$$

Therefore, the commutation relation between multi-j pair creation operator and the above tensor can be obtained as

$$[S^+, T_\kappa^{(k)}] = \sum_{j'} [S_{j'}^+, T_\kappa^{(k)}]$$

$$= \frac{1}{(2k+1)} \sum_{jj',mm'} (j||T^{(k)}||j')(jmj'm'|jj'k\kappa)$$

$$\times a_{jm}^+ (-1)^{j'+m'} [S_{j'}^+, a_{j',-m'}]$$

$$= \frac{1}{(2k+1)} \sum_{jj'} (j||T^{(k)}||j')(a_j^+, \times a_j'^+)_\kappa^{(k)} \tag{10.49}$$

The $T_\kappa^{(k)}$ becomes a quasi-spin scalar, if

$$[S^+, T_\kappa^{(k)}] = 0 \qquad (10.50)$$

which is true when $j = j'$. For $j \neq j'$, the right-hand side of Eq. (10.49) can be expressed as

$$= \frac{1}{2k+1} \sum_{j<j'} [(j\|T^{(k)}\|j')(a_j^+ \times a_{j'}^+)_\kappa^{(k)} + (j'\|T^{(k)}\|j)(a_{j'}^+ \times a_j^+)_\kappa^{(k)}]$$

$$= \frac{1}{2k+1} \sum_{j<j'} [(j\|T^{(k)}\|j') + (-1)^{(j+j'-k+1)}(j'\|T^{(k)}\|j)] \times (a_j^+ \times a_{j'}^+)_\kappa^{(k)}$$

$$= \frac{1}{2k+1} \sum_{j<j'} [1 + (-1)^k](j\|T^{(k)}\|j')(a_j^+ \times a_{j'}^+)_\kappa^{(k)} \qquad (10.51)$$

where the last equality can be obtained by using the property $(J'\|T^{(k)}\|J)^* = (-1)^{J-J'}(J\|T^{(k)}\|J')$ for Hermitian operators. This means that the commutation $[S^+, T_\kappa^{(k)}] = 0$ for odd k values, that is, the odd Hermitian tensors are proved as quasi-spin scalars. The result is actually a simple generalization of the results of seniority scheme in single-j shell, as discussed in the previous section. However, the situation can change for anti-Hermitian tensors, where the commutation $[S^+, T_\kappa^{(k)}]$ finally results in

$$\frac{1}{2k+1} \sum_{j<j'} [1 + (-1)^{k+1}](j\|T^{(k)}\|j')(a_j^+ \times a_{j'}^+)_\kappa^{(k)} \qquad (10.52)$$

It means that $[S^+, T_\kappa^{(k)}] = 0$ for even k values. Hence, there exist another class of tensors in multi-j environment, which can behave as quasi-spin scalars.

Also, one can easily get the commutation relations between the multi-j generators S^+, S^-, S^0, and creation and annihilation operators $\sum a_{jm}^+$, $\sum \tilde{a}_{jm}$ as

$$[S^0, \sum a_{jm}^+] = \frac{1}{2} \sum a_{jm}^+ \qquad (10.53)$$

$$[S^+, \sum a_{jm}^+] = 0$$

$$[S^-, \sum a_{jm}^+] = \sum (-1)^{j-m} a_{j,-m} = -\sum \tilde{a}_{jm}$$

and

$$[S^0, \sum \tilde{a}_{jm}] = \frac{1}{2} \sum \tilde{a}_{jm} \qquad (10.54)$$

$$[S^+, \sum \tilde{a}_{jm}] = 0$$

$$[S^-, \sum \tilde{a}_{jm}] = -\sum a_{jm}^+$$

Hence, $\sum a_{jm}^+$ becomes the $+\frac{1}{2}$ component of a quasi-spin tensor with $\frac{1}{2}$ rank, and $-\sum \tilde{a}_{jm}$ becomes the $-\frac{1}{2}$ component. Note that each of a_{jm}^+ and $-\tilde{a}_{jm}$ behaves as the $+\frac{1}{2}$ and $-\frac{1}{2}$ components of a quasi-spin tensor with $\frac{1}{2}$ rank. Therefore, the tensor $\sum_{jj'} (a_j^+ \times a_{j'}^+)^{(k)}$ becomes a linear combination of quasi-spin scalars and quasi-spin vectors. This further implies that the Hermitian tensor $T^{(k)}$ behaves as a quasi-spin scalar for odd-k values, while the anti-Hermitian tensor $T^{(k)}$ behaves as the $\kappa = 0$ component of the quasi-spin vector for even-k values. Interestingly, the operator used in electromagnetic transitions, spherical harmonics Y^L, is a Hermitian tensor. So, we will mainly be interested in the results of Hermitian tensor at the moment.

Furthermore, the multi-j orbits may or may not have different parities in the real situations. So, the multi-j pair creation operator can be redefined as [21]

$$S_1^+ = \sum_j (-1)^{l_j} S_j^+ \tag{10.55}$$

where the orbital angular momentum l_j takes care of the parity of any j-orbit. The phase factor $(-1)^k$ mentioned in Eq. (10.51) now changes to $(-1)^{l+l'+k}$. For the electromagnetic transitions, the phase factor can be rewritten as $(-1)^{l+l'+L}$ due to involved spherical harmonics Y^L. L is the multipolarity, and l and l' are the parities of the initial and final states involved in the electromagnetic transition, respectively. So, the Y^L tensor behaves as a quasi-spin scalar, if sum $l + l' + L$ is odd. On the other hand, it becomes the $\kappa = 0$ component of the quasi-spin vector, if sum $l + l' + L$ is even. Interestingly, the sum $l + l' + L$ remains even for electric transitions irrespective of the nature of L value. The same happens for magnetic transitions where the sum $l + l' + L$ remains odd, irrespective of the nature of L value.

Hence, *we find that the magnetic transitions, for both even and odd L, become quasi-spin scalar.* This means that the matrix elements of magnetic transitions are particle number independent and conserve the seniority, in the multi-j configuration. If we simply define a multi-j configuration as $\tilde{j} = j \otimes j' \otimes ...$, then the matrix elements for magnetic transitions can be expressed as

$$< \tilde{j}^n vlJM|O(Mag.)_M^L|\tilde{j}^n v'l'J'M' > = < \tilde{j}^v vlJM|O(Mag.)_M^L|\tilde{j}^v v'l'J'M' > \delta_{v,v'} \tag{10.56}$$

where pair degeneracy $\Omega = \frac{2\tilde{j}+1}{2} = \sum_j \frac{2j+1}{2}$. The result looks similar to the single-j shell for magnetic dipole moments. The only difference arises in terms of an additional possibility, which says that the magnetic quadrupole (or octupole,....) moments may also behave particle number independent. Similarly, *we find that the electric transitions, for both even and odd L, now behave similar to each other and behave as $\kappa = 0$ component of the quasi-spin vector.* Therefore, the matrix elements of electric transitions between the same seniority states (seniority conserving $\Delta v = 0$ transitions) in multi-j $\tilde{j}^n$ configuration can be written as

$$\langle \tilde{j}^n vlJM|Y_\kappa^L|\tilde{j}^n vl'J'M' \rangle = \left[\frac{\Omega - n}{\Omega - v}\right] \langle \tilde{j}^v vlJM|Y_\kappa^L|\tilde{j}^v vl'J'M' \rangle \tag{10.57}$$

On the other hand, the matrix elements of electric transitions between states having different seniorities by 2 (seniority changing $\Delta v = 2$ transitions) in multi-j $\tilde{j}^n$ configuration can be written as

$$\langle \tilde{j}^n vlJM|Y_\kappa^L|\tilde{j}^n v \pm 2, l'J'M' \rangle = \left[\sqrt{\frac{(n - v + 2)(2\Omega + 2 - n - v)}{2(2\Omega + 2 - 2v)}}\right] \langle \tilde{j}^v vlJM|Y_\kappa^L|\tilde{j}^v v \pm 2, l'J'M' \rangle \tag{10.58}$$

The reduced transition probabilities $B(EL)$ between J_i and J_f states, simply by defining a multi-j configuration as $\tilde{j} = j \otimes j'...$ along with the corresponding total pair degeneracy $\Omega = \frac{1}{2}(2\tilde{j} + 1) = \frac{1}{2}\sum_j (2j + 1)$, can be written as

$$B(EL) = \frac{1}{2J_i + 1}|\langle \tilde{j}^n vlJ_f|| \sum_i r_i^L Y^L(\theta_i, \phi_i)||\tilde{j}^n v'l'J_i \rangle|^2 \tag{10.59}$$

This implies that the $B(EL)$ values show a parabolic behavior in the multi-j case, irrespective of the L values (even or odd). The similar behavior remains to be valid in single-j case, only for even L values. Figure 10.1 shows the schematic variation of both the electric and the magnetic reduced transition probabilities in single-j and multi-j shells. The schematic results are shown for both the seniority conserving $\Delta v = 0$ and seniority changing $\Delta v = 2$ transitions, depending upon the seniority selection rules. The generalized seniority reduction formulas of multi-j for the reduced matrix elements with $\Delta v = 0$ and $\Delta v = 2$ transitions can also be written as follows:

$$\langle \tilde{j}^n vlJ_f || \sum_i r_i^L Y^L(\theta_i, \phi_i) || \tilde{j}^n vl'J_i \rangle = \left[\frac{\Omega - n}{\Omega - v} \right]$$
$$\langle \tilde{j}^v vlJ_f || \sum_i r_i^L Y^L(\theta_i, \phi_i) || \tilde{j}^v vl'J_i \rangle \qquad (10.60)$$

$$\langle \tilde{j}^n vlJ_f || \sum_i r_i^L Y^L(\theta_i, \phi_i) || \tilde{j}^n v \pm 2, l'J_i \rangle = \left[\sqrt{\frac{(n - v + 2)(2\Omega + 2 - n - v)}{4(\Omega + 1 - v)}} \right]$$
$$\langle \tilde{j}^v vlJ_f || \sum_i r_i^L Y^L(\theta_i, \phi_i) || \tilde{j}^v v \pm 2, l'J_i \rangle \quad (10.61)$$

10.5   Seniority Isomerism

Isomers are the long-lived excited states of nuclei. They exist due to structural peculiarities of nucleons in nuclei, resulting in the hindrance to their respective decays. On the basis of hindrance mechanisms, the isomers can be classified into five broad categories: spin isomers, K isomers, shape isomers, fission isomers, and seniority isomers [22]. The seniority isomers can actually be clubbed with spin isomers, particularly in semi-magic nuclei, where seniority behaves as a good quantum number. We note that the odd tensors for the magnetic transitions behave as quasi-spin scalars in single-j shell, and lead to particle number-independent behavior and conserve the seniority, as shown in Figure 10.1. On the other hand, we note that the even tensors for the electric transitions behave as $\kappa = 0$ component of the quasi-spin vector. Therefore, the reduced electric transition probabilities between the same seniority states show a parabolic behavior with a minimum at the middle of the shell, while it shows a maximum at the middle of the shell for seniority changing transitions. This further implies that the extra longevity of isomers may correspond to the middle of the shell for even-tensor electric transitions. It is long known that only $E2$ transitions occur in the case of single-j shell as the parity change is not possible in this case, i.e. no magnetic even multipole transitions are possible. From this, one can obviously conclude that the $E2$ transitions mainly lead to the seniority isomerism in single-j shell.

However, as we move to the multi-j shell, the situation becomes quite different. We establish that the magnetic transitions now behave as quasi-spin scalars for both even and odd multipole tensors and lead to particle number-independent behavior and conserve the seniority, as shown in Figure 10.1. On the other hand, the electric transitions behave as $\kappa = 0$ component of the quasi-spin vector irrespective of even or odd L values. Therefore, the reduced electric transition probabilities between the same seniority states show a parabolic minimum at the middle of active-shell for both even and odd multipole tensors, while it shows a maximum at the middle for seniority changing transitions, see Figure 10.1. This implies that one should get longer half-lives corresponding to both the even- and odd-tensor electric transitions at the middle of the multi-j shell. The situation was obvious, but not at all studied until a recent discovery of a new kind of seniority isomers in Sn isotopes is made [23].

As an example, we hereby discuss the recent results of the multi-j generalized seniority approach to the high-spin and high-seniority isomers in Sn-isotopes. Sn isotopes are long known to present an interesting ground for pairing correlations of nucleons in both experimental and theoretical studies. Experimental groups [24,25] recently reported the measurements of the higher seniority $v = 4$, 15^- and 13^- isomers in the Sn-isotopes. In particular, the surprisingly similar trends of the experimental $B(E2)$ values in the 15^- isomers and the $B(E1)$ values in the 13^- isomers were highlighted by Iskra *et al.* [25] without any justification.

It has generally been believed that only $E2$ transitions can lead to the seniority isomers, an obvious conclusion from the seniority formalism in the case of single-j orbital. But the situation gets changed in multi-j generalized seniority scheme. As a result, the electric odd-tensor transitions may also lead to seniority isomers. Maheshwari *et al.* [23] applied these results to the recent measurements in the 13^- and 15^- isomers in $^{116-130}$Sn isotopes, and successfully showed that the trends of $B(E1)$ values in the 13^- isomers and $B(E2)$ values in the 15^- isomers are very similar to each other. Hence, the seniority isomers with odd-tensor electric transitions were identified for the very first time and interpreted in terms of the generalized seniority.

The calculations were done by assuming $\Delta v = 0$ in the decay of all the high-spin 13^- ($v = 4$) and 15^- ($v = 4$) isomers using Eqs. (10.59) and (10.60). The resulting $B(EL)$ should be proportional to the coefficient $((\Omega - n)/(\Omega - v))^2$. The $g_{7/2}$ and $d_{5/2}$ orbits are taken to be completely filled till ^{114}Sn, as a core. So, the remaining active orbits are $h_{11/2}$, $d_{3/2}$, and $s_{1/2}$ in the $50 - 82$ neutron valence space. Since these orbits lie very near to each other in single-particle energies, it would be good enough to consider them as approximately degenerate. The calculated results with $\Omega = 9$ and $\tilde{j} = h_{11/2} \otimes d_{3/2} \otimes s_{1/2}$ explained the experimental data reasonably well for both the generalized seniority $v = 4$ 13^-, and 15^- isomers, as shown in Figure 10.2. The numerical results were obtained by fitting one of the experimental data, where this single-point fitting includes the information of radial integrals and involved spherical harmonics in terms of 3j-symbol. It fulfills the demand of the seniority conserving interaction in terms of the constant nature of radial integrals. Interestingly, the usage of multi-j configuration is found to be essential. Predictions can also be made for the 13^- isomers at ^{118}Sn, ^{128}Sn, and 15^- isomer at ^{118}Sn, as shown in the generalized seniority calculated trend in Figure 10.2.

Similarly, one can observe the identical behavior of the $B(E2)$s in the high-spin isomers for various semi-magic chains like $Z = 82$ isotopes and $N = 82$ isotones [26,27]. The goodness of generalized seniority quantum number is also found to be remarkable for the first excited 2^+ and 3^- states in Sn isotopes [28,29]. More recently, the g-factor trends have been obtained and understood by clubbing the multi-j configuration (as suggested by generalized seniority) to the Schmidt model in Sn isotopes, N = 82 isotones and Pb isotopes [30]. The results are found to be in better agreement with the experimental trend than the pure Schmidt model. Such theoretical results can be improved if the real case of non-degeneracy of multi-j orbits is taken into account. Furthermore, the generalized seniority can be extended to include both kinds of nucleons by introducing isospin quantum number, which may help to test the goodness of generalized seniority beyond semi-magic nuclei. These studies may also highlight the crucial role of symmetries in nuclei and the competition involved in the single-particle picture and collective picture of nucleus, as we venture away from the magic regions.

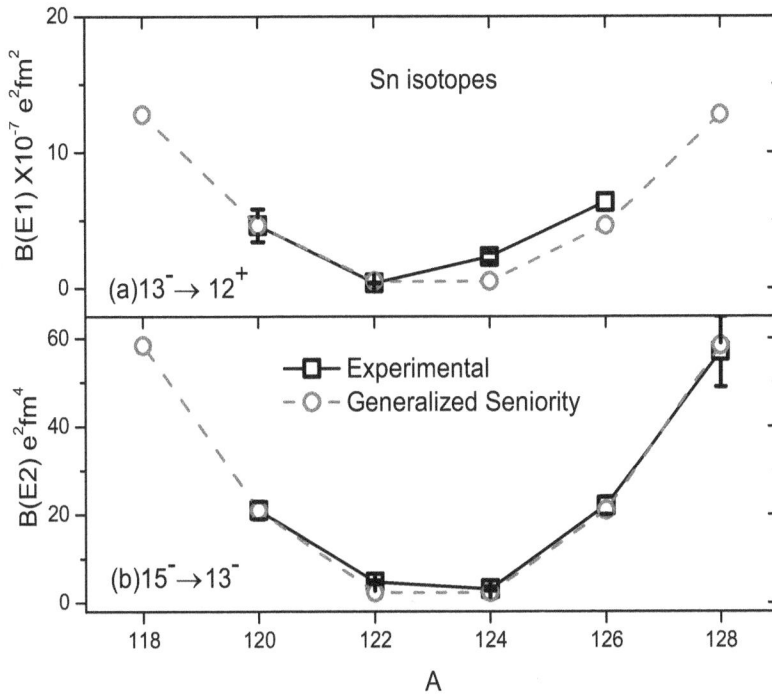

FIGURE 10.2
A comparison of experimental and generalized seniority calculated B(EL) values for both odd and even-tensor seniority isomers. As an example, (a) shows E1 decaying 13^- isomer, while (b) shows E2 decaying 15^- isomer for even-even Sn isotopes. The similar behavior of both the generalized seniority $v = 4$ isomers can be noticed, irrespective of the nature of involved tensor.

10.6   Summary

In this chapter, we have presented a brief review of the pairing interaction in quasi-spin scheme for both single-j and multi-j shells, in terms of seniority and generalized seniority, respectively. The selection rules for electromagnetic transition have been revisited in view of the generalized seniority scheme to explain the new kind of seniority isomers based on multi-j configurations. The results are recently extended for studying moments. However, the extension for both kinds of nucleons and many degenerate orbits is needed in future for a more detailed description of semi-magic nuclei and around.

Acknowledgment

BM would like to thank her mentor Prof. A. K. Jain, IIT Roorkee (currently at AINST, Amity University UP, Noida) for his guidance and support during this work.

Appendix: Quasi-Spin Algebra

The commutation relations for quasi-spin pair creation and pair annihilation operators can be obtained as

$$[S_j^+, S_j^-]_- = \sum_{m,m'>0} (-1)^{2j-m-m'} [a_{jm}^+ a_{j,-m}^+, a_{j,-m'} a_{jm'}]_-$$

$$[S_j^+, S_j^-]_- = \sum_{m,m'>0} (-1)^{2j-m-m'} (a_{jm}^+ a_{j,-m}^+ a_{j,-m'} a_{jm'} - a_{j,-m'} a_{jm'} a_{jm}^+ a_{j,-m}^+) \quad \text{(A.1)}$$

For simplicity leaving common j indices,

$$[S_j^+, S_j^-]_- = \sum_{m,m'>0} (-1)^{2j-m-m'} (a_m^+ a_{-m}^+ a_{-m'} a_{m'} - a_{-m'} a_{m'} a_m^+ a_{-m}^+) \quad \text{(A.2)}$$

Using Wick's theorem,

$$a_\alpha a_\beta a_\gamma^+ a_\delta^+ = \; : a_\alpha a_\beta a_\gamma^+ a_\delta^+ : + \sum_{singles} \overline{aa^+} : aa^+ : + \sum_{doubles} \overline{aa^+}\, \overline{aa^+} \quad \text{(A.3)}$$

$$: a_\alpha a_\beta a_\gamma^+ a_\delta^+ : = a_\gamma^+ a_\delta^+ a_\alpha a_\beta \quad \text{(A.4)}$$

The sum over singles and doubles corresponds to the single and double contractions, respectively. We consider the contractions between a and a^+ only, since, other contractions vanish:

$$\sum_{singles} \overline{aa^+} : aa^+ : = -\overline{a_\alpha a_\gamma^+} : a_\beta a_\delta^+ : + \overline{a_\alpha a_\delta^+} : a_\beta a_\gamma^+ :$$

$$+ \overline{a_\beta a_\gamma^+} : a_\alpha a_\delta^+ : - \overline{a_\beta a_\delta^+} : a_\alpha a_\gamma^+ : + \overline{a_\gamma^+ a_\delta^+} : a_\alpha a_\beta :$$

$$= +\delta_{\alpha\gamma} a_\delta^+ a_\beta - \delta_{\alpha\delta} a_\gamma^+ a_\beta - \delta_{\beta\gamma} a_\delta^+ a_\alpha + \delta_{\beta\delta} a_\gamma^+ a_\alpha \quad \text{(A.5)}$$

$$\sum_{doubles} \overline{aa^+}\,\overline{aa^+} = -\overline{a_\alpha a_\gamma^+}\, \overline{a_\beta a_\delta^+} + \overline{a_\alpha a_\delta^+}\, \overline{a_\beta a_\gamma^+}$$

$$= -\delta_{\alpha\gamma}\delta_{\beta\delta} + \delta_{\alpha\delta}\delta_{\beta\gamma} \quad \text{(A.6)}$$

From Eqs. (A.3–A.6)

$$a_\alpha a_\beta a_\gamma^+ a_\delta^+ = a_\gamma^+ a_\delta^+ a_\alpha a_\beta +$$

$$+ \delta_{\alpha\gamma} a_\delta^+ a_\beta - \delta_{\alpha\delta} a_\gamma^+ a_\beta - \delta_{\beta\gamma} a_\delta^+ a_\alpha + \delta_{\beta\delta} a_\gamma^+ a_\alpha$$

$$- \delta_{\alpha\gamma}\delta_{\beta\delta} + \delta_{\alpha\delta}\delta_{\beta\gamma} \quad \text{(A.7)}$$

So, the second term in Eq. (A.2) modifies to

$$a_{-m'} a_{m'} a_m^+ a_{-m}^+ = a_m^+ a_{-m}^+ a_{-m'} a_{m'}$$

$$+ \delta_{-m'm} a_{-m}^+ a_{m'} - \delta_{-m',-m} a_m^+ a_{m'} - \delta_{m'm} a_{-m}^+ a_{-m'}$$

$$+ \delta_{m',-m} a_m^+ a_{-m'} - \delta_{-m'm}\delta_{m',-m} + \delta_{-m',-m}\delta_{m'm} \quad \text{(A.8)}$$

Substituting (A.8) to (A.2), we get

$$[S_j^+, S_j^-]_- = \sum_{m,m'>0} (-1)^{2j-m-m'} (a_m^+ a_{-m}^+ a_{-m'} a_{m'} - a_{-m'} a_{m'} a_m^+ a_{-m}^+)$$

$$= \sum_{m,m'>0} (-1)^{2j-m-m'} (a_m^+ a_{-m}^+ a_{-m'} a_{m'} - a_m^+ a_{-m}^+ a_{-m'} a_{m'}$$

$$- \delta_{-m'm} a_{-m}^+ a_{m'} + \delta_{-m',-m} a_m^+ a_{m'} + \delta_{m'm} a_{-m}^+ a_{-m'}$$

$$- \delta_{m',-m} a_m^+ a_{-m'} + \delta_{-m'm} \delta_{m',-m} - \delta_{-m',-m} \delta_{m'm})$$

$$= \sum_{m,m'>0} (-1)^{2j-m-m'} (-\delta_{-m'm} a_{-m}^+ a_{m'} + \delta_{-m',-m} a_m^+ a_{m'}$$

$$+ \delta_{m'm} a_{-m}^+ a_{-m'} - \delta_{m',-m} a_m^+ a_{-m'}$$

$$+ \delta_{-m'm} \delta_{m',-m} - \delta_{-m',-m} \delta_{m'm}) \tag{A.9}$$

Since sum runs over $m, m' > 0$,

$$[S_j^+, S_j^-]_- = \sum_{m,m'>0} (-1)^{2j-m-m'} (-\underbrace{\delta_{-m'm}}_{=0} a_{-m}^+ a_{m'}$$

$$+ \delta_{-m',-m} a_m^+ a_{m'} + \delta_{m'm} a_{-m}^+ a_{-m'} - \underbrace{\delta_{m',-m}}_{0} a_m^+ a_{-m'}$$

$$+ \underbrace{\delta_{-m'm}}_{0} \delta_{m',-m} - \delta_{-m',-m} \delta_{m'm})$$

$$[S_j^+, S_j^-]_- = \sum_{m,m'>0} (-1)^{2j-m-m'} (+\delta_{-m',-m} a_m^+ a_{m'}$$

$$+ \delta_{m'm} a_{-m}^+ a_{-m'} - \delta_{-m',-m} \delta_{m'm})$$

$$= \sum_{m>0} (-1)^{2j-2m} (a_m^+ a_m + a_{-m}^+ a_{-m} - 1)$$

$$= \sum_{m>0} (a_m^+ a_m + a_{-m}^+ a_{-m} - 1)$$

$$= \sum_{m>0} (a_m^+ a_m + a_{-m}^+ a_{-m}) - \sum_{m>0} 1$$

$$[S_j^+, S_j^-]_- = \hat{N} - \Omega \tag{A.10}$$

$$[S_j^+, S_j^-]_- = 2S_j^0 \tag{A.11}$$

$$\boxed{[S_j^+, S_j^-] = 2S_j^0} \tag{A.12}$$

$$[S_j^+, S_j^0]_- = [S_j^+, \frac{\hat{N} - \Omega}{2}]$$

$$= \frac{1}{2} [S_j^+, \hat{N}]$$

$$= \frac{1}{2} [S_j^+, \sum_{m'>0} (a_{m'}^+ a_{m'} + a_{-m'}^+ a_{-m'})]$$

$$= \frac{1}{2} \sum_{m,m'>0} (-1)^{j-m} [a_m^+ a_{-m}^+, (a_{m'}^+ a_{m'} + a_{-m'}^+ a_{-m'})]$$

$$= \frac{1}{2} \sum_{m,m'>0} (-1)^{j-m} ([a_m^+ a_{-m}^+, a_{m'}^+ a_{m'}] + [a_m^+ a_{-m}^+, a_{-m'}^+ a_{-m'}]) \tag{A.13}$$

So,

$$[a_m^+ a_{-m}^+, a_{m'}^+ a_{m'}] = a_m^+ a_{-m}^+ a_{m'}^+ a_{m'} - a_{m'}^+ a_{m'} a_m^+ a_{-m}^+ \tag{A.14}$$

The second term of Eq. (A.14) can be expanded using Wick's theorem. Since

$$\overbrace{a_{-m'}^+ a_{m'}} = \overbrace{a_m^+ a_{-m}^+} = 0 \tag{A.15}$$

The contribution from doubly contracted terms will vanish. Furthermore, in the single contractions, only the contraction between a and a^+ would be non-zero. Therefore, we have

$$a_{m'}^+ a_{m'} a_m^+ a_{-m}^+ = \; : a_{m'}^+ a_{m'} a_m^+ a_{-m}^+ : + \overbrace{a_{m'} a_m^+} : a_{m'}^+ a_{-m}^+ : - \overbrace{a_{m'} a_{-m}^+} : a_{m'}^+ a_m^+ :$$

$$= a_m^+ a_{-m}^+ a_{m'}^+ a_{m'} + \overbrace{a_{m'} a_m^+} : a_{m'}^+ a_{-m}^+ : - \underbrace{\overbrace{a_{m'} a_{-m}^+}}_{=0} : a_{m'}^+ a_m^+ :$$

$$= a_m^+ a_{-m}^+ a_{m'}^+ a_{m'} + \delta_{mm'} a_{m'}^+ a_{-m}^+ \tag{A.16}$$

Substituting Eq. (A.14) in Eq. (A.10),

$$[a_m^+ a_{-m}^+, a_{m'}^+ a_{m'}] = -\delta_{mm'} a_{m'}^+ a_{-m}^+ \tag{A.17}$$

Similarly,

$$[a_m^+ a_{-m}^+, a_{-m'}^+ a_{-m'}] = a_m^+ a_{-m}^+ a_{-m'}^+ a_{-m'} - a_{-m'}^+ a_{-m'} a_m^+ a_{-m}^+$$

$$= + \overbrace{a_{-m'} a_{-m}^+} : a_{-m'}^+ a_m^+ :$$

$$= -\delta_{-m,-m'} a_m^+ a_{-m'}^+ \tag{A.18}$$

$$[S_j^+, S_j^0]_- = \frac{1}{2} \sum_{m,m'>0} (-1)^{j-m} [[a_m^+ a_{-m}^+, a_{m'}^+ a_{m'}] + [a_m^+ a_{-m}^+, a_{-m'}^+ a_{-m'}]]$$

$$= -\frac{1}{2} \sum_{m,m'>0} (-1)^{j-m} [\delta_{mm'} a_{m'}^+ a_{-m}^+ + \delta_{-m,-m'} a_m^+ a_{-m'}^+]$$

$$= -\frac{1}{2} \sum_{m>0} (-1)^{j-m} [a_m^+ a_{-m}^+ + a_m^+ a_{-m}^+]$$

$$= -\sum_{m>0} (-1)^{j-m} [a_m^+ a_{-m}^+] = -S_j^+ \tag{A.19}$$

$$\boxed{[S_j^+, S_j^0] = -S_j^+} \tag{A.20}$$

$$[S_j^-, S_j^0]_- = \frac{1}{2} [S_j^-, \hat{N}]$$

$$= \frac{1}{2} \sum_{m,m'>0} (-1)^{j-m} [a_{-m} a_m, (a_{m'}^+ a_{m'} + a_{-m'}^+ a_{-m'})]$$

$$= \frac{1}{2} \sum_{m,m'>0} (-1)^{j-m} [[a_{-m}a_m, a_{m'}^+ a_{m'}]$$

$$+ [a_{-m}a_m, a_{-m'}^+ a_{-m'}]] \tag{A.21}$$

$$[a_{-m}a_m, a_{m'}^+ a_{m'}] = a_{-m}a_m a_{m'}^+ a_{m'} - a_{m'}^+ a_{m'} a_{-m}a_m$$

$$= : a_{-m}a_m a_{m'}^+ a_{m'} : + \overline{a_m a_{m'}^+} : a_{-m}a_{m'} : - a_{m'}^+ a_{m'} a_{-m}a_m$$

$$= a_{m'}^+ a_{m'} a_{-m}a_m + \delta_{mm'} a_{-m}a_{m'} - a_{m'}^+ a_{m'} a_{-m}a_m$$

$$= \delta_{mm'} a_{-m}a_{m'} \tag{A.22}$$

$$[a_{-m}a_m, a_{-m'}^+ a_{-m'}] = a_{-m}a_m a_{-m'}^+ a_{-m'} - a_{-m'}^+ a_{-m'} a_{-m}a_m$$

$$=: a_{-m}a_m a_{-m'}^+ a_{-m'} : - \overline{a_{-m}a_{-m'}^+} : a_m a_{-m'} :$$

$$- a_{-m'}^+ a_{-m'} a_{-m}a_m$$

$$= a_{-m}a_m a_{-m'}^+ a_{-m'} - \overline{a_{-m}a_{-m'}^+} : a_m a_{-m'} :$$

$$- a_{-m'}^+ a_{-m'} a_{-m}a_m$$

$$= a_{-m'}^+ a_{-m'} a_{-m}a_m + \overline{a_{-m}a_{-m'}^+} a_{-m'}a_m$$

$$- a_{-m'}^+ a_{-m'}a_{-m}a_{-m}$$

$$= \delta_{-m,-m'} a_{-m'}a_m \tag{A.23}$$

Substituting Eq. (A.22) and (A.23) in Eq. (A.21), we have

$$\boxed{[S_j^-, S_j^0] = S_j^-} \tag{A.24}$$

Alternatively,

$$[S_j^0, S_j^+] = S_j^+ \tag{A.25}$$

$$[S_j^0, S_j^-] = -S_j^- \tag{A.26}$$

The commutations of S_j^+, S_j^-, and S_j^0 suggest that these operators follow the angular momentum algebra and can be called as quasi-spin operators.

References

[1] G. Racah, Theory of Complex Spectra. III, *Phys. Rev.* **63**, 367 (1943).

[2] G. Racah, *Nuclear levels and Casimir operators*, Research Council of Israel, Jerusalem, **L. Farkas Memorial Volume**, 294 (1952).

[3] B. H. Flowers, Studies in jj-Coupling. 1st Classification of Nuclear and Atomic States, *Proc. Roy. Soc. (London) A* **212**, 248 (1952).

[4] R. F. Casten, *Nuclear Structure from a Simple Perspective*, Oxford University Press (1990).

[5] R. D. Lawson, *Theory of the Nuclear Shell Model*, Oxford University Press, New York (1980).

[6] K. L. G. Heyde, *Basic Ideas and Concepts in Nuclear Physics*, CRC Press (2004); *From Nucleons to the Atomic Nucleus*, CreateSpace (1998).

[7] I. Talmi, *Simple Models of Complex Nuclei*, Harwood Academic (1993).

[8] D. J. Rowe, and J. L. Wood, *Fundamentals of Nuclear Models - Foundational Models*, World Scientific Publishing, Singapore (2010).

[9] V. K. B. Kota and Y. D. Devi, *Nuclear Shell Model and The Interacting Boson Model: Lecture Notes for Practitioners*, Inter University Consortium for DAE Facilities, Calcutta Centre (1996).

[10] P. Van Isacker, and S. Heinze, Partial Conservation of Seniority and Nuclear Isomerism, *Phys. Rev. Lett.* **100**, 052501 (2008).

[11] D. J. Rowe, and G. Rosensteel, Partially Solvable Pair-Coupling Models with Seniority-Conserving Interactions, *Phys. Rev. Lett.* **87**, 172501 (2001).

[12] L. Zamick, and P. Van Isacker, Partial dynamical symmetries in the $j = 9/2$ shell: Progress and puzzles, *Phys. Rev. C* **78**, 044327 (2008).

[13] I. Talmi, Solvability of eigenvalues in j^n configurations, *Nucl. Phys. A* **846**, 31 (2010).

[14] A. K. Kerman, Pairing forces and nuclear collective motion, *Ann. Phys. (NY)* **12** 300 (1961).

[15] A. K. Kerman, R. D. Lawson, and M. H. Macfarlane, Accuracy of the superconductivity approximation for pairing forces in nuclei, *Phys. Rev.* **124**, 162 (1961).

[16] K. Helmers, Symplectic invariants and flowers' classification of shell model states, *Nucl. Phys.* **23**, 594 (1961).

[17] A. Arima and M. Ichimura, Quasi-Spin formalism and matrix elements in the shell model, *Prog. of Theo. Phys.* **36**, 296 (1966).

[18] I. Talmi, Generalized seniority and structure of semi-magic nuclei, *Nucl. Phys. A* **172**, 1 (1971).

[19] S. Shlomo, and I. Talmi, Shell-model hamiltonians with generalized seniority eigenstates, *Nucl. Phys. A* **198**, 82 (1972).

[20] R. Arvieu, and S. A. Moszokowski, Generalized seniority and the surface delta interaction, *Phys. Rev.* **145**, 830 (1966).

[21] I. M. Green, and S. A. Moszokowski, Nuclear coupling schemes with a surface delta interaction, *Phys. Rev.* **139**, B790 (1965).

[22] A. K. Jain, B. Maheshwari, S. Garg, M. Patial, and B. Singh, Atlas of nuclear isomers, *Nucl. Data Sheets* **128**, 1 (2015).

[23] B. Maheshwari, and A. K. Jain, Odd tensor electric transitions in high-spin Sn-isomers and generalized seniority, *Phys. Lett. B* **753**, 122 (2016).

[24] A. Astier, M.-G. Porquet, Ch. Theisen, D. Verney, I. Deloncle, M. Houry, R. Lucas, F. Azaiez, G. Barreau, D. Curien, O. Dorvaux, G. Duchâne, B. J. P. Gall, N. Redon, M. Rousseau, and O. Stezowski, High-spin states with seniority $v = 4$, 5, and 6 in $^{119-126}$Sn, *Phys. Rev. C* **85**, 054316 (2012), and references therein.

[25] L. W. Iskra, R. Broda, R. V. F. Janssens, J. Wrzesiński, B. Szpak, C. J. Chiara, M. P. Carpenter, B. Fornal, N. Hoteling, F. G. Kondev, W. Królas, T. Lauritsen, T. Pawłat, D. Seweryniak, I. Stefanescu, W. B. Walters, and S. Zhu, Higher-seniority excitations in even neutron-rich Sn isotopes, *Phys. Rev. C* **89**, 044324 (2014).

[26] A. K. Jain, and B. Maheshwari, Goodness of Generalized Seniority in Semi-magic Nuclei, *Nucl. Phys. Rev.* **34**, 73 (2017).

[27] A. K. Jain, and B. Maheshwari, Generalized seniority states and isomers in tin isotopes, *Physica Scripta* **92**, 074004 (2017).

[28] B. Maheshwari, A. K. Jain, and B. Singh, Asymmetric behavior of the B(E2; $0^+ \rightarrow 2^+$) values in $^{104-130}$Sn and generalized seniority, *Nucl. Phys. A* **952**, 62 (2016).

[29] B. Maheshwari, S. Garg and A. K. Jain, $\Delta v = 2$ seniority changing transitions in yrast 3^- states and B(E3) systematics of Sn isotopes, *Pramana - J. Phys. (Rapid Communication)* **89**, 75 (2017).

[30] B. Maheshwari, and A. K. Jain, Generalized seniority Schmidt model and the g-factors in semi-magic nuclei, *Nucl. Phys. A* **986**, 232 (2019).

11

Nuclear High-Spin Spectroscopy in the $A \sim 60$ Mass Region

J. Gellanki and S.K. Mandal

University of Delhi

Amritanshu Shukla

Rajiv Gandhi Institute of Petroleum technology

CONTENTS

11.1 Introduction

The investigation and understanding of fundamental symmetries are the principal goals of physics in general and nuclear structure physics in particular. In order to fully understand the nuclear structure, one needs to exhibit the nuclei at extreme conditions, such as high angular momentum and high excitation energies. In general, the nuclear angular momentum can be built either from single-particle excitations or from collective excitations, such as collective rotation. Some "important phenomena" in nuclear high-spin spectroscopy are, for example, shape coexistence, band termination, and superdeformation.

The present chapter describes the above-mentioned phenomena in detail on the $A \sim 60$ region.

11.1.1 Shape Coexistence

The observation that a particular atomic nucleus (N, Z combination) can exhibit eigenstates with different shapes appears to be a unique type of behavior in finite many-body quantum systems. Shape coexistence in nuclei has now been a feature of nuclear structure for over 50 years and has been found throughout the periodic table. From cluster structure in light nuclei to fission isomers in the heaviest nuclei, a wide variety of shapes have been theoretically predicted and experimentally observed in the same nuclide.

Shape coexistence occurs when two competing minima exist in the nuclear potential energy surface corresponding to different shapes of the nuclear mean field. Such effects often occur in nuclei with a near magic number of protons and neutrons. Here, a spherical configuration competes with a deformed shape corresponding to particle-hole excitations across the shell gaps. The study of very elongated or superdeformed (SD) nuclei is an extreme form of shape coexistence.

Possible nuclear shapes in the $A \sim 60$ mass region are shown in Figure 11.1. They range from spherical nuclei to quadrupole deformed, highly deformed, and SD nuclei. If the spherical symmetry is broken, then the nucleus becomes deformed. This allows an orientation in space to be defined, and the nucleus can rotate around a specific axis.

11.1.2 Band Termination

For specific configurations of collective rotational bands, the gradual loss of collectivity at their intermediate spin values which terminates in a non-collective state at the maximum spin within the configuration is called termination of collective bands [2]. At this termination state, rotation is forbidden quantum mechanically. A deformed nucleus can increase its angular momentum by collective rotation. However, because the nucleus is a many-body quantal system, such a regular rotational motion must have an underlying microscopic basis, with the angular momentum being generated by small contributions from a sizeable number of valence nucleons. There is a limiting angular momentum that can be generated from the finite valence particles (and holes) and their individual spin contributions. At this point, the rotational band loses its collectivity and is named as terminate.

The concept of band termination spectroscopy reveals the study of the balance and interplay between the two extremes of nuclear dynamics, namely collective and single-particle degrees of freedom. Such a "demise of the rotational band," which reveals the finite particle basis of the nuclear multi-fermion system, has been seen in several nuclei

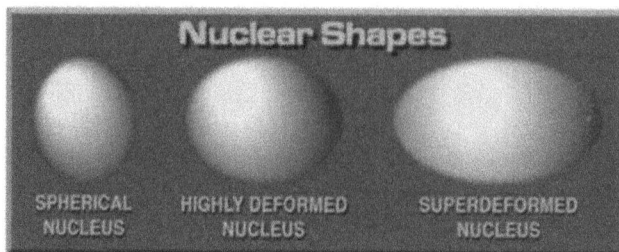

FIGURE 11.1
Possible nuclear shapes in the $A \sim 60$ mass region. (Picture taken from [1].)

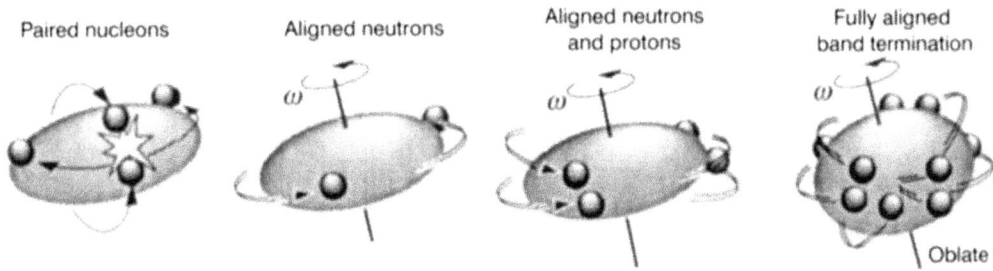

FIGURE 11.2
Schematic representation of the building of angular momentum in a rotational band from the collective rotation to a single-particle non-collective terminating state. (Picture taken from Ref. [2].) More details are given in the text.

throughout the periodic chart. The best examples of such bands known at present are in proton-rich nuclei in the $A \sim 60$ mass region, where terminating configurations involving proton particle-hole excitations across the $Z = 28$ gap can be observed over their entire spin range.

Figure 11.2 shows the construction of angular momentum in a rotational band ending up in a fully aligned non-collective terminating state. Here, the motion of nucleus is collective with paired nucleons at low spin. The maximum spin available will be determined by the maximal coupling of the single-particle spins for those particles outside the inert core. In order to generate higher angular momentum values, the core must be broken and the collective band structure should be terminated. Two different kinds of band terminations were mentioned in Ref. [3]. A favored termination is that at the termination point, the largest spins are obtained with least energy. On the other hand, in an unflavored termination, the largest spins are obtained with the most energy at the termination point.

11.1.3 Superdeformation

Superdeformation was first discovered nearly thirty years ago in the actinide fission isomers [4] and was explained a few years later as resulting from a secondary minimum at very large deformations [5].

Superdeformation occurs when quantum shell effects help stabilize a football shape in certain nuclei with a stable axial deformation with a major to minor axis ratio of $\sim$2:1. At the superdeformation state, the angular velocity of a nucleus can reach 10^{21} revolutions per second. Theoretically, superdeformation is explained through the existence of a second minimum of the nuclear potential at very large deformation, as shown in Figure 11.3 [6]. Here, the first and second minima of the nuclear potential are separated by an energy barrier. The regions where SD nuclei have been observed are close to those nuclei with proton and neutron numbers corresponding to large shell gaps (regions of low-level density) at large prolate deformations in the deformed shell model potential. The experimental evidence for a SD nucleus at high angular momentum is the emission of very fast rotational gamma-ray transitions with regular energy spacing. As shown in Figure 11.3, the states that belong to superdeformation bands are feeding back to the states in normal deformed bands through the green color marked linking transitions. Experimentally these liking transitions are very hard to observe due to their low intensity. Nearly 350 SD rotational bands have been observed and studied in different regions across the nuclear mass range. $A \sim 40$ [7], 60 [8], 80 [9], 130 [10], 150 [11], and 190 [12] are few examples of mass regions in which SD bands have been observed.

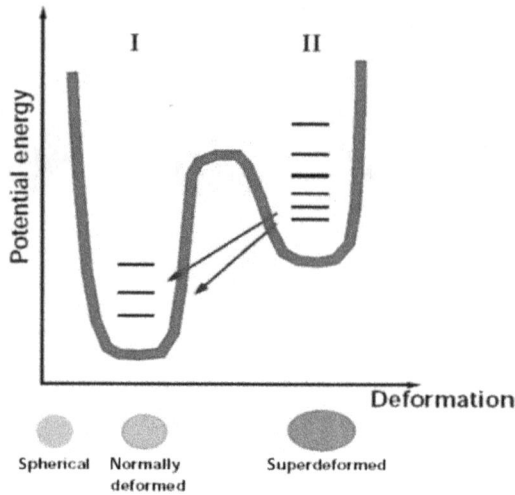

FIGURE 11.3

The potential energy of a nucleus versus the deformation. (Picture taken from [6].)

11.1.4 Interesting Facts in the $A \sim 60$ Mass Region

Because of their intermediate masses, the $A \sim 60$ nuclei are amenable to both microscopic and macroscopic theoretical treatments. Secondly, the proton and neutron shell effects act coherently in these nuclei, which results in a rich pattern of shape coexistence and shape transition. In the $A \sim 60$ region, within the same nuclei, they can exhibit various kinds of nuclear high-spin phenomena, such as shape coexistence, band termination, highly deformed bands, prompt proton decays (band decays by both gamma-ray transitions and prompt proton emission towards the daughter nucleus), and shape changes (see, e.g., Refs. [13–17]). In order to produce high angular moments in the $A \sim 60$ region, the doubly magic $N = Z = 28$ ^{56}Ni core must be broken, and the nucleons must be excited into the intruder $1g_{9/2}$ subshell starting at normal deformation (ND) with a few particles in the upper fp shell outside the core. The nuclei in this mass region become highly deformed and SD when the number of holes in the $1f_{7/2}$ orbital and the number of particles in the $1g_{9/2}$ orbital rapidly increase with spin and excitation energy. The large SD shell gaps at $N = Z = 30$ in the single-particle energy levels will provide several SD bands in this mass region. In addition to this, one can also expect "smooth band termination" with a smaller number of valence particles for the regular rotational sequence of the nuclei in this mass region. In contrast to the other mass regions, almost all SD bands in the second minima of the nuclear potential can be connected with the normal deformed or spherical states in the first minimum of the nuclear potential. There are some differences between the $A \sim 60$ mass region and other heavy mass regions. For example, in the $A \sim 190$ mass region, the observed decay-out is dominated by $E1$ transitions expected in a statistical decay process, whereas in the $A \sim 60$ mass region, the observed decay-out transitions are often of stretched $E2$ character with a non-statistical mechanism. The nuclei in this mass region are self-conjugate ($N \sim Z$) or nearly self-conjugate ($N \sim Z$), whereas the aspects like isospin ($T = 0$) pairing correlations might be important for their description, as seen in the $A \sim 100$ mass region [18].

All main features at high-spin states in the $A \sim 60$ mass region are overviewed in this chapter based mainly on the configuration-dependent cranked Nilsson–Strutinsky (CNS) approach.

11.1.5 Why Cranking Model Is Good at High Spin in the $A \sim 60$ Mass Region

In general, the models like large-scale shell-model accurately describe the spectroscopy of the low- to medium-spin states of $A \sim 60$ nuclei. However, this predictive power tends to decrease as excitation energy and spin increase. This is mainly because high-spin states in this mass region involve excitations of nucleons into the $1g_{9/2}$ intruder orbital. This orbital cannot be readily included in the above-mentioned effective interactions, because of both computational limits and problems attributed to monopole shifts. In addition, the $1g_{9/2}$ orbital is strongly shape driving, giving rise to well- and SD bands in a number of $A \sim 60$ nuclei.

11.2 Cranked Nilsson–Strutinsky (CNS) Model

Compared with the spherical shell-model calculations, the virtually unlimited model space and the physical understanding achieved are favorable properties of the CNS calculations. On the other hand, one of the drawbacks in CNS calculations is that not all aspects of the nuclear forces are taken into account. For instance, the pairing force is excluded from the calculations. Hence, a quantitative comparison between experimental and theoretical results is only relevant over some $I > 10\hbar$ in the $A \sim 60$ region. Before going in to more details of the CNS model, first we will discuss briefly the details of Nilsson and cranking models.

11.2.1 Nilsson Model

Nuclear shell model has failed to explain the properties of deformed nuclei (nuclei with N and Z far from the closed shells) and rapidly rotating nuclei. In order to explain the basic properties of nucleons moving in a deformed nucleus, a deformed shell model [19] was introduced by S.G. Nilsson in 1955 [20]. A small summary is given below.

In the Nilsson model, the modified oscillator potential [19,21] is used to describe the motion of the nucleons. For an axial symmetric deformation, the Nilsson Hamiltonian takes the following form:

$$H = -\hbar^2/2M\Delta + 1/2M(\omega_z^2 Z^2 + \omega\perp^2 (x^2 + y^2)) - 2k\hbar w_0 l.s - k\mu\hbar\omega_0 (l^2 - <l_2>_N). \quad (11.1)$$

Here, ω_0 is the harmonic oscillator frequency in the spherical limits, while W_z and $W_\perp$ are the frequencies of the anisotropic oscillator in the Z-direction and the directions $\perp^{\text{er}}$ to the Z-direction, respectively. The first term provides the K.E of the nucleons, and the second term defines a harmonic oscillator potential with a quadrupole deformation describing spheroidal shapes. To reproduce the empirical shell gaps for different N and Z values already for the spherical shell model [22,23], the third term, spin-orbit coupling, has been introduced. The last term, $(l^2 - <l_2>_N)$, is introduced to simulate the surface diffuseness depth, which leads to a proper single-particle ordering by lowering the energies of the large l orbitals within an N-shell. Here, the $<l_2>_N$ term is included [24] to keep the average energy of an N-shell unaltered by the l^2-term. The strength of the $l.s$ and l^2-terms is determined by the Nilsson parameters, K and μ. For small deformations, the single-particle states are defined by the approximate quantum numbers n, l, j, and Ω. The parameter Ω is the projection of the total angular momentum (j) on the Z-axis, n is the radial quantum number, and l is the orbital angular momentum. For large deformations, the single-particle states are described by the approximate "asymptotic quantum numbers," $[N, n_z, \Lambda]\Omega$. Here, $N = n_z + n_\perp$ is the principal quantum number, n_z and $n_\perp$ are the numbers of oscillator quanta in the Z-direction and in the $\perp^{\text{er}}$ direction, respectively, while Λ is the projection of the orbital angular momentum on the symmetry axis, i.e., the Z-axis.

11.2.2 Cranking Model

In order to explain the single-particle motion in a rotation nucleus, the cranking model was introduced by Inglis in 1954 [25,26]. In the cranking model, a nucleus with momentum $I > 0$ is described as rotating with a fixed frequency, ω, around a cranking axis (usually the x-axis), which is perpendicular to the symmetry axis. In this model, rotation is from the intrinsic frame of reference, and nucleons experience an additional potential caused by Coriolis and centrifugal forces.

The cranking Hamiltonian is then given by

$$h^\omega = h - \omega j_x,$$

where h^ω is the Hamiltonian in the body-fixed rotating system and h is the single-particle Hamiltonian in the laboratory system. j_x is the x-component of the single-particle angular momentum and is defined by $j_x = l_x + s_x$. The term $-\omega j_x$ is analogous to the Coriolis and centrifugal forces in classical mechanics. The eigenvalues of h^ω are referred to as the single-particle energies in the rotating system, which are generally noted as Routhians, e_i^ω. The total single-particle energy is calculated as the sum of the expectation values of the "single-particle energies" in the laboratory system, $E_{\text{tot}} = \Sigma_{\text{occ}} < h_i >$ and the total angular momentum I is calculated as $I = \Sigma_{\text{occ}} < J_x >_i$. Here, Σ_{occ} refers to the sum over the occupied proton and neutron orbitals.

In this model, the cranking Hamiltonian is invariant under a rotation of $180°$ around the x-axis, and the signature quantum number is used to classify the single-particle orbitals, $\alpha_{\text{tot}} = (\Sigma_{\text{occ}} \alpha_i) \bmod 2$ and $\alpha = I \bmod 2$. Thus, for an even number of nucleons, we have $\alpha = 0$ for $I = 0, 2$ and $\alpha = 1$ for $I = 1, 3, 5$, where for systems with an odd particle number, $\alpha = +1/2$ for $I = 1/2, 5/2, 9/2$ and $\alpha = -1/2$ for $I = 3/2, 7/2, 11/2$. Furthermore, we will only consider shapes that are invariant under reflection, which gives parity also a good quantum number.

11.2.3 Cranked Nilsson–Strutinsky (CNS)

In this model, the cranked Nilsson (modified oscillator) Hamiltonian is used to explain the properties of nucleon in the rotating nucleus:

$$h^\omega = h_{\text{osc}}(\varepsilon_2, \gamma) - V' - \omega j_x + 2\hbar\omega_0 \rho^2 \varepsilon_4 V_4(\gamma)$$

Here h_{osc} is the deformed harmonic-oscillator Hamilton, and is defined with different oscillator frequencies ω_x, ω_y, and ω_z expressed as $h_{\text{osc}} = p^2/2m + 1/2m \left(\omega_x^2 x^2 + \omega_y^2 y^2 + \omega_z^2 z^2\right)$.

Generally, these three frequencies are expressed in terms of quadrupole deformation coordinates ε_2 and γ, corresponding to the different ellipsoidal shapes:

$$\omega_i = \omega_0(\varepsilon_2, \gamma)\left[1 - 2/3\varepsilon_2 \cos(\gamma + i2\pi/3)\right], \quad i = x, y, z.$$

The parameter ε_2 specifies the degree of the deformation of the nucleus, while γ gives its degree of axial asymmetry. Figure 11.4 shows the values of ε_2 and γ, which are described as the all possible ellipsoidal shapes with rotation around the three principal axes. The spherical nucleus is at $\varepsilon_2 = 0$, and $\gamma = 0°$ and $60°$ axes are refer to as prolate shape and

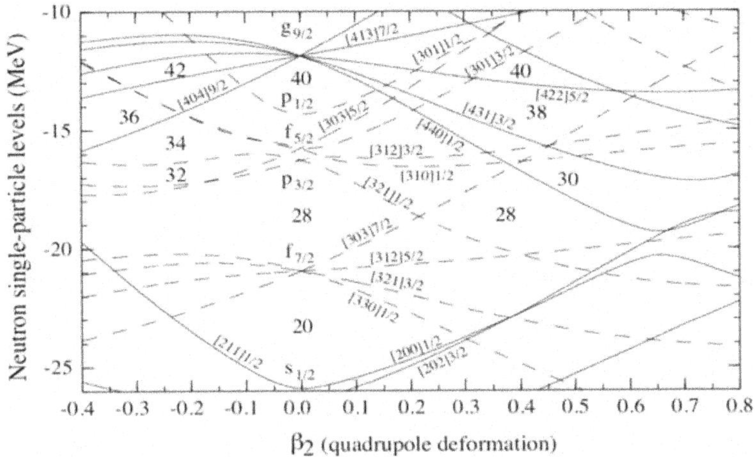

FIGURE 11.4
The neutron single-particle levels for the $A \sim 60$ mass region from calculation using a Wood-Saxon potential versus quadrupole deformation, β_2 [28].

oblate shape, respectively. These two shapes are axially symmetric with collective rotation around the perpendicular axis. At the border line, the nucleus rotates around the oblate ($\gamma = 60°$) or prolate ($\gamma = 120°$) symmetry axis corresponding to the non-collective limit. Away from the axes, ($0° < \gamma < 60°$), ($0° < \gamma < -60°$), and ($-60° < \gamma < -120°$), the nuclear shapes are triaxial with rotation around the shorter, intermediate, and longer principal axes, respectively. The term V' was already discussed in connection with the Nilsson model. A higher-order hexadecapole deformation, $2\hbar\omega_0\rho^2\varepsilon_4 V_4(\gamma)$, is also included, where ρ is the radius in the stretched coordinate system. In this model, the Hamiltonian is diagonalized using the eigenfunctions of the rotating oscillator [27], $|n_x n_2 n_3 \Sigma>$, as basis states. In this model, $N_{\rm rot}$ can be treated as a good quantum number and defined as $N_{\rm rot} = n_x + n_2 + n_3$. The main advantage of these calculations over the other, for example, spherical shell-model calculations is that the model space and the number of particle-hole excitations are practically unlimited. Few features of this model are as follows:

1. The total energy is then minimized with respect to the shape parameters for each spin state.

2. No virtual crossings.

3. By adding cranking term, the modified oscillator describes the fluctuating part of shell energy as a function of deformation, particle number, and the spin.

4. Configuration dependent.

In this model, configurations are specified by the number of particles with signature $\alpha = 1/2$ and $\alpha = -1/2$, respectively, in each $N_{\rm rot}$ shell. In addition to this, within each $N_{\rm rot}$-shell, it is generally possible to distinguish between orbitals which have their main amplitudes in the high-j intruder shell and the low-j orbitals which have their main amplitudes in the other j-shells. In the $A \sim 60$ mass region, the involved shells are $N_{\rm rot} = 3$ and 4. For $N_{\rm rot} = 3$ shell, the $1f_{7/2}$ and for the $N_{\rm rot} = 4$ shell, the $1g_{9/2}$ orbitals are considered as high-j. The leftover orbitals in the respective shells are considered as low-j.

The configuration for $Z = 30$ has one neutron hole in $1f_{7/2}$ and one neutron particle in the $1g_{9/2}$ subshell as shown in Table 11.1. The Nilsson scheme for the single-particle

TABLE 11.1

Occupation of Different High- and Low-j $N_{\rm rot}$ Orbitals of Different Signature α for a ^{60}Zn Neutron Configuration with One Hole in $1f_{7/2}$ and One Particle in $1g_{9/2}$

	$N_{\rm rot} =$	0	1	2	3	4	5
$\alpha = -1/2$	Low-j	0	1	3	1	0	0
	High-j	1	2	3	3	0	0
$\alpha = 1/2$	Low-j	0	1	3	1	0	0
	High-j	1	2	3	4	1	0

energies of the neutron orbit versus the quadrupole deformation for nuclei in the $A \sim 60$ mass region [28] is shown in Figure 11.4.

For spherical nuclei with a magic number of nucleons at 20, 28, and 40, i.e., at no deformation, one can see that the energy gaps are large. At a quadrupole deformation of $\beta_2 \sim 0.2$, the gaps disappear. The Nilsson diagram for single-particle energies of a deformed nucleus is very different from that of a spherical nucleus. Moving towards a larger deformation of $\beta_2 \sim 0.4$, a new distinct energy gap is formed corresponding to 28 neutrons [29]. This corresponds to the spherical nucleus $N = Z = 28$ ^{56}Ni, which thus is doubly magic both at spherical and deformed shapes. At an even larger deformation of $\beta_2 \sim 0.5$, another gap is formed at neutron number 30. This is the SD "doubly magic" core for the $N = Z = 30$ system, i.e., for ^{60}Zn [28]. Some other interesting points can be made when looking into the above diagram, the deformation, which causes low-Ω orbitals to be pushed down in energy for prolate nuclei and up for oblate nuclei, allows the intrusion of higher-shell orbitals into lower-shell orbitals. An example of these so-called "intruder" states is the $g_{9/2}$ orbital, which intrudes into the fp shell and is believed to drive collectivity in the Z, $N = 28$ region. Deformation causes the shell gaps at spherical magic numbers to disappear with new, small subshell gaps opening up at different deformation values.

More details on CNS calculations are given in Refs. [30,31].

11.3 Results on Few Nuclei

In the present chapter, the comparison results between experimentally observed bands and the theoretical calculations using the CNS model on few different nuclei in the $A \sim 60$ region have been discussed. The selected nuclei are $^{60}_{28}$Ni$_{32}$, $^{58}_{29}$Cu$_{29}$, $^{59}_{29}$Cu$_{30}$, $^{61}_{30}$Zn$_{31}$, and $^{62}_{30}$Zn$_{32}$.

11.3.1 ^{60}Ni

With a magic number of protons $Z = 28$, the Ni isotopes exhibit different kinds of high-spin rotational band structures with an increasing neutron number. The most important orbitals for a theoretical descriptions of Ni isotopes include the $N = 3$ high-j $1f_{7/2}$ and the upper fp-shell orbits $1f_{5/2}$, $2p_{3/2}$, and $2p_{1/2}$, and finally the $N = 4$ orbit $1g_{9/2}$. In the deformed rotating potential considered here, these j-shells will mix, i.e., the wave functions of the single-particle orbitals will have amplitudes in several j-shells. However, it turns out that if the deformation is not too large, these orbitals can be classified as having their main amplitudes in either of the other shells with lower j-values, [30]. For example, in the $N = 3$ shell, the orbitals will be dominated either by the high-j $1f_{7/2}$ shell or by the other j-shells, which means that they can be characterized as being of $1f_{7/2}$ character or of $1f_{5/2}$, $2p_{3/2}$ character. The classification into high-j and low-j orbitals is only possible if the so-called virtual crossings between weakly interacting orbitals are first removed. The classification is made easier by the use of the eigenstates of the rotating oscillator potential as basis states where the weak coupling between the different shells in this potential is neglected; see Ref. [31] for details.

With four neutrons outside the double-magic ^{56}Ni core, for ^{60}Ni nucleus, the experimentally observed high-spin level scheme is given in Ref. [32]. For ^{60}Ni, the Fermi surface will fall far below the $2p_{1/2}$ shell, so this orbital will not contribute significantly to the wave functions of the occupied orbitals. The observed level scheme comprises 270 gamma-ray transitions with seven high-spin band structures [32].

The calculated single-particle orbitals for neutrons as a function of the rotational frequency for typical deformation parameters $\varepsilon_2 = 0.26$ and $\gamma \sim 20°$ are shown in Figure 11.5. In this figure, the spherical origin of the different orbitals is indicated, but, as explained above, it is not really possible to make a distinction between $1f_{5/2}$ and $2p_{3/2}$ orbitals. Note also that two orbitals of $1f_{7/2}$ and fp characters, respectively, come close in the lower part of the diagram, which probably means that they will mix strongly. Our interest is, however, mainly the high-spin region, and it is then satisfactory to note that these orbitals split part for frequencies larger than 0.5 MeV, where the up-sloping orbital has the expected properties of the highest $1f_{7/2}$ orbital, while the lower orbital can be characterized as being of fp character. The single-particle orbitals for protons and neutrons are almost identical due to low Coulomb effects. The main difference is the Fermi energy, which is higher for neutrons than for protons. The large signature splitting of the lowest high-j $1g_{9/2}$ intruder orbital means that for configurations with one $1g_{9/2}$ neutron (or proton), one should expect only to observe the favored signature. A large energy gap is formed at particle number 31 at $\hbar\omega$ 1 MeV in the Routhian diagram. Favored configuration for $N = 32$ at this deformation will then be formed with all orbitals below the $N = 31$ gap filled and with the 32nd neutron in either signature of the (fp) orbital situated above the $N = 31$ gap.

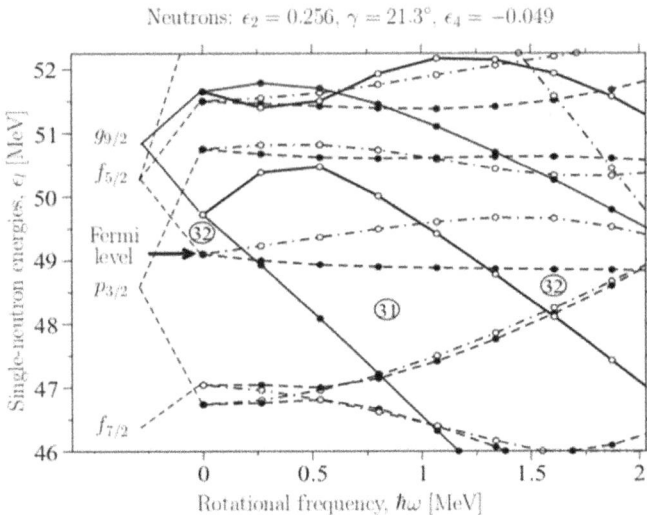

FIGURE 11.5

Calculated Routhians for neutrons for a typical deformation parameters at $\varepsilon_2 = 0.26$ and $\gamma \sim 20°$ in ^{60}Ni [32].

The ^{60}Ni configurations can then be written as Π $[(1f_{7/2})^{-p1}(1g_{9/2})^{p2}(fp)^{p3}] \otimes v$ $[(1f_{7/2})^{-n1} (1g_{9/2})^{n2}(fp)^{n3}]$ configuration can also be labeled as [p1p2, n1n2], where p1(n1) number of holes in $1f_{7/2}$ and p2(n2) is number of particles in $1g_{9/2}$.

Most of the observed dipole bands [32] were agreed with configurations, which have a hole in $1f_{7/2}$ orbitals, while the negative parity is created with one $1g_{9/2}$ particle. Then, because of the neutron excess, the only reasonable interpretation is that the hole is created in a proton orbital, while the $1g_{9/2}$ particle must be a neutron; see Ref. [32] for more details.

A comparison of the observed well-deformed bands is shown in Figure 11.6.

The experimentally observed WD1 band structures, which are "nonmagnetic," suggests that they have an even number of protons and neutron holes in $1f_{7/2}$ and an odd number of particles in $1g_{9/2}$ orbitals. As shown in Figure 11.6, a good agreement is made for the WD1 band structures with the configuration [20,01]. The observed band with the highest spin $I = 23^+$ is in a good agreement with the configuration which is having 3 proton holes in $1f_{7/2}$ and one proton and neutron particles in $1g_{9/2}$ orbital. The structure WD3, which is similar to structure WD2 and has a positive parity, is assigned to the calculated configuration [21,01]. More details on configuration assignments of remaining observed bands are reported in Ref. [32].

11.3.2 ^{59}Cu

The ^{59}Cu nucleus has three particles outside the doubly magic ^{56}Ni core: two neutrons and one proton. The most important involved orbitals for getting high-spin states are $N = 3$ high-j $1f_{7/2}$ and the $N = 4$ and $N = 5$ high-j orbits $1g_{9/2}$ and $1h_{11/2}$. Twelve different structures have been observed up to high spin in ^{59}Cu [34].

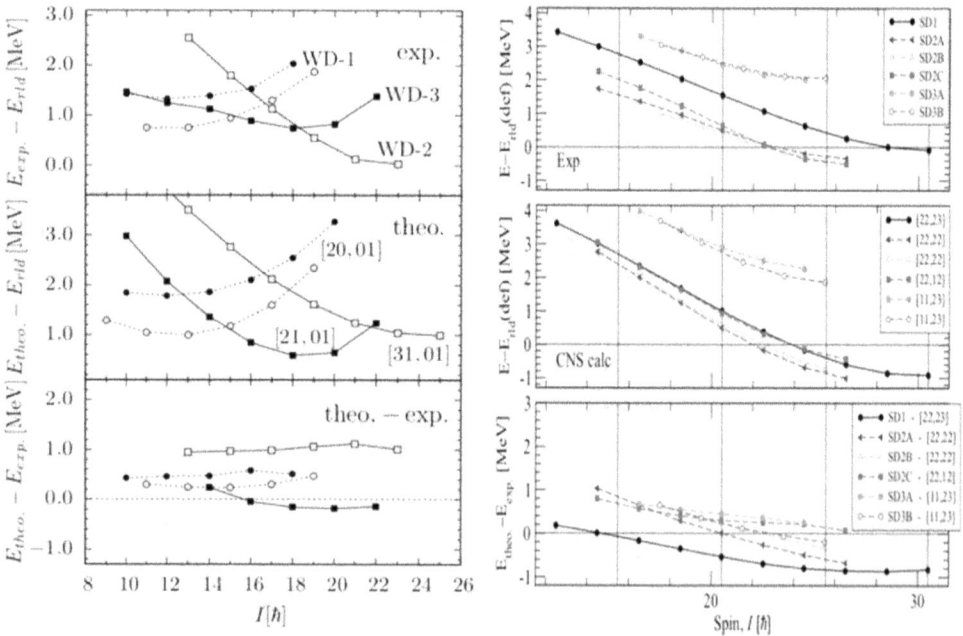

FIGURE 11.6

Experimental (top panel) and CNS calculated (middle panel) energies and their difference (lower panel) for the well-deformed structures in ^{60}Ni. The energies in the two upper panels are shown relative to the rotating liquid drop (rld) energies [33].

The upper panel in the left side of Figure 11.7 shows the experimental energies relative to an $I(I+1)$ reference for the negative-parity states in ^{59}Cu[34], while the lower panel shows the corresponding CNS calculations. The panels in the right side describe the same as the left-side panels but are drawn for positive-parity states in ^{59}Cu. The higher-spin states of negative-parity bands are formed when two particles are excited to the lowest $1g_{9/2}$ orbit. It is always energetically most favorable to excite one proton and one neutron, both with signature $\alpha = \frac{1}{2}$. Such an excitation can be combined with zero, one, two, three, and four $1f_{7/2}$ holes. With zero holes, i.e., [01,01] configurations, energetically very favored terminating states. More details on the bands are given in Ref. [34].

The calculated relative energies for configurations [11,11] and [21,11] which were assigned to bands 7, 8, and 6, respectively, are off by 0.5–1.0 MeV. This suggests that the spherical gap at particle number 28 should be increased by at least 0.5. MeV, because the three bands differ in their $1f_{7/2}$ hole contents. This modification can be done by a change of the Nilsson parameters, which will only marginally affect the relative energies within the bands.

The upper panel in the right side of Figure 11.7 shows the experimental excitation energies relative to a rigid rotor reference for the positive-parity states in ^{59}Cu. The lower panel shows the corresponding CNS calculations. As shown in the figure, the positive-parity bands were formed with the 1 or 3 nucleons in the $1g_{9/2}$ orbital. In order to create energetically favored high-spin positive-parity states in ^{59}Cu, three particles must be excited

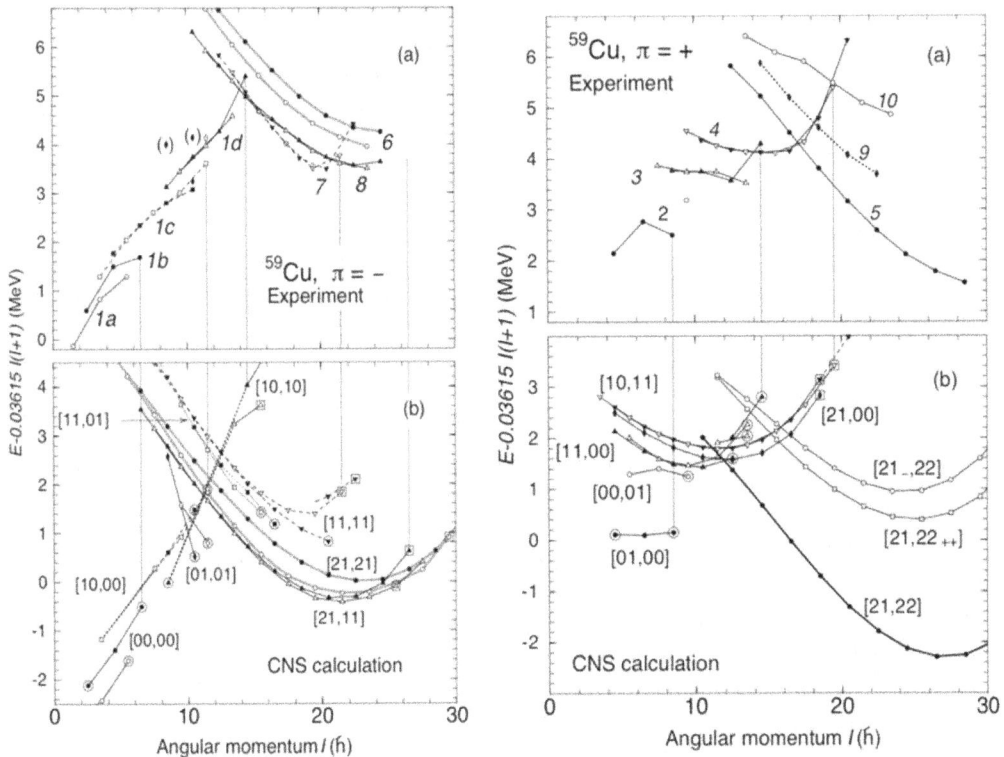

FIGURE 11.7
The upper panel in the left side shows the experimental energies relative to an $I(I+1)$ reference for the negative-parity states in ^{59}Cu, while the lower panel shows the corresponding CNS calculations. The panels in the right side describe the same as the left-side panels but are drawn for positive-parity states in ^{59}Cu.

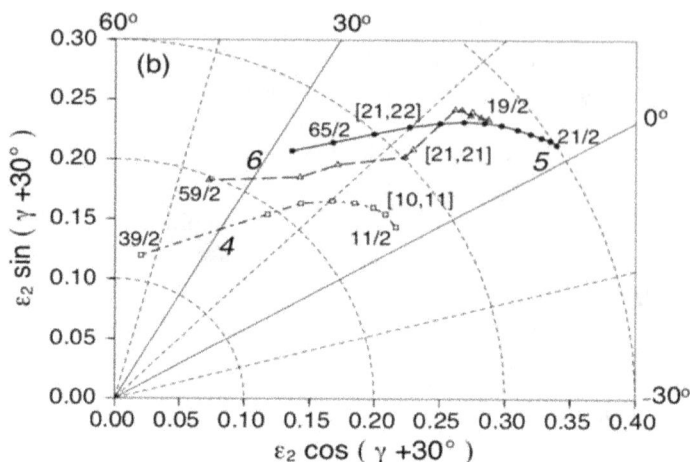

FIGURE 11.8
Calculated shape evolution of bands 4, 5, and 6 [34] in the ε_2-γ deformation plane.

to the $1g_{9/2}$ orbit. More details on the configuration assignments on the observed bands are given in Ref. [34]. The all observed bands involve different numbers of particle-hole excitations across the $N = Z = 28$ shell gap and different numbers of particles in the $1g_{9/2}$ orbit. The $1g_{9/2}$ orbit is strongly deformation driving. Together with an increasing number of particle-hole excitations, it is expected to induce an increasing amount of deformation or collectivity in the different structures. Calculated shape trajectories of bands 4, 5, and 6 [34] in the ε_2-γ deformation plane are shown in Figure 11.8. Starting at $\gamma = 0°$ with prolate shapes, the bands gradually change their shapes with increasing γ values. At intermediate high-spin values, they show the triaxiality and finally achieve non-collectivity at $\gamma = 60°$ with the highest spin states.

At high spin and medium to high excitation energy, it has been possible to follow the evolution of shapes from spherical to moderately-, well-, and SD shapes along the level scheme.

11.3.3 ^{61}Zn

Starting at ND with a few particles in the upper fp shell outside the soft doubly magic $N = Z = 28$ ^{56}Ni core, ^{61}Zn, like the other nuclei in the region, becomes well deformed and eventually SD by increasing the number of holes in $1f_{7/2}$ and particles in $1g_{9/2}$ orbitals, respectively. SD bands are defined as those with at least some four particle-holes in the $1f_{7/2}$ and four particles in the $1g_{9/2}$ orbital.

The experimental results [35] have built the level scheme with nearly about 120 excited states connected via some 180 gamma-ray transitions. In total, seven rotational structures up to $I \sim 25$ along with the three SD bands and six low-spin band structures have been identified. As we mainly focus on high-spin states, the predictions for the high-spin states are given in Figure 11.9. In the present CNS calculations, not only the relative energies between the bands but also the absolute scale can be compared. Low-spin states are generally predicted too high in energy because of the neglected pairing interaction. The high-spin states, on the other hand, should agree with the experimental results within a typical uncertainty of $\sim$1 MeV. Ideally, the observed transition energies will be predicted correctly by the calculations for each transition within a band. If this is the case, the band in the

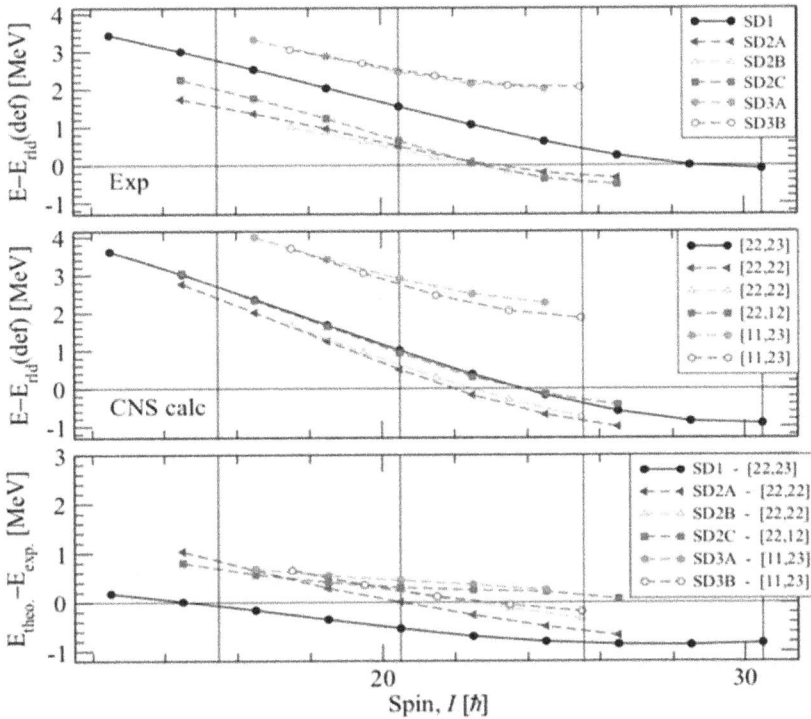

FIGURE 11.9

Comparison between experimentally observed superdeformed (SD) band structures and corresponding CNS predictions in ^{61}Zn. The top panel illustrates the experimental results where the bands are labeled according to Fig 1 of Ref. [35]. The middle panel shows the chosen predicted bands. The bottom panel plots the energy difference between the predictions and observations.

bottom panel will have a constant energy difference for all states of a given configuration or structure, i.e., it will be horizontal. Each band in the three panels is indicated according to its spin and parity assignment. Positive (negative) signature is drawn with filled (open) symbols. Solid (dashed) lines represent positive (negative) parity. The predictions for the SD band structures in ^{61}Zn are shown in Figure 11.9. In the top panel, the notation of each band corresponds to experimental bands; see Ref. [35]. The separation of the bands in the figure is based on their general behavior as a function of spin. In this figure, the bands that are favored in energy at high spins are illustrated.

The structure SD1 is assigned to the [22,23] configuration. In the bottom panel, the resulting band is in good agreement with the experimental data. The three bands in structure SD2 are matched with a [22,22] configuration or SD2B with either band A or C being its signature partner. The configuration [22,12] has been assigned to the band which is not a signature partner to SD2B. The configuration [11,23] was good in agreement with the signature partner bands of SD3. Experimentally observed signature splitting indicates an odd $1f_{7/2}$ hole, which agrees well with the matched configuration. The theoretically calculated signature splitting is, however, a little larger than what is experimentally observed. From these results, it is suggested that the standard Nilsson parameters κ and μ should be modified to improve the agreement between predictions and the observed structures in this mass region.

11.3.4 ^{62}Zn

With six valence nucleons outside the ^{56}Ni core, the ground state of ^{62}Zn is formed. The main orbitals involved in theoretical predictions of ^{62}Zn nucleus are $N = 3$ high-j $1f_{7/2}$ shell, the upper low-j (fp) shell $1f_{5/2}$, $2p_{3/2}$, and $2p_{1/2}$ and finally the $N = 4$ shell $1g_{9/2}$. For a spherical nucleus, these j shells are pure. In the deformed rotating potential, these j-shells will mix, i.e., the wave functions of the single-particle orbitals will have amplitudes in all the j-shells of a specific N_{rot}-shell. The low-j $N_{rot} = 3$ shells will only be active in the ground state of ^{62}Zn. More deformed, higher-spin configurations are formed by making particle excitations into the deformation-driving $1g_{9/2}$ orbitals and/or generating holes in the $1f_{7/2}$ orbitals.

Figure 11.10 shows the single-neutron orbitals plotted as a function of rotational frequency, $\hbar\omega$, at a specific deformation for the SD bands, $\varepsilon_2 = 0.41$, $\gamma = 0°$, and $\varepsilon_4 = 0.04$. As mentioned earlier, the single-particle orbitals for protons and neutrons are almost identical. For $Z = 30$, the $\pi[22]$ configuration, with two proton holes in $1f_{7/2}$ and proton particles in $1g_{9/2}$ orbitals, is favored in a large frequency range up to $\hbar\omega \sim 2.3$ MeV. At $\hbar\omega \leq 1.5$ MeV for $N = 32$, the $\alpha = \pm 1/2$ $(fp)_2$ are the favored orbitals to fill for the two neutrons above the 30 gap. Therefore, the γ [22] configuration is favored for $\hbar\omega \leq 1.5$ MeV. At $\hbar\omega \leq 1.5$ MeV, the $\alpha = 1/2$ $(fp)_2$ and the favored $(1g_{9/2})_2$ orbitals cross, which means that the occupation of the $\alpha = 1/2$ $(fp)_2$ and the favored $(1g_{9/2})_2$ neutron orbitals are lowest in energy at frequencies, and the occupation of the $\alpha = -1/2$ $1h_{11/2}$ orbital together with the $\alpha = 1/2$ $(1g_{9/2})_2$ orbital is favored for $N = 32$. At $\hbar\omega \approx 2.6$ MeV for $Z = 30$, the $\alpha = -1/2$ $1h_{11/2}$ orbital crosses the $(1f_{7/2})_3$ orbitals, so at higher frequencies in this deformation, it is favorable to fill one $1h_{11/2}$ orbital also for the protons.

The calculated total potential energy surfaces for the configuration [11,02] with one proton hole in $1f_{7/2}$ and one proton and two neutron particles in $1g_{9/2}$ orbitals are shown in Figure 11.11.

The configuration is with negative parity and signature $\alpha = 1$ with calculated maximum spin $I_{max} = 23^-$. The contour line separation is 0.2 MeV. The lowest energy minima are

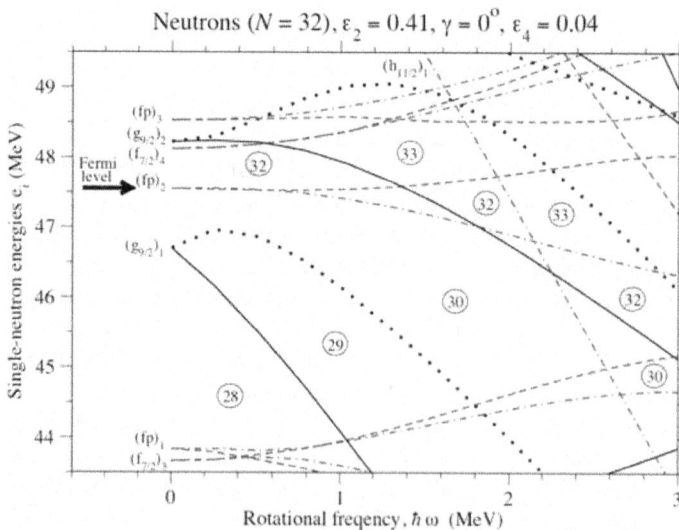

FIGURE 11.10

The neutron single-particle levels for the ^{62}Zn, at deformation parameters $\varepsilon_2 = 0.41$, $\gamma = 0°$, and $\varepsilon_4 = 0.04$.

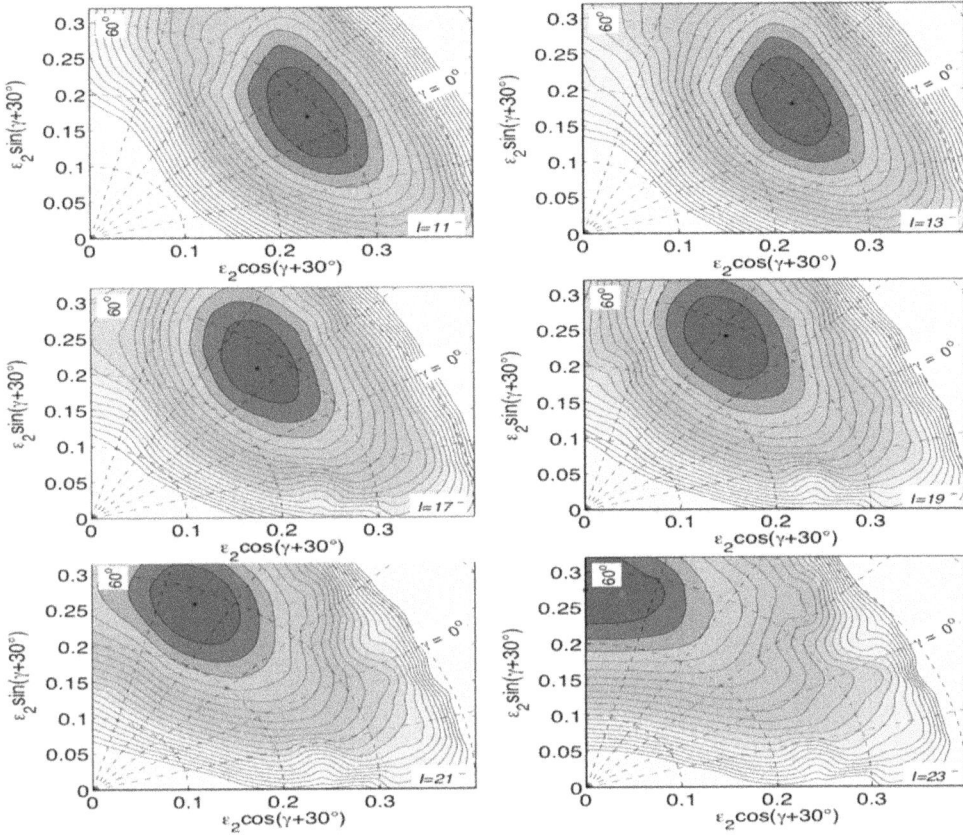

FIGURE 11.11

Calculated potential energy surfaces for the [11,02] configuration with negative parity and $\alpha = 1$ signature. The contour line separation is 0.2 MeV. See the text for details.

indicated with filled black circles. For the lowest spin values $I = 11^-$, 13^-, the minimum energy is found at $\gamma = 0°$, $-20°$, which indicates a large collectivity at these spin values. As going towards higher spin values $I = 17^-$, 19^-, 21^-, the collectivity reduces at $\gamma = 20°$, $-40°$. The maximum spin value $I = 23^-$ for the [11,02] configuration ended with a non-collective state at $\gamma = 60°$.

11.4 Configuration Assignment on Superdeformed Bands

Figure 11.12 illustrates the configuration assignments for the observed SD bands. The experimentally observed SD bands are shown in the top panel with the configurations assigned to them in the middle panel and the difference in energy between the two in the bottom panel. In the top and middle panels, the y-axis corresponds to the energy with the rld energy subtracted [33]. The positive (negative) parity bands have an even (odd) number of $1g_{9/2}$ particles, and it is instructive to see how the bands split up in groups with different I_{max} values depending on the number of $1g_{9/2}$ particles.

The signature partner configurations [22,2(−)3] and [22,2(+)3] are in good agreement with the observed bands SD1 and SD2. SD3 band is well fitted with the configuration [22,24].

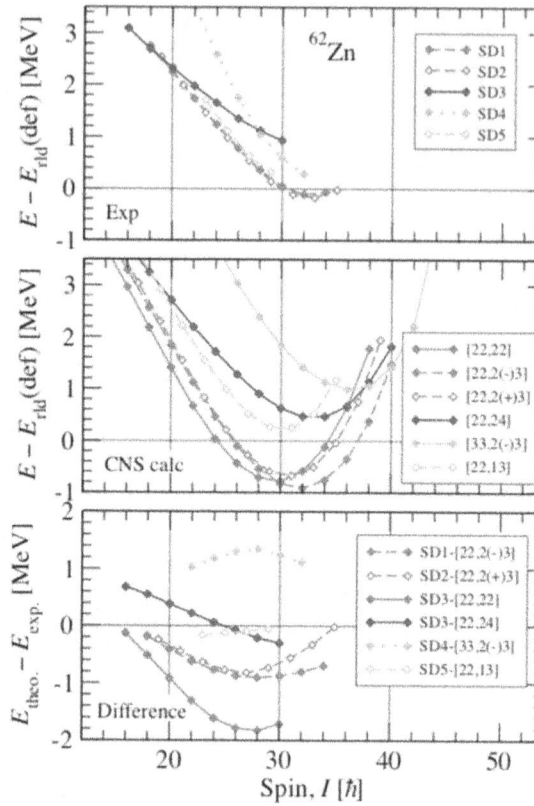

FIGURE 11.12

Comparison between experimentally observed structures and CNS predictions for observed SD bands [37] in ^{62}Zn. The top panel illustrates experimental energies relative to the rld energy, where the bands are labeled according to Ref. [37]. The middle panel shows the selected calculated bands. The bottom panel shows the energy difference between the predictions and observations. Signature $\alpha = 0$ is represented by filled symbols, and $\alpha = 1$ by open symbols.

The lowest energy calculated band with the configuration [22,22] has not been observed experimentally. The new SD4 band spin is uncertain, and the parity is undetermined [37]. The possible configuration assignment is [33,2(−)3] for SD4 as shown in the lower panel of Figure 11.12, within ∼1.4 MeV energy difference. In conclusion, the configuration of SD4 is unclear. The new SD5 band might be assigned to the negative signature partner of the [22,13] configuration. More details on the remaining observed band's configuration assignment have been given in Ref. [37].

11.5   New Nilsson Parameters

For a good agreement between the experimentally observed bands and the corresponding assigned calculated configurations using CNS codes with the standard Nilsson parameters, a large number of rotational bands in the $A \sim 60$ mass region have been selected. A new set

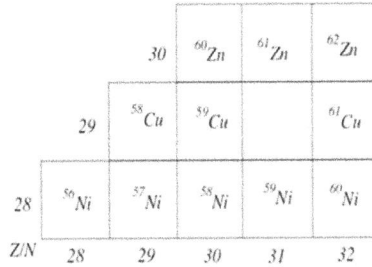

FIGURE 11.13
Selected nuclei for finding out the new Nilsson parameters in the $A \sim 60$ mass region [38].

of Nilsson parameters are determined, which is based on the comparison between theoretical predictions from CNS calculations and existing experimental data.

The chosen nuclei are the Ni, Cu, and Zn isotopes, as shown in Figure 11.13. The considered rotational bands from these isotopes have confirmed spin, parity, and well-established configuration assignments in the spin range $I \geq 15\hbar$. The details of the selected bands are given in Ref. [38]. The selected rotational bands from different nuclei are shown in Table 11.2. The involved orbitals for the selected bands are $N = 3$, $1f_{7/2}$ and $N = 4$, $1g_{9/2}$.

TABLE 11.2
Selected Bands for Finding Out the New Nilsson Parameter Set from Different Nuclei in the $A \sim 60$ Mass Region [38]

Nucleus	Bands	$[q_1, q_2]$	Maximum Spin, Parity, Signature	Configuration
^{56}Ni	SD2	[4,1]	$I^\pi = 17$, $\alpha = 1$	[2(+)1,20], [20,2(+)1][1]
^{57}Ni	SD1	[4,2]	$I^\pi = 47/2^-$, $\alpha = 1$	[2(+)1,21]
^{58}Ni	DSa, 3b	[3,1]	$I^\pi = 18^-$, 15^-, $\alpha = 0$, I	[20,11]
	Qla, lb	[3,2]	$I^\pi = 22^+$, 23^+, $\alpha = 0$, 1	[2(+)1,11]
	Q2a, 2b	[3,2]	$I^\pi = 20^+$, 21^+, $\alpha = 0$, 1	[2(−)1,11]
	Q3	[4,2]	$I^\pi = 22^+$, $\alpha = 0$	[20,22], [2(+)1,2(+)1][1]
	B1, B2	[4,3]	$I^\pi = 31^-$, $\alpha = 1$	[2(+)1,22], [31,22]
	B3	[5,3]	$I^\pi = 32^-$, $\alpha = 0$	[31,22]
^{59}Ni	Band 1	[2,1]	$I^\pi = 37/2^+$, $\alpha = +1/2$	[20,01]
	Band 2	[2,2]	$I^\pi = 43/2^-$, $\alpha = -1/2$	[2(+)1,01]
^{60}Ni	M2	[1,2]	$I^\pi = 17^+$, $\alpha = 0$, 1	[11,0(+)1], [1(+)0,02]
	WDla, b	[2,1]	$I^\pi = 18^-$, 19^-, $\alpha = 0$, 1	[20,0(∓)1]
	WD2	[3,2],[2,3]	$I^\pi = 23$, $\alpha = 1$	[31,0(+)1], [2(+)1,02][2]
	WD3	[2,2]	$I^\pi = 22^+$, $\alpha = 0$	[2(+)1,0(+)1]. [20,02][3]
^{58}Cu	SD	[4,2]	$I^\pi = 23^+$, $\alpha = 1$	[21,21]
^{59}Cu	Band 5	[4,3]	$I^\pi = 57/2^+$, $\alpha = + \frac{1}{2}$	[21,22]
	6A, 6B	[4,2]	$I^\pi = 4s/2^-$, $47/2^-$, $\alpha = +1/2, -1/2$	[21,2(∓)1]
	8A, 8B	[3,2]	$I^\pi = 47/2^-$, $49/2^-$, $\alpha = -1/2, +1/2$	[21,11]

(*Continued*)

TABLE 11.2 (*Continued*)
Selected Bands for Finding Out the New Nilsson Parameter Set from Different Nuclei in the $A \sim 60$ Mass Region [38]

Nucleus	Bands	$[q_1, q_2]$	Maximum Spin, Parity, Signature	Configuration
^{61}Cu	D3a, D3b	[1,2]	$I^\pi = 37/2^-, 35/2^-, \alpha = +1/2, -1/2$	$[1(+)1,0(+)1]$
	Q4	[2,3]	$I^\pi = 53/2^+, \alpha = +1/2$	$[21,02]$
	Q5	[4,3]	$I^\pi = 53/2^+, \alpha = +1/2$	$[21,22]$
	Q7a, b	[3,3]	$I^\pi = 53/2^+, 55/2^+, \alpha = 1/2, -1/2$	$[21,1(\mp)2]$
^{60}Zn	SD	[4,4]	$I^\pi = 30^+, \alpha = 0$	$[22,22]$
^{61}Zn	sD1	[4,5]	$I^\pi = 57/2^+, \alpha = +1/2$	$[22,23]$
	ND5a, b	[1,2]	$I^\pi = 37/2^-, 39/2^-, \alpha = +1/2, -1/2$	$[11,01]$
	SD2A, B	[3,4]	$I^\pi = 55/2^-, 57/2^-, \alpha = +1/2, -1/2$	$[22,12]$
	SD2C	[4,4]	$I^\pi = 55/2^-, \alpha = -1/2$	$[22,2(-)2]$
^{62}Zn	SD1, SD2	[3,5]	$I^\pi = 34^-, 35^-, \alpha = 0, 1$	$[22,13]$
	SD3	[4,6]	$I^\pi = 30^+, \alpha = 0$	$[22,24]$
	SD5	[4,5]	$I^\pi = (29^-), \alpha = 1$	$[22,2(+)3]$
	WD1	[2,4]	$I^\pi = 30^+, \alpha = 0$	$[22,02]$
	WD2a, b	[2,3]	$I^\pi = 27^-, 28^-, \alpha = 1, 0$	$[11,1(+)2]$
	WD3	[2,3]	$I^\pi = 27^-, \alpha = 1$	$[22,0(+)1],$ $[2(+)1,02]^3$
	WD4	[1,4]	$I^\pi = 27^-, \alpha = 1$	$[1(+)2,02]$
	WD5	[3,4]	$I^\pi = 31^+, \alpha = 1$	$[22,1(+)2]$
	WD6-7	[2,3]	$I^\pi = 27^-, \alpha = 1$	$[11,1(+)2]$
	WD7-6	[2,3]	$I^\pi = 27^-, \alpha = 1$	$[2(+)1,02],$ $[22,0(+)1]^3$
	TBla, lb	[1,2]	$I^\pi = 20^+, 21^+, \alpha = 0, 1$	$[11,0(+)1]$
	TB2a, 2b	[1,3]	$I^\pi = 23^-, 24^-, \alpha = 1, 0$	$[11,02]$
	ND9	[0,3]	$I^\pi = 19^-, \alpha = 1$	$[0(+)1,02]$
	ND8	[0,2]	$I^\pi = 16^+, \alpha = 0$	$[00,02]$
	ND7	[0,2]	$I^\pi = 17^+, \alpha = 1$	$[0(+)1,0(-)1]$
	ND6a, b	[0,2]	$I^\pi = 15, 16^+, \alpha = 1, 0$	$[0(\pm)1,0(-)1]$

The method called effective alignment [39–42] has been used to analyze the relative properties of the rotational bands in neighboring nuclei. With this method, the consistency of the spin, parity, and theoretical configuration assignments of the existing experimental bands can be investigated. The effective alignment, denoted by i_{eff}, is defined as the difference of spin values at constant rotational frequency which measures the contribution from different Nilsson orbitals. The main contribution to i_{eff} comes from the alignment of the orbital which is being occupied when going from the A to $(A + 1)$ nucleus, but differences between these two nuclei in deformation, pairing, etc. will also affect i_{eff}. To extract the effective alignments, one should know the relative spins. In the case of SD bands, the two compared bands maintain an almost constant difference of the equilibrium deformation between the bands over a wide range of rotational frequencies. This gives an i_{eff} which is predominately defined by the alignment properties of the particle in the single-particle orbital by which the two bands differ. In the case of smooth terminating bands, the sudden shape changes at the terminating state make the i_{eff} value to differ from the alignment of the single-particle orbital. The transition depopulating the terminating state links the states having the largest difference in equilibrium deformation within the band, where the terminating state has $\gamma = 60°$ and the state with $I_{\max} - 2$

has typically $\gamma \sim 30°$ and a larger ε_2 than the terminating state. Thus, from the i_{eff} approach, one can also measure how well theory describes the shape changes close to termination also.

The effective alignment between bands A and B, which is calculated at a constant rotational frequency $\hbar\omega$, is defined [39–42] as $i_{ef}^{A,B}(w) = I_B(w) - I_A(w)$. Band A in the lighter nucleus is taken as a reference. In a standard plot of spin I versus transition energy E_γ (or rotational frequency $\omega \approx \leftarrow E_\gamma/2$), an increase of I with particle number will be seen, if the added particle has a positive spin contribution. This is illustrated in Figure 11.15, which shows the total spin versus rotational frequency for the observed bands SD [21,21] in ^{58}Cu, band 5 [21,22] in ^{59}Cu, and SD [22,22] in ^{60}Zn.

From Figure 11.14, one can measure the i_{eff} of the second $1g_{9/2}$ neutron and proton. For rotational frequencies below 1 MeV, the three bands in the figure align differently, indicating different pairing correlations. These correlations disappear at high spin values, leading to an almost constant i_{eff} for $\hbar\omega \geq 1$ MeV.

The CNS calculations were performed to the selected bands mentioned in Table 11.2, with the standard Nilsson parameters in the $A \sim 60$ mass region. The $N = 4$, $1g_{9/2}$, and $N = 3$, $1f_{7/2}$ orbitals are purer than the other orbitals ($1f_{5/2}$, $2p_{3/2}$ and $2p_{1/2}$), and the position of the corresponding subshells gives a direct measure of gaps at $Z = N = 28$ and 40. Therefore, we introduce four cases from which one can find out revised energy gaps and thereby determine the new Nilsson parameters:

Case 1. $1g_{9/2}$ neutron particle excitation

Case 2. $1g_{9/2}$ proton particle excitation

Case 3. $1f_{7/2}$ neutron hole excitation

Case 4. $1f_{7/2}$ proton hole excitation

All these four cases are tested considering the bands in neighboring nuclei, whose configurations are identical, except for the particle in the $1g_{9/2}$ or $1f_{7/2}$ orbital.

In addition, bands in the same nucleus, which differ by a $1f_{7/2} \rightarrow fp$ or a $fp \rightarrow 1g_{9/2}$ excitation, are considered as a measure of the position of the $1f_{7/2}$ and $1g_{9/2}$

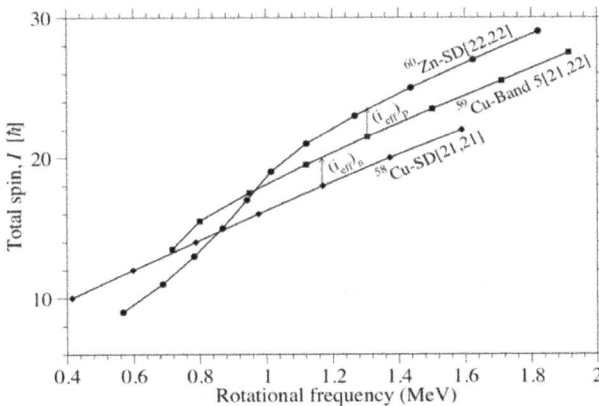

FIGURE 11.14

The total spin vs. rotational frequency ω for the experimental bands SD [21,21] from ^{58}Cu, band 5 [21,22] from ^{59}Cu, and SD [22,22] from ^{60}Zn. The value of i_{eff} is illustrated at specific ω-values.

TABLE 11.3

Standard and Derived New Nilsson Parameters for $N = 3$ and $N = 4$ Shells in the $A \sim 60$ Mass Region

	κ_{old}	μ'_{old}	κ_{new}	μ'_{new}
$N = 4$				
Protons	0.065	0.0370	0.0560	0.0319
Neutrons	0.070	0.0273	0.0804	0.0313
$N = 3$				
Protons	0.090	0.0270	0.090	0.0370
Neutrons	0.090	0.0225	0.090	0.0325

subshells, respectively. However, these values are more uncertain because they are also sensitive to the position of the fp orbitals.

After performing several calculations with the CNS model, for a better agreement between experimental and calculated band energies, the following changes are made:

1. The $N = 4$ proton and neutron $1g_{9/2}$ orbitals should be shifted by a value which is somewhat larger than 0.5 MeV in different directions. After performing some tests, it appears that 0.7 MeV $\approx 0.066\hbar\omega_0$ (for the $A \sim 60$ mass region) is an appropriate value, where the neutron $1g_{9/2}$ subshell is shifted downwards, while the proton $1g_{9/2}$ subshell is shifted upwards.

2. For $N = 3$ shell, the $1f_{5/2}$ orbital should be lowered by ≈ 1 MeV with respect to the $2p_{3/2}$ orbital.

3. The $N = 3$ shell, the $1f_{7/2}$ proton, and neutron orbital should be lowered, or the $Z = 28$ and $N = 28$ gaps should both be increased by ~ 0.5 MeV.

4. For a real 28-gap increment, the $1f_{7/2}$ orbital must be lowered more, say ~ 1 MeV. If both the $1f_{7/2}$ and the $1f_{5/2}$ shells are lowered by 1 MeV, it means that the spin-orbit splitting between these two shells is unchanged, and it is not necessary to introduce any l-dependence for κ. The derived new and old Nilsson parameters in the $A \sim 60$ region are given in Table 11.3.

11.5.1 Results with the New Nilsson Parameters

With the standard and the derived new Nilsson parameters, the energy gaps for $N = 3$ and $N = 4$ shells in the $A \sim 60$ mass region are shown in Figure 11.15.

The comparable values between two bands are taken from the differences between the experimental and calculated band energies which are drawn from $E - E_{\text{rld}}$ plots.

From Table 11.4, the average energy difference for all four cases is close to zero, i.e., the new parameters give a much improved description of the relative properties of bands with similar configurations in neighboring nuclei and in the same nuclei.

11.6 Conclusion

The $A \sim 60$ mass region is a very complex region, and a large variety of nuclear structure phenomena can be experimentally investigated, explained, and compared with the theoretical models.

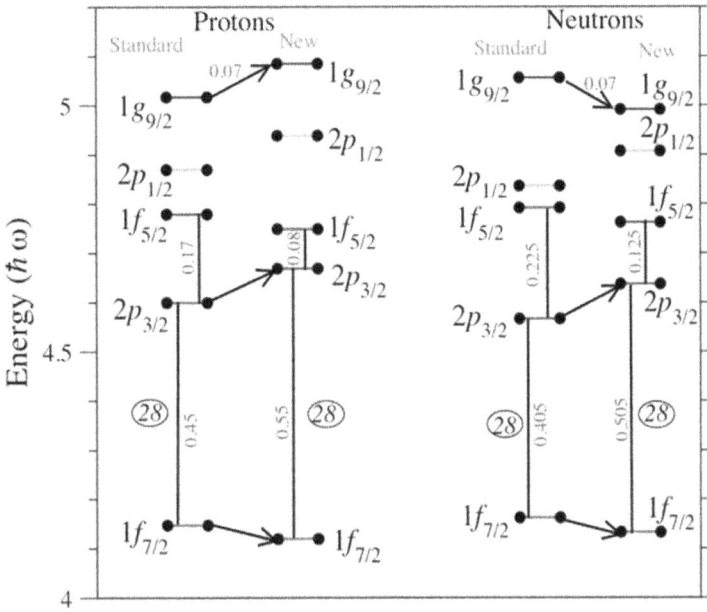

FIGURE 11.15
For $N = 4$, the $1g_{9/2}$ shell has been lifted by 0.07 $\hbar\omega_0$ for the protons and lowered by 0.07 $\hbar\omega_0$ for the neutrons, while for $N = 3$, the $l = 3$ shells have been lowered by 0.1 $\hbar\omega_0$ relative to the $l = 1$ shells for both neutrons and protons. More details are given in Ref. [38].

It is possible in the $A \sim 60$ region to follow the configurations with an increasing particle number by adding one particle in a specific orbital [42–44]. A special quality in this mass region is to follow the configurations from close-to-spherical to large deformations from successive excitations of holes and particles.

The experimental characteristics of the rotational bands are compared with the results from CNS calculations. Most of the observed high-spin structures are concluded that the low-spin normal deformed structures have no holes in the $1f_{7/2}$ orbitals, the well-deformed bands have two holes, and the SD bands have three to four holes in the $1f_{7/2}$ orbitals. With more particles in the high-j $1g_{9/2}$ orbitals, the terminating sequences become more down-sloping, when drawn relative to the rld reference, i.e., they are more favored in energy at high-spin states. More holes in the high-j $1f_{7/2}$ orbitals, on the other hand, lead to terminating sequences with a large curvature, corresponding to small moments of inertia. The effective alignment method has been applied to check out the spin, parity, and theoretical configuration assignments of the existing experimental bands and thereby identify the behavior of the band structures in the $A \sim 60$ mass region. Confirmed experimental bands and their theoretical configuration assignments are used to investigate an improved Nilsson parameter set in the $A \sim 60$ mass region. These new Nilsson parameters gave an improved overall description of data.

The way of determining the gap from crossing frequencies between orbitals at high frequencies has an advantage that it is calculated from essentially pure single-particle properties and therefore independent of pairing correlations.

TABLE 11.4

Energy Differences between the Bands of Different Neighboring Nuclei and within the Same Nuclei Are Calculated for Four Different Cases [38]

	Nuclei	Spin Range $\hbar$	Band 1 – Band 2	Energies Calculations with Old parameters	Diff. MeV	Energies Calculations with New parameters	Diff. MeV
$\nu(g_{9/2})$	^{57}Ni– ^{58}Ni	~ 17–23	SD1[2(+)1,21]-[2(+)1,22]B1a	−1.1, −0.8	0.3	−1.2, −1.3	−0.1
	^{58}Cu– ^{59}Cu	~ 17–23	SD[21,21] – [21,22] Band 5	−1.7, −1.3	0.4	−1.35, −1.35	0.0
$1g_{9/2}$ neutron particle excitation	^{60}Zn– ^{61}Zn	~ 22–30	SD[22,22] – [22,23] SD1	−1.6, −0.8	0.8	−1.2, −1.2	0.0
	^{61}Zn– ^{62}Zn	~ 14–20	ND5b[11,01] – [11,02]TB2	−0.3,0.2	0.5	−0.7, −0.6	0.1
		~ 22–28	SD1[22,23] – [22,24]SD3	−0.8, −0.3	0.5	−0.9, −0.3	0.6
			Avg. Diff.		**0.5**		**0.12**
$\pi(g_{9/2})$	^{58}Ni– ^{59}Cu	~ 14–18	D3a[20,11] – [21,11]8B	0.3, −0.8	−1.1	−0.1, −0.65	−0.55
	^{59}Cu– ^{60}Zn	~ 22–28	Band5[21,22]-[22,22]SD	−1.3, −1.6	−0.3	−1.5, −1.3	0.2
$1g_{9/2}$ proton particle excitation	^{61}Cu– ^{62}Zn	~ 20–26	Q4[21,02] – [22,02]WD	0.0, −0.5	−0.5	−0.6, −0.43	0.17
		~ 20–27	Q7b[21,1(+)2]– [22,1(+)2]WD5	0.0, −1.0	−1.0	−0.6, −0.6	0.0
			Avg. Diff.		**−0.7**		**−0.045**
$\nu(f_{7/2})$	^{59}Ni– ^{58}Ni	~ 15–19	Band 1[20,01] – [20,11]D3	0.5,0.3	−0.2	−0.15, −0.2	−0.05
		~ 14–22	Band2[2(+)1,01]-[2(+)1,11]Q1b	0.1, −0.2	−0.3	−0.4, −0.55	−0.1
$1f_{7/2}$ neutron hole excitation	^{59}Cu– ^{58}Cu	~ 16–24	8A, 8B[21,11] – [21,21]SD	0.0, −1.0	−0.4	−0.9, −1.3	−0.4
			Avg. Diff.		**−0.3**		**−0.11**

(*Continued*)

TABLE 11.4 (*Continued*)
Energy Differences between the Bands of Different Neighboring Nuclei and within the Same Nuclei Are Calculated for Four Different Cases [38]

	Nuclei	Spin Range $\hbar$	Band 1 – Band 2	Energies Calculations with Old Parameters	Diff. MeV	Energies Calculations with New Parameters	Diff. MeV
$\pi(f_{7/2})$	^{62}Zn–^{61}Cu	∼18-24	TB2[11,02] – [21,02]Q4	0.2,-0.2	0.0	-0.7, -0.7	0.0
1f$_{7/2}$ proton hole excitation	^{59}Cu–^{58}Ni	∼22-28	Band5[21,22] – [31,22]B3	-1.3, -1.5	-0.2	-1.5, -1.5	0.0
			Avg. Diff.		**-0.1**		**0.0**
$\nu(g_{9/2})$	^{59}Cu	∼18-24	6A[21,21] – [21,22]Band5	-1.3, -1.3	0.0	-1.0, -1.4	-0.4
	^{62}Zn	∼12-18	TB1a[11,01] – [11,02]TB2b	-0.7,0.2	0.9	-0.5, -0.4	0.1
			Avg. Diff.		**0.45**		**-0.15**
$\pi(g_{9/2})$	^{59}Ni	∼14-18	Band1[20,01]-[21,01]Band2,3	0.6,0.1	-0.5	-0.1, -0.1	0.0
	^{62}Zn	∼19-23	TB2[11,02]-[1(+)2,02]WD4	0.1,05	-0.4	-0.7, -0.2	0.5
			Avg. Diff.		**-0.45**		**0.25**
$\nu(f_{7/2})$	^{61}Cu	∼20-26	Q7[21,12]-[21,22]Q5	0.0, -0.4	-0.4	-0.5, -0.7	-0.2
	^{62}Zn	∼20-30	WD1[22,02]-[22,12]WD5	-0.4, -1.0	-0.6	-0.5,-0.5	0.0
			Avg. Diff.		**-0.5**		**-0.1**
$\pi(f_{7/2})$	^{62}Zn	∼15-19	ND9[01,02]-[11,02]TB2	0.0, 0.0	0.0	-0.6, -0.6	0.0
	^{58}Ni	∼20-25	WD4[12,02]-[22,02]WD1	0.3, -0.5	-0.8	-0.2, -0.3	-0.1
			Avg. Diff.		**-0.4**		**-0.05**

References

[1] http://www.aip.org/png/html/shapes.htm.

[2] A.V. Afanasjev, D.B. Fossan, G.J. Lane, and I. Ragnarsson, *Phys. Rep.* 322, 1 (1999).

[3] I. Ragnarsson, *Acta Physics Polonica B* 32, 2597 (2001).

[4] S.M. Polikanov et al., *Sov. Phys. JETP* 15, 1016 (1962).

[5] V.M. Strutinsky, *Nucl. Phys.* A95, 420 (1967); A122, 1 (1968).

[6] C. Andreiou, PhD thesis, Lund University, ISBN 91-628-5308-2(KFS AB, Lund, 2002).

[7] C.E. Svensson et al., *Phys. Rev. Lett.* 85, 2693 (2000).

[8] C.E. Svensson et al., *Phys. Rev. Lett.* 79, 1233 (1997).

[9] C. Baktash et al., *Phys. Rev. Lett.* 74, 1946 (1995).

[10] P.J. Nolan et al., *J. Phys. G* 11, L17 (1985).

[11] P.J. Twin et al., *Phys. Rev. Lett.* 57, 811 (1986).

[12] E.F. Moore et al., *Phys. Rev. Lett.* 63, 360 (1989).

[13] E.K. Johansson et al., *Phys. Rev. C* 80, 014321 (2009).

[14] C. Andreoiu et al., *Eur. Phys. J.* A14, 317 (2002).

[15] L.-L. Andersson et al., *Eur. Phys. J.* A36, 251 (2008).

[16] L.-L. Andersson et al., *Phys. Rev. C* 79, 024312 (2009).

[17] J. Gellanki et al., *Phys. Rev. C* 86, 034304 (2012).

[18] B. Cederwall et al., *Nature* 469, 68−71 (2011).

[19] S.G. Nilsson, I. Ragnarsson, *Shapes and Shells in Nuclear Structure* (Cambridge University Press, 1995).

[20] S.G. Nilsson, *Dan. Mat.Fys. Medd.* 29, no. 16 (1955).

[21] S.G. Nilsson et al., *Nucl. Phys.* A131, 1 (1969).

[22] O. Haxel, J.H.D. Jensen, and H.E. Suess, *Phys. Rev.* 75, 1766 (1949).

[23] M.G. Mayer, *Phys. Rev.* 75, 1969 (1949).

[24] C. Gustafsson, I.L. Lamm, B. Nilsson and S.G. Nilsson, *Ark, Fys.* 36, 613 (1967).

[25] D.R. Inglis, *Phys. Rev. C* 96, 1059 (1954).

[26] D.R. Inglis, *Phys. Rev. C* 103, 1786 (1956).

[27] V.G. Zelevinskii et al., *J. Nucl. Phys.* 22, 565(1975).

[28] R. Wyss, Private Communication.

[29] D. Rudolph et al., *Eur. Phys. J.* A4, 115 (1999).

[30] A. Afanasjev, D. Fossan, G. Lane, and I. Ragnarsson, *Phys. Rep.* 322, 1 (1999).

[31] T. Bengtsson and I. Ragnarsson, *Nucl. Phys.* A436, 14 (1985).

[32] D.A. Torres et al., *Phys. Rev. C* 78, 054318 (2008).

[33] B.G. Carlsson and I. Ragnarsson, *Phys. Rev. C* 74, 011302 (R) (2006).

[34] C. Andreiou et al., *Eur. Phys. J. A* 14, 317−348 (2002).

[35] L.-L. Andersson et. al., *Phys. Rev. C* 79, 024312 (2009).

[36] C.-H. Yu et al., *Phys. Rev. C* 60, 031305(R) (1999).

[37] J. Gellanki et al., *Phys. Rev. C* 86, 051304(R) (2009).

[38] J. Gellanki, *Comprehensive Gamma-Ray Spectroscopy of 62Zn and studies of Nilsson Parameters in the Mass A = 60 Region*. Lund University, Department of Nuclear Physics (2013)

[39] A.V. Afanasjev et al., *Phys. Rev. C* 59, 3166 (1999).

[40] I. Ragnarsson, *Phys. Lett.* B264, 5 (1991).

[41] A.V. Afanasjev and I. Ragnarsson, *Nucl. Phys.* A628, 508−596 (1998).

[42] I. Ragnarsson, *Nucl. Phys.* A557, 167c (1993).

[43] C.E. Svensson et al., *Phys. Rev. Lett.* 82, 3400 (1999).

[44] W. Nazarewicz and I. Ragnarsson, *Handbook of Nuclear Properties,* edited by D.N. Poenaru and W. Greiner (Clarenndon Press, Oxford, 1996), p. 80.

12

Nuclear Structure Aspects of Bubble Nuclei

Akhilesh Kumar Yadav and Amritanshu Shukla

Rajiv Gandhi Institute of Petroleum Technology

CONTENTS

12.1 Introduction

The bubble effect, i.e. the possibility of stable nuclei having central density depletion at the interior of nuclear surfaces in the exotic nuclei, was firstly proposed and investigated by H. A. Wilson for spherical nuclei (Wilson 1946). It was further investigated by Wong, considering the shell correction energy and liquid drop model (Wong 1972). These studies suggested that the stability of spherical bubble nuclei with the bubble at the interior may occur as such shape with fissibility parameter $x \geq 1.0$ makes it stable against fission by preserving the symmetry, which was termed as breathing mode of deformation (Wong 1973). The spheroidal deformation has also been explained by considering the fundamental volume conservation principle of torus nuclei where the inner and outer surfaces of the torus are constrained to move in such a manner that the ratio of outer and inner surface radii remain constant. The similar phenomenon of breathing deformation was also studied by Krishan and co-workers in the effective density functional theory. They showed that the large fissibility parameter minimizes the energy of bubble shape (Krishnan and Pu 1973). Later, the phenomena of Coulomb energy and surface energy are also observed to explain the survival of heavy nuclei by assuming bubble- or torus-type geometrical structure of nuclei which formed void deep inside the nucleus in heavy nuclei by preserving the volume conservation. The formed void increases the surface area, and hence, binding energy decreases at the same time an average distance between protons also increases and Coulomb energy decreases. These two phenomena balance each other, and a new geometry shows the tendency to observe a more stable heavy-nuclei system.

The bubble nuclei may be formed in nuclear reactions when the collision with high-energy antiproton and antideuterium annihilation takes place in the deep of nuclei (Bulgac 2003).

Also, the possibility of the formation of nuclei with bubble structure has been explored in the stellar matter on finite-temperature neutron stars. Since then, there exist various models to explore the possibility of bubble structure in the various mass regions as well as deformed nuclei. The independent particle model predicted by Siemens and Bethe (1967) and the Hartree-Fock model (Davies et al. 1972) concluded that the single-particle state low-density region exists for different sizes of the bubble. The phenomenon of the low-density single-particle state is also supported by the Strutinsky shell correction energy and explained as local deformation minima for ^{36}Ar, ^{84}Se, ^{138}Ce, ^{174}Yb, ^{200}Hg, and ^{250}Rf. The existence of bubble or semi-bubble effect, i.e. vanishing or decreased central density in light-, medium-, and superheavy nuclei have been studied by Grasso (Khan et al. 2008) and was suggested that shell effect in lighter nuclei and the Coulomb repulsion due to the movement of proton towards the nuclear surface in heavier nuclei appeared as main reason for the central density depletion. The shell effect in lighter nuclei is forced by s-orbit, and it contributes to central density in the nonrelativistic picture. The unoccupied 1s-orbit of proton in ^{34}Si and ^{46}Ar nuclei and unfilled 2s state in heavy-nuclei ^{206}Hg caused central density depression. In the absence of centrifugal barrier in exotic nuclei, s-orbit has peaked radial distribution in the interior of nuclei with their wave function extending to surface and contributing to central density. For non-zero angular momentum, wave function peaks are suppressed into interior of nuclei and do not contribute to central density. The deformation and temperature correlation with bubble effect have been recently studied by Kumawat et al. (2017) who suggested that the highly deformed nuclei changed the shell structure of nuclei, and under super- or hyper-deformed condition, bubble effect disappears; also, temperature plays an important role in bubble structure, i.e. at temperature equivalent to 3 MeV, the transition from superconducting state to normal state vanishes the bubble effect. The first experimental evidence of proton bubble nuclei in doubly magic nuclei ^{34}Si has renewed the vast field of nuclear structure to explore the understanding of bubble structure and anomaly depletion of central density (Mutschler et al. 2017).

For the study of the potential bubble effect in nuclei, unoccupied s-orbit is a necessary condition. In addition, if the unoccupied s-orbit near the Fermi surface is surrounded by higher angular momentum values and well separated from occupied energy level, i.e. there exists a weak dynamical correlation, then it can be a potentially most favored condition. The bubble structure in magic nuclei was studied separately for proton and neutron magic numbers in the RMF+BCS model, and the effect of temperature was compared from ground-state temperature (0 K) to transition temperature equivalent to 3 MeV in the statistical theory (Saxena et al. 2019). The bubble effect is strongly affected by the spin-orbit coupling; therefore, the choice of model for the study is very crucial. The RMF model, which contains the spin-orbit term and pairing effect based on self-consistency, is very useful to identify the relation between bubble formation and predicted mean-field potential. This model, with effective nucleon meson coupling and nonlinear nucleon meson coupling, has successfully explained many nuclear structural properties of nuclei over the entire nuclear landscape. For the reliability of this model, ground-state observables have been calculated in various parameter sets and have been compared with experimentally observed bubble potential cases ("Systematic study of bubble nuclei in relativistic mean field model," 2016). The bubble effect in nuclei is defined by depletion fraction (F) and quantified as

$$F = \frac{\rho_{\max} - \rho_c}{\rho_{\max}} \times 100 \qquad (12.1)$$

where ρ_c and $\rho_{\max}$ are central and maximum charge densities of nuclei, respectively. Duguet suggested that the bubble effect in ^{34}Si is driven by only protons, and the direct measurement of proton density distribution by electron scattering can prove it. But this is not possible for unstable light nuclei like ^{34}Si to perform electron scattering with a sufficient intensity (Dugu et al. 2017).

12.2 Bubble Effect in Light- and Medium-Mass Nuclei

The RMF model calculations are used to explain many nuclear phenomena for nuclei lying on the beta stability line as well as far from island of stability, i.e. exotic nuclei under various parameter sets, namely, NL-SH, NL065, NL3*, and NL3. For the reliability of model prediction, we have compared the ground-state observables such as binding energy, charge radii, matter radii, neutron radii, and proton radii and found that results are almost similar to experimentally observed cases. In this section, we show the results for potentially bubble light nuclei ^{20}O, ^{22}O, ^{24}O, ^{34}Si, ^{36}Si, ^{36}S, ^{46}Ar, and ^{48}Ca on the basis of NL-SH parameter set calculation. The neighboring nonbubble nuclei have also been studied to identify the signature of bubbleness in central density. Figure 12.1 shows the neutron and proton linear density plot. In Figure 12.2, a single-particle shell energy diagram is provided, and a spin-orbit potential is shown in Figure 12.3. From density plot, one can suppose to $2s_{1/2}$ orbital populated in the following pairs: e.g. ^{20}O, ^{22}O vs ^{24}O and ^{34}Si, ^{36}Si vs ^{36}S, and ^{46}Ar vs ^{48}Ca in ground-state configuration. In Figure 12.1a, one can see the depletion in central density for ^{20}O, ^{22}O for neutron and proton numbers. In Figure 12.1b, the density is plotted for ^{34,36}Si, and compared with ^{36}S nuclei, the depletion in proton density clearly indicates that ^{34,36}Si are the cases of potentially proton bubble nuclei. In Figure 12.1c, we show the comparative density plot for ^{46}Ar and ^{48}Ca, and the proton density distribution shows that ^{46}Ar is the case for proton bubble nuclei. In nuclei ^{24}O, the $2s_{1/2}$ orbital state ($l = 0$) lies at the Fermi level (Figure 12.2), so the nucleon gets filled s-orbital state, and no depletion is observed for neutron density. While in ^{46}Ar, the $2s_{1/2}$ state lies above the Fermi level and unoccupied s-orbital state shows depletion in central density for both neutron and proton,

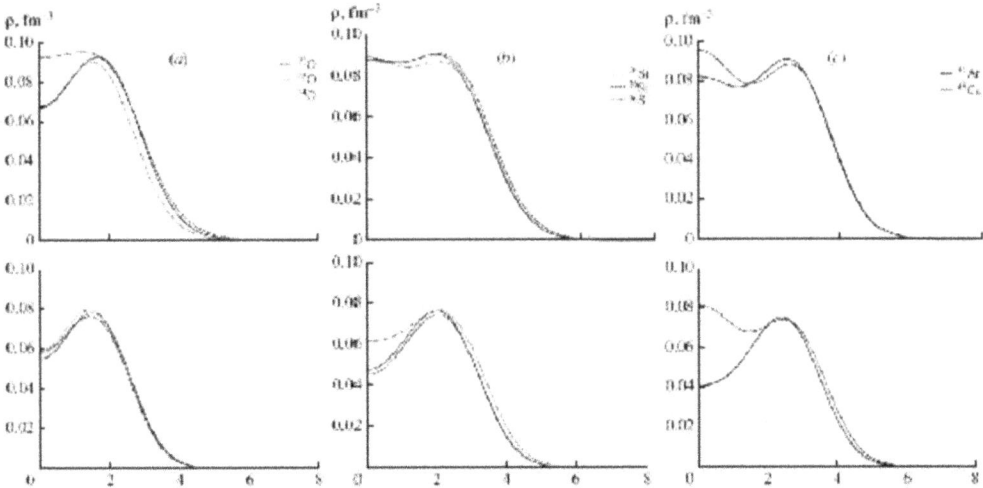

FIGURE 12.1
Comparison of linear density ρ for proton and neutron in nuclei ^{20,22,24}O, ^{34,36}Si, ^{46}Ar, and ^{48}Ca. The upper panel shows neutron density, and the lower panel shows proton density.

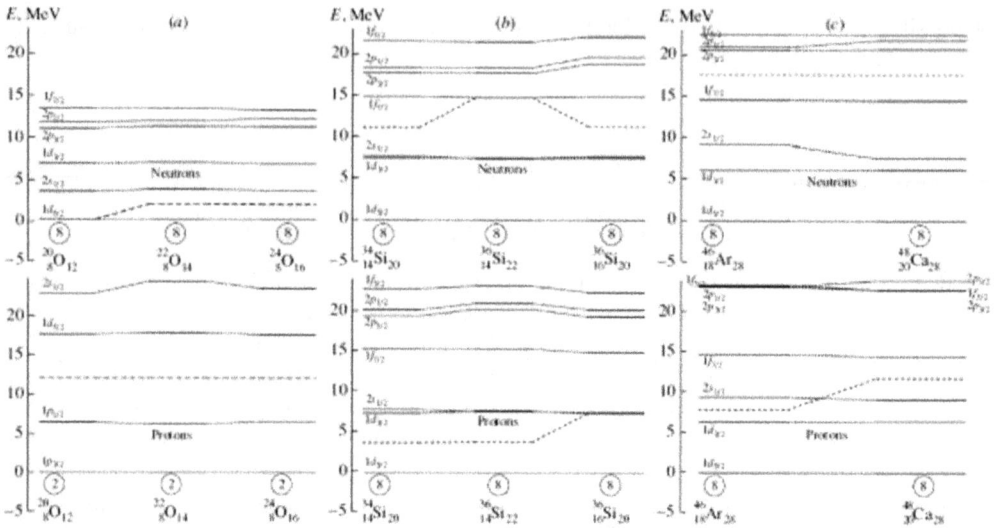

FIGURE 12.2
Comparison of single-particle energy diagrams for ^{20,22,24}O, ^{34,36}Si, ^{46}Ar, and ^{48}Ca nuclei. The upper panel shows for neutrons, and the lower panel shows for protons. The dashed lines in the single-particle spectra represent the position of the Fermi level.

in ^{48}Ca nuclei, the $2s_{1/2}$ orbital lies below the Fermi level get filled by nucleon and no density depletion is observed. As discussed in Section 12.2, the spin-orbit potential depends on the density profile of nucleon and isospin, which affects the spin-orbit coupling and shell splitting. In Figure 12.3, the spin-orbit potential is shown for the nuclei under consideration. In Figure 12.3a, one can see that ^{20,22}O shows more attractive spin-orbit potential nature at the outer surface of nuclei than ^{24}O. In Figure 12.3b and c, the shell effect can be seen in the interior of nuclei for ^{34,36}Si and ^{36}S, and 2p shell splitting in ^{46}Ar and ^{48}Ca. In Figures 12.2 and 12.3, one can see that the occupancy of $2s_{1/2}$ orbital varies with varying positions of the Fermi level, which plays a significant role in central density depletion. The s-d shell inversion arising due to pairing and tensor effects causes $1d_{3/2}$ level drop below $2s_{1/2}$ level and thus $1d_{3/2}$ gets filled before $2s_{1/2}$ orbital and hence the unoccupied $2s_{1/2}$ orbital shows the semi bubble effect. The bubble effect in potentially bubble nuclei using RMF (NL3*) and non-relativistic (SHF) theory for ^{22}O, ^{23}F, ^{34}Si, ^{36}S, ^{36}Ar, ^{46}Ar, ^{84}Se, ^{134}Ce, ^{174}Yb, and ^{200}Hg nuclei was studied and observed that density distribution for proton

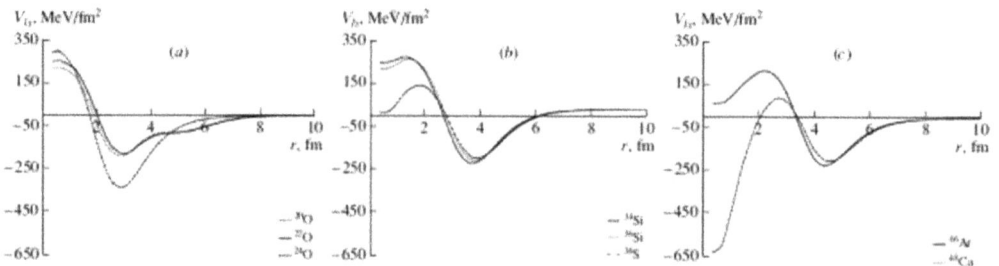

FIGURE 12.3
Comparison of spin-orbit potential as a function of radius r for ^{20,22,24}O, ^{34,36}Si, ^{46}Ar, and ^{48}Ca nuclei.

shows more depletion than neutron in central density as due to Coulomb repulsion particle occupies higher orbital state and s-orbit remains vacant ("Systematic study of bubble nuclei in relativistic mean field model," 2016).

In a recent work, Kay and Hoffmann observed that in ^{34}Si, the proton bubble density is due to a different sign of interior and outer surface spin-orbit potential that causes the 1p neutron shell gap reduction by a factor of 2, and they have also shown the weak binding energy effect in Woods-Saxon potential by considering the cases of ^{35}Si–^{41}Ca nuclei where the spin-orbit splitting decreased by 1 MeV in 1p shell, i.e. 1p$_{3/2}$ and 1p$_{1/2}$ (Kay et al. 2017). The major part of the decrement occurs between ^{35}Si and ^{37}S; 1p shell becomes closer with decreasing Z; and at ^{35}Si, 1p$_{1/2}$ state reaches the threshold value. The abrupt change of spin-orbit splitting in 1p neutron shell was also studied by Hamamoto and Sagawa in the HFB model, and they explained that weakly bound neutron with low orbital angular momentum has a certain probability that neutron lies outside the core of the nucleus, which reduces the strength of well-bound nucleon potential and reflected as halo phenomenon. The nucleon that participates in paring with the nearby unbound state is responsible for the bubble effect (Hamamoto and Sagawa 2004). Since this nucleon is weakly coupled to the core nucleus, it has an almost negligible contribution to single-particle potential and very small to many-body pairing correlation. The pairing strength of 2s$_{1/2}$ orbital neutron was observed on various aspects, and it was found that single nucleon binding energy approaches to zero, and hence, pairing gap also becomes zero and the strength of spin response approaches to possible hole particle coupling.

In Tables 12.1 and 12.2, one can see that how the occupation probabilities under the ground-state configuration change from bubble (unfilled or half-filled 2s$_{1/2}$) to nonbubble (filled 2s$_{1/2}$) nuclei for both proton and neutron numbers ("Systematic study of bubble nuclei in relativistic mean field model," 2016). The depletion fraction can be quantified using Equation 1 in the NL-SH model for proton, neutron, and total nucleon numbers. The results are given in Table 12.3.

From Table 12.3, it is clear that ^{22}O is an ideal case to observe the bubble effect for proton and neutron numbers and ^{34,36}Si and ^{46}Ar are the potential cases to observe the proton bubble nuclei. ^{20}O and ^{36}S also show the bubble characters. Using mean field with

TABLE 12.1

The Proton Occupation Probability Calculated in the NL-SH Model for Potentially Bubble Nuclei.

Shell	The Proton Occupancy in the Shell Model to Predict Potential Bubble Nuclei in Ground-State Configuration							
	^{20}O	^{22}O	^{24}O	^{34}Si	^{36}Si	^{36}S	^{46}Ar	^{48}Ca
1s$_{1/2}$	2.0	2.0	2.0	2.0	2.0	2.0	2.0	2.0
1p$_{3/2}$	3.96	3.98	3.99	3.99	3.99	3.99	4.0	4.0
1p$_{1/2}$	1.85	1.92	1.95	1.99	1.99	1.99	2.0	2.0
1d$_{5/2}$	0.10	0.06	0.04	5.63	5.72	5.86	5.95	5.96
1d$_{3/2}$	0.02	0.01	0.01	0.17	0.15	1.23	3.53	3.86
2s$_{1/2}$	0.01	0.01	0.00	0.10	0.07	0.67	0.38	1.70
2p$_{3/2}$	0.01	0.01	0.00	0.01	0.01	0.03	0.01	0.02
2p$_{1/2}$	0.00	0.00	0.00	0.01	0.00	0.01	0.01	0.03
1f$_{7/2}$	0.02	0.01	0.01	0.05	0.04	0.12	0.10	0.40
1f$_{5/2}$	0.00	0.00	0.00	0.01	0.01	0.02	0.02	0.01

Note: In order to see the variation of occupancy of s-orbital state over proton density, we have shown the results for neighboring non-bubble nuclei too.

TABLE 12.2

The Same as Table 12.1, but for Neutron and Neutron Density Variation

	The Neutron Occupancy in the Shell Model to Predict Potentially Bubble Nuclei in Ground-State Configuration							
Shell	^{20}O	^{22}O	^{24}O	^{34}Si	^{36}Si	^{36}S	^{46}Ar	^{48}Ca
$1s_{1/2}$	2.0	2.0	2.0	2.0	2.0	2.0	2.0	2.0
$1p_{3/2}$	3.99	4.0	4.0	4.0	4.0	4.0	4.0	4.0
$1p_{1/2}$	1.99	2.0	2.0	2.0	2.0	2.0	2.0	2.0
$1d_{5/2}$	3.61	5.59	5.95	5.96	5.99	5.95	5.99	5.99
$1d_{3/2}$	0.07	0.07	0.16	3.71	3.96	3.68	3.99	3.99
$2s_{1/2}$	0.13	0.23	1.79	1.87	1.98	1.82	1.99	1.99
$2p_{3/2}$	0.03	0.02	0.02	0.05	0.11	0.05	0.06	0.12
$2p_{1/2}$	0.01	0.01	0.01	0.02	0.04	0.02	0.03	0.05
$1f_{7/2}$	0.04	0.02	0.02	0.23	1.72	0.31	7.62	7.62
$1f_{5/2}$	0.01	0.01	0.01	0.03	0.04	0.04	0.18	0.08

TABLE 12.3

Depletion Fraction for Neutron, Proton, and Total Nucleon Numbers in ^{20,22,24}O, ^{34,36}Si, ^{46}Ar, and ^{48}Ca Nuclei.

Nuclei	Depletion Fraction in Neutron No.	Depletion Fraction in Proton No.	Total Depletion Fraction in Nuclei
^{20}O	21.71	21.06	21.40
^{22}O	24.54	25.32	24.90
^{24}O	2.61	19.34	9.54
^{34}Si	3.26	34.20	17.50
^{36}Si	3.42	36.69	18.47
^{36}S	0.00	18.78	8.04
^{46}Ar	9.79	43.44	25.17
^{48}Ca	0.00	0.00	0.00

pairing approach, Phuc et al. (2018) have shown that half-filled $2s_{1/2}$ in ^{22}O, ^{34}Si, and ^{46}Ar nuclei and half-filled $3s_{1/2}$ orbital in ^{206}Hg are responsible for central density depletion, and their shell closure takes place in such a way that low orbital state lies near the Fermi surface. In the case of bubble nuclei, the energy level of low angular momentum state increases as compared to higher orbital state, and remains unoccupied and creates a conventional magic bubble number such as 18, 34, 50, 58, 70, 80, and 120.

12.3 Bubble Effect in Superheavy Nuclei

The phenomena of bubble structure in heavy and superheavy nuclei have been studied in various models, which suggest that the bubble structure in superheavy nuclei occurs due to large Coulombic repulsion at extreme density. The density profile of many nuclei on the basis of various models also suggests that the nuclear forces favor the saturated nuclear density of 0.16 nucleon/fm^3. Therefore, finding how the superheavy nuclei achieve their stability against large Coulombic repulsion has always been the challenge. At the initial stage, the stability of superheavy nuclei is explained by assuming the bubble- or torus-type structure which creates a void (hole) in the interior of nuclei that pushes further the nuclear surface and nucleons, and hence, surface energy increases while the distance between the protons decreases due to large surface. This reflected to a decrease in Coulomb repulsion, so the balancing between the two energies provides the sufficient binding energy for the survival of superheavy nuclei. The α-particle model of nuclei has been discussed since the very beginning of nuclear theories, and the binding energy explained for lighter nuclei has also used this model to explain bubble structure in superheavy nuclei by considering the packed fullerene-type structure of $Z = 120$ nuclei formed by 60 α-clusters, as shown in Figure 12.4 (Hess et al. 2018).

The idea of fullerene-type structure comes from the RMF theory, and the nuclear density also suggests depletion in central density. The fullerene-type structure is hollow inside, and 60 α-clusters are distributed at the vertices of 20 hexagons and 12 pentagons. The fullerene structure of $^{284}120$ nuclei is explained by assuming that 120 protons and 120 neutrons form 60 α-clusters, and 60 neutrons provide an extra bonding to nucleus as 1 α-per neutron, and rest of the four neutrons are neglected and founded two different bond lengths which explain the hollow structure of $^{304}120$ nuclei similar to other RMF calculations performed by Greiner for $^{292}120$, $^{240}120$, and $^{120}120$ nuclei and hollow structure reflected as depletion in central density (Greiner 2002). Recently, the bubble effect in superheavy nuclei was also explained by considering Coulombic repulsion (Nazarewicz 2018), which arises by extreme particle densities that try to minimize the repulsion either by pushing the nuclear matter towards the surface or by deforming the shape of nuclei. In the process of pushing the nucleon, a hole is created at the interior of nuclei which stimulates the redistribution process of the nucleon, and the nuclear surface increases which is reflected as depletion of central density, while in the case of deformed nuclei, the deformation creates pocket barriers which change the orbital state of nucleon and provide extra binding energy, thus surviving against Coulomb repulsion. The heaviest element synthesized in the lab is ^{294}Og, and the various study models suggest that the density profiles of triaxially deformed ^{294}Og, spherical ^{302}Og, and prolately deformed ^{326}Og show central depletion in nucleonic density. Singh et al. (2013) studied the ground-state properties of superheavy nuclei $Z = 105-118$ and $Z = 120$ using

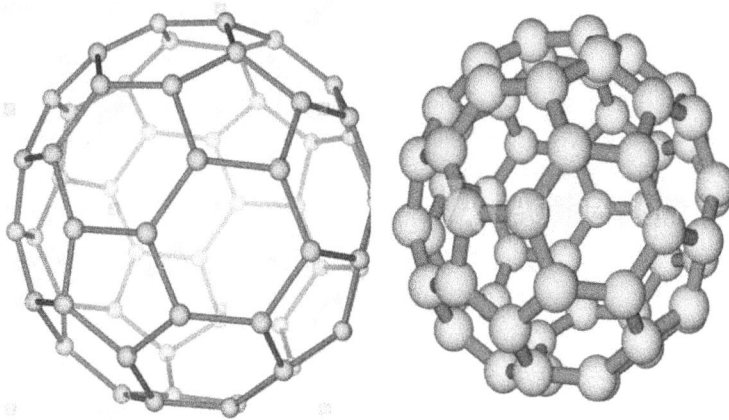

FIGURE 12.4
The packed α-cluster structure of $Z = 120$ nuclei consisting of 20 hexagons and 12 pentagons is shown in the left panel, and in the right panel, the hollow structure at the center of $Z = 120$ nuclei is shown. (Courtesy: www.google.com.)

SHF (SKI4) and RMF (NL3*) models in spherical, prolate, and oblate deformed shapes, and they observed that by increasing N/Z ratio from proton dripline to neutron dripline, the density profile of most of the nuclei follows the general pattern of peaking at center and decreasing to zero at the surface, while few of them show depletion at the center and slowly decreasing to zero at the surface, which are known as bubble nuclei. The phenomenon of bubble nuclei was observed in spherical symmetry for $^{285}113$, $^{294}117$, and $^{292,293}120$ nuclei, and oblate and prolate shapes also show the bubble effect in various superheavy nuclei under different quadrupole deformations.

12.4   Roles of Tensor Forces, Pairing, and Deformation in Bubble Nuclei

The exotic nuclei are unstable nuclei whose structural properties are different from stable nuclei. The phenomenon of bubble effect occurs in exotic nuclei, and it has been observed on the basis of shell effect, pairing effect, tensor effect, and deformation in various mass regions from lighter to the heavier and then superheavy region in literature. It has been found that tensor force plays an important role in the structure of bubble nuclei. The theoretical study and literature suggest that the tensor force can be easily verified in the lighter mass region compared to heavy nuclei. It was initially observed by Grasso (Grasso et al. 2009) in spherically bubble nuclei ^{34}Si in the framework of Skyrme Hartree-Fock approach in Sly4 parameter set. Latter, it was observed in bubble nuclei (^{34}Si and ^{46}Ar) by Zhao (Wang et al. 2011) who predicted that tensor force does not have much effect on central density of spherically symmetric well-paired nuclei ^{34}Si in which the neutron and proton shell is completely filled and lies well below the Fermi level, and hence, only zero orbital ($2s_{1/2}$) wave function contributes to central density depletion. In the case of ^{46}Ar, pairing and tensor force modified $1d_{3/2}$ orbit, while $2s_{1/2}$ did not show any change; hence, $s_{1/2}$-$1d_{3/2}$ (s-d) inversion is responsible for the bubble effect. The pairing effect and deformation is also observed in light, heavy, and superheavy nuclei (namely—^{34}Si, ^{46}Ar, and isotonic chain of $N = 82$, 126, 284), using the density functional theory in SV-min functional (Klüpfel et al. 2009). Schuetrumpf et al. (2017) observed that superheavy nuclei ^{294}Og shows proton bubble density in triaxial symmetry, while ^{326}Og nuclei show bubble in prolate deformed symmetry, and hence concluded that deformation does not affect the proton density depression. In ^{206}Hg, the lighter isotone 82, 2s orbital shell is partially filled and responsible for depletion. Shukla and Åberg studied the axially deformed bubble nuclei (Shukla and Åberg 2014) in light mass region, and suggested that the depletion is the result of mixing between low and high angular momentum states, and it is impossible to ignore the contribution of s-wave, which can be minimized due to deformation (Yao et al. 2015). The calculations were performed in the NL065 parameter set in an axially deformed harmonic oscillator basis. For $N/Z = 14$ nucleon distribution in spherical bubble nuclei is explained by an extreme shell model that has fully occupied $0d_{5/2}$ and empty $1s_{1/2}$, the density is plotted in three different sets of (s-d) orbital occupation number, as shown in Figure 12.5.

In Figure 12.5, one can see how the occupation probability of 1s orbital changes the central density distribution in deformed nuclei. As the deformation decreases the central neutron density depletion increases along with neutron number, to identifying this change the isotonic chain 10, 12, 14, 16, 18, 20 has been studied as a function of quadrupole deformation parameter. In Figure 12.6, the bubble size is observed as a function of deformation, and the figure also shows how the increasing prolate and

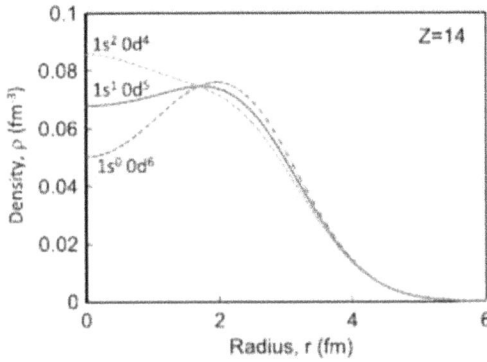

FIGURE 12.5
(Color online) Densities for the extreme single-particle picture for 14 protons in a spherical well. The graph shows how the bubble structure gradually increases with the increasing emptiness of 1s orbital.

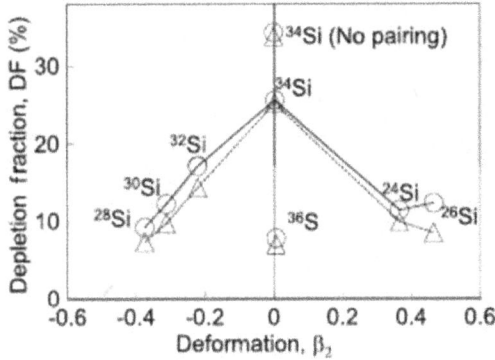

FIGURE 12.6
Proton depletion fractions for Si isotopes ($Z_{\bullet} = 14$) as functions of calculated deformations. The triangle shows the depletion along the perpendicular axis, ω, and the circles show depletion along the symmetry axis, Z. For comparison, the depletion of ^{34}Si is shown when pairing is excluded. In the graph, the drastic change in depletion is shown by adding two protons in ^{34}Si nuclei, i.e. ^{36}S. The lines connecting Si isotopes are drawn to guide the eye.

oblate deformations reduce the depletion fraction which is analogous to the increasing occupation of 1s orbital state in deformed nuclei. In order to compare the depletion fraction in ^{34}Si nuclei mirror nuclei ($Z = 20$ and $N = 14$), ^{34}Ca has also been studied (see Figure 12.7).

In literature, it has been discussed that ^{34}Si and mirror nuclei ^{34}Ca show proton bubble structures in the spherical symmetry, while the bubble structure is also observed in oblately deformed nuclei ^{32}Si and ^{32}Ar, and prolately deformed ^{24}Ne (Shukla and Åberg 2014).

In Figure 12.8, the proton and neutron densities are shown for ^{32}Si. The central region of proton density clearly shows that it is less dense in comparison to the peripheral region and hence shows depletion, while in neutron density, there is no density variation seen. Therefore, the above discussion and the density profile clearly indicate that ^{32}Si, ^{32}Ar, and ^{24}Ne show bubble structures in axially deformed symmetry.

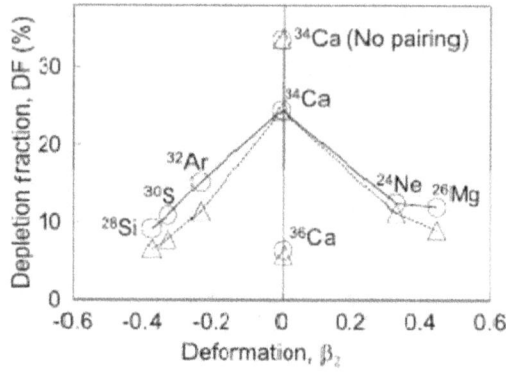

FIGURE 12.7

The same as Figure 12.2, but for neutron depletions of $N = 14$ isotones.

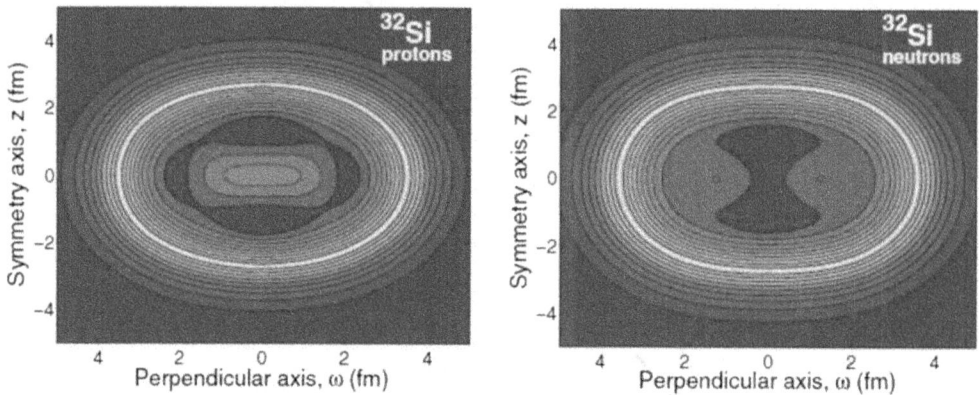

FIGURE 12.8

(Color online) Contour plots for proton and neutron densities for ^{32}Si along symmetry axis Z, and perpendicular axis ω. The white line marks the nuclear shape, defined as half maximum density. Maximal densities for proton and neutron are 0.0805 and 0.0920 fm^{-3}, respectively, and the distance between equidensity lines is 5% of the respective maximal density.

12.5 Roles of Shell Gaps and Spin-Orbit Splitting in the Formation of Bubble Nuclei

The spin-orbit coupling is an essential tool to understand the phenomenon of shell structure in mean-field approximation. The nucleons are considered as an independent particle to move inside the nucleus under different potentials which are derived from density functional theory and the spin-orbit term naturally considered in covariant density functional. In nonrelativistic mean field (MF) calculations like Skyrme (Bender et al. 2003) and Gogni-type (Vautherin and Brink 1972) interaction, the form of spin-orbit coupling is different from the relativistic mean field (RMF) models where it emerges directly from the Dirac equation. The spin-orbit potentials are given for MF and RMF by the following equation (Sorlin and Porquet 2013):

$$\emptyset_n^{ls} = -\left[\mu_1 \frac{\partial \rho_n(r)}{\partial r} + \mu_2 \frac{\partial \rho_p(r)}{\partial r}\right] \vec{l}.\vec{s} \tag{12.2}$$

where $\emptyset_n^{ls}$ denotes the spin-orbit potential and parameter μ_1 and μ_2 contain the isospin and density dependence. The ratio $\frac{\mu_1}{\mu_2}$ determining the isospin dependence has a fixed value equal to 2 in the case of nonrelativistic functionals (MF) and for RMF its value is not fixed and is close to one ($\frac{\mu_1}{\mu_2} = 1$, therefore no isospin dependence). The density dependence of spin-orbit potential is given by the following equations:

$$\text{For HF/Skyrme} \qquad \emptyset_{ls}^n \alpha \frac{1}{r} \frac{\partial}{\partial r} [2\rho_n(r) + \rho_p(r)] \vec{l}.\vec{s} \tag{12.3}$$

$$\text{For RMF} \qquad \emptyset_{ls}^n \alpha \frac{1}{r} \frac{\partial}{\partial r} [\rho_n(r) + \rho_p(r)] \vec{l}.\vec{s} \tag{12.4}$$

In the relativistic mean-field and nonrelativistic model, there is a difference between the isospin (Eq. 12.2) and density dependence (Eqs. 12.3 and 12.4) which results in the form of different spin-orbit potentials and shell splitting (Lalazissis et al. 2017). In order to remove the discrepancy in two different functional approaches, Mutschler (Mutschler et al. 2017) tried to set an experimental constraint for the spin-orbit term. For this purpose, he studied ^{34}Si experimentally and explained that neutron density to proton density asymmetry caused the bubble structure of ^{34}Si. Burgunder experimentally studied the effect of spin-orbit term on shell splitting and showed that the central density depletion in bubble nuclei reduced the splitting size of $2p_{1/2}$, and $2p_{3/2}$ of $2p$ shell state (Burgunder et al. 2014). The outcomes of these two experiments are used to evaluate the isospin and density dependence of the spin-orbit term in various models.

The bubble effect in lighter nuclei is significantly contributed by the zero orbital angular momentum ($l = 0$) state, i.e. s-orbital state, as the s-orbital wave function is the only non-zero wave function having a radial distribution peaked at the center of the nucleus. The higher angular momentum non-zero l wave function which is packed at the interior of the nuclei do not contribute to central density; therefore, a vacancy in s ($l = 0$) state near the Fermi surface depressed nuclear density near the center ($r = 0$) and led to depletion in central density. The single-particle shell structure has also predicted the bubble effects in ^{22}O and ^{34}Si nuclei (Grasso et al. 2009).

In RMF model calculation, the bubble effects resulted in the low angular momentum states are not favored by mean-field potential, pushing the nucleon to higher angular momentum state and hence formed a bubble which further pushed the nuclear matter towards surface; thus, the bubble formation in nuclei increases the matter radii (Shukla and Åberg 2014). In a similar study of bubble nuclei, Wang et al. (2011) suggested that weak spin-orbit pairing and small deformation are required to ensure the low occupation probability of s-orbital state. In superheavy nuclei, the stability of nuclei is due to quantal shell effect, while the bubble effect arises due to strong Coulomb repulsion between nucleons, which pushed nucleons from interior to outer surface of nuclei. The single-particle level structure near the Fermi energy surface controls the density fluctuation; therefore, the depletion in central density is also the function of shell effect. The nuclear density is approximated by electron scattering in the case of proton density, and for neutron density, it is measured by parity violation in superheavy nuclei. The density distribution of nuclei is the direct tool to quantify the nuclear size evaluated by the root mean square radii of neutron and proton, i.e. r_n and r_p. The nuclear structure can be identified by neutron skin. Pei et al. (2005) measured nuclear skin in superheavy nuclei, ^{208}Pb, as 0.15, which was further increased by asymmetry.

12.6   Anti-bubble Effect in Magic Nuclei

Saxena et al. (2019) studied the anti-bubble effect in conventional magic nuclei in neutron numbers N = 8, 20, 28, 40, 50, 82, 126, 164, 184, 226 and proton numbers Z = 8, 20, 28, 40, 50, 82 using the relativistic mean-field and BCS approach in ground-state temperature ($T = 0$ K) and excited-state temperature (T equivalent to 3 MeV). They showed that the bubble effect decreases with an increase in temperature, and at transition temperature (T equivalent to 3 MeV), the bubble effect in central density disappears. The phenomenon of the anti-bubble effect is similar to the BCS theory of superconductivity, in which at the transition temperature, the superconducting state transforms to the normal state. In some of the nuclei, the superfluid behavior of nuclei vanishes at $T = 3$ MeV and behave as normal fluid nuclei. The comparative results for conventional magic nuclei are shown in Tables 12.4 and 12.5 (Saxena et al. 2019).

TABLE 12.4

Depletion Fractions Observed in Various Models, Namely, NL3*, NL-SH, PK1, and TMA for Conventional Magic Bubble Nuclei for Proton

Nuclei	Bubble Effect Observed in Proton No.	Depletion Fraction (F): $F = \frac{\rho_{max} - \rho_c}{\rho_{max}} \times 100$			
$_Z X^A$	Z	NL3*	NL-SH	PK1	TMA
$_8 O^{22}$	8	17.16	32.64	26.91	31.37
$_{20} Ca^{34}$	20	34.51	40.32	39.20	40.61
$_{28} Ni^{80}$	28	14.18	14.33	14.22	14.22
$_{40} Zr^{90}$	40	15.12	19.45	13.84	13.38
$_{50} Sn^{106}$	50	20.36	20.33	20.46	20.42
$_{82} Pb^{240}$	82	15.12	19.34	14.70	18.64

TABLE 12.5

The Same as Table 12.2, but for Neutron

Nuclei	Bubble Effect Observed in Proton No.	Depletion Fraction (F): $F = \frac{\rho_{max} - \rho_c}{\rho_{max}} \times 100$			
$_Z X^A$	N	NL3*	NL-SH	PK1	TMA
$_{14} Si^{22}$	8	7.45	15.30	10.10	14.38
$_{14} Si^{34}$	20	20.10	24.91	23.63	25.32
$_{18} Ar^{46}$	28	20.30	37.67	33.11	42.00
$_{16} S^{56}$	40	14.58	31.13	26.32	29.63
$_{18} Ar^{58}$	40	3.79	23.50	18.32	26.03
$_{58} Ce^{184}$	126	14.47	14.03	14.40	14.39
$_{87} Fr^{251}$	164	22.50	34.16	23.59	36.45
$_{115} Mc^{299}$	184	17.93	21.23	21.39	20.57
$_{118} Og^{302}$	184	15.62	18.82	18.01	18.07
$_{120}{}^{292}$	172	15.36	15.89	15.96	17.77
$_{113} Nh^{341}$	228	19.24	17.31	21.17	20.31
$_{119}{}^{347}$	228	20.04	21.22	22.38	21.78

12.7 Summary and Conclusion

In this chapter, we have discussed the bubble nuclei on the basis of the various models used in theoretical study, namely, non-relativistic and relativistic mean-field models with different parameter sets in lighter, heavier, and superheavy mass regions. We have shown that bubble effects in lighter nuclei ^{22}O, ^{34}Si, and ^{46}Ar are due to unoccupied or partially filled s-orbital, i.e. shell effect is responsible, and in heavy nuclei, due to inversion of s-orbital state in higher orbital state, i.e. tensor effect. We have also discussed the deformation effect on bubble structure and, in superheavy nuclei the Coulomb repulsion forces the bubble type structure to be observed for ^{294,302,326}Og, ^{120,240,292}120. We have also discussed the effect of temperature (anti-bubble), shell structure, spin-orbit coupling, deformation, and pairing forces. The bubble effect in axially deformed, i.e. prolately (^{24}Ne) and oblately deformed (^{32}Si, ^{32}Ar), nuclei are also discussed. The phenomena of bubble nuclei are unique, and many more theoretical and experimental observations are underway to explore the potential bubble nuclei over the entire nuclear landscape.

Acknowledgments

The research scholar (Akhilesh Kumar Yadav) is very thankful to CSIR-UGC for providing financial support under the Junior Research Fellowship Scheme (Ministry of Human Resource Development Group) and also thankful to Rajiv Gandhi Institute of Petroleum Technology, Jais, Amethi, U.P., India for providing a platform to carry out the research work.

References

[1] Bender, M., Heenen, P., & Reinhard, P.(2003). Self-consistent mean-field models for nuclear structure, *Rev. Mod. Phys.75*(January), 121–180. doi:10.1103/RevModPhys. 75.121

[2] Bulgac, A. (2003). Neutron stars, bubble nuclei, and quantum billiards, AIP Conference Proceedings *597*, 415–417. doi:10.1063/1.1427491

[3] Burgunder, G., Sorlin, O., Nowacki, F., Giron, S., Hammache, F., Moukaddam, M., ... Thomas, J. C. (2014). Experimental Study of the Two-Body Spin-Orbit Force in Nuclei, *042502*(January), 1–5. doi:10.1103/PhysRevLett.112.042502

[4] Davies, K. T. R., Wong, C. Y., & Krieger, S. J. (1972). Hartree-Fock calculations of bubble nuclei. *Physics Letters B, 41*(4), 455–457. doi:10.1016/0370-2693(72)90673-9

[5] Duguet, T., Somà, V., Lecluse, S., Barbieri, C., & Navrátil, P. (2017). Ab initio calculation of the potential bubble nucleus Si 34. *Physical Review C, 95*(3), 1–17. doi:10.1103/PhysRevC.95.034319

[6] Grasso, M., Gaudefroy, L., Khan, E., Nikšić, T., Piekarewicz, J., Sorlin, O., ... Vretenar, D. (2009). Nuclear "bubble" structure in Si34. *Physical Review C – Nuclear Physics, 79*(3), 1–8. doi:10.1103/PhysRevC.79.034318

[7] Greiner, W. (2002). Superheavies and beyond. *Acta Physica Hungarica New Series Heavy Ion Physics, 16*(1–4), 3–17. doi:10.1556/APH.16.2002.1-4.2

[8] Hamamoto, I., & Sagawa, H. (2004). Weakly bound p 3/2 neutrons and spin-response function in the many-body pair correlation of neutron drip line nuclei, *034317*, 1–8. doi:10.1103/PhysRevC.70.034317

[9] Hess, P. O., Stöcker, H., Mişicu, Ş., & Mishustin, I. N. (2018). The Fullerene-like Structure of Superheavy Element Z = 120 (Greinerium) – A Tribute to Walter Greiner. *Walter Greiner Memorial Volume*, 263–274. doi:10.1142/9789813234284_0018

[10] Kay, B. P., Hoffman, C. R., & Macchiavelli, A. O. (2017). Effect of Weak Binding on the Apparent Spin-Orbit Splitting in Nuclei, *182502*(November), 1–4. doi:10.1103/PhysRevLett.119.182502

[11] Khan, E., Grasso, M., Margueron, J., & Van Giai, N. (2008). Detecting bubbles in exotic nuclei, *800*, 37–46. doi:10.1016/j.nuclphysa.2007.11.012

[12] Klüpfel, P., Reinhard, P. G., Bürvenich, T. J., & Maruhn, J. A. (2009). Variations on a theme by Skyrme: A systematic study of adjustments of model parameters. *Physical Review C – Nuclear Physics, 79*(3), 1–23. doi:10.1103/PhysRevC.79.034310

[13] Krishnan, R. M., & Pu, W. W. T. (1973). Bubble nuclei. *Physics Letters B, 47*(3), 225–226. doi:10.1016/0370-2693(73)90715-6

[14] Kumawat, M., Saxena, G., Agrawal, B. K., & Aggarwal, M. (2017). Proton bubble and robustness of neutron skin thickness, *Proceedings of the DAE symp. on Nucl. Phys. 62*, 218–219.

[15] Lalazissis, G. A., Karakatsanis, K., Ring, P., & Litvinova, E. (2017). Density and Isospin Dependence of Spin-Orbit Splittings in N = 20 Nuclei within Relativistic Mean-Field Models, Bulg. J. Phys. *44(4)*, 334–344.

[16] Mutschler, A., Lemasson, A., Sorlin, O., Bazin, D., Borcea, C., Borcea, R., . . . Wimmer, K. (2017). A proton density bubble in the doubly magic 34 Si nucleus. *Nature Physics, 13*(2), 152–156. doi:10.1038/nphys3916

[17] Nazarewicz, W. (2018). The limits of nuclear mass and charge. *Nature Physics, 14*(June), 537–541. doi:10.1038/s41567-018-0163-3

[18] Pei, J. C., Xu, F. R., & Stevenson, P. D. (2005). Density distributions of superheavy nuclei. *Physical Review C – Nuclear Physics, 71*(3), 1–7. doi:10.1103/PhysRevC.71.034302

[19] Phuc, L. T., Hung, N. Q., & Dang, N. D. (2018). Bubble nuclei within the self-consistent Hartree-Fock mean field plus pairing approach. *Physical Review C, 97*(2), 1–13. doi:10.1103/PhysRevC.97.024331

[20] Saxena, G., Kumawat, M., Kaushik, M., Jain, S. K., & Aggarwal, M. (2019). Bubble structure in magic nuclei. *Physics Letters, Section B: Nuclear, Elementary Particle, and High-Energy Physics, 788*, 1–6. doi:10.1016/j.physletb.2018.08.076

[21] Schuetrumpf, B., Nazarewicz, W., & Reinhard, P. G. (2017). Central depression in nucleonic densities: Trend analysis in the nuclear density functional theory approach. *Physical Review C, 96*(2), 1–6. doi:10.1103/PhysRevC.96.024306

[22] Shukla, A., & Åberg, S. (2014). Deformed bubble nuclei in the light-mass region. *Physical Review C – Nuclear Physics*, *89*(1), 1–5. doi:10.1103/PhysRevC.89.014329

[23] Siemens P. J. and Bethe, H. A. (1967). Unpublished Notes, *Phys. Rev. Lett. 18*, 704.

[24] Singh, S. K., Ikram, M., & Patra, S. K. (2013). Ground state properties and bubble structure of synthesized superheavy nuclei. *International Journal of Modern Physics E*, *22*(1). doi:10.1142/S0218301313500018

[25] Sorlin, O., & Porquet, M. G. (2013). Evolution of the N = 28 shell closure: A test bench for nuclear forces. *Physica Scripta*, *014003*(T152). doi:10.1088/0031-8949/2013/T152/014003

[26] Systematic study of bubble nuclei in relativistic mean field model. (2016). *Physics of Atomic Nuclei*, *79*(1), 11–20. doi:10.1134/S1063778816010191

[27] Vautherin, D., & Brink, D. M. (1972). Hartree-Fock Interaction with Skyrme's Interaction. I.Spherical Nuclei. *Physical Review C – Nuclear Physics*, *5*(3). doi:10.1103/PhysRevC.5.626

[28] Wang, Y. Z., Gu, J. Z., Zhang, X. Z., & Dong, J. M. (2011). Tensor effect on bubble nuclei. *Chinese Physics Letters*, *28*(10), 2–5. doi:10.1088/0256-307X/28/10/102101

[29] Wilson, H. A. (1946). A Sphericl Shell Nuclear Model. *Phys. Rev.*, *69*-538. doi:10.1103/PhysRev.69.538

[30] Wong, C. Y. (1972). Hartree-Fock calculations of bubble nuclei, *Physics Letters B*, *41*(4), 451–454. doi:10.1016/0370-2693(72)90672-7

[31] Wong, C. Y. (1973). Toroidal and spherical bubble nuclei. *Annals of Physics*, *77*(1–2), 279–353. doi:10.1016/0003-4916(73)90420-X

[32] Yao, J. M., Zhou, E. F., & Li, Z. P. (2015). Beyond relativistic mean-field approach for nuclear octupole excitations. *Physical Review C – Nuclear Physics*, *92*(4), 1–5. doi:10.1103/PhysRevC.92.041304

13

Correlation of Nuclear Structure Observable with the Nuclear Reaction Measurable in the Aspect of Astrophysical P-Process

Awanish Bajpeyi, Ajeet Singh, and Amritanshu Shukla

Rajiv Gandhi Institute of Petroleum Technology

CONTENTS

13.1 Introduction

When we look at the sky in the night, the stars we see are within our own galaxy. But there are some fuzzy patches too, and these are other galaxies like our own—they are much and much further away than the stars. Almost all of these galaxies are moving away from us—some at speeds of several hundreds or even thousands of kilometers per second. It means that the universe is expanding, and if so, then in the past, it must have been much smaller. Rigorous studies on the subject suggest that there was a moment when all the matter in the universe was packed into a point and expanded outwards. That moment is termed as the Big Bang, which happened about 14 billion years ago! It is largely agreed through such astronomical studies that about 14 billion years ago, matter and energy, time and space came into existence through a big explosion, and about 300,000 years after it, matter and energy started to coalesce into complex structures called atoms. These atoms or chemical elements have played a very important role in human life. We are surrounded by these in the day-to-day life in and out. The periodic table, extended recently, now accommodates elements up to $Z = 118$, containing the basic constituents of matter, with which the world around us is constructed. However, when the universe began with a Big Bang, it started out with no elements at all, and many of the elements that make up the Earth and the people on it today were created in the nuclear furnaces inside stars and were released once the star reached the end of its life. In fact, the ordinary matter in our universe (known as baryonic matter) is made up of only 94 naturally occurring elements (rest are artificially synthesized in the lab), the familiar beasts of the periodic table. Moreover, approximately 73% of the mass of the visible universe is in the form of hydrogen. Helium makes up about 25% of the mass, and everything else represents only about 2%. While the abundance of these more massive ("heavy", $A > 4$) elements seem to be quite low, it is important to remember that most of the atoms in our bodies and the Earth are a part of this small portion of the matter of the universe only. The birth, life, and death of a star are described in terms of nuclear reactions. The chemical elements that make up the matter we observe throughout the universe were created in these reactions. Detailed understanding of the formation and abundance of these nuclei remains a standing challenge and is quite significant for our understanding of nucleosynthesis and nuclear astrophysics in general and particularly for the formation of heavy nuclei, ranging from neutron drip line to proton drip line.

13.1.1 Nucleosynthesis: What Is the Origin of Chemical Elements? How Are They Formed?

The process of formation of nuclei in the universe from pre-existing nuclei is called nucleosynthesis. Broadly, nucleosynthesis can be categorized into Big Bang nucleosynthesis, stellar nucleosynthesis, supernova explosion, and cosmic-ray spallation. The origin of all the naturally occurring elements falls into two phases: Big Bang or primordial nucleosynthesis (i.e. the origin of the "light" elements) and the stellar nucleosynthesis (i.e. the origin and production of the "heavy" elements) [1]. When astronomers refer to the "light elements", they refer mainly to hydrogen and helium and their isotopes. Nuclei of hydrogen (protons) are believed to have formed as soon as the temperature had dropped enough to make the existence of free quarks impossible. The low-mass elements, hydrogen, and helium, were produced in the hot, dense conditions of the birth of the universe itself. Big Bang nucleosynthesis refers to the process of element production during the early phases of the universe, shortly after the Big Bang. It is thought to be responsible for the formation of hydrogen (H) and its isotope deuterium ^{2}H, helium (He) and its varieties ^{3}He and ^{4}He, and lithium (Li) and its isotope of ^{7}Li.

At the beginning of the 20th century, Rutherford summarized the role of nuclear reactions in the formation of heavier elements in the energy production of stars. Afterwards, the major breakthrough was made by Gamow in 1928 which made it possible to understand reactions much better, as he formulated the quantum mechanical tunneling effect to describe the occurrence of nuclear reactions at energies lower than the coulomb barrier. Renowned physicists George Gamow, Enrico Fermi, Edward Teller, Maria Mayer, and many others of that era suggested that neutron capture with primordial nucleons, electron, and radiation is responsible for the formation of all the nuclei. But this hypothesis could not work because if all the nuclei do not have the same abundance, then how their origins can be the same? This leads to abandoning the said hypothesis. At the same time, George Gamow and his team continued the research to find the processes of formation of nuclei and came up with the idea of hot Big Bang universe hypothesis, which is very popular now. Fred Hoyle in 1946 came up with a different approach to resolve the problem, and his finding was the real beginning of the stellar nucleosynthesis. He mentioned that there is a peak in the solar system abundance curve near iron and proposed that the process responsible for the origin of nuclei exists in a very high temperature $\sim3 \times 109$ K, which means there is a connection of origin of nuclei with supernovae and stars. Hence, all the nuclei were made up of the stars. Until then, understanding the formation of elements, especially heavier ones, and their abundance was not very clear and was one of the most puzzling problems among the researchers involved in astronomy and astrophysics. The subject gained critical momentum with the remarkable work of Burbidge et al. [2] and Cameron [3]; in these two separate but almost simultaneous works, they postulated a series of energy-generating processes in stars for explaining the origin of chemical elements, and these research papers explained how much stars could contribute to the synthesis of nuclei heavier than hydrogen. E. M Burbidge, G. R. Burbidge, W. A. Fowler, and F. Hoyle (BBFH) published a paper titled "Synthesis of elements in stars", in which they cleared the picture for the evolution of the nuclei and recommended a series of processes that are responsible for the origin of nuclei. They suggested after the initial couple of seconds, fusion and fission processes happened including eight processes, i.e. H burning, He burning, α, e, x, r, s, and p processes, and these processes were supposed to produce almost all known elements in stars. They described the formation process for the elements heavier than the above-said lightest elements in a proper manner. This was the beginning of production of the elements in stars. If this does not happen, today there would be no heavier elements. After 40 years, Wallerstein et al. [4] updated this work and included the available current knowledge about nucleosynthesis. For the sake of completeness, we give an outline of these processes below in brief. Since the pioneering work of BBFH in 1957, the field has flourished tremendously and great achievements have been made, especially in past few decades, particularly due to consistent experimental advancements related to radioactive ion beams.

13.1.1.1   Nucleosynthesis of Elements $Z > 26$

As discussed, the synthesis of elements up to iron is possible by thermonuclear fusion reactions and quasi-equilibrium processes that happen during either explosive phases or phases of stellar evolution with consecutive burning phases via a number of charged particle reactions. For heavier nuclei ($Z > 26$), the situation is quite different, because fusion process cannot take place due to enhancement in the height of coulomb barrier, as one moves to higher atomic numbers. So, cross sections for such fusion nuclear reactions are expected to be very small for the nuclei above iron, and one cannot explain the observed abundance of such chemical elements.

The formation of heavy nuclei beyond iron is possible only via photodisintegration processes which occur at the very high temperature in the order of Giga Kelvin. BBFH and Cameron suggested three distinct nucleosynthesis processes that are responsible for the formation of heavy nuclei beyond iron: the rapid neutron capture process (r-process), slow neutron capture process (s-process), and proton capture process (p-process). The nuclei formed by these processes are referred to as r-nuclei, s-nuclei, and p-nuclei, respectively [5]. Most of the nuclei heavier than iron are formed by s- and r-process. The p-nuclei, however, cannot be produced by neutron capture reactions, and it was supposed that these are produced by proton capture process. It is important to note that the natural abundances of these p-nuclei are much less as compared to other isotopes produced by neutron capture processes. In following subsections, we will give a brief description of these processes for the sake of the continuity and completeness of the discussion.

The r-process

The r-process is a nucleosynthesis process that produces elements above iron and is responsible for the formation of about half of the nuclei above iron. The conditions necessary for the process are higher temperature ($\sim$1–3 $\times$ 109 K), higher neutron density ($\geq$1,020 cm^{-3}), and smaller time scales (0.01–10 seconds) that are less than the mean time for β-decay ($\sim$1 s) [6], host for the cause of supernova explosions of type II. The exact environment for this process, however, is still unconfirmed. It is assumed that the r-process takes place between the surface of the neutron star and the shock wave that propagates outward [7], and in this region, the entropy is very high. Due to this high entropy, most nucleons are in the form of free neutrons or bound into α-particles rather than heavy nuclei. This mode of synthesis is responsible for the production of a large number of isotopes in the range of 70 < 3 < 209, and also for the synthesis of uranium and thorium. This process may also be responsible for some light element synthesis, e.g., S^{36}, Ca^{46}, Ca^{48}, and perhaps Ti^{47}, Ti^{49}, and Ti^{50}.

The S-Process

This process of neutron capture occurs with the emission of gamma radiation (n, γ) which takes place during the burning phase of Asymptotic Giant Branch (AGB) stars at relatively low temperatures ($T \sim$ 1–3 $\times$ 10^8 K) and relatively low neutron densities (10^6 cm^{-3}). This slow neutron capture takes at large time scales, ranging from 10^2 years to 10^5 years for each neutron capture [8]. Due to long time scale in slow neutron capture, the s-process path follows the valley of beta stability. The s-process is also responsible for the formation of almost half of the nuclei above iron, and it has three components: main s component, weak s component, and strong s component. The main s component happens in the environment of low-mass AGB stars at solar-like metal content and is responsible for the formation of nuclei between ^{90}Zr and ^{209}Bi. The strong s component happens in the environment of AGB stars at low metal content and is responsible for the nuclei between ^{208}Pb and ^{209}Bi. The weak s component, which is responsible for the production of nuclei between iron and yttrium (56 < A < 90), takes place during convective core-He burning in massive stars, where temperatures reach (2.2 − 3.5) $\times$ 108 K [9].

The P-Process

The thirty-five nuclei (listed in Table 13.1) lie between ^{74}Se and ^{196}Hg on the neutron-deficient side of the stability chart of nuclides, and their formation is not possible by any of the above-mentioned processes. The nuclei are supposed to be formed by the proton capture

TABLE 13.1

Comparison of the Parameters of Effective Interactions NL3 and NL3* in the RMF Theory

Parameter	NL3	NL3*
M	939 (MeV)	939 (MeV)
m_σ	508.194 (MeV)	502.574 (MeV)
m_ω	782.501 (MeV)	782.60 (MeV)
m_ρ	763.000 (MeV)	763.00 (MeV)
g_2	−10.431 (fm^{-1})	−10.809 (fm^{-1})
g_σ	10.217	10.094
g_ω	12.868	12.806
g_ρ	4.474	4.575
g_3	−28.885	−30.1486

process (p-process) and are commonly referred as p-nuclei [10]. The natural abundance of these p-nuclei is very less (order of 0.01%–1%) in comparison to the nuclei produced through neutron capture processes, with the exception of only molybdenum and ruthenium isotopes. Also, almost all the p-nuclei are even-even nuclei except the four nuclei (^{113}In, ^{115}Sn, ^{138}La, and ^{180}Ta). These even-even nuclei are very stable due to pairing force, and the only decay mode of these is β^+ β^+, β^+EC, $ECEC$ decay which also has very large half-lives.

To understand the formation of p-nuclei, information obtained from stellar spectra, meteoritic data, and nuclear reaction experiments, as well as nuclear reaction calculations, is put together, and it has been found that the abundance curve of p-nuclei is almost parallel to the r- and s-nuclei. This suggests that r- and s-process seeds are responsible for the origin of p-nuclei but fail to produce p-nuclei in various mass ranges [11]. The possible sites and scenarios for the production of p-nuclei were first proposed by BBFH and Cameron. Additionally, alternative sites have also been proposed for the evidence related to the formation of these nuclei. These studies proposed that photodisintegration is an alternative way to produce p-nuclei and it plays a very important role in the synthesis of p-nuclei either by directly producing them through devastation of their neutron-rich neighbor isotopes through sequences of (γ, n) reactions or predominant photodisintegration process occurring via heavier, unstable nuclides by (γ, p) or (γ, α) reactions and subsequent beta decays. There are three main constituents that are proposed as the scenarios for the p-process. The first one is the time scale (temperature variation as a function of time) of the process and the peak temperature; the second one is proton density as proton captures can prevail over (γ, p) reactions at high temperatures; and the last one is seed abundances, i.e. the number and composition of nuclei on which photodisintegrations act ab initio [12]. Recently, two alternative processes, rp-process and vp-process, have also been suggested for synthesis of p-nuclei. The rp-process is supposed to occur in a hydrogen-rich environment at high temperatures, which is responsible for the synthesis of higher mass nuclei. The vp-process proceeds in a type II supernova, and nuclei up to mass 152 are produced in these conditions. Actually, these rp- and vp-processes can synthesize only a handful of p-nuclei, and it was suggested by later studies that p-nuclei are synthesized from the devastation of s- and r-nuclei via (γ, n), (γ, p), and (γ, α) photodisintegration reactions, which take place in the neon-oxygen layers of massive stars during type II supernovae explosions at the very high temperature range of 2–3 billion degrees Kelvin. It can produce all the p-nuclei, and this is supposed that Type 2 supernovae explosions are the best candidates for the occurrence of p-process. In these reactions flow, neutron emission channels are most prominent for the abundance of p-nuclei in the neutron-deficient side, but with the increase in neutron separation energies, these (γ, n) reactions stop after five to ten steps, and then at the same time, proton and alpha reactions will start and feed another isotopic chain. After certain time, with the decrease in temperature, these reactions slow down and β+ decays of unstable nuclei take over the whole process. Finally, photodisintegration terminates quickly, and all unstable nuclei achieve stability [11]. So, an extended network of nuclear reactions for stable and unstable nuclei is required for the p-process modeling. Approximately, twenty thousand nuclear reactions on about two thousand nuclei are part of this network, and the major nuclear reactions involved in the modeling are the (γ, n), (γ, p), and (γ, α) reactions, their inverse capture reactions, and β+decays [12]. Despite the extensive studies, it is yet to be confirmed about the exact site and scenario for the production of the said 35 p-nuclei.

13.1.2 Abundance of Elements

To date, approximately four thousand possible isotopes of nuclei are present in the universe in which nearly three hundred are stable nuclei and remaining are unstable [13]. In Figure 13.1, a schematic representation of the nuclear landscape (chart of nuclides) has

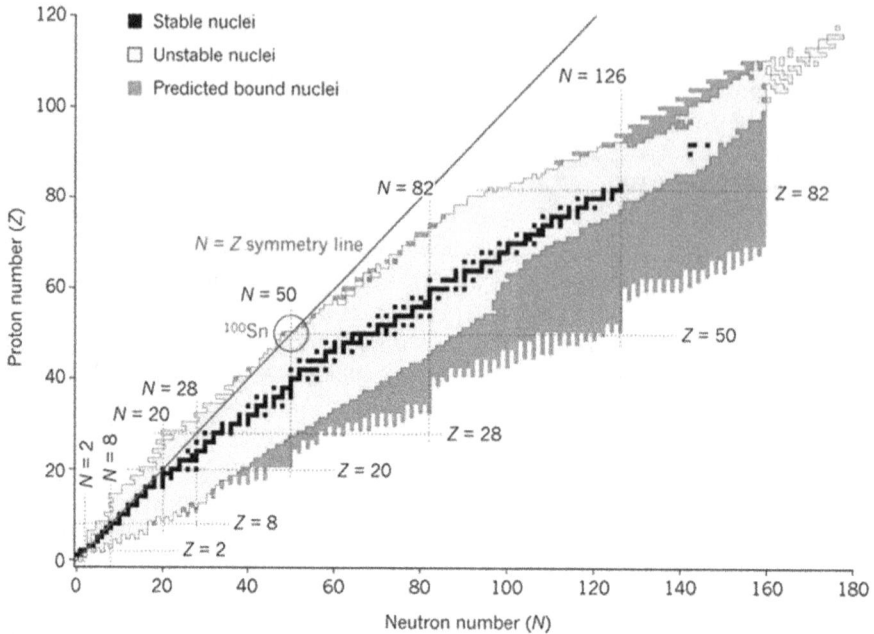

FIGURE 13.1
Nuclear chart with proton numbers in the *y*-axis and neutron numbers in the *x*-axis. Stable nuclei are indicated with black boxes. The yellow shed covers the known nuclei, and the green shed covers the unknown nuclei [14].

been given. Every nucleus with its unique proton and neutron numbers occupies a spot on the nuclear landscape. Here, the region where stable nuclei (in black squares) lie is called the valley of stability. The nuclei spotted in the yellow region are unstable nuclei, and the predicted nuclei that have never been observed are shown in the green region.

In order to understand the formation of elements, it is necessary to know in which condition the elements exist in the universe, and it is also important to know their relative quantity with each other. In addition to the formation, it is also quite crucial to know how much abundant these nuclei are? The abundance of a nucleus is driven by the number and abundance of its stable isotopes, which in turn rely on the stability of the nuclei during thermonuclear reactions in stellar interiors.

In 1910, Harkins and Oddo predicted that nuclei with even atomic numbers are more abundant compared to their odd atomic number neighbors [15]. The abundance of elements can be described mainly in two ways: the first is solar abundance, and the second one is isotopic abundance. The solar abundances are deduced by the analysis of meteorites through Cosmo chemistry and observations of the Sun by astronomers. In the field of Cosmo chemistry, researchers determine elemental abundances by chemical analysis of meteorites. The astronomers determine the solar abundances from stars by identifying and analyzing the electromagnetic waves emitted at unique frequencies by nuclei near the stellar surface. From the studies made by researchers with respect to abundance, in the early stage, it was taken into account that only spectral lines are used to calculate the abundances with stellar atmosphere and interior models, but later studies rely upon many other sources such as solar winds, solar flares, solar energetic particles, and propagation of wave oscillations, particularly acoustic pressure waves to calculate the abundances. The isotopic abundance of a specific element is calculated using the isotopic ratios of the specific element found on the Earth. A plot of relative abundance of nuclei versus mass number has been shown in

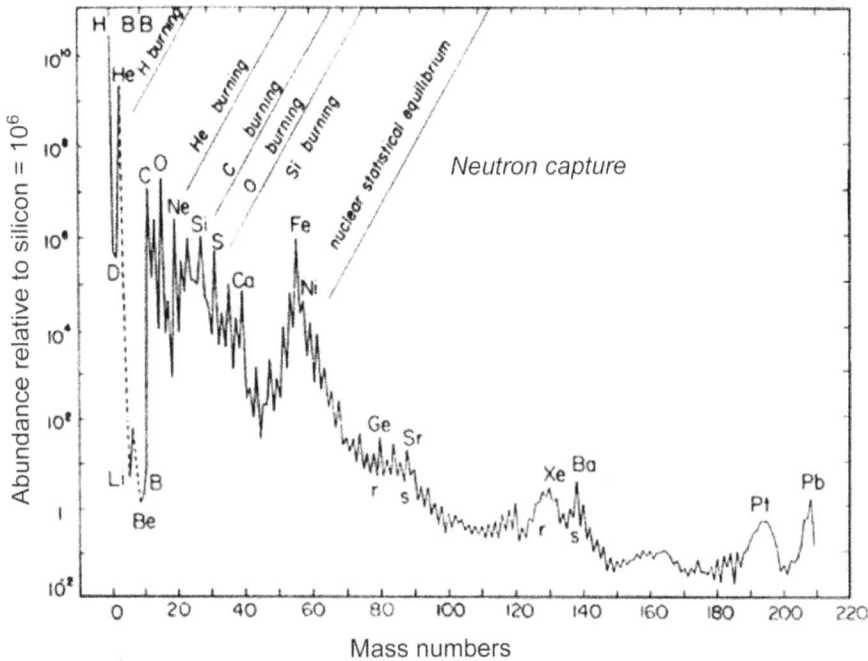

FIGURE 13.2
Abundance of nuclei in the solar system in respect of Si abundance [16].

Figure 13.2. The abundances are normalized with respect to the Si atoms. It is evident from the figure that the lightest elements H and He are the most abundant elements compared to the next lightest elements Li, Be, and B, which have low abundances in the solar system. The elements from C to Ca have a trend of decreasing abundances with increasing mass numbers, and as we move towards higher mass numbers, the abundance curve rises near the Fe region. Beyond the iron peak, again elements have a general trend of decreasing abundances with increasing mass numbers. The gradual decrement in abundance with increasing mass number makes the topic of research of the study of heavy nuclei beyond iron.

13.1.3   Theoretical Research Advancements in the Study of P-Process

The experimental cross-section data are scarce because of two major drawbacks of experiments. The first one is, in the laboratory, the target nucleus is always in its ground state, while target nucleus lies in its thermally excited state under stellar conditions, and in the case of heavy nuclei, high nuclear level density and high coulomb barriers are also the hurdles for the experimental efforts. Secondly, typical p-process reaction networks require a large network calculation involving nearly two thousand nuclei and approximately twenty thousand reactions, and most of the nuclei in this reaction network are unstable, which gives rise to many challenges that constrain the direct determination of reaction rates, i.e. tiny cross sections at astrophysically relevant energies, different sensitivities of the cross sections inside and outside the astrophysical energy window. Thus, it is very difficult to perform a large number of experiments in the laboratory to understand the p-process nucleosynthesis solely through experimental measurements. Hence, theoretical considerations are highly desired to supplement experimental studies and provide input for the network calculations to meet the shortfall of cross-section data. Even if experimental cross-section data are available, very often these data are not fully commensurable with the energy regime of the stellar condition. In such cases, cross sections obtained with theoretical models can be normalized to the available experimental data. If the uncertainty of the energy dependence of the calculated cross section is small, this method provides a safe extrapolation to relevant stellar energies from the measured energies.

In the theoretical framework, reaction cross-section results are mainly obtained by the Hauser-Feshbach [17] statistical model calculations, and these cross sections are used to calculate the astrophysical reaction rates, which are required for p-process network studies based on theoretical models. The input parameters, i.e. nuclear level density, optical potential, γ-ray strength function, and nuclear masses, energies, isospin effects and widths of the giant dipole resonances (GDR), enter as an input for the calculation of cross sections of the statistical model code. However, the particle and γ-transmission coefficients and level densities of excited states are the most important input quantities of the Hauser-Feshbach model. The theoretical calculations depend highly on the reliability and accuracy of these inputs, and hence, the uncertainty in calculated cross-section results depends on the uncertainties of these inputs. Therefore, it is very important to ensure the reliability of the model inputs by calculating as many experimental observables as possible and compare them with the experimental observations. The statistical model codes like NON-SMOKER [18] and TALYS [19] have been widely used for nuclear reaction calculations. TALYS is quite preferred due to its versatility in handling different nuclear reactions into one single software packages, and it contains many special features. Also, its source code is open for the user, and modification in any part of the code is possible. Moreover, it covers most of the major reaction mechanisms that are considered important for light particle-induced nuclear reactions, i.e. direct, pre-equilibrium, compound mechanisms to calculate the total reaction probability. It also includes the possibility of adding different inputs from varying sources (models and/or databases). Additionally, TALYS is optimized for a large projectile energy range from 1 KeV to 200 MeV, and such a large energy range for the calculation of cross-section reaction has not taken in any other code.

13.2   Formalism

13.2.1   Nuclear Structure Formalism

There are a number of theoretical approaches being currently used for solving the nuclear quantum many-body problem to calculate nuclear density/wave function, and it is a separate area of interest for nuclear physics researchers. One of the most successful approaches used for a wide range of nuclei across the nuclear mass table is the effective mean-field approach based on mean-field assumptions. Effects beyond the mean field are taken into account in an effective manner through the nucleon-nucleon interaction, whose parameters are chosen to reproduce the values of some basic observables of a wide set of nuclei. The most successful theories for the microscopic description of the ground-state properties of the finite nuclei are the conventional shell model or the Hartree-Fock calculations with effective density-dependent interactions. These theories are able to reproduce nuclear densities, binding energies for the finite nuclear matter and spherical doubly magic nuclei with proper pairing correlations in single close shell nuclei and also the proper size of ground-state deformations in doubly open-shell nuclei, which is one of the great achievements of these theories. However, in the recent past, the relativistic models have attracted much attention and describe the nucleus as a system of Dirac nucleons interacting via meson fields.

The relativistic mean-field (RMF) theory has proved to be very successful among the relativistic models to describe the ground-state properties of the nuclei over the entire range of the periodic table, from light doubly magic nuclei to medium heavy spherical superfluid nuclei and, furthermore, to heavy deformed nuclei in the rare earth and actinide regions up to superheavy nuclei, and this model has to compete with its nonrelativistic similitude, such as the Skyrme force which has been the most successful in describing nuclear structure and low-energy dynamics within the Hartree-Fock calculations. The RMF theory was first introduced as a full-fledged quantum field theory by Walecka [20]. The RMF theory is a phenomenological theory and is conceptually similar to the density-dependent Hartree-Fock theory with Skyrme forces, but the relativistic theory of nuclear matter has had some success in clearing up the long-standing problems encountered in the nonrelativistic theory, and this RMF theory works well for better results for the ground states of even-even nuclei. In the RMF models, the nuclear many-body system is described in terms of interacting baryons and mesons, and the parameters are phenomenologically fitted to the saturation properties of nuclear matter. In the past three decades, many investigations have been devoted to a relativistic description of the ground-state properties of nuclei. In the past three decades, the RMF model has been successfully used to explain the properties of finite nuclei as well as infinite nuclear matter. The ground-state properties such as binding energy, matter radii, charge radii, deformation density profile, and nuclear halo have been calculated using the RMF theory, and an excellent agreement has been found with the experimental results. The RMF method has the advantage that with proper relativistic kinematics and with the mesons and their properties already known or fixed from the properties of a few nuclei, it gives excellent results for many ground-state properties such as binding energy, root mean square (rms) radii, quadrupole and hexadecapole deformation, not only of spherical, but also of deformed nuclei lying close to the beta stability line as well as for the nuclei lying far from the stability line. The major attractive features of the RMF formalism are the spin-orbit strength, other associated spin properties, and associated nuclear shell structure automatically arising from meson-nucleon interaction [21].

The details of the RMF model and solutions to the equations for finite nuclei and nuclear matter can be found in the literature [22,23]. The nucleus is described as a system of point-like nucleons, Dirac spinors, coupled to mesons and photons in the RMF theory. The nucleons interact by the exchange of a scalar meson σ and vector particles, i.e. ω meson, ρ meson, and the photon. The scalar mesons provide a strong intermediate-range attraction between the nucleons. But for the vector particles, we have to distinguish the spatial components and time-like components. For the ω mesons, time-like components provide a very strong repulsion at short distances for all the combination of like and unlike particles (pp, nn, and pn). Similarly, for the ρ meson the time-like components provide the short-range repulsion only for like particles (pp and nn) and short-range attraction for unlike particles (pn), and with that, the spatial components of the ω and ρ mesons provide attraction for all like, unlike combinations and attractive for like but repulsive for unlike particles, respectively. The basic ingredient for the RMF is the Lagrangian density for nucleon-meson many body systems [24,25].

$$L = \overline{\Psi}_i \{i\gamma^\mu \partial_\mu - M\}\psi_i + \frac{1}{2}\partial^\mu\sigma\partial_\mu\sigma - \frac{1}{2}m_\sigma^2\sigma^2 - \frac{1}{3}g_2\sigma^3 - \frac{1}{4}g_3\sigma^4 - g_s\overline{\Psi}_i\Psi_i\sigma - \frac{1}{4}\Omega^{\mu\nu}\Omega_{\mu\nu}$$
$$+ \frac{1}{2}m_w^2 V^\mu V_\mu + \frac{1}{4}c_3(V_\mu V^\mu)^2 - g_w\overline{\Psi}_i\gamma^\mu\Psi_i V_\mu - \frac{1}{4}B^{\mu\nu}B_{\mu\nu}$$
$$+ \frac{1}{2}m_\rho^2 R^\mu R_\mu - g_\rho\overline{\Psi}_i\gamma^\mu\tau\overline{\Psi}_i R^\mu - \frac{1}{4}F^{\mu\nu}F_{\mu\nu} - e\overline{\Psi}_i\gamma^\mu\frac{(1-\tau_{3i})}{2}\Psi_i A_\mu \quad (13.1)$$

The field for the σ meson is denoted by σ, that for the ω meson by V_μ, and for the isovector ρ meson by R_μ. A^μ denotes the electromagnetic field. ψ_i is the Dirac spinors for the nucleons whose third component of isospin is denoted by τ_{3i}. Here g_s, g_w, g_ρ, and $e^2/(4\pi) = 1/137$ are the coupling constants for σ, ω, and ρ mesons. M is the mass of the nucleon, and m_σ, m_ω and m_ρ are the masses of the σ, ω, and ρ mesons, respectively. $\Omega^{\mu\nu}, B^{\mu\nu}$, and $F^{\mu\nu}$ are the field tensors for the V^μ, R^μ and the photon fields, respectively.

13.2.2 Pairing Approach in RMF

Pairing is a very crucial quantity for open shell nuclei in determining the nuclear properties, though it does not contribute significantly to the lighter-mass region. The constant gap, BCS-pairing approach is reasonably applicable in this mass region. In order to take care of the pairing effects in the present work, we use the constant gap for proton and neutron, as given in [26,27] $\Delta_p = RB_s e^{sI-tI^2}/Z^{1/3}$ and $\Delta_n = RB_s e^{-sI-tI^2}/A^{1/3}$ with $R = 5.72$, $s = 0.118$, $t = 8.12$, $B_s = 1$, and $I = (N - Z)/(N + Z)$. This type of prescription for pairing effects, both in RMF and Skyrme-based approaches, has already been used by many authors [28,29].

13.2.3 RMF Parametrization

There are a number of parameter sets for solving the standard RMF Lagrangians. We have used the NL3* parameter set, which is a slightly modified version of the NL3 parameter set. For the RMF results of the nuclear structure, the Lagrangian parameters are usually obtained by a fitting procedure to some bulk properties of a set of spherical nuclei. The NL1 and NL-SH were the most frequently used parameter sets among the existing parametrization.

These parameter sets gave good results for binding energies, charge radii, etc. for most of the cases along the beta stability line. However, the results were less satisfactory for the nuclei lying away from beta stability. So, a new parametrization NL3 was proposed to solve effective non-linear Lagrangian density of the RMF theory. It provided excellent results not only for the nuclei lying along the stability line but also for the nuclei far away from the stability line. However, with increasing experimental data getting accumulated, the NL3 parameter set also could not suffice specially in describing some properties of Hg and Pb isotopes. Recently, Lalazissis et al. proposed a new parameter set by improving NL3 which was obtained through a new global fit of ground-state properties of spherical nuclei and infinite nuclei matter. The new parametrization obtained in this fit is termed as NL3*, and it contains six phenomenological parameters. It is able to improve the description of nuclear masses, correct nuclear sizes, and overcome the shortcomings observed previously with the NL3 force. A calculation performed with NL3* provides results for the ground-state properties that match excellently with experiments and have been used in a number of studies [29,30]. A comparison between the NL3 and NL3* parameters is given in Table 13.1.

13.2.4 Nuclear Reaction Formalism

Astrophysical reaction rates involving charged particles are based mainly on theoretical cross sections obtained from the Hauser-Feshbach statistical model calculations. The ground of the Hauser-Feshbach formalism is the assumption that the nuclear reactions are taking place through the compound nucleus state, and in general, nucleosynthesis of the elements heavier than $A \sim 25$ happens via the formation of compound nucleus state.

13.2.4.1 Reaction Cross Section

The cross section (σ) is the measurement of probability that a nuclear reaction will occur, and it is defined as follows:

$$\sigma = \frac{(N_R/t)}{[N_b/(tA)]N_t} \tag{13.2}$$

where N_R/t = number of interactions per time, N_b/tA = number of incident particles per area per time, and N_t = number of target nuclei within the beam.

Simply, an interaction area between the incident particle and target nucleus can be termed as a cross section. Consider a nuclear reaction $i^\mu(j,o)m^\nu$ and $\sigma^{\mu\nu}(E_{ij})$ is the cross section of the reaction from the target state i^μ to the excited state m^ν of the final nucleus which can be written as follows:

$$\sigma^{\mu\nu}(E_{ij}) = \frac{\pi\hbar^2/(2\mu_{ij}E_{ij})}{(2J_i^\mu + 1)(2J_j + 1)} \sum_{J,\pi}(2J+1) \tag{13.3}$$

$$\times \frac{T_j^\mu(E,J,\pi,E_i^\mu,J_i^\mu,\pi_i^\mu)T_o^\mu(E,J,\pi,E_m^\nu,J_m^\nu,\pi_m^\nu)}{T_{tot}(E,J,\pi)}$$

where J denotes the angular momentum, E_{ij} is the corresponding excitation energy, and π is the parity of excited states. Equation (13.3) is called the Hauser-Feshbach equation. The total transmission coefficients $T_{tot} = \sum_{v,o}T_o^\nu$ describe the transitions to all possible bound and unbound states ν in all energetically accessible exit channels o including the entrance channel i. If the target states μ in an astrophysical plasma of temperature T^* are thermally populated, the astrophysical cross section $\sigma*$ is given by

$$\sigma*(E_{ij}) = \frac{\Sigma_\mu(2J_i^\mu + 1)\exp(-E_i^\mu/kT^*)\Sigma_\nu\sigma^{\mu\nu}(E_{ij})}{\Sigma_\mu(2J_i^\mu + 1)\exp(-E_i^\mu/kT^*)} \tag{13.4}$$

where k is the Boltzmann constant. The summation over ν replaces $T_o^\nu(E,J,\pi)$ in Eq. (13.3) by the total transmission coefficient

$$T_o(E,J,\pi) = \sum_{\nu=0}^{\nu_m}T_o^\nu(E,J,\pi,E_m^\nu,J_m^\nu,\pi_m^\nu) \tag{13.5}$$

$$+ \int_{E_m^{\nu_m}}^{E-S_{m,o}} \sum_{J_m,\pi_m} T_o(E,J,\pi,E_m,J_m,\pi_m)\rho(E_m,J_m,\pi_m)dE_m$$

where $S_{m,o}$ is the channel separation energy and the summation over excited states above the highest experimentally known state ν_m are changed to integration over the level density ρ. The important ingredients for the statistical model calculations are the particle and gamma transmission coefficients and the level density of excited states. Hence, the reliability of statistical model calculations depends on the accuracy with the ingredients for such calculation can be evaluated. Equation (13.5) is the total transmission coefficient, which is dependent on radiative transmission coefficients, and nuclear level density. Using the Schrodinger equation, particle transmission coefficient can be determined by solving the Schrodinger equation for optical nuclear potential $U(r)$. This can be solved via radial equation:

$$\left[\frac{\hbar^2}{2\mu}\left(\frac{d^2}{dr^2} + \frac{l(l+1)}{r^2} + \frac{Z_1Z_2e^2}{r} + U(r)\right)\right]\phi_l(r) = E\phi_l(r) \tag{13.6}$$

13.3   Results and Discussion

We have calculated important reaction observables that are used for the astrophysical models [11]. The study of molybdenum isotopes is essential from the nuclear structure perspective, as ^{92}Mo ($Z = 42$, $N = 50$) contains the magic neutron number and a number of protons are also near semi-closed shell. We have performed the detailed nuclear structure calculations for the nuclei relevant to the proton and alpha capture reactions for Mo isotopes. To ensure the model prediction reliability regarding shape and density, we have studied ground-state properties (binding energy, rms matter radii, and rms charge radii), coherently with the nuclear reaction rates.

13.3.1   Ground-State Properties: Nuclear Densities, Binding Energy, rms Matter Radii, and rms Charge Radii

In Table 13.2, we present the calculated values along with experimental values, if available, for ground-state properties namely, the binding energy per nucleon, neutron radii, proton radii, rms matter radii, and charge radii for all the nuclei under consideration to have a comparable look. The rms charge radius (r_c) is obtained from the point proton rms radius through the following relation [25]:

$$r_c = (r_p^2 + 0.64)^{1/2}$$

considering the size of the proton radius as 0.8 fm. It can be noticed from Table 13.1 that in some cases, values of binding energy, charge radius, and deformation calculated in the present work for ^{92}Mo, ^{95}Tc, ^{99}Rh, ^{98}Mo, and ^{104}Ru match perfectly with the experimental results, and overall our calculated results for ground-state properties account fairly well in comparison to the experimental/theoretical results, as available. It is important to note here that since the charge radius is obtained from the density profile (as shown in Figures 13.3 and 13.4) and our RMF results for r_c match excellently with the experimental results, we can reliably comment on the density profiles as well as density-dependent properties.

The $A \sim 100$ mass region provides a unique opportunity from the nuclear structure perspective as the interplay between single-particle and collective degrees of freedom leads to shape phase transitions along isotopic and isotonic chains giving rise to the possibility of nuclear shape coexistence. In Mo isotopes particularly, starting from $N = 50$, the nuclear shape gradually evolves from being spherical, and driven by the enhanced proton-neutron residual interaction, sets to a large deformation at $N = 60$. $_{42}$Mo$_{56}$ lying in between the chain is quite interesting, where proton cross-shell excitations from the $Z = 28$–40 pf shell to the $g_{9/2}$ orbital play an important role to make it a potential case for shape coexistence. This specific behavior of ^{98}Mo is reflected in nuclear density distribution and also neighboring nuclei ^{99}Tc, for which density distribution is slightly different from the other nuclei studied in the present work. In Figures 13.3 and 13.4, we present neutron and proton linear density plots for all the nuclei under consideration in this chapter. In Figure 13.3a and b, we present comparative results of ^{92,94,98}Mo for proton density distribution and neutron density distribution, respectively. In the case of ^{92}Mo and ^{94}Mo, one can clearly see from the density plots (Figure 13.3a, b) that there is a consistent region of density depletion at the center for protons as well as neutrons, while ^{98}Mo density distribution is different due to the above-said reason. In Figure 13.3c and d, we present the comparative results of ^{93,95,99}Tc and ^{102}Ru for proton density distribution and neutron density distribution, respectively. In the case of ^{93}Tc and ^{95}Tc, one can clearly see from the density plots (Figure 13.4c and d) that there is a consistent region of density depletion at the center

TABLE 13.2

Comparison of Calculated Ground-State Properties for ^{92,94,98}Mo, ^{96,98,104}Ru, ^{93,95,99}Tc, ^{97,99,105}Rh, ^{102}Ru, and ^{100,102,108}Pd with the Data Available

Nucleus	B.E./A (in MeV)		r_{rms} (in fm)		r_{charge} (in fm)		Deformation	
	Calc.	Expt.	Calc.	Expt.	Calc.	Expt.	Calc.	Expt.
^{92}Mo	−8.657	−8.657	4.284	—	4.318	4.316(0.001)	0.000	0.109(0.003)
^{94}Mo	−8.640	−8.662	4.327	—	4.333	4.352(0.001)	0.000	0.151(0.002)
^{98}Mo	−8.594	−8.635	4.439	—	4.406	4.409(0.001)	0.224	0.168(0.001)
^{96}Ru	−8.595	−8.609	4.344	—	4.375	4.393(0.004)	0.003	0.154(0.002)
^{98}Ru	−8.597	−8.620	4.401	—	4.412	4.423(0.005)	0.155	0.205(0.015)
^{104}Ru	−8.538	−8.587	4.533	—	4.484	4.510(0.002)	0.222	0.274(0.002)
^{93}Tc	−8.627	−8.608	4.294	—	4.340	—	0.009	—
^{95}Tc	−8.620	−8.622	4.335	—	4.354	—	0.009	—
^{99}Tc	−8.595	−8.613	4.444	—	4.424	—	0.212	—
^{97}Rh	−8.567	−8.560	4.353	—	4.396	—	0.003	—
^{99}Rh	−8.579	−8.580	4.408	—	4.431	—	0.147	—
^{105}Rh	−8.545	−8.573	4.535	—	4.498	—	0.200	—
^{102}Ru	−8.570	−8.607	4.493	—	4.465	4.482(0.002)	0.219	0.170(0.002)
^{100}Pd	−8.553	−8.564	4.415	—	4.449	—	0.135	—
^{102}Pd	−8.564	−8.580	4.461	—	4.475	4.484 (0.004)	0.175	0.139 (0.005)
^{108}Pd	−8.525	−8.567	4.575	—	4.530	4.556 (0.003)	0.182	0.243 (0.004)

Note: The experimental values along with their uncertainties given in parentheses for *B.E./A* and deformation are taken from [31] and for r_{charge} are taken from [32].

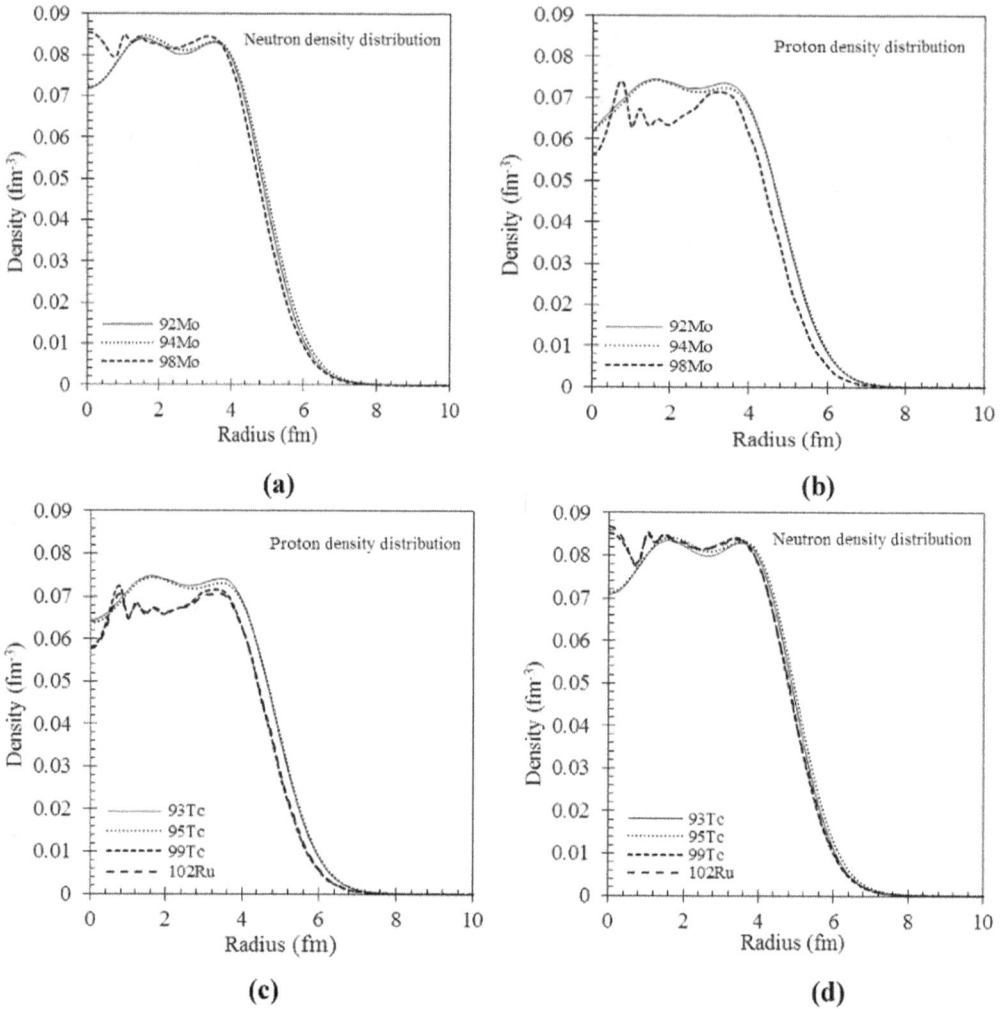

FIGURE 13.3

(a & b) Linear proton and neutron density distribution (as a function of r) for ^{92,94,98}Mo isotopes as calculated using the RMF model. (c & d) Linear proton and neutron density distribution (as a function of r) for ^{93,95,99}Tc and ^{102}Ru isotopes as calculated using the RMF model.

for protons as well as neutrons, while density distribution plot for ^{99}Tc and ^{102}Ru has a different trend. In Figure 13.4a and b, we present the comparative results of ^{96,98,104}Ru for proton density distribution and neutron density distribution, respectively. One can see that both proton and neutron density distribution plots have different trends. In Figure 13.4c and d, we present the comparative results of ^{97,99,105}Rh and ^{100,102,108}Pd for proton density distribution and neutron density distribution, respectively. One can see that the proton density distribution plot for ^{97,99,105}Rh and ^{100,102,108}Pd has a similar trend, while in neutron density plot, only ^{97}Rh shows a distinct change. Overall, one can see that the RMF calculations provide sufficiently good results for ground-state nuclear observables, and hence, the nuclear density distribution obtained from RMF can be reliably employed to calculate the optical potential and subsequently the nuclear reaction rates at low projectile energy.

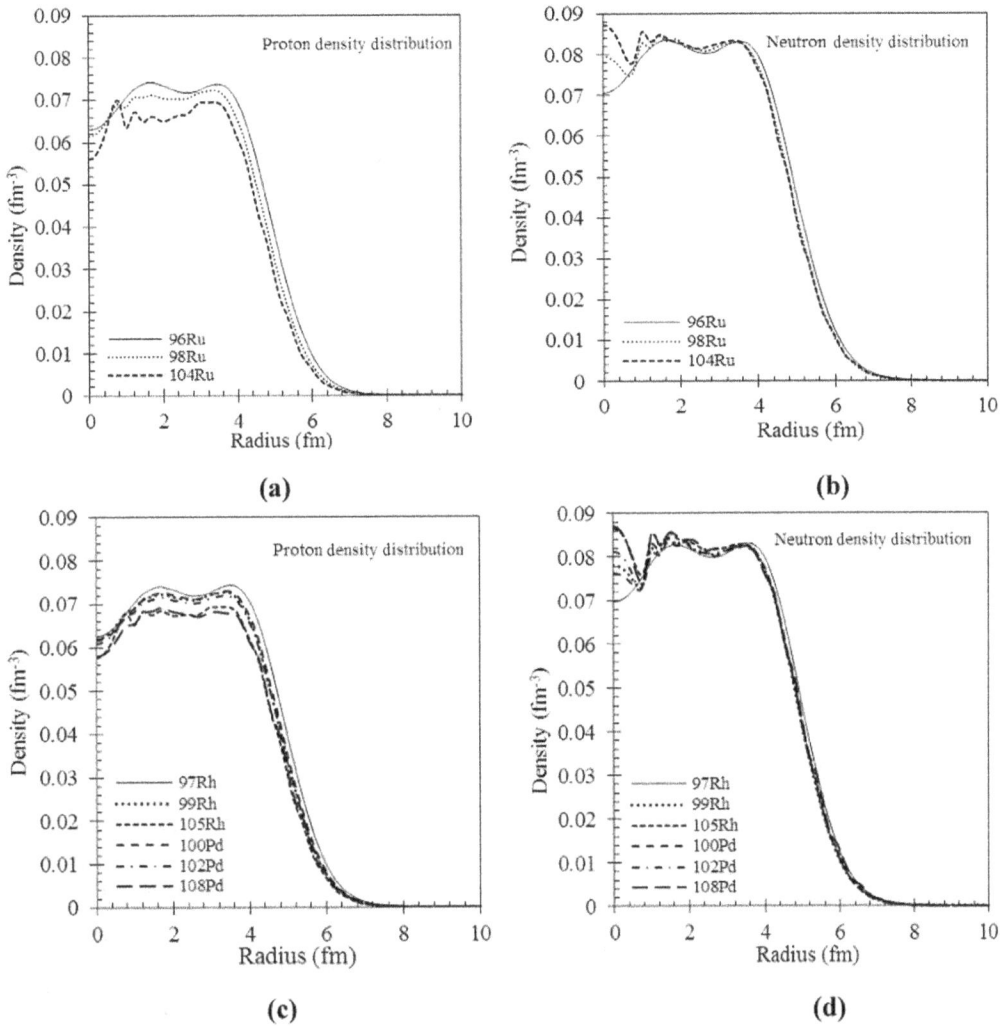

FIGURE 13.4
(a & b) Linear proton and neutron density distribution (as a function of r) for ^{96,98,104}Ru isotopes as calculated using the RMF model. **(c & d)** Linear proton and neutron density distribution (as a function of r) for ^{97,99,105}Rh and ^{100,102,108}Pd isotopes as calculated using the RMF model.

13.3.2 The Cross Section of ^{92,94,98}Mo (p, γ) ^{93,95,99}Tc

The calculation of cross section for proton capture reactions is performed using two different optical potentials through TALYS code, i.e. phenomenological KD03 potential and microscopic JLM potential (calculated using nuclear densities obtained from the RMF model). These nuclear densities (RMF) were calculated using the RMF formalism and added to TALYS source files manually for calculating the potential and cross section. The reaction rate depends on the models of the level density and optical potential adopted in the calculation of cross sections. The proton capture cross sections of the stable ^{92,94,98}Mo isotopes were measured first time using the activation method in the proton energy range between 1.5 and 3 MeV with a calibrated HPGe detector by Sauter and Käppeler. In the experiment, thin layers of natural molybdenum were irradiated at the Karlsruhe 3.75 MV

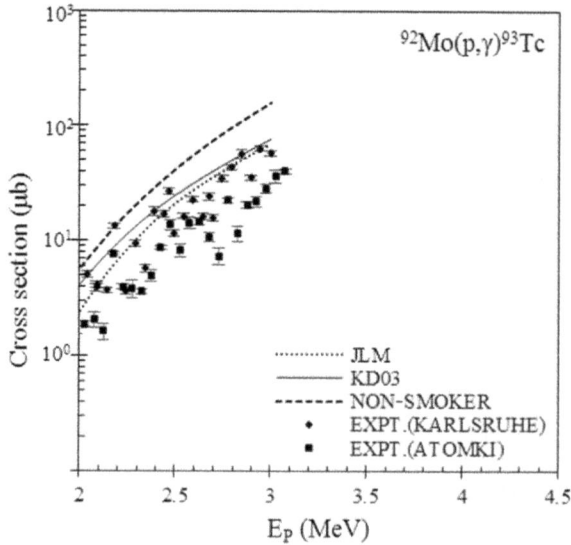

FIGURE 13.5

Comparison of the present calculations (JLM and KD03) for proton capture cross section (as a function of energy in the center of mass frame) for ^{92}Mo(p, γ)^{93}Tc calculated with NON-SMOKER results and experimental data as available.

Van de Graaff accelerator with proton beams of 20–55 μA. There were strong fluctuations in the ^{92}Mo cross-section results using the above technique, so keeping this in mind, ^{92}Mo(p, γ)^{93}Tc reaction was studied again by the Cologne group with both activation and in-beam γ-spectroscopy methods [12]. In this experiment, the Cologne group tried to reduce the uncertainty in the cross-section results by using thick target instead of the thin target layer. However, the results obtained still have much error bar, and there is still a strong need to have more cross-section data for ^{92}Mo(p, γ)^{93}Tc.

In Figures 13.5–13.7, we show the results of our calculated cross section in the energy range of 2–4.5 MeV and compare them with experimentally and theoretically available values.

The cross-section results for ^{92}Mo(p, γ)^{93}Tc reaction are displayed in Figure 13.5 where we show our results along with experimental results obtained by the experiment performed at KARLSRUHE, ATOMKI, and theoretical (NON-SMOKER) results [12,33]. We see that the NON-SMOKER results overpredict the experimental results, while the JLM and KD03 results lie within the experimental error bar.

It is important to mention, here, that there is much uncertainty in the measured cross section for ^{92}Mo, in comparison to other cases. This large fluctuation may be attributed to a very low-level density in ^{93}Tc, as a result of its magic neutron number $N = 50$. Since one of the basic assumptions of the statistical model is that level density is high enough so that a statistical treatment is possible, the applicability of the statistical model for such cases requires further modifications to provide reliable reaction rates for γ-process networks [33].

The cross-section results for ^{94}Mo(p, γ)^{95}Tc reaction are shown in Figure 13.6, where we show the experimental results [12] of the experiment performed in the KARLSRUHE along with NON-SMOKER and our calculated JLM and KD03 results, respectively. We find that the NON-SMOKER results overpredict the experimental results, while our JLM and KD03 results are close to experimental results.

The cross-section results for ^{98}Mo(p, γ)^{99}Tc reaction are illustrated in Figure 13.7, where we show the experimental results [33] by the experiments performed in ATOM KI

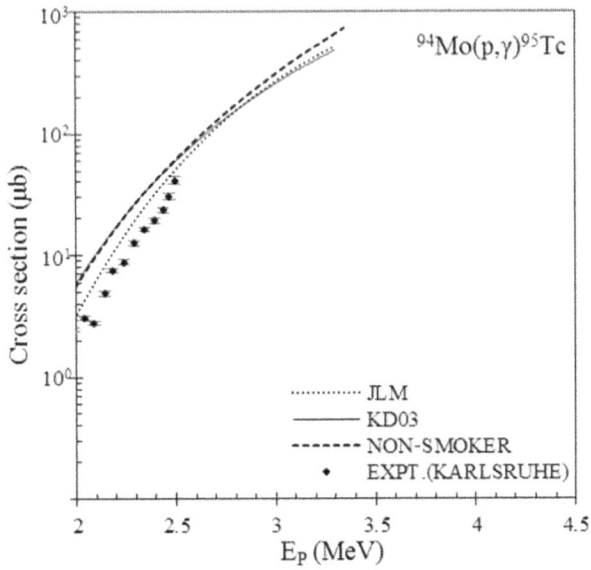

FIGURE 13.6
Comparison of the present calculations (JLM and KD03) for proton capture cross section (as a function of energy in the center of mass frame) for ^{94}Mo$(p, \gamma)^{95}$Tc calculated with NON-SMOKER results and experimental data as available.

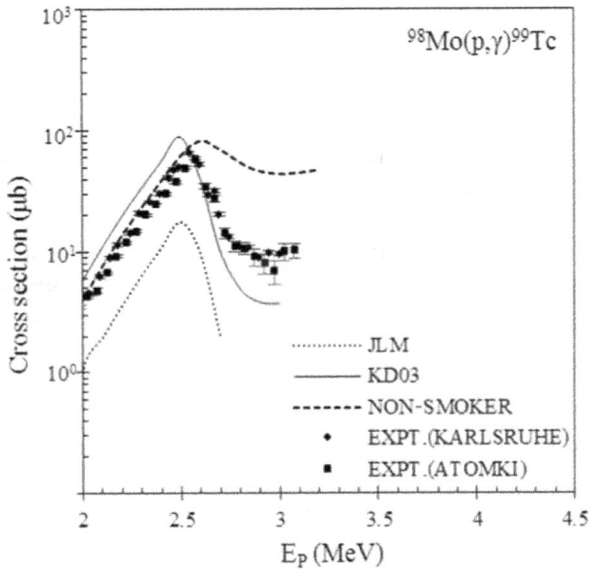

FIGURE 13.7
Comparison of the present calculations (JLM and KD03) for proton capture cross section (as a function of energy in the center of mass frame) for ^{98}Mo$(p, \gamma)^{99}$Tc calculated with NON-SMOKER results and experimental data as available.

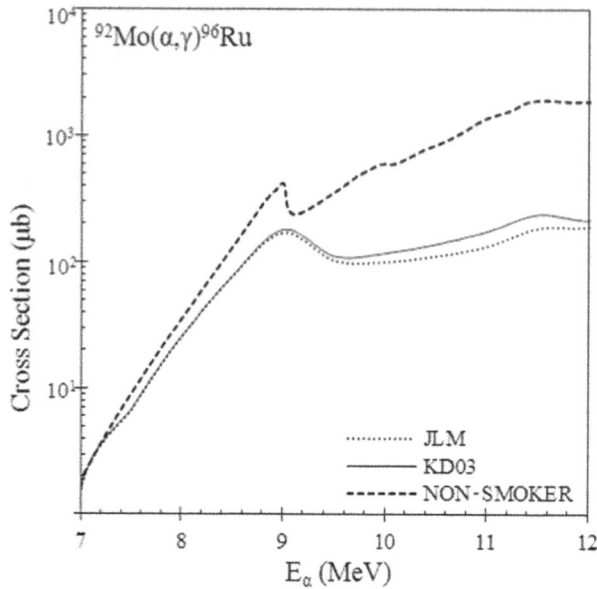

FIGURE 13.8

Comparison of the present calculations (JLM and KD03) for alpha capture cross section (as a function of energy in the center of mass frame) for ^{92}Mo$(\alpha, \gamma)^{96}$Ru calculated with the NON-SMOKER results.

with NON-SMOKER and our calculated JLM and KD03 results. We find that the NON-SMOKER and KD03 results are close to the experimental results, but the JLM results underestimate the experimental data.

One can see that for all cases, except for ^{98}Mo, our calculated results are very close to experimental results and lie within the experimental error bars. Also, one can see that the results calculated using both potentials are qualitatively very similar, except in the case of ^{98}Mo, where the results calculated using the JLM potential slightly underestimate the cross section, and the results using the KD03 potential slightly overestimate the experimentally observed cross section for the low-energy region. Our calculated results are also found to be very close to the calculations for ^{98}Mo from Ref. [19] using the Talys 1.0 code.

13.3.3 Cross Section of ^{92,94,98}Mo (α, γ) ^{96,98,102}Ru

The results of alpha capture cross sections for the molybdenum isotopes are plotted in Figures 13.8–13.10 obtained using the same formalism as discussed above. The results for cross section and S-factors are calculated in the energy range (Gamow's window) of 7−12 MeV and reaction rates are calculated in the temperature range of 2−5 ($\times 10^9$K), which is relevant to the p-process temperatures. The experimental results are not available for the alpha capture reactions of molybdenum isotopes. The results in this section are given as prediction for future experimental activities and are also compared to theoretical calculations using NON-SMOKER code for checking the results qualitatively as well as enriching data to the nuclear reaction data libraries. One can see from Figures 13.8–13.10 that the JLM and KD03 results are almost identical and have a similar trend with NON-SMOKER results.

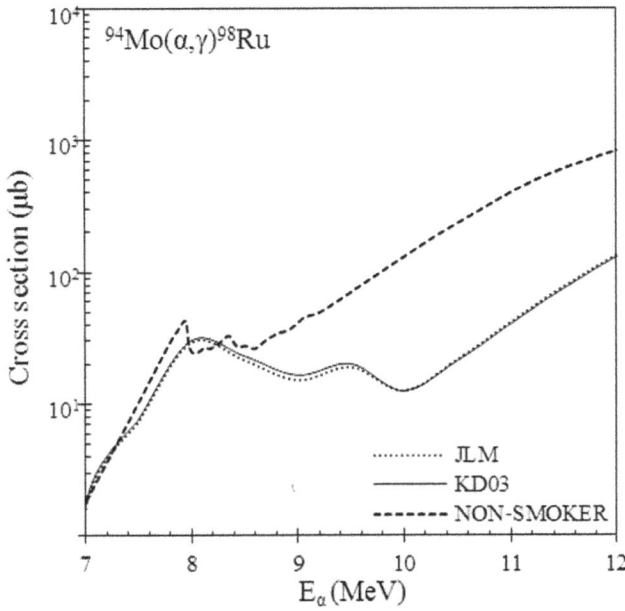

FIGURE 13.9
Comparison of the present calculations (JLM and KD03) for alpha capture cross section (as a function of energy in the center of mass frame) for ^{94}Mo(α, γ)^{98}Ru calculated with the NON-SMOKER results.

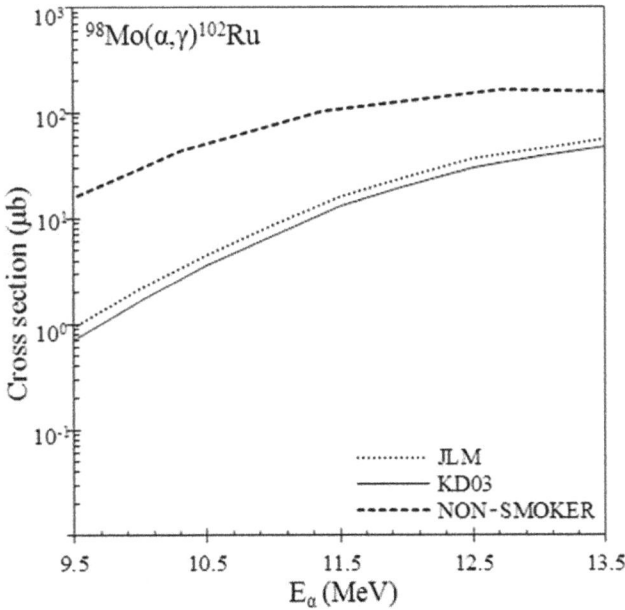

FIGURE 13.10
Comparison of the present calculations (JLM and KD03) for alpha capture cross section (as a function of energy in the center of mass frame) for ^{98}Mo(α, γ)^{102}Ru calculated with the NON-SMOKER results.

13.4   Conclusion

The proton and alpha capture reactions calculation has been performed in the Hauser-Feshbach statistical model with the help of microscopic optical potential obtained using the densities calculated from the RMF approach and phenomenological optical potential. The reliability of the calculated nuclear densities (through the RMF model), which are further used as input for reaction studies, has been performed by comparing ground-state properties of the nuclei, i.e. binding energy, rms matter radii, and rms radii, with the corresponding experimental observations. Hence, as is evident from the results, it is difficult to establish the unique reliability of the present calculations with experimental data, in comparison to other calculations. However, the results provide fair agreement. Moreover, it is important to emphasize here that the present work is an attempt to consistently study the nuclear structure and the nuclear reaction observables. It seems that one can successfully employ the approach to calculate reaction rates using the optical potential (JLM and KD03) using RMF densities, though finer corrections may be required to enhance the reliability and predictability of the approach. The success of these calculations can be taken as a proof of principle of using the RMF densities/inputs for performing nuclear reaction calculations. This approach opens a door to study proton and alpha capture reactions much reliably and in a consistent manner even for the reactions that are experimentally inaccessible.

References

[1] Arnett W. D., *Supernovae and Nucleosynthesis*, Princeton University Press (1996).

[2] Burbidge E. M., Burbidge G.R., Fowler W. A. and Hoyle F., *Rev. Mod. Phys.* **29**, 547 (1957).

[3] Cameron A. G. W., *Chalk River Rep. CRL*, Vol. **41** (Atomic Energy of Cande Ltd. (1957).

[4] Wallerstein G., Iben I., Parker P., Boesgaard A. M., Hale G.M., Champagne A. E., Barnes C. A., Käppeler F., Smith V. V., Hoffman R. D., Timmes F. X., Sneden C., Boyd R. N., Meyer B. S. and Lambert D. L., *Rev. Mod. Phys.* **69**, 995 (1997).

[5] Mayer B. S., *Annu. Rev. Astron. Astrophys.* **32**, 153 (1994).

[6] Cowan J. J., Thielemann F. -K. and Truran J. W., *Phys. Rep.* **208**, 267 (1991).

[7] Meyer B. S., Mathews G. J., Howard W. M., Woosley S. E. and Hoffman R. D., *Astrophys. J.* **399**, 656 (1992).

[8] Käppeler F., EPJ Web of Conferences **21**, 03006 (2012).

[9] Raiteri C. M., Gallino R., Busso. M., Neuberger D., Käppeler F., *Ap. J.* **419**, 207 (1993).

[10] Bajpeyi A., Koning A. J., Shukla A. and Aberg S., *Eur. Phys. J. A* **51**, 157 (2015).

[11] Rauscher T., Dauphas N., Dillmann I., Fröhlich, C., Fülöp, Z., and Gyürky, G. *Rep. Prog. Phys.* **76**, 066201 (2013).

[12] Sauter T., Käppeler F., *Phys. Rev. C* **55**, 6 (1997).

[13] Magill, J., Pfennig G., Dreher R. and Sti Z., Karlsruher Nuklidkarte/Chart of the Nuclides. 9th edition. Nucleonica GmbH, Eggenstein-Leopoldshafen 2015.

[14] Bazin D., *Nature* **486**, 330 (2012).

[15] Asplund M., Grevesse N., Sauval A. J. and Scott P., *Annu. Rev. Astron. Astrophys.* **47**, 481 (2009).

[16] Cameron A. G. W., Essays in Nuclear Astrophysics, ed. Barnes C. A., Clayton D. D. and Schramm D. N. (Cambridge University Press, Cambridge), 1982.

[17] Hauser W. and Feshbach H., *Phys. Rev.* **87**, 366 (1952).

[18] Goriely S., in: Wender S. (Ed.), Proc. Capture Gamma-Ray Spectroscopy and Related Topics, in: IOP Conference Proceedings., **529**, 287 (2000).

[19] Koning A. J., Hilaire S., Duijvestijn M. C., "TALYS-1.0", Proceedings of the 27, 2007, International Conference on Nuclear Data for Science and Technology - ND2007, April 22-Nice, France, edited by Bersillon O., Gunsing F., Bauge E., Jacqmin R and Leray S (EDP Sciences, 2008), 211.

[20] Walecka J. D., *Ann. Phys. (N.Y.)* **83**, 491 (1974).

[21] Horowitz C. J. and Serot B. D., *Nucl. Phys. A* **368**, 503 (1981).

[22] Reinhard P. -G., *Z. Phys. A* **329**, 257 (1993).

[23] Rufa M., Reinhard P. G., Maruhn J., Greiner W. and Strayer M. R., *Phys. Rev. C* **38**, 390 (1988).

[24] Gambhir Y. K., Ring P. and Thimet A., *Ann. Phys.* **198**, 132 (1990).

[25] Patra S. K. and Praharaj C. R., *Phys. Rev. C* **44**, 2552 (1991).

[26] Madland D. G. and Nix J. R., *Nucl. Phys. A* **476**, 1 (1981).

[27] Möller P. and Nix J. R., *At. Data Nucl. Data Tables* **39**, 213 (1988).

[28] Shukla A., Sharma B. K., Chandra R., Arumugam P. and Patra S. K., *Phys. Rev. C* **76**, 034601 (2007).

[29] Shukla A., Aberg S. and Bajpeyi A., *Phys. Ato. Nucl.* **79**, 11 (2016).

[30] Shukla A., Aberg S. and Bajpeyi A., *J. Phys. G: Nucl. Part. Phys.* **44**, 025104 (2017).

[31] http://www.nndc.bnl.gov/masses/.

[32] Angeli I., *At. Data Nucl. Data Tables* **87**, 185206 (2004).

[33] Gyürky Gy., Vakulenko M., Fulop Z., Halasz Z., Kiss G. G., Somorjai E. and Szucs T., *Nucl. Phys. A* **922**, 112 (2014).

14

Constraining the Nuclear Matter EoS from the Properties of Celestial Objects

Bijay Kumar Agrawal

Saha Institute of Nuclear Physics
Homi Bhabha National Institute

Tuhin Malik

BITS-Pilani

CONTENTS

14.1 Introduction

Understanding the behavior of EoS of dense matter relevant to neutron stars (NSs) is one of the main objectives of both nuclear physics and astrophysics to date [1,2]. The typical densities encountered in the core region of an NS are a few times the nuclear saturation density ($\rho_0 \approx 0.16$ fm^{-3}). Till date, the theoretical knowledge of nuclear interactions at typical densities of the NS interior is quite uncertain [3]. Theoretically, the problem boils down to treating nuclear matter equation of state (EoS) with underlying physical principles in the domain of neutron-rich, highly asymmetric matter. Here, the confluence of properties of finite nuclei and those of infinite nuclear matter should be reflected in the current observational data. To gain the knowledge of the EoS, one should also revert to the indirect methods such as the correlations of finite nuclei and the NSs observable with various nuclear matter properties [4,5].

Astrophysical observations are the complementary probe for the information in the region of the dense-matter EoS which is not experimentally accessible in the laboratory [1]. The NSs are massive and compact astrophysical objects, and the coalescence of binary NS systems is one of the most promising sources of gravitational waves (GWs) observable by ground-based detectors [6–11]. The GW signals emitted during an NS merger depend on the behavior of NS matter at high densities. Therefore, its detection opens the possibility of constraining the nuclear matter parameters (NMPs) characterizing the EoS [12,13]. During the last stages of the inspiral motion of the coalescing NSs, the strong gravity of each of them induces a tidal deformation in the companion star. Decoding the gravitational wave phase evolution caused by that deformation allows the determination of the dimensionless tidal deformability parameter [14–18]. It is a measure of the response to the gravitational pull on the NS surface correlating with pressure gradients inside the NS, and therefore, it has been proposed as an effective probe of the EOS of nuclear matter relevant for NSs [5,19,20].

In August 2017, the Advanced LIGO and Advanced Virgo gravitational-wave observatories detected for the first time GWs emitted from the inspiral of two low-mass compact objects which are consistent with a binary neutron star (BNS) merger [21], and subsequently the electromagnetic counterparts are also observed [22–28]. This discovery opens a new window to look into the nuclear matter theories relevant for NS. The tidal deformations of the NSs comes into the analysis of this GW observation data provide relevant constraints on the uncertainty of parameters of nuclear equation of matter EoS. In the recent studies, possible correlations between the tidal deformability of an NS and the properties of infinite nuclear matter such as the slope and curvature of the symmetry energy coefficient have been explored widely [5,20,29–32].

In view of the recent observation GW170817, in this chapter, we will study the dependence of the tidal deformability Λ on the NMPs describing the EoS. In other words, we shall examine if there is a unique dependence of Λ on the various nuclear matter properties obtained for a diverse set of nuclear models. One of the simplest approaches is to find the correlation between NS properties and nuclear matter properties in a diverse set of nuclear models that are constrained by the bulk properties of finite nuclei [33,34] and the observed $\approx 2M_\odot$ NS maximum mass [35]. We have employed a diverse set of relativistic and non-relativistic mean-field models. For all of these models, both the NS and nuclear matter properties vary over a wide range [4].

The present chapter is organized as follows. In Section 14.2, we outline very briefly the energy density functionals obtained for the relativistic and non-relativistic mean-field models. The expressions for the various nuclear matters and the tidal deformability for a given EoS are also described. In Section 14.3, we discuss our choice for a diverse set of mean-field models and present our results. Finally, the conclusions are presented in Section 14.4.

14.2   Theoretical Framework

The properties of NSs are calculated using nuclear energy density functional based on the relativistic and non-relativistic mean-field models. In the following subsections, we briefly outline the effective Lagrangian for the RMF models and the Hamiltonian density for the non-relativistic model derived from the Skyrme force.

14.2.1   The Relativistic Mean-Field Theory

The relativistic mean-field (RMF) model describes the interaction of nucleons via the exchange of σ, ω, and ρ mesons. [36–39]. The σ mesons create a strong attractive central force and influence the spin-orbit nuclear force; on the other hand, ω mesons are responsible for the repulsive central force. The protons and neutrons only differ in terms of their isospin projections; the ρ mesons are included to distinguish between these baryons.

The effective Lagrangian for the RMF models we have employed in the present work can be grouped into two categories, namely, (i) in which the coupling constants are independent of the density, the mesonic field couple linearly to the nucleonic degree of freedoms describing their interactions with the nucleons and non-linear terms for mesons describe their self and mixed interactions, and (ii) contains density dependent coupling constant which account for the non-linear contributions of mesonic field and the interactions with the nucleons are described by the linear terms for the mesons. The nucleonic-mesonic interaction can be written as

$$\mathcal{L} = \bar{\psi}\left[\gamma^{\mu}\left(i\partial_{\mu} - g_{v}V_{\mu} - g_{\rho}\vec{\tau}.\vec{b_{\mu}}\right) - (M_{n} + g_{s}\phi)\right]\psi. \tag{14.1}$$

where the parameters g_{σ}, g_{ω}, and g_{ρ} describe the strength of the couplings of baryon(ψ) with σ, ω, and ρ mesons. The interaction part of the effective Lagrangian of the RMF model belonging to the category (i) can be written as

$$\mathcal{L}_{int} = \bar{\psi}\left[g_{\sigma}\sigma - \gamma^{\mu}\left(g_{\omega}\omega_{\mu} + \frac{1}{2}g_{\rho}\tau.\rho_{\mu} + \frac{e}{2}(1 + \tau_{3})A_{\mu}\right)\right]\psi - \frac{\kappa_{3}}{6M}g_{\sigma}m_{\sigma}^{2}\sigma^{3}$$
$$- \frac{\kappa_{4}}{24M^{2}}g_{\sigma}^{2}m_{\sigma}^{2}\sigma^{4} + \frac{1}{24}\zeta_{0}g_{\omega}^{2}(\omega_{\mu}\omega^{\mu})^{2} + \frac{\eta_{2\rho}}{4M^{2}}g_{\omega}^{2}m_{\rho}^{2}\omega_{\mu}\omega^{\mu}\rho_{\nu}\rho^{\nu} \tag{14.2}$$

where the symbols have a usual meaning and the details can be found in Refs. [37,38,40,41].

For the density-dependent meson-exchange model [42], interaction part of the Lagrangian does not contain any non-linear term, but the meson-nucleon strengths g_{σ}, g_{ω}, and g_{ρ} have an explicit density dependence in the following form:

$$g_{i}(\rho) = g_{i}(\rho_{sat})f_{i}(x), \quad \text{for } i = \sigma, \omega \tag{14.3}$$

where the density dependence is given by

$$f_{i}(x) = a_{i}\frac{1 + b_{i}(x + d_{i})^{2}}{1 + c_{i}(x + e_{i})^{2}} \tag{14.4}$$

in which x is given by $x = \rho/\rho_{sat}$, and ρ_{sat} denotes the baryon density at saturation in symmetric nuclear matter. For the ρ meson, density dependence is of exponential form and given by

$$f_{\rho}(x) = exp(-a_{\rho}(x - 1)) \tag{14.5}$$

The coupling strengths in Eqs. (14.1) and (14.2) are usually calibrated to reproduce the measured binding energies, charge radii. and the properties of giant resonances.

14.2.2 Skyrme Energy Density Functional

The total energy E of the system can be expressed as

$$E = \int \mathcal{H}(r)d^{3}r \tag{14.6}$$

where the Skyrme energy density functional $\mathcal{H}(r)$ is given by [43,44]

$$\mathcal{H} = \mathcal{K} + \mathcal{H}_{0} + \mathcal{H}_{3} + \mathcal{H}_{\text{eff}} + \mathcal{H}_{\text{fin}} + \mathcal{H}_{\text{so}} + \mathcal{H}_{\text{sg}} + \mathcal{H}_{\text{Coul}} \tag{14.7}$$

where $\mathcal{K} = \frac{\hbar^{2}}{2m}\tau$ is the kinetic energy term, $\mathcal{H}_{0}$ is the zero-range term, $\mathcal{H}_{3}$ is the density-dependent term, $\mathcal{H}_{\text{eff}}$ is an effective-mass term, $\mathcal{H}_{\text{fin}}$ is a finite-range term, $\mathcal{H}_{\text{so}}$

is a spin-orbit term, $\mathcal{H}_{sg}$ is a term due to tensor coupling with spin and gradient, and $\mathcal{H}_{Coul}$ is the contribution to the energy density due to the Coulomb interaction. For the Skyrme interaction, we have

$$\mathcal{H}_0 = \frac{1}{4}t_0\left[(2+x_0)\rho^2 - (2x_0+1)(\rho_p^2 + \rho_n^2)\right], \tag{14.8}$$

$$\mathcal{H}_3 = \frac{1}{24}t_3\rho^\alpha\left[(2+x_3)\rho^2 - (2x_3+1)(\rho_p^2 + \rho_n^2)\right], \tag{14.9}$$

$$\mathcal{H}_{eff} = \frac{1}{8}\left[t_1(2+x_1) + t_2(2+x_2)\right]\tau\rho + \frac{1}{8}\left[t_2(2x_2+1) - t_1(2x_1+1)\right](\tau_p\rho_p + \tau_n\rho_n), \tag{14.10}$$

$$\mathcal{H}_{fin} = \frac{1}{32}\left[3t_1(2+x_1) - t_2(2+x_2)\right](\nabla\rho)^2$$
$$- \frac{1}{32}\left[3t_1(2x_1+1) + t_2(2x_2+1)\right]\left[(\nabla\rho_p)^2 + (\nabla\rho_n)^2\right], \tag{14.11}$$

$$\mathcal{H}_{so} = \frac{W_0}{2}\left[\mathbf{J}\cdot\nabla\rho + \mathbf{J_p}\cdot\nabla\rho_p + \mathbf{J_n}\cdot\nabla\rho_n\right], \tag{14.12}$$

$$\mathcal{H}_{sg} = -\frac{1}{16}(t_1x_1 + t_2x_2)\mathbf{J}^2 + \frac{1}{16}(t_1 - t_2)\left[\mathbf{J_p}^2 + \mathbf{J_n}^2\right]. \tag{14.13}$$

Here, $\rho = \rho_p + \rho_n$, $\tau = \tau_p + \tau_n$, and $\mathbf{J} = \mathbf{J_p} + \mathbf{J_n}$ are the particle number density, kinetic energy density, and spin density with p and n denoting the protons and neutrons, respectively. We have used the value of $\hbar^2/2m = 20.734$ MeVfm^2 in our calculations. For the case of uniform matter, $\mathcal{H}_{fin}, \mathcal{H}_{so}, \mathcal{H}_{sg}$, and $\mathcal{H}_{Coul}$ do not contribute.

Furthermore, the $\mathcal{H}(\nabla)$ depends only on the ρ_p and ρ_n and is independent of position coordinate r, $\mathcal{H}(r) \rightarrow \mathcal{E}(\rho_p, \rho_n)$. The energy per nucleon, $\epsilon(\rho_p, \rho_n), = \frac{\mathcal{E}(\rho_p, \rho_n)}{\rho}$ can be expressed in terms of the total density ρ and the asymmetry parameter $I = \frac{\rho_n - \rho_p}{\rho}$ as

$$\epsilon(\rho, I) = \frac{3}{5}\frac{\hbar^2}{2m}\left(\frac{3\pi^2}{2}\right)^{2/3}\rho^{2/3}F_{5/3} + \frac{1}{8}t_0\rho[2(x_0+2) - (2x_0+1)F_2]$$
$$+ \frac{1}{48}t_3\rho^{\sigma+1}[2(x_3+2) - (2x_3+1)F_2]$$
$$+ \frac{3}{40}\left(\frac{3\pi^2}{2}\right)^{2/3}\rho^{5/3}\left\{[t_1(x_1+2) + t_2(x_2+2)]F_{5/3}\right.$$
$$\left. + \frac{1}{2}[t_2(2x_2+1) - t_1(2x_1+1)]F_{8/3}\right\} \tag{14.14}$$

$$F_m(I) = \frac{1}{2}[(1+I)^m + (1-I)^m] \tag{14.15}$$

Once the Skyrme parameters are known, the EoS for nuclear matter at any given asymmetry I can be obtained using Eq. (14.16). To a good approximation, the EoS can also be expressed in terms of various nuclear matters calculated at the saturation density ρ_0.

14.2.3 Properties of Nuclear Matter and Tidal Deformability

The EoS to a good approximation can be decomposed into the EoS for the symmetric matter and the density-dependent symmetry energy as

$$\epsilon(\rho, I) = \epsilon(\rho, I = 0) + J(\rho)I^2 \tag{14.16}$$

The EoS for the symmetric nuclear matter $\epsilon(\rho, I = 0)$ and symmetry energy $J(\rho)$ can be further expressed in terms of the nuclear matter properties at the saturation density ρ_0 as

$$\epsilon(\rho, 0) = \epsilon_0 + \frac{1}{2}K_0 \left(\frac{\rho - \rho_0}{3\rho_0}\right)^2 + \frac{1}{6}Q_0 \left(\frac{\rho - \rho_0}{3\rho_0}\right)^3, \tag{14.17}$$

$$J(\rho) = J_0 + L_0 \left(\frac{\rho - \rho_0}{3\rho_0}\right) + \frac{1}{2}K_{\text{sym},0} \left(\frac{\rho - \rho_0}{3\rho_0}\right)^2. \tag{14.18}$$

$$\epsilon_0 = \epsilon(\rho_0, I = 0) \tag{14.19}$$

The quantities appearing in Eqs. (14.17) and (14.19) are the incompressibility coefficient (K_0), skewness parameter (Q_0), symmetry energy coefficient (J_0), it's slope (L_0), and the curvature parameters ($K_{\text{sym},0}$). They are defined as follows:

$$K_0 = 9\rho_0^2 \left(\frac{\partial^2 \epsilon(\rho, 0)}{\partial \rho^2}\right)_{\rho_0}$$

$$Q_0 = 27\rho_0^3 \left(\frac{\partial^3 \epsilon(\rho, 0)}{\partial \rho^3}\right)_{\rho_0}$$

$$J_0 = \frac{1}{2} \left(\frac{\partial^2 \epsilon(\rho, I)}{\partial I^2}\right)_{I=0}$$

$$L_0 = 3\rho_0 \left(\frac{\partial J(\rho)}{\partial \rho}\right)_{\rho_0}$$

$$K_{\text{sym},0} = 9\rho_0^2 \left(\frac{\partial^2 J(\rho)}{\partial \rho^2}\right)_{\rho_0}$$

The tidal deformability parameter λ is defined as [14,15,18,45]

$$Q_{ij} = -\lambda \mathcal{E}_{ij}, \tag{14.20}$$

where Q_{ij} is the induced quadrupole moment of a star in a binary due to the static external tidal field $\mathcal{E}_{ij}$ of the companion star. The parameter λ can be expressed in terms of the dimensionless quadrupole tidal Love number k_2 as

$$\lambda = \frac{2}{3}k_2 R^5, \tag{14.21}$$

where R is the radius of the NS. The value of k_2 is typically in the range $\simeq 0.05 - 0.15$ [15,18,46] for NS and depends on the stellar structure. It can be calculated using the following expression [15]:

$$k_2 = \frac{8C^5}{5}(1 - 2C)^2 [2 + 2C(y_R - 1) - y_R]$$

$$\times \left\{ 2C(6 - 3y_R + 3C(5y_R - 8)) \right.$$

$$+ 4C^3 [13 - 11y_R + C(3y_R - 2) + 2C^2(1 + y_R)]$$

$$\left. + 3(1 - 2C)^2 [2 - y_R + 2C(y_R - 1)] \log(1 - 2C) \right\}^{-1}, \tag{14.22}$$

where C ($\equiv \frac{M}{R}$) is the compactness parameter of the star of mass M. The quantity y_R ($\equiv y(R)$) can be obtained by solving the following differential equation:

$$r\frac{dy(r)}{dr} + y(r)^2 + y(r)F(r) + r^2 Q(r) = 0, \tag{14.23}$$

with

$$F(r) = \frac{r - 4\pi r^3\left(\epsilon(r) - p(r)\right)}{r - 2M(r)}, \tag{14.24}$$

$$Q(r) = \frac{4\pi r\left(5\epsilon(r) + 9p(r) + \frac{\epsilon(r)+p(r)}{\partial p(r)/\partial\epsilon(r)} - \frac{6}{4\pi r^2}\right)}{r - 2M(r)}$$
$$- 4\left[\frac{M(r) + 4\pi r^3 p(r)}{r^2\left(1 - 2M(r)/r\right)}\right]^2. \tag{14.25}$$

For a given EOS, together with boundary conditions $p(0) = P_c$ and $M(0) = 0$, the Tolman-Oppenheimer-Volkoff(TOV) equations [47] can be integrated to obtain $\epsilon(r), p(r)$, and $M(r)$, which can be employed to solve Eq. (14.23) with boundary condition $y(0) = 2$. Then, one can define the dimensionless tidal deformability:

$$\Lambda = \frac{2}{3}k_2 C^{-5}. \tag{14.26}$$

14.3   Results and Discussions

The correlations between properties of finite nuclei and neutron star, and the properties of nuclear matter at the saturation density have been studied extensively [4,5,48–52]. The existence of strong correlations between the neutron-skin thickness, dipole polarizability, NS radii, and the tidal deformability, and few selected properties of the nuclear matter indicates that the accurate knowledge of these observables could constrain the equation of state in narrower bounds. We provide here a comprehensive analysis of correlations of tidal deformability with NMPs obtained using a diverse set of nuclear mean-field models which can be broadly classified as relativistic mean-field (RMF) models and non-relativistic Skyrme Hartree-Fock (SHF). These models are constrained by the bulk properties of finite nuclei and the observed lower bound on the NS maximum mass. Yet, the nuclear matter properties vary over a wide range for the representative set of models consisting of 18 RMF and 24 SHF models as summarized in Tables 15.4 and 15.5. Further details of the models can be found in Ref. [4] and references their in.

In Figure 14.1, we plot the distributions of e_0, ρ_0, K_0, and Q_0 pertaining to the symmetric nuclear matter obtained using the models listed in Tables 15.4 and 15.5. Similar plots for J_0, L_0, and $K_{\text{sym},0}$ are shown in Figure 14.2. The mean (μ) and the standard deviation (σ) for each of the NMPs are indicated in their respective plots. For the comparison of the spread in the values of these nuclear matter parameters relative to each other, their values in the figures are transformed as $\frac{x-\mu}{\mu}$, where x denotes a NMP. It is evident that the values of e_0, ρ_0, and K_0 are better constrained relative to other nuclear matter parameters. The values of Q_0, L_0, and $K_{\text{sym},0}$ are poorly constrained. The later quantities are crucial in determining the behavior of the EoS for the asymmetric nuclear matter. In what follows, we shall demonstrate how the accurate knowledge of tidal deformability for the NSs is essential to constrain these NMPs which would enable one to understand the behavior the EoS for the asymmetric nuclear matter.

TABLE 14.1

The Nuclear Matter Properties for a Representative Set of RMF Models Calculated at the Saturation Density ρ_0

Model	ρ_0 (fm^{-3})	e_0 (MeV)	K_0 (MeV)	Q_0 (MeV)	M_0 (MeV)	J_0 (MeV)	L_0 (MeV)	$K_{\text{sym},0}$ (MeV)
BSR2	0.149	−16.03	240	−52.1	2829	31.4	62.2	−3.4
BSR3	0.15	−16.09	230.6	−119.4	2648	32.6	70.5	−7.8
BSR6	0.149	−16.13	235.9	−11.4	2820	35.4	85.6	−47.8
DD2	0.149	−16.02	242.2	167.4	3076	31.7	55	−93.4
DDHδ	0.153	−16.25	240.2	−539.8	2343	25.6	48.6	80.7
DDHδMod	0.153	−16.25	240.2	−539.8	2343	31.9	57.5	80.3
DDME1	0.152	−16.23	243.9	316.2	3249	33.1	55.4	−101.3
DDME2	0.152	−16.14	251.3	479	3493	32.3	51.3	−87.5
FSU2	0.15	−16.28	237.8	−156.1	2698	37.6	112.7	25.4
GM1	0.153	−16.3	300.1	−215.1	3387	32.5	93.9	18
NL3	0.148	−16.25	271.6	205.5	3464	37.4	118.5	100.9
NL3$\sigma\rho$4	0.148	−16.25	271.6	205.5	3464	33	68.3	−26.8
NL3$\sigma\rho$6	0.148	−16.25	271.6	205.5	3464	31.5	55.4	25
NL3$\omega\rho$02	0.148	−16.25	271.6	205.5	3464	33.1	68.2	−53.1
NL3$\omega\rho$03	0.148	−16.25	271.6	205.5	3464	31.7	55.3	−7.5
TM1	0.145	−16.26	281.2	−286.3	3088	36.9	110.8	33.6
TM1-2	0.145	−16.26	281.2	−199.3	3175	36.9	111.4	41.9
TW	0.153	−16.25	240.2	−539.8	2343	32.8	55.3	−124.8

The quantities listed below are the energy per nucleon e_0, incompressibility coefficient (K_0), and skewness parameter(Q_0) for the symmetric nuclear matter. The symmetry energy coefficient (J_0), it's slope (L_0), and the curvature ($K_{\text{sym},0}$) determining the density dependence of the symmetry energy.

14.3.1 Nuclear Matter Equation of State

All the models considered are consistent with the observational constraint provided by the existence of 2 M$_\odot$ NS [53,54]. Moreover, the SHF models employed do not become acausal for masses below 2 M$_\odot$. The inner crust of all the models are obtained in unified manner [53] and for the outer crust we have used Baym-Pethick-Sutherland [55]. The mass of the merging stars in the GW170817 are in the range of $\approx 1.2 - 1.6 M_\odot$, and on average, the masses of merging stars are $\approx 1.4 M_\odot$. Here, we present our results for the NS with mass $1.4 M_\odot$. Therefore, other non-nucleonic degrees of freedom are ignored [56,57].

The variation of pressure (p) with the energy density (ε) in the left panel and that of $dp/d\varepsilon$ in the right panel are plotted in Figure 14.3, for NS matter using our representative set of models. The central density corresponding to the NS maximum mass for each EoS represented by black circles. The dashed line indicates the causality limit (i.e. $dp/d\varepsilon = 1$). The values of $dp/d\varepsilon$ for SHF models are larger at higher densities ($\rho \gg \rho_0$) than those for the RMF models. The EoS for only BSk20 and BSk26 models are marginally acausal, whereas rest of all the models indicate maximum mass NS configurations within the causality limit.

14.3.2 Tidal Deformability

GW170817 observation provides an upper bound on the dimensionless tidal deformability parameter Λ of an NS. The extracted bound on Λ for an NS with canonical mass is $<$ 800(400) for 90%(50%) confidence limit for the low spin binary. A lower bound on the tidal deformability $\sim$ 400 is also set from the analysis of the UV-optical-infrared counterpart of GW170817 complemented with numerical relativity results [58]. Similar analysis, but

TABLE 14.2

The Same as Table 14.1, but for the SHF Models

Model	ρ_0 (fm^{-3})	e_0 (MeV)	K_0 (MeV)	Q_0 (MeV)	M_0 (MeV)	J_0 (MeV)	L_0 (MeV)	$K_{sym,0}$ (MeV)
SKa	0.155	−15.99	263.2	−300.3	2858	32.91	74.62	−78.46
SKb	0.155	−16	263	−300.3	2856	23.88	47.6	−78.5
SkI2	0.1575	−15.77	241	−340	2552	33.4	104.3	70.7
SkI3	0.1577	−15.98	258.2	−303.7	2795	34.83	100.5	73
SkI4	0.16	−15.95	247.95	−329	2646	29.5	60.39	−40.56
SkI5	0.156	−15.85	255.8	−302.1	2768	36.64	129.3	159.5
SkI6	0.159	−15.89	248.17	−327.8	2650	29.9	59.24	−46.77
Sly2	0.161	−15.99	229.92	−370.3	2389	32	47.46	−115.13
Sly230a	0.16	−15.99	229.9	−364.2	2394	31.99	44.3	−98.3
Sly4	0.159	−15.97	230	−362.9	2397	32.04	46	−119.8
Sly9	0.151	−15.8	229.84	−355.6	2402	31.98	54.86	−81.42
SkMP	0.157	−15.56	230.87	−342.7	2428	29.89	70.31	−49.82
SKOp	0.16	−15.75	222.36	−390.8	2277	31.95	68.94	−78.82
KDE0V1	0.165	−16.23	227.54	−384.9	2346	34.58	54.69	−127.12
SK255	0.157	−16.33	254.96	−350.2	2709	37.4	95	−58.3
SK272	0.155	−16.28	271.55	−305.2	2953	37.4	91.7	−67.8
Rs	0.158	−15.53	236.7	−348.3	2492	30.58	85.7	−9.1
BSk20	0.16	−16.08	241.4	−282.1	2615	30	37.4	−136.5
BSk21	0.158	−16.05	245.8	−274.1	2676	30	46.6	−37.2
BSk22	0.1578	−16.09	245.9	−275.4	2675	32	68.5	13
BSk23	0.1578	−16.07	245.7	−274.9	2674	31	57.8	−11.3
BSk24	0.1578	−16.05	245.5	−274.4	2672	30	46.4	−37.6
BSk25	0.1587	−16.03	236	−316.3	2516	29	36.9	−28.5
BSk26	0.1589	−16.06	240.8	−282.8	2607	30	37.5	−135.6

with a larger number of models, pushes the lower bound to ∼200 [59]. Recently [5], we have investigated the correlation of tidal deformability with NMPs. The tidal deformability is usually assumed to be dependent on R^5 with R being the NS radii (Eq. 14.26). We have shown that the Love number k_2 is not model independent, which makes the tidal deformability to be dependent on $R^{6.13}$.

We calculate the values of the coefficients for the correlation of Λ, k_2, and R with the nuclear matter saturation parameters K_0, Q_0, M_0, J_0, L_0, $K_{sym,0}$, and with several linear combinations of two parameters. In our analysis, the correlation between a pair of quantities is quantified in terms of Pearsons correlation coefficient, denoted as $\mathcal{R}$ [60]. A pair of quantities is said to be fully correlated to each other if the magnitude of $\mathcal{R}$ is unity and is weakly correlated if $|\mathcal{R}| < 0.5$.

In Figure 14.4, we plot the values of $\Lambda_{1.4}$ for the NS with the canonical mass as the function of K_0 and M_0 for RMF, SHF, and all the models together. Similar plots for the L_0 and $K_{sym,0}$ are shown in Figure 14.5. The correlations of $\Lambda_{1.4}$ with K_0 and M_0 are moderate for RMF, SHF, and all the models; however, it is marginally higher in the case of RMF models. The correlations of $\Lambda_{1.4}$ with L_0 and $K_{sym,0}$ for RMF are at variance with those of SHF models. The correlations are stronger for the case of SHF models. For the case of RMF models, $\Lambda_{1.4}$ shows the moderate correlation with L_0 and weaker correlations with $K_{sym,0}$. Consequently, the correlations of $\Lambda_{1.4}$ with L_0 and $K_{sym,0}$ become moderate when all the models are considered together. The results presented in Figures 14.4 and 14.5 also revel that symmetric nuclear matter part of the EoS may be some what similar; however, the density dependence of the symmetry energy for the RMF and SHF models seems to differ.

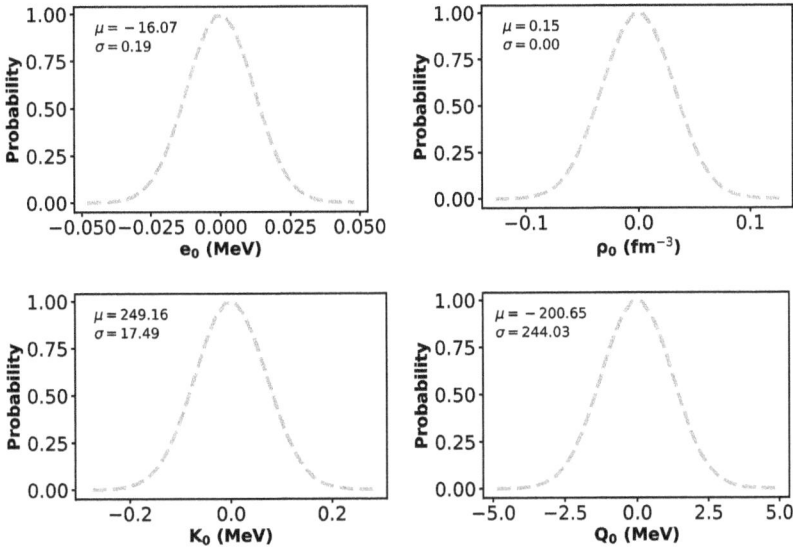

FIGURE 14.1
(Color online) Probability distributions of energy per nucleon e_0 , saturation density (ρ_0), incompressibility coefficient $(K_0$), and skewness parameter$(Q_0$) corresponding to the symmetric nuclear matter obtained for a representative set of RMF and SHF models. The values of mean (μ) and the standard deviations (σ) for each of the NMP are also given. The NMPs as displayed along the abscissa are obtained by appropriate transformation so that the spread in their values can be compared. (See the text for details.)

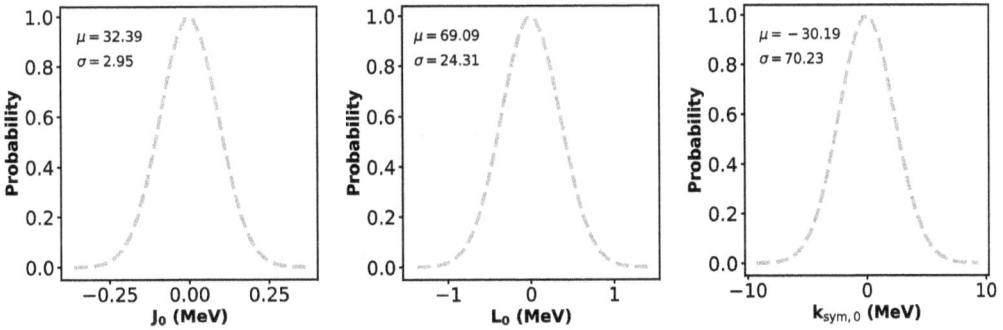

FIGURE 14.2
(Color online) The same as Figure 14.1, but for symmetry energy coefficient $(J_0$), it's slope $(L_0$), and the curvature $(K_{\mathrm{sym},0}$) which determine the density dependence of the symmetry energy coefficient.

The correlations of $R_{1.4}$ and $k_{2,1.4}$ with several NMPs are also calculated. The results are summarized in Table 14.3. None of the nuclear matter parameter individually seems to be strongly correlated to $\Lambda_{1.4}$, $R_{1.4}$, and $k_{2,1.4}$, when all the 42 models are considered together.

One may expect that the properties of NSs should depend simultaneously on the behavior of symmetric nuclear matter and the symmetry energy. Because these properties are calculated by solving the TOV equations which require the EoS for the asymmetric nuclear matter as the main input. Such an EoS can be decomposed into the EoS for the symmetric nuclear matter and the density-dependent symmetry energy as given by

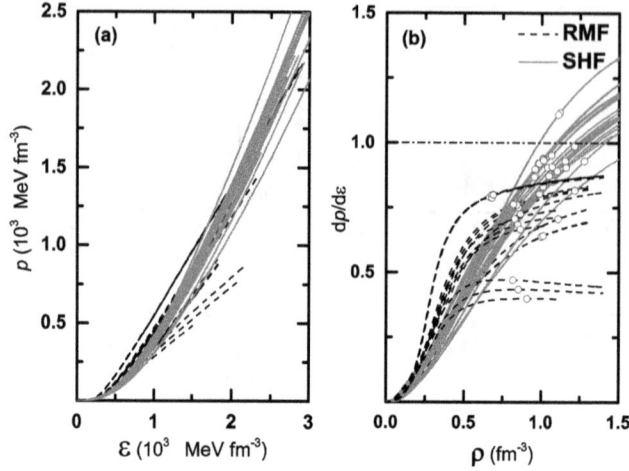

FIGURE 14.3
(Color online) Variations of (a) the pressure p with the energy density ϵ and (b) $dp/d\varepsilon$ with the nucleonic density for the NS matter in beta equilibrium. The black dashed and red lines depict the the results for RMF and SHF models, respectively. The circles in the right panel correspond to the central densities and the slopes $dp/d\varepsilon$ at the maximum NS mass for each of the EoS. The EoSs for BSk20 and BSk26 are marginally acausal at the NS maximum masses $\sim 2.2\ M_\odot$ [53,54].

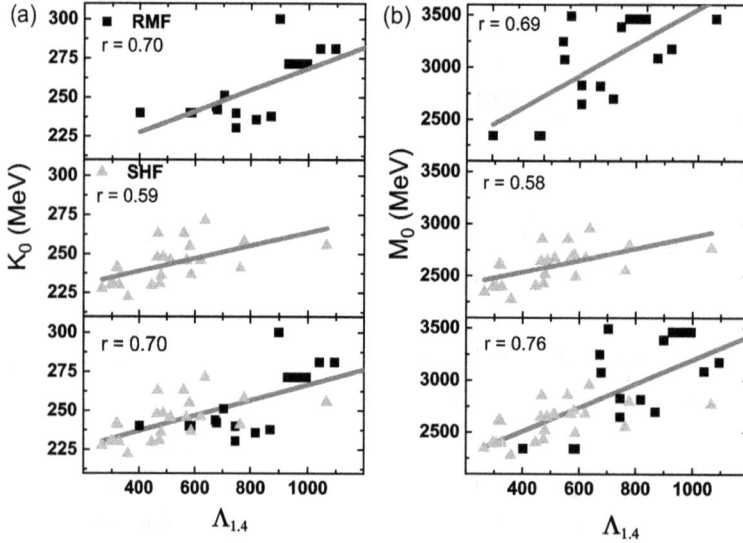

FIGURE 14.4
(Color online) Plots for K_0 (a) and M_0 (b) as a function of $\Lambda_{1.4}$ for the RMF, SHF, and all the models. The correlations are weak and exhibit only marginal model dependence.

Eq. (14.14). We consider various combinations involving the NMPs K_0 and M_0 with the L_0 and $K_{\text{sym},0}$. Namely, these combinations are $K_0 + \alpha L_0$, $M_0 + \beta L_0$, and $M_0 + \eta K_{\text{sym},0}$. The values of α, β, and η are obtained so that, for given NS mass, they yield optimum correlations. The variations of $k_{2,1.4}$ and $\Lambda_{1.4}$ with these linear combinations are displayed in Figure 14.6. The tidal deformability depends strongly on the linear combinations considered

TABLE 14.3

Pearson's Correlation Coefficients $\mathcal{R}$ Obtained for the Correlations between Various NS and Nuclear Matter Properties

	K_0	Q_0	M_0	J_0	L_0	$K_{\mathrm{sym},0}$
			RMF models			
$\Lambda_{1.4}$	0.70	0.40	0.69	0.59	0.7	0.47
$R_{1.4}$	0.63	0.38	0.64	0.71	0.77	0.42
$k_{2,1.4}$	0.64	0.51	0.74	-0.09	0.11	0.40
			SHF models			
$\Lambda_{1.4}$	0.59	0.37	0.58	0.48	0.86	0.94
$R_{1.4}$	0.62	0.32	0.59	0.50	0.88	0.82
$k_{2,1.4}$	0.57	0.48	0.61	-0.08	0.51	0.86
			All models			
$\Lambda_{1.4}$	0.70	0.57	0.76	0.53	0.71	0.71
$R_{1.4}$	0.67	0.54	0.73	0.59	0.75	0.69
$k_{2,1.4}$	0.64	0.51	0.73	0.06	0.39	0.70

The values of tidal deformability Λ, radius R, and the Love number k_2 are evaluated for the NS with canonical mass 1.4 $M_\odot$. The NMPs considered are the nuclear matter incompressibility K_0, the skewness Q_0, slope of incompressibility M_0, symmetry energy J_0, slope of symmetry energy L_0, and the curvature parameters $K_{\mathrm{sym},0}$ evaluated at the saturation densities.

TABLE 14.4

The Values of Pearson's Correlation Coefficients $\mathcal{R}$ Obtained for the Correlations of $\Lambda_{1.4}$, $R_{1.4}$, and $k_{2,1.4}$ with Various Linear Combinations of EoS Parameters

	$K_0 + \alpha L_0$		$M_0 + \beta L_0$		$M_0 + \eta K_{\mathrm{sym},0}$	
	$\mathcal{R}$	α	$\mathcal{R}$	β	$\mathcal{R}$	η
$\Lambda_{1.4}$	0.83	0.74	0.92	13.68	0.95	4.83
$R_{1.4}$	0.84	0.98	0.93	16.62	0.91	5.00
$k_{2,1.4}$	0.65	0.16	0.75	4.81	0.92	5.31

as displayed in the figure even when all the models are considered together. This dependence is much stronger than those observed in the case of individual NMP. The correlations of $k_{2,1.4}$ and $\Lambda_{1.4}$ with $M_0 + \eta K_{\mathrm{sym},0}$ are quite strong in comparison to those for the individual NMPs or their other linear combinations. Therefore, if one knows the value of $\Lambda_{1.4}$ with greater precision, the values of few selected NMPs can constrained. The values of α, β, and η in the case of $\Lambda_{1.4}$, $R_{1.4}$, and $k_{2,1.4}$ are listed in Table 15.7 along with corresponding correlation coefficients. In Figure 14.7, we plot the probability distribution for the correlations of $\Lambda_{1.4}$ with $M_0 + \beta L_0$ and $M_0 + \eta K_{\mathrm{sym},0}$ calculated using bootstrap method. The bootstrap method is a statistical tool to build a sufficiently large number of data set by resampling the original one with random replacement. It was introduced by Efron in 1979 [61]. It is based on a simple assumption that the data set of independent observations contains information on its parent distribution. To evaluate the confidence intervals of any estimated quantity from the data set, one can generate a large number of data sets and find the distribution of the estimated quantity. It has been successfully introduced in nuclear physics [62].

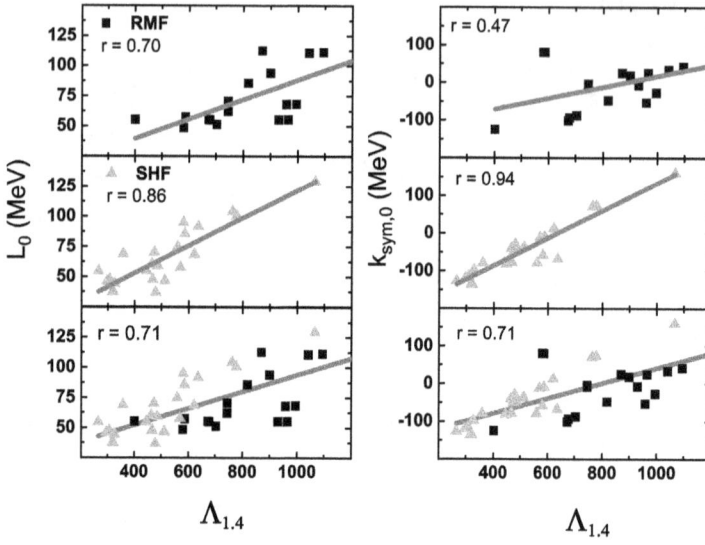

FIGURE 14.5
(Color online) The same as Figure 14.4, but for L_0 and $K_{sym,0}$. The $\Lambda_{1.4}$ depends only moderately on L_0 and weakly on $K_{sym,0}$ for the RMF models. The SHF models display stronger dependence on L_0 and $K_{sym,0}$ for $\Lambda_{1.4}$.

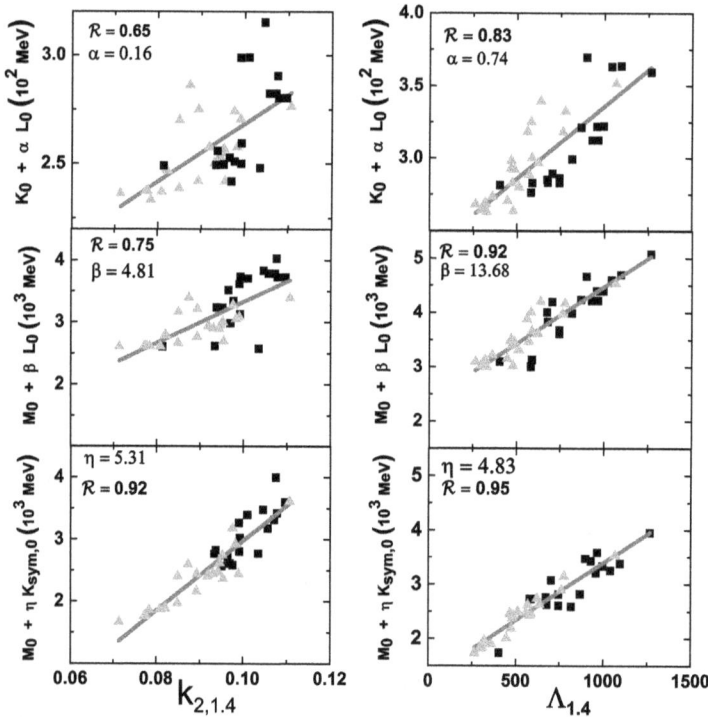

FIGURE 14.6
(Color online) The plots for the linear combinations for the NMPs versus the Love number $k_{2,1.4}$ and tidal deformability $\Lambda_{1.4}$ for the NS with the canonical mass obtained for all the models. The correlations of $k_{2,1.4}$ and $\Lambda_{1.4}$ with the $M_0 + \eta K_{sym,0}$ are the strongest one.

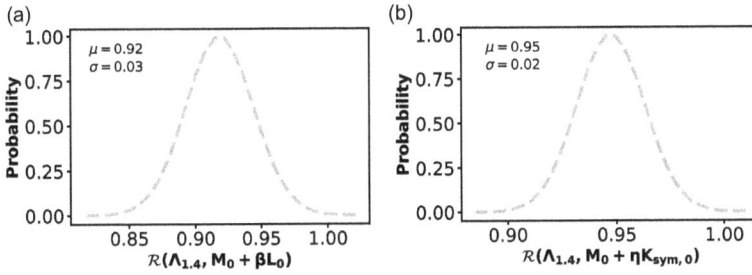

FIGURE 14.7
(Color online) The probability distribution for the correlations of $\Lambda_{1.4}$ with $M_0 + \beta L_0$ (a) and $M_0 + \eta K_{\text{sym},0}$ (b) calculated using bootstrap method. (See the text for details.)

14.4 Conclusions

The correlations of recently observed tidal deformability, in merging NSs, with various properties of nuclear matter have been investigated in detail for a diverse set of relativistic and non-relativistic mean-field models. The dependence of tidal deformability on the properties of symmetric nuclear matter such as incompressibility and its density derivative is found to be moderate for both the relativistic and non-relativistic mean-field models. However, dependence of tidal deformability on the slope and the curvature of the symmetry energy coefficient is at variance for the relativistic and non-relativistic models. In particular, the non-relativistic models display stronger correlations of tidal deformability with the symmetry energy coefficients. In contrast, the relativistic models display only weak correlations between the tidal deformability and the curvature of the symmetry energy coefficient. The correlations become stronger and appear to be model independent if the linear combinations of the slopes of the incompressibility coefficient and the symmetry energy coefficient are considered. In particular, tidal deformability depends linearly on the combinations of slope of incompressibility coefficient and the curvature of symmetry energy coefficient. Once the values of tidal deformability are known with greater precision, our results can be utilized to constrain the values of nuclear matter properties to narrower bounds which may not be possible with the terrestrial data on the finite nuclei.

References

[1] L. Rezzolla, P. Pizzochero, D. I. Jones, N. Rea, and I. Vidaña, *Astrophys. Space Sci. Libr.* **457**, pp. (2018).

[2] P. Haensel, A. Y. Potekhin, and D. G. Yakovlev, *Astrophys. Space Sci. Libr.* **326**, pp. 1 (2007).

[3] *Progr. Part. Nucl. Phys.* **109**, 103714 (2019), ISSN 0146–6410.

[4] N. Alam, B. K. Agrawal, M. Fortin, H. Pais, C. Providência, A. R. Raduta, and A. Sulaksono, *Phys. Rev.* **C94**, 052801 (2016a), 1610.06344.

[5] T. Malik, N. Alam, M. Fortin, C. Providência, B. K. Agrawal, T. K. Jha, B. Kumar, and S. K. Patra, *Phys. Rev. C* **98**, 035804 (2018).

[6] J. H. Taylor and J. M. Weisberg, *Astrophys. J.* **253**, 908 (1982).

[7] R. W. P. Drever, *Gravitational Radiation* ((North-Holland, Amsterdam, 1983) p. 321, 1983).

[8] K. Thorne, *Three Hundred Years of Gravitation* (Cambridge, UK: Univ. Pr. (1987) 684 p, 1987).

[9] A. Brillet and et al. (Virgo Collaboration), Technical Report No. VIR-0517A-15 (1989).

[10] J. Hough and et al., MPQ Technical Report No. 147 [GWD/137/JH(89)] (1989).

[11] C. Cutler, T. A. Apostolatos, L. Bildsten, L. S. Finn, E. E. Flanagan, D. Kennefick, D. M. Markovic, A. Ori, E. Poisson, G. J. Sussman, et al., *Phys. Rev. Lett.* **70**, 2984 (1993).

[12] J. Faber, *Class. Quant. Grav.* **26**, 114004 (2009).

[13] M. D. Duez, *Class. Quant. Grav.* **27**, 114002 (2010).

[14] E. E. Flanagan and T. Hinderer, *Phys. Rev. D* **77**, 021502 (2008).

[15] T. Hinderer, *Astrophys. J.* **677**, 1216 (2008).

[16] T. Damour and A. Nagar, *Phys. Rev. D* **80**, 084035 (2009).

[17] T. Binnington and E. Poisson, *Phys. Rev. D* **80**, 084018 (2009).

[18] T. Hinderer, B. D. Lackey, R. N. Lang, and J. S. Read, *Phys. Rev. D* **81**, 123016 (2010).

[19] S. De, D. Finstad, J. M. Lattimer, D. A. Brown, E. Berger, and C. M. Biwer, *Phys. Rev. Lett.* **121**, 091102 (2018), [Erratum: Phys. Rev. Lett.121,no.25,259902(2018)], 1804.08583.

[20] T. Malik, B. K. Agrawal, J. N. De, S. K. Samaddar, C. Providência, C. Mondal, and T. K. Jha, *Phys. Rev.* **C99**, 052801 (2019), 1901.04371.

[21] B. Abbott et al., *Phys. Rev. Lett.* **119**, 161101 (2017).

[22] B. P. Abbott et al., *Astrophys. J.* **848**, L13 (2017), 1710.05834.

[23] A. Goldstein et al., *Astrophys. J.* **848**, L14 (2017), 1710.05446.

[24] B. P. Abbott et al., *Astrophys. J.* **848**, L12 (2017).

[25] D. A. Coulter et al., *Science* (2017), [Science358,1556(2017)], 1710.05452.

[26] E. Troja et al., *Nature* **551**, 71 (2017), [Nature551,71(2017)], 1710.05433.

[27] D. Haggard, M. Nynka, J. J. Ruan, V. Kalogera, S. Bradley Cenko, P. Evans, and J. A. Kennea, *Astrophys. J.* **848**, L25 (2017), 1710.05852.

[28] G. Hallinan et al., *Science* **358**, 1579 (2017), 1710.05435.

[29] F. J. Fattoyev, J. Piekarewicz, and C. J. Horowitz, *Phys. Rev. Lett.* **120**, 172702 (2018).

[30] N.-B. Zhang and B.-A. Li, *Eur. Phys. J.* **A55**, 39 (2019), 1807.07698.

[31] N.-B. Zhang and B.-A. Li, *J. Phys.* **G46**, 014002 (2019), 1808.07955.

[32] Z. Carson, A. W. Steiner, and K. Yagi, *Phys. Rev.* **D99**, 043010 (2019), 1812.08910.

[33] M. Dutra, O. Lourenço, J. S. Sá Martins, A. Delfino, J. R. Stone, and P. D. Stevenson, *Phys. Rev. C* **85**, 035201 (2012).

[34] M. Dutra, O. Lourenço, S. S. Avancini, B. V. Carlson, A. Delfino, D. P. Menezes, C. Providência, S. Typel, and J. R. Stone, *Phys. Rev. C* **90**, 055203 (2014).

[35] J. Antoniadis and *et al.*, *Science* **340** (2013).

[36] J. D. Walecka, *Annals Phys.* **83**, 491 (1974).

[37] J. Boguta and A. R. Bodmer, *Nucl. Phys.* **A292**, 413 (1977).

[38] J. Boguta and H. Stoecker, *Phys. Lett.* **120B**, 289 (1983).

[39] B. D. Serot and J. D. Walecka, *Int. J. Mod. Phys.* **E6**, 515 (1997), nucl-th/9701058.

[40] R. J. Furnstahl, B. D. Serot, and H.-B. Tang, *Nucl. Phys.* **A615**, 441 (1997), [Erratum: Nucl. Phys.A640,505(1998)], nucl-th/9608035.

[41] B. G. Todd-Rutel and J. Piekarewicz, *Phys. Rev. Lett.* **95**, 122501 (2005).

[42] G. A. Lalazissis, T. Nikšić, D. Vretenar, and P. Ring, *Phys. Rev. C* **71**, 024312 (2005).

[43] D. Vautherin and D. M. Brink, *Phys. Rev. C* **5**, 626 (1972).

[44] E. Chabanat, P. Bonche, P. Haensel, J. Meyer, and R. Schaeffer, *Nucl. Phys. A* **627**, 710 (1997).

[45] T. Damour, A. Nagar, and L. Villain, *Phys. Rev. D* **85**, 123007 (2012).

[46] S. Postnikov, M. Prakash, and J. M. Lattimer, *Phys. Rev. D* **82**, 024016 (2010).

[47] S. Weinberg, *Gravitation and Cosmology* (Wiley, New York, 1972).

[48] M. Centelles, X. Roca-Maza, X. Viñas, and M. Warda, *Phys. Rev. C* **82**, 054314 (2010).

[49] M. Centelles, X. Roca-Maza, X. Viñas, and M. Warda, *Phys. Rev. Lett.* **102**, 122502 (2009).

[50] J. Piekarewicz, B. K. Agrawal, G. Colò, W. Nazarewicz, N. Paar, P.-G. Reinhard, X. Roca-Maza, and D. Vretenar, *Phys. Rev. C* **85**, 041302(R) (2012).

[51] P.-G. Reinhard, J. Piekarewicz, W. Nazarewicz, B. K. Agrawal, N. Paar, and X. Roca-Maza, *Phys. Rev. C* **88**, 034325 (2013).

[52] T. Malik, B. K. Agrawal, J. N. De, S. K. Samaddar, C. Providência, C. Mondal, and T. K. Jha, *Phys. Rev. C* **99**, 052801 (2019).

[53] M. Fortin, C. Providência, A. R. Raduta, F. Gulminelli, J. L. Zdunik, P. Haensel, and M. Bejger, *Phys. Rev. C* **94**, 035804 (2016).

[54] N. Alam, B. K. Agrawal, M. Fortin, H. Pais, C. Providência, A. R. Raduta, and A. Sulaksono, *Phys. Rev.* **C94**, 052801 (2016).

[55] G. Baym, C. Pethick, and P. Sutherland, *Astrophys. J.* **170**, 299 (1971).

[56] S. K. Dhiman, R. Kumar, and B. K. Agrawal, *Phys. Rev. C* **76**, 045801 (2007).

[57] M. Fortin, S. S. Avancini, C. Providência, and I. Vidaña, *Phys. Rev. C* **95**, 065803 (2017).

[58] D. Radice, A. Perego, F. Zappa, and S. Bernuzzi, *Astrophys. J. Lett.* **852**, L29 (2018).

[59] A. Bauswein, Talk delivered at International School of Nuclear Physics (2018).

[60] S. Brandt, *Statistical and Computational Methods in Data Analysis* (Springer, New York, 3rd English edition, 1997).

[61] B. Efron, *Ann. Stat.* **7**, 1 (1979), ISSN 00905364, http://dx.doi.org/10.2307/2958830.

[62] A. Pastore, *J. Phys.* **G46**, 052001 (2019), 1810.05585.

15

Weak Interactions and Nuclear Structure

Naftali Auerbach
Tel Aviv University
Michigan State University

Vladimir Zelevinsky
Michigan State University

Bui Minh Loc
Tel Aviv University
Ho Chi Minh City University of Education

CONTENTS

15.1   Introduction: Weak Interactions, Standard Model, and Nuclear Physics

Weak interactions entered nuclear physics earlier than the very idea of nuclear physics was born [1]. In 1900, E. Rutherford was able to separate α- and β-parts of the products of radioactive decays discovered in 1896 by H. Becquerel who in 1901 identified the beta particles with electrons introduced earlier by J.J. Thomson. The radioactive displacement law (F. Soddy and K. Fajans, 1913) established that the nuclear charge Z changes by one unit in beta decay, while the nuclear mass number A does not change. The natural radioactivity was observed only for the β^- decay when the nucleus changes its charge $Z \to Z + 1$.

The next important step in understanding the nature of weak interactions is related to the idea of neutrino (W. Pauli, 1930) that restores conservation laws of energy and angular momentum [2]. A rich bibliography on all aspects of neutrino physics and its history can be found in the resource letter [3]. After the discovery of the neutron, the β^+ decay was found experimentally for the unstable phosphorus isotope with the neutron and positron in the final state while $Z \to Z - 1$ (Frederic and Irene Joliot-Curie, 1934). The new weak process, the electron capture from the atomic K-shell, was discovered by L. Alvarez (1937). The experimental evidence for interactions induced by the neutrino was produced by F. Reines and C. Kowan only in the end of the 1950s [4,5].

Weak interactions from the very beginning served as a natural laboratory for studies of symmetries. Chen-Ning Yang and Tsung-Dao Lee (1956) performed a comprehensive analysis of existing data and showed that there is no proof that the weak interaction conserves spatial parity [6]. The immediate experiment by Chien-Shiung Wu and collaborators [7] demonstrated that parity is violated in the beta decay of ^{60}Co where there is a complete correlation of outgoing electrons with the spin polarization of the cobalt nucleus. Such a correlation between polar (electron momentum) and axial (nuclear spin) vectors shows parity ($\mathcal{P}$) nonconservation. As the electron comes in pair with the particle called the electron antineutrino, the last one is fully right-polarized (helicity $+1$), while this implies that the electron in this decay is left-polarized. The situation is opposite in the positron decay. The particle that comes along with the positron, the electron neutrino, is left-polarized, and the positron is right-polarized.

This seems to imply that as parity is completely violated, it is compensated by the full violation of charge conjugation symmetry $\mathcal{C}$, while the combined inversion, $\mathcal{CP}$, is still a good symmetry (Landau hypotheses [8]). However, later experiments with meson decays [9,10] found the $\mathcal{CP}$-violation which means that the time reversal symmetry $\mathcal{T}$ is also violated; only the full product $\mathcal{CPT}$ still appears as an exact fundamental symmetry. The $\mathcal{CP}$-violation is one of the conditions [11] necessary for understanding the particle-antiparticle asymmetry of the Universe. The important direction of symmetry studies is the ongoing experimental search for the static electric dipole moment (EDM) of elementary particles or individual nuclei, atoms or molecules. This discovery would show the simultaneous violation of $\mathcal{P}$ and $\mathcal{T}$ symmetries [12,13].

The weak nuclear transformations make up a large chapter of nuclear physics, medicine, and technology. Their knowledge is necessary for nuclear technological processes in all spheres of their application. The use of isotopes is on the rise, and weak interactions appear in all stages, including their harvesting, practical application, and decay. The astrophysical weak processes take place in various stellar objects, at different epochs of their life, and finally determine the abundances of stable elements in the Universe. The rates of weak transformations carry plentiful information on nuclear structure, including violations of fundamental symmetries. The collective and stochastic features of nuclear dynamics are successfully studied with the use of weak interactions. Theoretical predictions of rare observables are invaluable for the search of phenomena outside of the current standard model.

The traditional description of the mechanism of weak processes was based on the Fermi theory of the contact four-fermion interaction, for example for the neutron decay into proton, electron, and electron neutrino. This theory is still extremely useful for nuclear weak interactions. The great step in understanding the nature of weak interactions was in the discovery that this description is just a low-energy limit substituting the underlying interaction moderated by heavy mesons ($W^{\pm}$ and Z^0) by the effective contact four-fermion process. The new theory (S. Weinberg, S. Glashow, and A. Salam [14]) confirmed by the discovery of those mesons unified electromagnetism and weak interactions as a part of the full standard model of elementary particles and their interactions. The standard model can allow for only extremely small values for the EDM which are beyond the experimental reach. This brings a special interest to the EDM search as such a discovery would be the important phenomenon outside of the standard model.

Another very active direction in physics of weak interactions is the neutrino physics [15]. Now it is known that there are (at least) three types of neutrino and corresponding antineutrino (the problem of the sterile neutrino is still not solved). They belong to families labeled by their charged fermions (electron, muon, or tauon). However, these families are interrelated as corresponding neutrinos are not stationary quantum states—they are mixed and oscillate among themselves. This is possible only if the neutrinos are not massless, but their masses are still not known even if it is certain that they are smaller than 1 eV. Because of the finite mass, neutrinos are not always polarized with the fixed helicity. The parity violation in the weak processes is therefore not the property of the massless neutrino, but a feature of the left-handed nature of the weak interaction itself. Now one can imagine that the neutrino and antineutrino are actually the same particle, Majorana fermion. If so, there is no ban for a neutrinoless double beta decay which, for example, can proceed in two steps when virtual neutrinos cancel each other (another field of the current active experimental and theoretical search). Great interest to the physics of neutrino, theoretical and observational, including such an exciting development as the Ice Cube underground laboratory in Antarctica [16], is inherently related to the role of the neutrino in many astrophysical processes, including the life and explosions of supernova, physics of neutron stars, the above-mentioned problem of the charge asymmetry of the Universe, etc.

Our further text is not intended to give a full review of the weak interaction physics. It will be devoted to a much more modest task—to discuss the role of the weak interaction in nuclear structure and its use for the problems of fundamental symmetries when nuclei, atoms, and molecules can serve as natural laboratories where the complex many-body physics, in spite of its inevitable role as a complicated background, can enhance the manifestations of basic laws of nature. Of course, currently the area of nuclear astrophysics is one of the most actively developing branches of physics. There are many relatively fresh review papers, almost each of them is longer than our whole text. In our short section on weak interactions and astrophysics, we limit ourselves by a very concise introduction to this area; here, the reader is referred to a good book [17] and many references in our short section.

15.2 Beta Decay in Nuclei

15.2.1 Weak Processes in Nuclei

According to the current theory of electroweak interactions [14,18], the weak processes are mediated by the heavy vector mesons, charged $W^\pm$ and neutral Z^0. W-mesons with mass 80.4 GeV are responsible for charged currents and Z-mesons with mass 90.2 GeV for neutral currents. The typical example of the charged weak current, Figure 15.1, is the decay of the free neutron

$$n \to p + e^- + \bar{\nu}_e, \tag{15.1}$$

into proton, electron, and electron antineutrino. The electron carries away the negative electric charge, while the antineutrino is necessary for energy and angular momentum conservation. The term "antineutrino" is introduced to be able to prescribe the lepton charge which is conserved if these quantum numbers are opposite for the electron and the neutrino-like particle that appears in this process. On the quark level, the first act of this process is the virtual decay of one of the d-quarks in the neutron into the u-quark and W^- meson, followed by the decay of the latter into electron and electron antineutrino. The neutron lifetime, about 15 min, is that large because of a very big virtuality of the intermediate meson with its Compton wavelength $\hbar/(M_W c) \approx 10^{-16}$ cm and quite small energy released in this process. This allows one to return to the old four-fermion picture of this decay, Figure 15.2, using the universal Fermi constant:

$$\frac{G_F}{\sqrt{2}} = \frac{e^2}{8 M_W^2 \sin^2 \theta_W}. \tag{15.2}$$

Here, the Weinberg angle θ_W is one of the empirical parameters of the standard model given, at low energies of beta decay, by the value $\sin^2 \theta_W = 0,2223 \pm 0.0021$ (it is actually

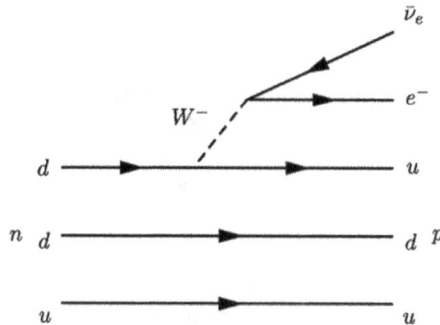

FIGURE 15.1
The process of neutron beta decay in quark picture.

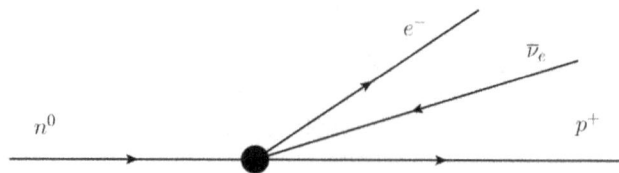

FIGURE 15.2
The process of neutron beta decay for the four-fermion point interaction picture.

determined in the standard model by the Higgs condensate). Frequently the constant $g^2 = e^2/\sin^2\theta_W$ is used instead of (15.2).

The problem of the neutron lifetime during past years acquired a special interest. The two approaches to the experimental measurement of this time turn out to give slightly different numbers. Roughly speaking, the *bottle* experiments measure neutron disappearance as a function of time, while the *beam* measurements use the appearance of protons. The beam measurement [19] at National Institute of Standards and Technology (NIST) gave the mean neutron lifetime $\tau_n = 887.7 \pm 1.2[\text{stat}] \pm 1.9[\text{syst}]$ s. The result of the bottle Grenoble experiment [20] was $878.5 \pm 0.7[\text{stat}] \pm 0.3[\text{syst}]$ s. The difference of 9 s seems to be a real experimental puzzle. One idea of explanation prescribes a longer lifetime counted by protons to the presence of another, until now not known, decay channel which can be speculatively given to the appearance of a not registered dark matter particle. There are also very weak decay channels with an extra gamma quantum or with the electron+proton in a bound hydrogen atomic states, but they could not resolve the puzzle. Some weak effects can also come from the interaction of the neutron magnetic moment with the parts of the experimental equipment, but the problem is still present while several laboratories in various countries are working in this direction.

The analogs of beta decay (15.1) are frequent in nuclei defining the so-called β^- branch of nuclear decays,

$$^A(Z)_N \Rightarrow {}^A(Z+1)_{N-1} + e^- + \bar{\nu}_e. \tag{15.3}$$

In this way, the neutron-rich nuclei can return to the more energetically beneficial relation between the numbers of neutrons and protons. Opposite to that, the nuclei with a surplus of protons can get rid of extra energy performing the β^+ decay:

$$^A(Z)_N \Rightarrow {}^A(Z-1)_{N+1} + e^+ + \nu_e. \tag{15.4}$$

Both decays are really allowed if the nuclear product has lower energy (Q-value, or the thermal effect of the reaction, is positive). The energy threshold for the β^+ process is even lower by $2m_ec^2$ if it goes as orbital capture of an atomic electron:

$$^A(Z)_N + e^- \Rightarrow {}^A(Z-1)_{N+1} + \nu_e. \tag{15.5}$$

The hole in the atomic structure is later filled in either by an outer electron with the X-ray emission or by a direct emission of an *Auger electron*. The recent GSI (Gesellschaft fuer Schwerionenforschung, currently Helmholz Center for Heavy Ion Research, Darmstadt, Germany) experiments [21] used almost completely ionized atoms when the only decay channel is allowed. The decay curve had oscillations which, however, most probably are not related to neutrino oscillations.

There are several cases when the usual beta decay of an even-even nucleus is energetically forbidden as the neighboring odd-odd nucleus has the ground state not accessible by energy (essentially the pairing effect). This is the case, for example, in $^{76}_{32}\text{Ge}_{44}$. The abundance of neutrons here would naturally lead to β^- decay to $^{76}_{33}\text{As}_{43}$, which, however, has the ground state higher in energy and undergoes its own β^- decay to the next even-even nucleus $^{76}_{34}\text{Se}_{42}$. As the ground-state energy of ^{76}Se is lower than that of ^{76}Ge, energetically allowed is the direct *double beta decay*; it is enough energy even for the beta decay to the two excited states of ^{76}Se. The double decay as the second-order process in weak interaction certainly is very slow; for the ^{76}Ge, the half-life is 1.8×10^{21} years. The fastest double decay case is observed for ^{100}Mo with half-life 7×10^{18} years, while the longest (half-life $(1.8\pm0.5) \times 10^{22}$ years) double decay case measured directly [22] is that of electron capture in ^{124}Xe found as a byproduct in the work of the xenon-based dark matter detector. A historian of weak interactions [23] comments the first direct observation [24] of the double decay of ^{82}Se

with the half-life of 10^{20} years that *"This result was very important, particularly from the psychological point of view."*

In all the observed cases of double beta decay [25], the process releases two antineutrinos:

$$^A(Z)_N \Rightarrow {}^A(Z+2)_{N-2} + 2e^- + 2\bar{\nu}_e. \tag{15.6}$$

There are serious efforts in the search of a *neutrinoless* double beta decay. This is possible if the neutrino is a *Majorana fermion* identical to its antiparticle. Then, the antineutrino released in the first act of beta decay could induce the second act which, for Dirac fermions, would require the neutrino. Moreover, the same conclusion about the Majorana type would follow from the very fact of the neutrinoless double beta decay even if the intrinsic mechanism would be different (not containing a virtual neutrino) [26]. Until now, there are no observations of the neutrinoless double beta decay ($0\nu\beta\beta$) so that the question of the nature of the neutrino remains open. The last result belongs to the EXO collaboration [27] that gives no statistically significant evidence for $0\nu\beta\beta$ in the case of ^{136}Xe, leading to a lower limit on the half-life of 1.8×10^{25} years at the 90% confidence level.

As in the neutron decay, also in nuclei, there are rare examples of beta decay with the electron remaining in one of the atomic states of the daughter nucleus. Then, almost the whole decay energy is carried off by the antineutrino. This was observed at GSI (Darmstadt) in fully ionized atoms ^{163}Dy^{66+} [28] and ^{187}Re^{75+} [29]. In both cases, the ionization works in favor of the beta decay (the neutral ^{163}Dy is stable, while in the ^{187}Re case, the decay time is shortened by nine orders of magnitude). At the same time, there are no statistically significant effects of chemical environment or season variations [30].

We can also mention rare cases when the neutron undergoing "quasi-free" beta decay produces the proton in the continuum. This channel was observed [31] in the decay of the halo neutron nucleus ^{11}Be into ^{10}Be+p with the branching ratio 8×10^{-6} compared to the decays into the bound states of ^{11}B. Even the processes with the deuteron emission are known when the deuteron is formed in beta decay of the two-neutron halo pair in ^{6}He [32] and ^{11}Li [33].

15.2.2 Fermi and Gamow-Teller Transitions

Among the numerous examples of the nuclear beta decay, the special role belongs to the so-called *superallowed* transitions. In the four-fermion Fermi picture, those are the cases where the coordinate wave function of the nucleon undergoing the decay does not change (only the nucleon spin can change), while its isospin projection flips transforming neutron to proton or vice versa. The quantum amplitude of this transformation has to be integrated over the nuclear volume with the actual many-body wave function of the initial and final nucleus as a whole. The wave functions of the leptons [electron or positron and (anti)neutrino] with the wavelengths exceeding the size of the nucleus are taken as plane waves equal to 1 inside the nuclear volume. Then, the nuclear matrix element may contain only an isospin part, t^+ or t^-, and spin operators $\vec{\sigma}$ summed over all nucleons.

The corresponding operators inducing the beta process are therefore the simplest sums over all nucleons being vectors with respect to isospin: the *Fermi operator* which is just the increasing or decreasing total isospin component,

$$F_{\pm} = \sum_a t_{\pm}(a) \equiv T_{\pm}, \tag{15.7}$$

and the *Gamow-Teller (GT) operator*, vector in isospin and in spin of nucleons,

$$(GT)_{\nu;\pm} = \sum_a \sigma_\nu(a) t_{\pm}(a), \tag{15.8}$$

with the obvious selection rules $\Delta S = 0, \Delta T_3 = \pm 1$ in the Fermi case and $\Delta S = 0, \pm 1; \Delta T_3 = \pm 1$ in the GT case. Both operators do not carry any radial dependence. However, acting inside a complex mesoscopic system of interacting constituents, they may change the state of the nucleus in a nontrivial way.

The Fermi operator (15.7) moves the nucleus inside the isobaric multiplet to the *isobaric analog* state (IAS). However, the daughter state has a different energy, at least due to the mass difference and Coulomb energy [34], although strong nuclear forces also have small charge-dependent components. For the GT operator, the isospin can change, and due to the possible additional spin-flip, several final states are possible, typically at energies higher than IAS. Because of energy conservation, an actual beta process can populate only a part of formally allowed final states. The states inaccessible in beta decay can be studied in charge-exchange reactions [35], which is important for understanding stellar processes, as well as for physics of double beta decay [36].

The structures of operators (15.7) and (15.8) are so simple that allow us to find the exact operator *sum rules*. For the Fermi operator, the total operator strength summed over final states,

$$B^{\pm}(F) = \sum_f |\langle f|F_{\pm}|i\rangle|^2 = \langle i|T_{\pm}T_{\mp}|i\rangle, \qquad (15.9)$$

has, for an initial pure isospin state $|TT_3\rangle$, the universal value

$$B^{\pm}(F) = (T \mp T_3)(T \pm T_3 + 1). \qquad (15.10)$$

In a nucleus with neutron excess, the ground state typically has $T_3 = T \geq 0$, and $B^{+}(F)$ vanishes as a new neutron born in $p \rightarrow n$ decay would occupy the orbital already filled by a neutron. But the opposite process is allowed for any of $N - Z$ excess neutrons, so that $B^{-}(F) \rightarrow 2T = 2T_3 = N - Z$. In the consideration of nuclei close to the proton drip line, $Z > N$, the roles of neutrons and protons are inverted. For example, the nucleus $^{18}_{10}\text{Ne}_8$ has a Fermi decay branch with probability 7.7% from its ground 0^+ state with isospin $T = 1$ into its IAS, the first excited state of $^{18}_{9}\text{F}_9$, which also has quantum numbers 0^+ while the GT decay $J = 0 \rightarrow J = 0$ is forbidden. The lifetime of $^{18}_{10}\text{Ne}_8$ is determined by the GT transition to the ground state with $J^{\Pi} = 1^+$ (branching 92%).

Using the Hermiticity properties of the operators expressed in spherical components, we obtain for the GT strength

$$B^{\pm}(GT) = \sum_{f;\nu} |\langle f|(GT)_{\nu;\pm}|i\rangle|^2 = \sum_{\nu}(-)^{\nu}\langle i|(GT)_{-\nu;\mp}(GT)_{\nu;\pm}|i\rangle, \qquad (15.11)$$

or, with the summation over particles,

$$B^{\pm}(GT) = \sum_{ab}\langle i|(\vec{\sigma}_a \cdot \vec{\sigma}_b)t_a^{\mp}t_b^{\pm}|i\rangle. \qquad (15.12)$$

Applying the commutation properties of isospin operators, this leads to the *Ikeda sum rule* [37]:

$$B^{-}(GT) - B^{+}(GT) = 6\sum_a (t_a)_3 = 6T_3 = 3(N - Z). \qquad (15.13)$$

15.2.3 Lifetime in Beta Decay

Using the perturbation theory with respect to weak interactions, we come to the allowed transitions if the lepton (electron and antineutrino in the β^- decay) wave functions can be taken constant inside the nucleus. As the antineutrino energy is defined by the conservation law, the corresponding probability contains the density of electron states in the continuum

$$\rho(E_e) = \frac{(E - E_e)E_e\sqrt{[(E - E_e)^2 - m_\nu^2 c^4](E_e^2 - m_e^2 c^4)}}{4\pi^4 \hbar^6 c^6}, \tag{15.14}$$

where E is the total decay energy shared by the electron and the antineutrino; this expression can be complicated by the presence of the neutrino oscillations. In principle, the small but non-vanishing mass of the (anti)neutrino can be extracted from the shape of the energy spectrum. The corresponding experiments already for many years try to use the cases with the smallest total decay energy, such as tritium decay ${}^3\text{H} \to {}^3\text{He} + e^- + \bar{\nu}_e$ with total energy release of just 18.6 keV and the half-life of 12.3 years. In reality, this is extremely hard, and the most advanced Karlsruhe Tritium Neutrino Experiment (*KATRIN*) goes on for many years. The available data on neutrino oscillations determine only differences of squared masses of neutrinos of three generations, $e, \mu,$ and τ, but not the absolute values.

The lifetime with respect to the beta decay is an empirical quantity. In order to understand the underlying physics, we have to correctly evaluate the level density of outgoing electrons and neutrinos in the continuum. In the simplest kinematical approximation, this quantity integrated over (anti)neutrino momentum is given by Eq. (15.14). The neutrino mass here might be important only for the purpose of its measurement at the very end of the spectrum; then, also corrections for the possible oscillation effect and appearance of a neutrino with different mass can be introduced. The difficulty of the direct measurement of the neutrino mass is related to the fact that even in the tritium beta decay, thinking about $m_\nu < 1$ eV, the probability of the electron being registered within 1 eV from the endpoint is just 10^{-13}.

The transition rate $\dot{w}$ is determined by the nuclear matrix element for fixed initial and final wave functions and the integral over the electron spectrum. With the four-fermion Hamiltonian, two nucleons and two leptons are taken at the same point in the nuclear matrix element, see Figure 15.2. But the Coulomb interaction distorts the electron (positron) wave function inside the nucleus. This generates the so-called *Fermi function* $F(Z, p)$, where p is the electron (positron) momentum, that was tabulated long ago with the aid of realistic atomic calculations [38] accounting for relativistic effects. Going to integration over p and collecting all constant factors, the beta transition rate can be presented in the form

$$\dot{w} = \frac{m_e^5 c^4}{2\pi^3 \hbar^7} G^2 |M|^2 f(Z, p_{\max}), \tag{15.15}$$

where M is the nuclear matrix element, and f is the dimensionless integral over electron momentum of the Fermi function and the density of states. Here G is the fundamental constant of the four-fermion weak interaction. With the half-life time $t_{1/2}$, we have $\dot{w} = \ln(2)/t_{1/2}$, and the standard characteristic of the beta decay is given by the product

$$ft_{1/2} = \frac{K}{G^2 |M|^2}, \tag{15.16}$$

where we define the universal constant K,

$$\frac{K}{(\hbar c)^6} = 2\pi^3 \ln(2) \frac{\hbar}{(m_e c^2)^5} = (8120.2776 \pm 0.0009) \times 10^{-10}\,\text{s GeV}^{-4}. \tag{15.17}$$

Traditionally, the $\log_{10}(ft_{1/2}) \equiv \Lambda$ is used as a simple number characterizing the specific beta decay. In the case of both Fermi and GT decays allowed, we have instead of (15.16)

$$ft_{1/2} = \frac{K}{G_V^2} \frac{1}{B(F) + (G_A/G_V)^2 B(GT)}, \tag{15.18}$$

with the numerical values

$$\frac{K}{G_V^2} = 6144.2 \pm 1.6\,\text{s}, \quad \frac{G_A}{G_V} = -1.2694 \pm 0.0028. \tag{15.19}$$

Figure 15.3, taken from the book [39], shows the distribution of Λ over known beta decays.

15.2.4 Briefly about Actual Data

In the first approximation, the β^- transition $0^+ \to 0^+$ inside a given isobaric multiplet is a pure Fermi decay with the simple matrix element (15.9). In particular, for $T = 1$, $N - Z = 2$, we expect the universal result

$$ft_{1/2} = \frac{K}{2G_V^2}. \tag{15.20}$$

These are the so-called *superallowed* transitions encountered in relatively light nuclei. The result has to be a universal number close to 3070 s which is long ago known to be the case. Of course, there are small effects not accounted for in Eq. (15.20). They are coming from isospin violation in nuclear forces and from electromagnetic interactions (radiative corrections). Here the high precision is important for defining the constant G_V and checking the unitarity of the quark mixing matrix, see in more detail in Section 3.

Traditionally [1], the beta transitions in nuclei are classified by the magnitudes of $\lg(ft)$ determined by the nuclear matrix elements of weak interaction between mother and daughter nuclei (for the same pair of nuclei, there could be several transitions into different final states with their partial values of $\lg(ft)$). In the case of superallowed transitions, the initial and final states are strongly overlapping (up to a change of isospin). A mirror transition $^{17}_9\text{F}_8 \to {}^{17}_8\text{O}_9$ with both, initial and final states $5/2^+$ being members of the same isotopic doublet, proceeds as electron capture with half-life time 66 c and $\lg(ft) = 3.38$. The fastest mirror transitions take place if they are pure Fermi-type. In certain cases, the spin-parity quantum numbers do not allow Fermi transition: the simplest example is for ^6He (its only bound state 0^+, "alpha + neutron pair", decays by the GT transition into

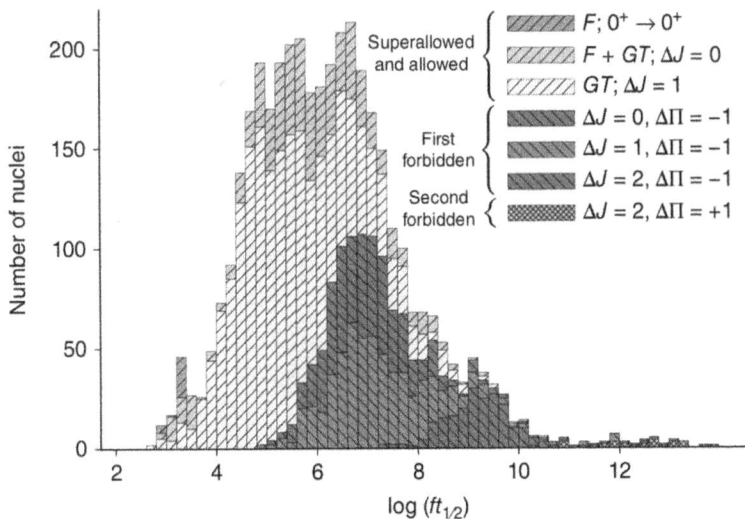

FIGURE 15.3
The distribution of Λ over known beta decays.

the "alpha + deuteron" state 1^+ of ^{6}Li with the quite small $\lg(ft)=2.95$). This process is sensitive to the spatial structure of the nuclear wave functions [40]. The transitions $0^+ \to 0^+$ between different isobaric multiplets ($\Delta T = \pm 1$) are much slower than inside the multiplet, and the corresponding quantity ft is greater by 3-6 orders of magnitude. Such a suppression by isospin tells us that even in heavy nuclei, isospin is still a good quantum number. For example, three $0^+ \to 0^+$ transitions from ^{234}Np, $T = 24$, to different levels in ^{234}U, $T = 25$, have ft from 3.4×10^8 to 1.5×10^9 [41]. For allowed transitions into a certain state of the daughter nucleus, the experimentally defined function of electron energy E and decay intensity $N(E)$,

$$\sqrt{\frac{N(E)}{f(Z,E)p_e E}} = \text{const}\,(E_0 - E),\qquad(15.21)$$

is given by the straight line going down up to the intersection with the horizontal axis at the full decay energy E_0. This is the so-called Kurie plot [42] which indeed is a straight line as can be seen from Figure 15.4 for tritium beta decay, where the small deviations at the spectrum end, if measured, would define the neutrino mass.

A classical example of an allowed transition is the pure GT decay ^{60}Co$\to^{60}$Ni where the parity nonconservation was found for the first time, Figure 15.5. The transition can proceed to two low-lying collective states in the daughter nucleus, 4^+ at 2.495 MeV, the corresponding $\lg(ft) = 7.5$, and 2^+ at 1.332 MeV; the second possibility is realized only in 10^{-4} fraction of decays because of a big spin mismatch as the mother state is 5^+. Another famous case is the GT decay of the only radioactive carbon isotope ^{14}C to stable ^{14}N with the half-life of 5730 years ($\lg(ft) = 9$). Here, the energy release is quite small, only 156 keV.

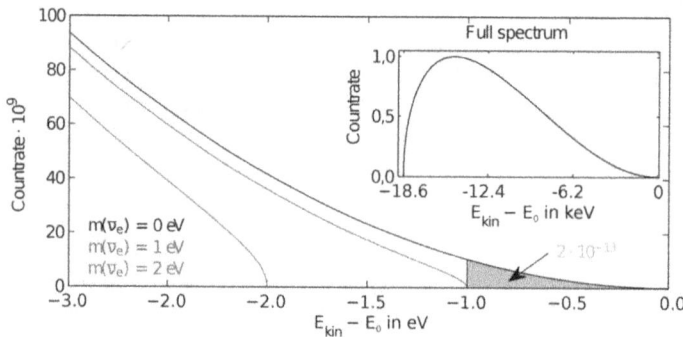

FIGURE 15.4
Kurie plot for the beta spectrum of tritium.

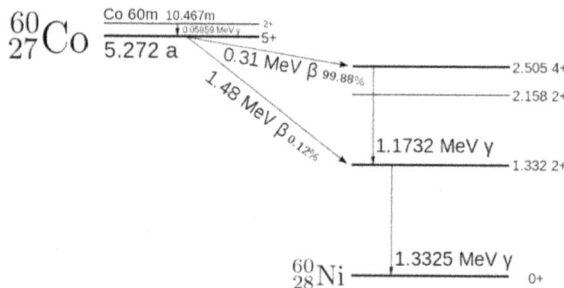

FIGURE 15.5
Scheme of the beta decay ^{60}Co$\to^{60}$Ni.

The intensity of this decay is used in radioactive dating [43] for the old samples, up to 6×10^4 years. In some cases, only the spin-orbit coupling makes the transition allowed.

For the so-called forbidden transitions, usually we have either opposite parities of the initial and final nuclear states or the nuclear spin difference $\Delta J > 1$. The simple Fermi and GT transitions are then forbidden. The first effect that allows them comes from higher orbital components of the lepton [electron and (anti)neutrino] wave functions. This, formally seen as the centrifugal barrier for the leptons, can be interpreted as retardation. As a result, the selection rules for nuclear spin and parity change. Other new effects include relativistic corrections for nuclear wave functions and operators also influencing electron wave functions. Every next step in this direction typically increases $\lg(ft)$ by 2–3 units. In fact, the ladder of consecutive approximations in the process of applying the higher-order terms is similar to the mutipole expansion for the electromagnetic processes. For forbidden decays, the Kurie plot deviates from the straight line.

New physics can appear in beta decay of well-deformed nuclei. The low-lying stationary states are grouped in rotational bands with intraband energy growing with spin as $J(J+1)$. With an axial symmetry of deformation, the bands can be labeled by an additional quantum number of $|K|$, where K is a total spin projection onto the symmetry axis. Then, in addition to the standard selection rules for parity and total spin J, the new selection rule for the Fermi and GT transitions appears, $|\Delta K| = 0, \pm1$. This is the source of isomerism in many heavy nuclei. In the odd-odd nucleus ^{176}Lu, the ground state has spin-parity 7^- and $K = 7$. According to the J selection rules, its GT decay proceeds into the excited state 6^+ of the ground-state band $K = 0$ of the even-even nucleus ^{176}Hf. The large K difference makes this decay almost improbable: the half-life is 7.4×10^{10} years with $\lg(ft) = 18.7$.

The ^{180}Ta^m nucleus is the rarest naturally occurring isotope [45]. It exists in an isomeric state 9^- with a half-life longer than 1.2×10^{15} years at an excitation energy of 77 keV, while the ground state of ^{180}Ta has quantum numbers 1^+ and its half-life is 8.15 hours. The ground state starts the $K = 1$ band of collective rotation of the odd-odd nucleus (an almost pure Nilsson configuration $\{(p_{7/2}[404]) \times (n_{9/2}[624])\}_{K=1}$ around the axis perpendicular to the symmetry axis), while the higher-K band is due to a noncollective rotation of only a few unpaired nucleons around the symmetry axis (the main configuration is $\{(p_{9/2}[504]) \times (n_{9/2}[624])\}_{K=9}$). The gamma-radiation between these states is practically improbable. The weak transition between the two rotational schemes is strongly hindered because of the very different geometry of these two rotations so that a large rearrangement is required in order to go from one form of motion to the other.

With measurement of the angle between the decay electron and recoil nucleus, one can study the angular correlation between electron and neutrino that has a form

$$W(\theta) = 1 + a \frac{v}{c} \cos\theta, \tag{15.22}$$

where v is the electron velocity and θ is the angle between momenta of electron and neutrino. The angle between the nuclear recoil and electron momentum is $\theta' = (p_\nu/p_{\text{recoil}}) \sin\theta$. Such measurements were instrumental in establishing the $V - A$ form of the weak interaction. For a pure GT decay ^{6}He$\rightarrow^6$Li, the experimental result [44] is $a = -0.0334 \pm 0.003$. In the neutron decay, the Fermi and GT transitions contribute almost equally, and the asymmetry coefficient a is small being the worst known quantity in the neutron decay (except for the total lifetime puzzle mentioned earlier).

15.2.5 Gamow-Teller Transitions and the Quenching Problem

This is the part of the physics of weak interactions in nuclei that, being very important for understanding many problems, seems to be less well understood. Even the Ikeda sum rule (15.13) is not saturated experimentally being noticeably quenched. We note here that this exact operator identity is certainly fulfilled in the complete isospin-conserving shell-model diagonalization for a finite orbital space that includes appropriate neutron and proton orbitals, while in approximate approaches, such as RPA, one needs to make special efforts [46] for the fulfillment of this sum rule. The experimental quenching is known for a long time [47–49] on the level of 60–70%.

Here we deal with the general problem of spin-isospin nuclear excitations. The standard experimental tool and the main source of experimental information are the use of charge-exchange reactions $(p, n), (n, p)$ or similar processes with heavy ions, generated by the operators $\sigma_i \tau^\pm$. The isovector part of effective nucleon-nucleon (NN) forces is the most pronounced at kinetic energies of few hundred MeV. Experimentally, the GT giant resonance [50] was found at zero degrees (selection of the orbital momentum $L = 0$ component). The systematics of charge-exchange reactions [51] indicates also the presence of resonance structures at nonzero angles corresponding to $L \neq 0$.

Basically, two types of theories were introduced to discuss the quenching of GT and M1 strength in nuclei.

A. The quenching of the GT strength in charge-exchange reactions refers to the main peaks where the GT strength is concentrated. So, it was suggested that the remaining strength is fragmented and spread out at several tens of MeV above the main peaks [52–55]. The nuclear force (and especially the tensor component) causes a mixing of the one particle-one hole $(1p - 1h)$ components with $2p - 2h$ excitations, causing the strength to be spread out. There are attempts to locate experimentally this fragmented strength, but it is quite difficult to reach a conclusive result. This mechanism could also affect other giant resonances, but no such effect was clearly observed so far.

B. The other approach taken in the past to explain the missing GT and M1 strength was to consider the influence of internal nucleon degrees of freedom that couple to the nucleus degrees of freedom and remove the strength to very high excitation energies, about 300 MeV above the ground state [56–62]. The GT operator can act upon the quarks inside each nucleon exciting the closest nucleon resonances, first of all Δ_{33} with spin and isospin 3/2. Simple estimates of such nucleon polarizability [60] show that the effect is local inside the nucleus and therefore is not sensitive to the details of nuclear structure, contrary to the usual shell-model interactions. It should also influence nuclear magnetic moments and the M1 isovector giant resonance. The partial transfer of the GT strength to the 300 MeV excitation energy would give a natural explanation to the quenching of the observed strength at lower energies. Effectively, all nuclear matrix elements of the spin-isospin symmetry are to be renormalized. In general, there are no direct visible indications of the non-trivial quark structure of nucleons in low-energy nuclear structure. Even the fact that neutron and proton as bound states of quark and gluons are, strictly speaking, not fermions but compound objects satisfying more complicated commutation relations [63], seems not to lead to considerable observable consequences in low-energy nuclear phenomena. Therefore, the idea of high-energy structures excited by the simple nucleon operators and visible in low-energy experiments is quite attractive.

And then, it is quite possible that both mechanisms A and B contribute to this puzzle of the missing GT strength. None of these possibilities can be ruled out at present.

Some charge-exchange experiments, especially covering a wider energy interval, show a smaller quenching effect. In the reaction $^{90}\text{Zr}(n, p)$ over the broad energy range up to 70 MeV [64], the quenching factor is reported as 0.88 ± 0.06. It is stressed that the noticeable strength is hidden at relatively high (but still pure nucleonic) energy in the continuum, where there are experimental problems with uncertainties and normalization. Another question is the elimination of the isovector spin-monopole resonance with the same spin-parity quantum numbers (excited by the operator $r^2 \vec{\sigma} \tau^\pm$). Assuming these results, the coupling to the Δ degrees of freedom should be taken weaker than accepted earlier.

The shell-model analysis of the GT physics [65] stressed a strong anticorrelation of the GT strength and spin-orbit splitting noticed in [66]. This is accompanied by the strong enhancement of low-energy quadrupole collectivity. These two possible collective behaviors are opposite with respect to underlying physics, see Figure 15.6. It follows from the shell-model consideration that the GT strength considerably grows for the processes started in excited states. The earlier analysis for the ^{24}Mg nucleus [67] showed a steady increase of this strength as a function of excitation energy of the initial state. Apart from the statistical effect of the level density, a considerable part of this increase comes from the suppression of spatial symmetry and corresponding progress towards the Wigner $\mathcal{SU}(4)$ symmetry. However, for the final states of the daughter nucleus in the higher part of the shell-model space, the growth of the summary GT strength (for the excitation of the ground state of the mother nucleus) comes through many small increments, see Figures 15.7 and 15.8, which creates additional experimental problems. Another consequence of the shell-model analysis is that the presence of the spin-orbit coupling mixes the orbital momenta $L = 0$ and $L = 2$, while

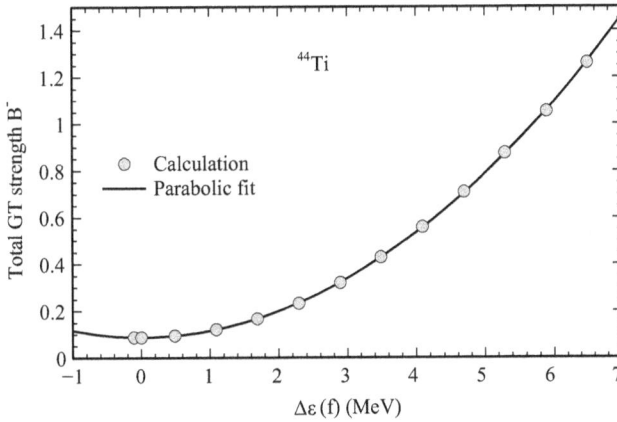

FIGURE 15.6
The parabolic dependence of the GT strength on the spin-orbit splitting in ^{44}Ti.

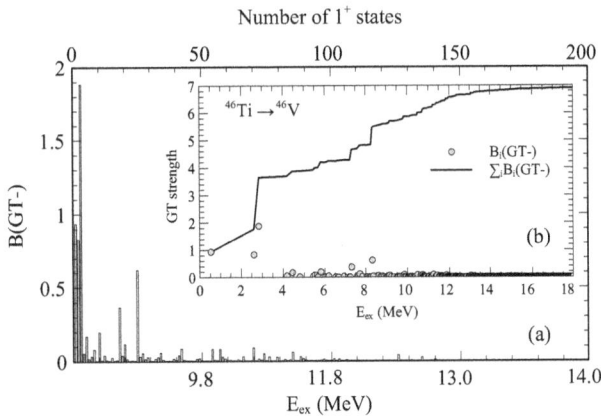

FIGURE 15.7
(a) Distribution of the GT strengths from the ground state of ^{46}Ti along the excitation energy in the daughter state ^{46}V. (b) Cumulative sum of the GT strengths growing as a function of the excitation energy in the daughter nucleus ^{46}V.

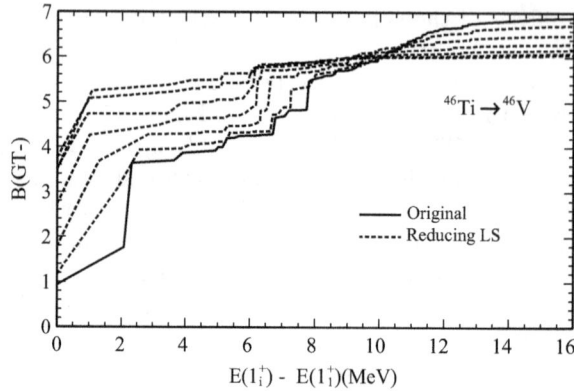

FIGURE 15.8
Cumulative sum of the GT strengths growing as a function of the excitation energy in the daughter nucleus ^{46}V. The consecutive lines illustrate the accumulation process for several values of the spin-orbit splitting.

the reaction analysis with the help of the multipolar expansion typically extracts only the $L = 0$ component.

The recent publication [68] announced that the quenching problem is resolved by the new set of advanced calculations based on the effective field theory. The main change is related with the introduction of two-body terms in the GT operator. The additional operator terms which are lowering the full GT matrix element are not discussed explicitly in this short text, but they do not require anymore comparison with the Ikeda sum rule. It would be still premature to judge these results as the charge-exchange reactions are excluded from consideration [68], while practically all experimental results are extracted from such reactions that do not involve the weak interaction and are due to the strong interaction. All these charge-exchange reactions show a quenching in the main GT peaks of the same size as in the beta decays.

15.3   Relation to Fundamental Theory

15.3.1   Quark Mixing (CKM) Matrix

As now we are interested in the high-precision results, we need here a more detailed description of nuclear weak processes. Weak interaction is not diagonal with respect to quark flavors. The propagation of free quarks preserves the flavors, but they can be mixed in weak processes. This idea was suggested by Cabibbo [69] for two quark generations and extended by Kobayashi and Maskawa for three generations [70]. Instead of flavor states, for example for down quarks, d, s and b, the weak interactions select their linear combinations which are transformed in weak processes to up quarks, u, c and t. The Cabibbo-Kobayashi-Maskawa (CKM) matrix formally parameterizes the relation between the mass and weak bases,

$$\begin{pmatrix} d' \\ s' \\ b' \end{pmatrix} = \begin{pmatrix} V_{ud} & V_{us} & V_{ub} \\ V_{cd} & V_{cs} & V_{cb} \\ V_{td} & V_{ts} & V_{tb} \end{pmatrix} \begin{pmatrix} d \\ s \\ b \end{pmatrix}. \tag{15.23}$$

Unprimed are quark mass eigenstates. In terms of the matrix elements $V_{f'f}$ of the matrix (15.23), the part of the weak Lagrangian responsible for the processes with the quark flavor changes $f \to f'$ and corresponding lepton production has the form

$$\mathcal{L} = -\frac{1}{\sqrt{2}} G_F V_{f'f} (\bar{u}_{f'} \gamma_\mu (1 - \gamma^5) d_f)(\bar{\ell} \gamma^\mu (1 - \gamma^5) \nu) + (\text{H.c.}), \qquad (15.24)$$

where it is usually used $G_F/\sqrt{2} = g^2/(8m_W^2)$, and g is the parameter of the standard model, compare Eq. (15.2).

If there are only three generations, the unitarity requirement determines that this matrix contains three mixing angles and only one possible complex phase (for two generations, the phases always can be removed by the redefinition of basis vectors). This phase is responsible for the effects of time reversal ($\mathcal{T}$) violation, and therefore, due to the full $\mathcal{CPT}$ invariance, for $\mathcal{CP}$-violation [71]. The search for the sources of the $\mathcal{CP}$-violation is a necessary step for solving the problem of baryon-antibaryon asymmetry of the Universe. There exist various parameterizations of the CKM matrix, for example, an explicitly unitary form in terms of angles (sines $s_{1,2,3}$ and cosines $c_{1,2,3}$) and the $\mathcal{CP}$-violating phase δ,

$$(\text{CKM}) = \begin{pmatrix} c_1 & -s_1 c_3 & -s_1 s_3 \\ s_1 c_2 & c_1 c_2 c_3 - s_2 s_3 e^{i\delta} & c_1 c_2 s_3 + s_2 c_3 e^{i\delta} \\ s_1 s_2 & c_1 s_2 c_3 + c_2 s_3 e^{i\delta} & c_1 s_2 s_3 - c_2 c_3 e^{i\delta} \end{pmatrix}. \qquad (15.25)$$

In terms of the representation (15.23), the unitarity leads to many relations between the matrix elements. Nuclear experiments with weak interaction processes can give a way of checking some of them, first of all

$$|V_{ud}|^2 + |V_{us}|^2 + |V_{ub}|^2 = 1. \qquad (15.26)$$

The matrix element V_{ud} can be extracted from beta decays and has the largest absolute value among three terms in Eq. (15.26), $|V_{ud}| \approx 0.97$, while from hyperon decays $|V_{us}| \approx 0.2$, and $|V_{ub}| \approx 0.003$ is quite small. The problem is to account for many secondary effects and enhance the precision of the fulfillment of Eq. (15.26) which would tell us that the standard model is complete and we do not expect more quark generations involved in weak interactions. The matrix element V_{ud} works as the main ingredient in the neutron decay, pion decay and all superallowed $0^+ \to 0^+$ nuclear decays. In all known cases, the experimental value of the $ft_{1/2}$ is close to 3040 s, independently of isospin of nuclear states involved in the process. The comprehensive work on the problem of unitarity (15.26) was performed by I. Towner and J. Hardy, see the review [72] and the later publication [73]. The whole problem of the search for the manifestations of new physics in weak processes is reviewed in [74,75].

The γ^μ and $\gamma^\mu \gamma^5$ terms in the quark matrix element (15.24) give the Fermi and GT decays, respectively. The Fermi matrix elements as a function of the quark momenta can be parameterized using the idea of the conserved vector current (CVC) [76,77]. This current is the charge-changing component of the general vector current where the charge-conserving component is the electromagnetic current. Then, the Fermi matrix elements, as a function of the quark momentum transfer $q = p_1 - p_2$, have two types of form-factors, vector $g_V(q^2)\gamma^\mu$ and the so-called weak magnetism usually written as $(1/2m)f_M(q^2)i\sigma_{\mu\nu}q^\nu$ where the proton mass m is used for correct dimension. With the assumption of the G-parity [78], which is a combination of the charge-transformation and isospin rotation, the GT transitions are also characterized by two form-factors, axial $g_A(q^2)\gamma^\mu\gamma^5$ and pseudoscalar $(1/2m)f_P(q^2)q_\mu\gamma^5$. An important difference between the Fermi and GT transitions is that due to the conservation of the vector current, $g_V(0) = 1$, while $g_A(0) \neq 0$, and in beta decay we just need the limit $q^2 \to 0$. Finally, the relation between those form-factors (taken at $q^2 = 0$) and nuclear transition matrix element T_{fi} is given [72] by

$$T_{fi} = -\frac{1}{\sqrt{2}} G_F V_{ud} (g_V M_F \delta_{\mu,0} - g_A M_{GT} \delta_{\mu,i}) (\bar{\ell}\gamma^\mu (1 - \gamma^5)\nu), \qquad (15.27)$$

where M_F is the matrix element of $(\bar{u}\gamma_0 d)$ and $M_{GT;i}$ is the matrix element of $(\bar{u}\gamma_i \gamma^5 d)$. In the general case when the phase space integrals f_V and f_A can be slightly different for the Fermi and GT transitions (the so-called shape correction function [79]), the analog of Eq. (15.18) is

$$t = \frac{K}{G_F^2} \frac{1}{V_{ud}^2[f_V |M_F|^2 + f_A |M_{GT}|^2]}. \qquad (15.28)$$

15.3.2   Isospin Violation

Corrections to the universal description of the Fermi transitions are mainly related to the isospin violation. Being important for many problems of nuclear structure and reactions, it comes, in particular, from the Coulomb interaction that breaks symmetry between protons and neutrons [62]. Traditionally, the change of the Fermi matrix element in Eq. (15.28) is written as

$$|M_F|^2 = |M_F^\circ|^2 (1 - \delta_C), \qquad (15.29)$$

where the correction δ_C, usually on the level of 1% or even smaller, is especially needed for those decays where the experimental precision is high. The "naked" matrix element M_F° for the Fermi process is trivial, Eq. (15.10).

One approach [72,80] to the calculation of the correction δ_C uses the shell model and the mean-field description. Because of the too large orbital spaces necessary for the full account of Coulomb admixtures, the correction sought for is split into two parts. The first part results from the shell-model calculation in the practically accessible orbital space (no nodal mixing), and typically, at least in relatively light nuclei especially important for the unitarity problem, it is less than 0.1%. This result is rather sensitive to the details of the used shell model but can be related to very small branching ratios measured for the transitions into non-analog states in the daughter nuclei. The second part of the full correction comes from the radial differences of the proton and neutron single-particle functions (nodal mixing) not included into the first part in a smaller space. Here, the correction can reach 0.6% being noticeably different for different versions of the mean field used as a source for single-particle orbitals. The additional check includes the isobaric mass multiplet equations with the coefficients found from the shell model [81]. The authors [72] comment that "it is impossible to be definitive about any systematic errors that might be introduced by these methods".

There are alternative approaches [82,83] for finding the correction δ_C which avoid the subdivision of the Coulomb effects into two components. An operator part of the perturbing average Coulomb potential V_C in a reasonable approximation of the uniformly charged sphere of radius R excites the giant isovector monopole resonance (IVMR),

$$\langle IVMR|V_C|0\rangle = \frac{Ze^2}{2R^3} \langle IVMR| \sum_a r_a^2 t_3(a)|0\rangle. \qquad (15.30)$$

The IVMR is a superposition of different isospins (equal to that of the initial state and shifted by ± 1). They are split in energy by the Lane symmetry potential V_s [84],

$$V_s = \frac{v_1}{A} (\mathbf{t} \cdot \mathbf{T}), \quad v_1 \approx 100\,\text{MeV}, \qquad (15.31)$$

although the value of v_1 is not firmly defined. The centroid of the IVMR is located at about 3-4 shell-model oscillator frequencies, and the resonance has a considerable spreading width serving as a doorway state to more complicated configurations. Various model descriptions of the IVMR [82] lead to close values of the correction δ_C in average smaller than in [72] by a factor around 2.

The description of the IVMR [83] based on the sum rules weighted with the transition frequencies (or their inverse powers) allows the use of the random phase approximation (RPA) methods to take into account the single-particle continuum [85,86]. In a good approximation, the correction δ_C is related to the global properties of the IVMR and the nuclear charge radius being not too sensitive to the applied RPA versions. It agrees also with limited experimental data [87]. Again different specific assumptions lead to slightly different results for δ_C, in general in reasonable agreement with the predictions of [82] and smaller than comes out in [72] as the collectivity of the IVMR pushed it up by the repulsive particle-hole interaction in an appropriate channel.

15.3.3 To the Evaluation of Unitarity

The appearance of charged particles in the beta-decay processes involves the presence of accompanying electromagnetic processes, both virtual (radiative processes with gamma quanta and with W and Z bosons) and real (photon bremsstrahlung, including infrared ones). These corrections can be divided into a universal part and the part directly related to the nuclear structure. We do not discuss here the specific details just referring to the recent calculations [88] and their application to the unitarity problem [72]. The universal part of the radiation correction to the superallowed decays is close to 1.5% for all available cases, while the part dependent on nuclear structure is the largest for lighter nuclei (–0.345% for ^{10}C) and typically on the level of –0.04% for heavier nuclei.

The ft value corrected for isospin violation and nucleus-specific radiation corrections is traditionally denoted as $\mathcal{F}t$. It is currently a common way to introduce the universal radiation correction Δ_R^V as

$$\mathcal{F}t = \frac{K}{2G_V^2(1 + \Delta_R^V)} \tag{15.32}$$

where the constant K is as defined in Eq. (15.17), while G_V can now be determined, after introducing transition-dependent corrections in the left-hand side of (15.32), for each specific case under study and, due to the CVC, be a universal constant. As tabulated in [72] for thirteen precisely measured $0^+ \rightarrow 0^+$ Fermi transitions, with very small and well-known branching ratios to other final states, it is universally equal to 3071 ± 0.83 s. The corresponding value for the element of the CKM matrix for superallowed transitions is given in [72] as $|V_{ud}| = 0.97425\pm0.00022$. Being extracted [89] from the transitions between $T = 1/2$ isospin doublets (here we have to deal with mixed Fermi and GT processes) with the use of the CVC hypothesis, this value is $|V_{ud}| = 0.9719\pm0.0017$. This number, as well as the result $|V_{ud}| = 0.9742\pm0.0026$ from the rare (branching 10^{-8}) pion decay $\pi^+ \rightarrow \pi^0 e^+ \nu_e$, does not contradict to that from the superallowed beta decay but has bigger uncertainties.

The values of V_{us} and V_{ub} are extracted from several kaon decays with greater uncertainties. The result given in [72] is $|V_{us}| = 0.22521\pm0.00094$. Finally, the small matrix element V_{ub} is found from B-boson decays as $|V_{ub}| = (3.93 \pm 0.36) \times 10^{-3}$, but a worse precision here does not influence much the result for the unitarity which is given in [72] as

$$|V_{ud}|^2 + |V_{us}|^2 + |V_{ub}|^2 = 0.99990 \pm 0.00060, \tag{15.33}$$

and 95% of this sum comes from V_{ud}.

The latest review [73] analyzes, evaluates, and combines the results of 222 independent measurements for twenty superallowed transitions (Figure 15.9). In fact, only three

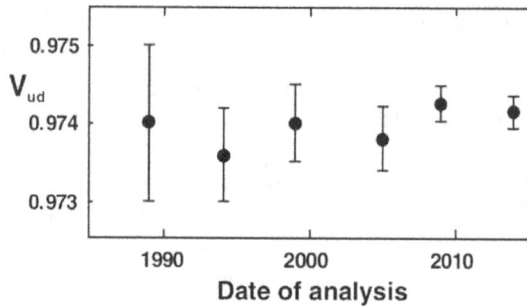

FIGURE 15.9
Values of V_{ud} as determined from superallowed $0^+ \to 0^+$ β decays plotted as a function of analysis date, spanning the past two and a half decades.

quantities from each transition are necessary: Q-value, lifetime $t_{1/2}$, and the branching ratio for the specific transition. Many Q-values are now measured with high precision with the use of a Penning trap [90], where the special attention should be given to distinguish the isomeric nuclear states. In some cases, the earlier-mentioned problem of a multitude of weak GT transitions [91] introduces additional uncertainty of the branching ratio. New data, the slightly modified procedure for the isospin violation corrections, and better statistics provided a new value for the average quantity $\overline{\mathcal{F}t} = 3072 \pm 0.72$ s. This can be interpreted as a confirmation of the CVC hypothesis (the constancy of G_V).

The situation with the isospin violation is still not fully resolved as the methods including dynamical correlations give smaller corrections, while the mean-field estimates do not include RPA correlations and continuum effects. The static tests based on the Wood-Saxon or Hartree-Fock mean field (the authors of [73] certainly prefer the Wood-Saxon method) cannot account for dynamical and collective effects. The new review slightly changed the value of V_{ud}, and now it is given as $|V_{ud}| = 0.97417 \pm 0.00021$. The values of V_{us} and V_{ub} entering the unitarity line (15.26) depend on the data provided by the Particle Data Group and partly rely on the lattice calculations. Stressing that this is not yet "the final word," the authors of [73] give the resulting sum (15.26) as 0.99978 ± 0.00055 with the main uncertainty coming from Δ_R^V in Eq. (15.32). Indeed, the recent reevaluation of this quantity [92] with the use of dispersion relations brings the result to $|V_{ud}| = 0.97370 \pm 0.00014$ and the sum (15.26) to 0.9984 ± 0.0004. The discussion is still continuing; we can mention the reevaluation [93] of the radial overlap of proton and neutron radial functions. The violation of the CVC principle could come in the form of the induced scalar in the vector part of the $p - n$ weak current, $H_S = if_S q_\mu (\bar{p}n)(\bar{e}\gamma_\mu(1 + \gamma_5)\bar{\nu}_e)$. This would bring the so-called Firz term in the neutron beta decay [75]. Here q_μ is the momentum transfer vector between hadrons and leptons. The limit on this term given in [73] is $|m_e f_s / g_V| < 0.0035$.

15.4   Weak Interactions and Astrophysics

A significant fraction of all experimental and theoretical efforts in nuclear physics is currently directly related to astrophysical problems [94–96]. As the review paper [95] starts, *nuclear structure physics and visible Universe are intimately related*. Numerous nuclear reactions at different conditions (site, density, temperature, various phases of stellar life, availability of nuclear species, etc.) determine the fate of a star, time scales of its evolution, outgoing fluxes of matter and radiation, and resulting abundances of various chemical elements in the Universe. We will not discuss here the earliest high-temperature (and non-equilibrium) stage where the $\mathcal{CP}$ symmetry was violated creating a world populated by particles and without antiparticles. The whole topic is too large so that we refer mainly to the review articles and limit ourselves by few examples directly related to weak interactions. We have to mention that many astrophysical reactions take place at a relatively low temperature (compared to the nuclear scale), and therefore, their description requires a detailed account for underlying nuclear structure. In many cases, the whole machinery of the shell model seems to be necessary. Still, this might be not sufficient for the description of neutrino-induced reactions at higher energy. In fact, the importance of deep knowledge of nuclear structure was stressed long ago, see the classical review article [97].

The first evaluation of the nucleosynthesis was made in the famous α, β, γ paper by Alpher, Bethe, and Gamow [98], while the new calculations were done in [99]. The basic features of the currently accepted picture were given by Zel'dovich [100] and Peebles [101]. A very detailed book [102] summarizes the knowledge to the end of the 20th century, after the discovery and study of the microwave background radiation. The predictions of this blackbody radiation [103] with a correct order of magnitude for its temperature were related to the abundance of the lightest chemical elements. The weak interaction process at temperature about 3 MeV, $e^- + p \to \nu_e + n$, leads to the thermal equilibrium between neutron and proton concentrations $\sim \exp(-\Delta m_{np}/T)$. Then, the radiative recombination of neutrons and protons brings in the first nuclei, deuterons, which are accumulated as the temperature falls below the deuteron binding energy. The sun-like low-mass stars have their energy source provided by the weak interaction fusion $p + p \to d + e^+ + \nu$ avoiding the unbound diproton; this reaction determines the star lifetime [104]. The observed deficit of neutrino eventually brought the current understanding of neutrino oscillations [105]. This deficit is compensated by the neutral current events produced by neutrino of all flavors.

Next the helium isotopes are produced by nuclear synthesis processes, and with falling temperature, the process stops (there are no stable nuclei with $A = 5$ or 8). The weak interaction β^+ decays are an important component of the CNO cycle. After several alpha-captures, the synthesis of heavier elements proceeds through fusion, photonuclear processes, and the slow neutron capture, the so-called s-process [106]. The slowness here is determined by the neutron component that is followed by beta decay in the direction of the valley of stability. This competition is the main source of production below the iron peak (the nucleus ^{62}Ni has the largest binding energy per nucleon). However, the s-process continues further, probably until lead isotopes. The known peaks for the s-process are related to the closed nucleon shells for stable nuclei [107], $N = 50$ and $N = 82$. In many cases, there is a strong dependence on temperature of the medium. For example, the lifetime of the earlier-mentioned rarest isotope ^{180}Ta is cut by many orders of magnitude through electromagnetic excitation in the s-process conditions [108].

Weak interactions provide the effective instrument for further evolution [109]. Different varieties of the processes, electron capture, Fermi and GT transitions, all play their roles in the chemical history, see a very clear description in Ref. [107]. The detailed quantitative understanding of those processes and resulting abundances requires the whole machinery of nuclear structure going beyond simple independent particle schemes [110]. Here, the observed quenching of the GT strength has to be accounted for. The realistic shell-model analysis with a reasonable account for the nucleonic interactions may give results in many cases noticeably different from the simple non-interacting shell model [109]. The detailed nuclear information then has to be made a part of a huge reaction network. A relatively less understood fraction of the available nuclear physics input is related to the reactions, including weak processes, where the nuclei are involved in their excited states. Here, available empirical information is limited, as well as knowledge of local temperature and corresponding occupancies of various nuclear states. An important set of data refers to the competition between the electron capture and beta decay as it regulates the flux and energy of outgoing solar neutrino and therefore the resulting cooling as well as the electron concentration Y_e. It is important to preserve the self-consistency of the general picture: the actual nuclear conditions in the stellar medium generate certain reactions that release particles and energy in turn defining those conditions.

The transition to the stellar collapse phase is significantly influenced by all details of the neutrino propagation and their interactions with nucleons, nuclei, and electrons. With the growing neutron number, the electron capture changing one more proton to a neutron becomes, in the simple shell-model scheme, forbidden by the Pauli principle as the newborn neutron cannot occupy its orbital. Certainly, this is an over-simplified feature of the primitive shell model. In reality, the GT processes become possible due to the smoothing occupation numbers both because of the residual interaction and thermal excitations [111]. The fate of the star depends mainly on the neutrino interaction with matter, including the excitation of giant resonances, and on the stellar equation of state which we cannot discuss here. As stressed in Ref. [94], *the muon and tau neutrinos react with the proto-neutron star matter only by neutral-current reactions, while the spectra are usually assumed to be identical.*

After the iron-nickel region, the large fraction of the elements is created by the *rapid* processes. Usually called as the *r*-process, it develops as a sequence, sometimes alternating, of neutron capture acts and then beta decays bringing the neutron-rich unstable nuclei (up to 20–30 units from the valley of stability) to the allowed n/p ratio at the stability valley (the north-west direction on a standard nuclear chart). In the *rp* process, the characteristic time scale is determined by the waiting points with beta decays and slow two-proton capture reactions. The whole evolution mostly unravels through exotic nuclei with largely unknown main characteristics including masses, lifetimes, deformation, and magic numbers. This is the region where the new radioactive ion facilities are expected to provide the necessary information. Currently, the most popular locations for the *r*-process are the sites with strong neutron fluxes, like type II supernovae and neutron star mergers, although this problem cannot be considered completely solved. One of the interesting problems is the competition between beta decays and reactions (ν_e, e^-) driven by *neutrino wind* [112]. At the stage preceding the stellar collapse, the main fight is the competition between the gravity and weak processes, mainly electron capture, although, especially at increased temperature, the beta decays from excited states play a role.

The lifetimes along the *r*-process are expected to be mainly defined by the GT transitions. This is one of the hard current nuclear theory problems as for the nuclei in these regions the orbital space necessary for the successful shell-model calculations is too big while the quenching of the GT strength makes the results derived with the use of a smaller space with the functionals adjusted near the ground state, or with the RPA-type schemes, not fully reliable. There are also topics with beta-delayed neutrons and contributions of isomers and beta processes of higher forbiddenness. The so-called rapid proton (*rp*) process ends with the SnSbTe cycle where the time scale is defined by the weak beta decays of tin isotopes 103 and 104. Here, there is no clear understanding of the nuclear structure details because of the hidden alpha clustering in this region.

It is known that some objects in the Universe have very strong magnetic fields. The critical field

$$B_c = \frac{m_e^2 e^3 c}{\hbar^3} = 2.35 \times 10^9 \, \text{Gs} \tag{15.34}$$

corresponds to the situation when the magnetic cyclotron frequency $\hbar\omega_c$ becomes comparable to the Rydberg energy. At such fields, the atomic structure becomes highly anisotropic. For the pulsars, the typical field magnitude is $B \sim 10^{12}$ Gs, while for some neutron stars (*magnetars*), it can reach 10^{15} Gs or even higher. At the so-called Schwinger field,

$$B_S = \frac{m_e^2 c^3}{e\hbar} = 4.41 \times 10^{13} \, \text{Gs}, \tag{15.35}$$

the cyclotron frequency reaches the electron mass. We have no opportunity to discuss here this extremal physics in any detail and refer to broad available literature. see, for example, [113] and references therein, but we mention a couple of important aspects related to beta decay and other weak processes in such strong fields. One of the main effects that makes the standard results invalid is that the density of final states is radically changed becoming for the electron, instead of usual expressions (15.14), the density of Landau magnetic levels. This can strongly change the electron capture rates and, correspondingly, neutrino energy loss that typically greatly increases by magnetic fields [114,115]. The high matter density (the electron Fermi energy can reach 10 MeV) and possible Coulomb crystal structure of neutron stars make the whole situation even more complicated.

15.5 Parity Violation

15.5.1 Parity Violation and Neutrino

As mentioned in the Introduction, the nuclear beta decay of a polarized nucleus ^{60}Co demonstrated, for the first time in quantum physics, parity violation [7]. The experiment was prompted and preceded by the analysis [6] of available data where it was shown that the reflection invariance was not actually checked before. The violation of parity $\mathcal{P}$ is seen by the $\mathcal{P}$-odd correlation of the beta electron momentum $\mathbf{p}$ or velocity $\mathbf{v}$ with the spin polarization direction $\mathbf{I}$ of the mother nucleus. In the GT transition ^{60}Co(5^+) $\rightarrow ^{60}$Ni(4^+), the electron and electron antineutrino carry away spin 1 shared between them so that their spins have to be parallel. The experimental result describes the electron angular distribution $f(\theta)$ as a function of the angle θ between $\mathbf{p}$ and $\mathbf{I}$,

$$f(\theta) \propto 1 + A\alpha \frac{v}{c} \cos\theta, \tag{15.36}$$

where the polarization degree of the decaying nucleus was $\alpha \approx 0.65$. The result $A \approx -1$ shows the presence of both scalar (term 1) and pseudoscalar ($\mathbf{I} \cdot \mathbf{p}$) correlations as $\mathbf{I}$ is an axial vector. The relativistic electrons, $v/c \approx 1$, are moving mainly against the mother nucleus spin which means that for the momentum conservation, the antineutrinos are moving along that spin. Together the leptons carry away the difference of nuclear spins equal to 1. This tells us that the antineutrino is right polarized (spin along its motion), while the electron is obliged to be left polarized to conserve the total spin.

The antineutrino is always observed as a right-polarized particle with helicity ($\vec{\sigma} = 2\mathbf{s}$)

$$h = \frac{(\vec{\sigma} \cdot \mathbf{p})}{|\mathbf{p}|} = 1, \tag{15.37}$$

The neutrino, a particle that appears in β^+ processes along with the positron, is always left polarized, $h = -1$. The same properties are shared by muonic neutrino and antineutrino defined in a similar way decays. As for many years it was thought that all generations of the neutrino family are massless, they could be described by the Dirac equation with $m = 0$. Then, the helicity property allows to use, instead of the four-component Dirac spinors, just two-component Weyl spinors satisfying

$$(\vec{\sigma} \cdot \mathbf{n})\psi^\pm = \pm \psi^\pm, \quad \mathbf{n} = \frac{c\mathbf{p}}{E} = \frac{\mathbf{p}}{p} \, \text{sign} \, E. \tag{15.38}$$

In a four-component language, helicity for a massless particle is equivalent to *chirality* given by the 4-matrix γ^5. The polarization effects become especially pronounced in the situation with extra strong magnetic field [119] when both electrons and neutrino are highly relativistic. Then, the geometry of the process is essentially determined by the electron Landau levels in the plane perpendicular to the magnetic field with spin polarized along the field so that the whole picture is extremely anisotropic.

The presence of two variants of a particle like left-polarized neutrino ν and right-polarized antineutrino $\bar{\nu}$ by itself does not signal the violation of natural symmetries. The massless photons can be circularly, right or left, polarized (photon polarization is similar to helicity), and just the existence of these two analogous states converting into each other under spatial inversion completely restores the $\mathcal{P}$-symmetry of the electromagnetic field. But now we have a particle (neutrino) and an antiparticle (antineutrino). The spatial inversion transforms the left-polarized neutrino into right-polarized neutrino that does not exist experimentally if we define the neutrino as the lepton that appears along with the positron (we can describe this, and a combination of the antineutrino with the electron, as a conservation of the lepton charge). Then, we have to accept that parity is completely violated. At the same time, we see that here also charge symmetry $\mathcal{C}$ is completely violated as well: the charge conjugation of the left neutrino would produce the non-existent left antineutrino. Another direct evidence of $\mathcal{C}$ violation comes from pionic decays. In all cases, massless neutrino and antineutrino seem to satisfy Eq. (15.38) with sign minus. A natural idea is [8] that $\mathcal{P}$ and $\mathcal{C}$ symmetries are both violated, while the *combined* $\mathcal{CP}$ symmetry still does exist transforming the left neutrino into analogous right antineutrino.

This way of arguing assumes that neutrinos and their anti-analogs are massless. Now we know that, as proven by neutrino oscillations between their electron, muon and tau flavors, they have quite small but nonzero masses, even if the exact values of these masses, being less than 1 eV, are currently not known yet. This means that for a neutrino and its analogs, there exists a rest frame where the polarization is not defined, so it cannot serve as an eternal property. We have to accept that the polarization is a characteristic not for the nature of particles but for the *left currents* working in the weak, or more precisely electroweak, interactions. If so, the massive neutrino and antineutrino have to satisfy the four-component Dirac equation and the new question arises—are they Majorana or Dirac particles—which can be answered by observing (or not) a neutrinoless double beta decay.

As the neutrino in beta decay is practically polarized (the polarization would be complete for the massless neutrino), and the electron has a polarization of the order v/c, the daughter nucleus with nonzero spin also turns out to be partially polarized. If the final nucleus is born in an excited state, the polarization is partly transferred to gamma quanta of the next decay [116]. This circular polarization is a remnant of parity nonconservation in the weak decay. It allows (along with correlations appearing in beta decay of aligned nuclei) to separate the Fermi and GT contributions. For example, consider an electron beta decay from the mother nucleus with spin J_0 to the daughter nucleus with spin J_1 which then produces a photon of multipolarity L coming to the ground state of the daughter nucleus with spin J_2. In this case, the difference of intensities of quanta with left and right circular polarization is

$$I_- - I_+ = 2A' \frac{v}{c} \cos \vartheta, \qquad (15.39)$$

where v is the electron velocity, and ϑ is the angle between the gamma quantum and the electron. The asymmetry A' is expressed as

$$A' = \frac{\xi}{1 + x^2} (2x + \eta), \quad x = \frac{G_V M_F}{G_A M_{GT}}. \qquad (15.40)$$

Here, the geometric quantities ξ and η depend on quantum numbers $J_{0,1,2}$ of participating states and the multipolarity of the gamma radiation (vector coupling of participating spins).

The effect disappears for pure Fermi transitions, while for pure GT transitions, it is universal independently of matrix elements, $A' = \xi\eta$. If the transition contains both contributions, Fermi and GT, the observed asymmetry provides a chance to determine the ratio of these contributions (but such an experiment requires $\beta - \gamma$ coincidence) [117].

With no chance to discuss in detail, we still have to mention the known phenomenon in chemistry and biochemistry: a broken mirror symmetry of bioorganic objects noticed long ago by L. Pasteur. All biopolymers, including DNA, RNA, and enzymes, are *homochiral*. There are serious discussions concerning the mechanism of appearance and stability of such chiral asymmetry [118]. The relation to the parity violation in weak interactions was suggested long ago [120,121] and recently returned into discussions [122,123] when it was demonstrated a tiny difference in two minima of molecular energy (left and right *enantiomers*) sometimes ascribed to the parity violation in weak interactions. If so, this splitting between two equivalent potential wells due to quantum tunneling between them can be, on a very different scale, of the same nature as parity doublets in nuclei with a combination of quadrupole and octupole deformation that will be discussed later in relation to the search of the nuclear EDM.

15.5.2 Mixing of Neutron Resonances

The non-conservation of parity in weak interactions was discovered in 1956 by Lee and Yang [6] after analyzing the weak decay of K-mesons. Very soon after this discovery, an experiment was performed demonstrating that the effects of parity violation exist in complex nuclei [7]. In the subsequent decades, many more experiments on parity violation in nuclei have been performed. Most of these experiments involved the spin degree of freedom and used polarized nucleon beams [124].

The theoretical description of parity violation is based on the notion that the weak interaction among nucleons is due to the process of meson exchange. In the lowest order, the exchanged meson couples weakly to one nucleon and strongly to the second one, see Figure 15.10. The mesons exchanged are the charged pions $\pi^\pm$, the rho meson ρ, and the omega meson ω. Theoretical and experimental studies attempt to determine the values of the weak coupling constants [125,126].

In the end of the 20th century, there has been renewed interest in the subject of parity violation in nuclei. A new class of parity violation experiments were introduced: the measurements of parity-violating effects in the compound nucleus. First, these were experiments in Dubna [127], and a few years later, extensive studies were undertaken in Los Alamos [128]. The experiments and theory of this process combine two special phenomena: weak interaction, and chaos in nuclear dynamics. The basic premise of this

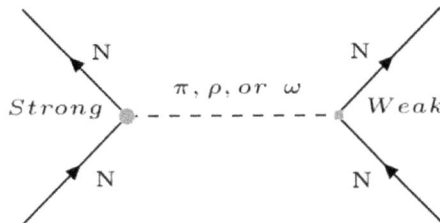

FIGURE 15.10
The theoretical description of parity violation is based on the notion that the weak interaction among nucleons is due to the processes of meson exchange. In the lowest order, the exchanged meson couples weakly to the first nucleon and strongly to the second.

approach is the idea that due to a high density of states in the compound nucleus, action of weak perturbations, including parity mixing, will be enhanced. In low-energy neutron resonances, the very small separation of s- and p-waves (typically $10-100$ eV) leads to large enhancements of parity-violating observables. As we will see, these enhancement factors could often reach 10^6.

Indeed, the first neutron scattering experiment [127] and later the experiments at Los Alamos [128] produced parity-violating effects of the order of $(1-10)\%$. In these experiments, one scatters polarized epithermal neutrons with helicities $\pm 1/2$ off unpolarized nuclear targets. One then measures the longitudinal asymmetry,

$$P = \frac{\sigma_+(E) - \sigma_-(E)}{\sigma_+(E) + \sigma_-(E)}, \tag{15.41}$$

where $\sigma_\pm(E)$ are total cross sections for neutrons with energy E and helicities ± 1. In the absence of parity violation $P = 0$, while a nonzero longitudinal asymmetry signifies the existence of parity non-conservation in the neutron plus target system. The measurements of P are made for p-wave resonances in the compound nucleus. Those p-wave resonances with $J = \frac{1}{2}^-$ may mix with $J = \frac{1}{2}^+$ (s-wave) resonances through a parity non-conserving force, and therefore give rise to nonzero values for P.

The polarization P is proportional to the parity-violating matrix element, and therefore measuring the asymmetry (15.41), one can obtain information about the parity-violating component of the nuclear force. In a number of theoretical papers [129–132], it was shown that the leading term contributing to P in resonant scattering is

$$P(E_\mu) = 2 \sum_q \frac{\langle \mu | V_{\mathrm{PV}} | q \rangle}{E_\mu - E_q} \frac{\gamma_q}{\gamma_\mu}, \tag{15.42}$$

where V_{PV} is the parity-violating interaction. The state $|\mu\rangle$ is a $J = \frac{1}{2}^-$ resonance excited in the reaction and E_μ its energy. The states $|q\rangle$ are $J = \frac{1}{2}^+$ states that mix with the resonance $|\mu\rangle$, γ_q, and γ_μ are the amplitudes of the neutron decay to the target ground state of the $|q\rangle$ and $|\mu\rangle$ resonances, respectively. For low neutron energies, due to the centrifugal barrier penetration factor, the ratio for $l = 0$ and $l = 1$ decays is

$$\frac{\gamma_q}{\gamma_\mu} \simeq \frac{\sqrt{3}}{kR}, \tag{15.43}$$

where k is the neutron wave number, and R is the nuclear target radius. For $E \simeq 1$ eV and a heavy nucleus, this ratio is about 10^3. This enhancement sometimes referred to as "kinematical" is very important in the observation of $P(E_\mu)$.

We may express the decay amplitudes γ_μ and γ_q in the following manner:

$$\gamma_\mu = a_p^{(\mu)} \langle \varphi_p | U | \chi_p^{(+)}(E_\mu) \rangle \tag{15.44}$$

and

$$\gamma_q = a_s^{(q)} \langle \varphi_s | U | \chi_s^{(+)}(E_\mu) \rangle, \tag{15.45}$$

where φ_p and φ_s are the single-particle bound states, $\chi_p^{(+)}(E)$ and $\chi_s^{(+)}(E)$ are the neutron continuum wave functions in p- and s-channels, and U is a one-body potential. The symbols $a_p^{(\mu)}$ and $a_s^{(q)}$ are the single-particle spectroscopic amplitudes of the bound p- and s-states in the wave functions of the $|\mu\rangle$ and $|q\rangle$ states, respectively. Denoting

$$\langle \varphi_p | U | \chi_p^{(+)}(E_\mu) \rangle = \gamma_p(E_\mu), \tag{15.46}$$

$$\langle \varphi_s | U | \chi_s^{(+)}(E_\mu) \rangle = \gamma_s(E_\mu), \tag{15.47}$$

we write Eq. (15.42) as

$$P(E_\mu) = 2 \frac{\gamma_p(E_\mu)}{\gamma_s(E_\mu)} \sum_q \frac{\langle \mu | V_{\mathrm{PV}} | q \rangle}{E_\mu - E_q} \frac{a_s^{(q)}}{a_s^{(\mu)}}. \tag{15.48}$$

In what follows, we will mainly deal with the quantity

$$A_\mu = \sum_q \frac{\langle \mu | V_{\mathrm{PV}} | q \rangle}{E_\mu - E_q} \frac{a_s^{(q)}}{a_s^{(\mu)}}. \tag{15.49}$$

Let us divide the full parity conserving nuclear Hamiltonian into

$$H = H_0 + V, \tag{15.50}$$

where H_0 represents the mean-field Hamiltonian, and V is a residual interaction. We now assume that we are in the energy regime where the nuclear states $|c\rangle$ which are eigenstates of H when expanded in the basis eigenstates of H_0 will contain a large number N of "principal" components $|\varphi_i\rangle$ that all have amplitudes of the same order of magnitude $1/\sqrt{N}$. The number N is proportional to the density of states in the excitation energy regime under discussion. The assumption about the large number of components is an expression of the fact that at this energy, the nucleus is in a stage of strong ("chaotic") mixing, and its wave functions, when expressed in terms of simple configurations, have amplitudes that are randomly distributed. In this case,

$$|c\rangle = \sum_{i=1}^N a_i^{(c)} |\varphi_i\rangle. \tag{15.51}$$

Let V_{sv} be a one- or two-body interaction that violates a symmetry of H. One can write an off-diagonal matrix element as

$$\langle c | V_{sv} | c' \rangle = \sum_{i,j=1}^N a_i^{(c)} a_j^{(c')} \langle \varphi_i | V_{sv} | \varphi_j \rangle. \tag{15.52}$$

Since the interaction V_{sv} is at most two-body, and in many cases (such as in parity violation for valence orbitals outside of a closed shell), its effective dominant part is one-body, one can connect a given configuration $|\varphi_i\rangle$ to one or very few states $|\varphi_i\rangle$. The double sum in Eq. (15.52) is thus reduced to a single one. For the sake of argument, let us write

$$\langle \varphi_i | V_{sv} | \varphi_j \rangle = p_i \delta_{ij} \tag{15.53}$$

(for a one-body V_{sv}, this is exact in the absence of radial excitations, while for a two-body V_{sv}, there is only a small number of j that will connect to given i). Then,

$$\langle c | V_{sv} | c' \rangle = \sum_{i=1}^N a_i^{(c)} a_i^{(c')} p_i. \tag{15.54}$$

Because of the chaotic nature of the states $|c\rangle$ and $|c'\rangle$, the coefficients a_i^c are independent random Gaussian variables, and as $\sum_i^N |a_i^{(c)}|^2 = 1$, each $a_i^{(c)}$ is on the average $a_i^{(c)} \simeq 1/\sqrt{N}$.

From a random walk argument, one then obtains $[\overline{\langle c|V_{sv}|c'\rangle^2}]^{1/2} \sim 1/\sqrt{N}$. The average energy spacing between adjacent levels $|c\rangle$ and $|c'\rangle$ is

$$D_{cc'} = E_c - E'_c = \frac{\epsilon}{N}, \tag{15.55}$$

and therefore, for two adjacent levels $|c\rangle$ and $|c'\rangle$,

$$\frac{\langle c|V_{sc}|c'\rangle}{E_c - E_{c'}} \sim \sqrt{N}. \tag{15.56}$$

We now take V_{sv} to be the parity-violating interaction V_{PV}. We obtain for the mixing amplitude

$$\frac{\langle \mu|V_{\text{PV}}|q\rangle}{E_\mu - E_q} \sim \sqrt{N}. \tag{15.57}$$

Thus, the left-hand side of Eq. (15.57) is proportional to $\sqrt{N}$ if $|q\rangle$ is a $J = \frac{1}{2}^+$ level adjacent to the $J = \frac{1}{2}^-$, $|\mu\rangle$ level. This proportionality to $\sqrt{N}$ is expected to persist in the sum of Eq. (15.42) because the contributions from other terms that come from more distant states $|q\rangle$ will average to zero due to the randomness of the sign of the matrix elements and decay amplitudes. Thus, only one or very few large terms that stem from levels $|q\rangle$ that are very close to $|\mu\rangle$ will contribute to the sum in Eq. (15.42). For a neutron of energy $E = 1$ eV incident on the ^{232}Th target, the excitation energy in the compound ^{233}Th nucleus is $E_x = 4.8$ MeV. At this energy, the density of states is about 10^5 $J = \frac{1}{2}$ levels per MeV. When compared to a single-particle spacing, this gives $N \simeq 10^6$, and therefore, the "dynamical" enhancement is $\sqrt{N} \simeq 10^3$. We should notice that in addition to the γ_s/γ_p ratio which is around 10^3, there is the factor 10^3 from the dynamical enhancement, altogether we come to an enhancement of about 10^6 in heavy nuclei.

In the paper [133], a numerical study of the parity violation enhancement in the compound nucleus was performed. It is very difficult to do this for a heavy nucleus, so a light nucleus ^{9}Be was used. By scaling the energy spacing, it was possible to extrapolate the results to the case of a heavy nucleus. The numerical calculations confirmed the idea of parity violation enhancement in the regime of large density of states.

One of the important characteristics of symmetry breaking is the symmetry violating spreading width [131,133]. In the case of parity violation, the spreading width for a given state $|\mu\rangle$ is

$$\Gamma^\downarrow_{\text{PV}} = \frac{2\pi\overline{\langle \mu|V_{\text{PV}}|q\rangle^2}}{\overline{D}_\mu}, \tag{15.58}$$

where in the numerator we have the mean square parity-violating matrix element averaged over the states $|q\rangle$, and $\overline{D}_\mu$ is again the average level spacing $|q\rangle$ in the vicinity of $|\mu\rangle$ states. In the study of [133], the average value and variance were: $\Gamma^\downarrow_{\text{PV}} = 7.5 \pm 3.5 \times 10^{-6}$eV. The spreading width in general, in the regime of chaos, is independent of the density of states N, and from this fact, one can find easily that there is the enhancement described in Eq. (15.56).

The "dynamical" enhancement is not limited to effects of parity violation in the compound nucleus. One expects a similar situation to occur for other symmetries weakly broken in nuclei, see, for example, [45] mentioned earlier in relation to nuclear isomers. Time reversal symmetry is another example. Time reversal violation observables that are linearly dependent on the mixing amplitudes will show such effects of enhancement when studied in the compound nucleus regime. Also, this "dynamical" enhancement should occur in many complex quantum systems in the situation when the density of states is high and the wave functions are complicated (chaotic). Thus, enhancement of parity violation is expected also in atoms and molecules.

15.5.3   Parity Violation in Fission

Low-energy fission of heavy nuclei is the most impressive process related to the consistent motion of big blobs of nuclear matter either from rest (spontaneous fission) or under the action of a relatively minor external agent, a photon or a neutron of low energy (even on thermal level). Although many regularities of this process are known in details, mainly due to the practical use of a large energy freed by the act of fission, we still do not have full microscopic description of fission. When the experimentalists [134] claimed that they observe the parity violation in the fission process, this was generally met with distrust. It was hard to believe that the weak forces (indeed, weak on the background of strong and electromagnetic forces which are responsible for almost macroscopic fission) can leave an observable imprint on the results of this process. However, again the ideas of quantum chaos can help to understand what is going on.

We have already seen from the resonance ("dynamical") enhancement of parity violation that the effects of the weak interaction, and actually other small perturbations, become effectively strong in the region of chaotic states and high-level density. The mixing of close compound states can be enhanced by 3–4 orders of magnitude. The low-energy fission is an almost adiabatic process when the shape degrees of freedom are slowly changing along collective trajectories during the long lifetime of the compound nucleus moving through potential barriers to the scission point. The parity violation enacted in a compound stage and resulting in a small admixture of opposite parity is kept through this adiabatic evolution [135]. The result of the parity violation can be visible in the final fragments.

If only momenta $\mathbf{p}_f$ of the fission fragments going back to back in the frame of the mother nucleus are available for registration, there are no pseudoscalar correlations. In the neutron-induced fission, you can try a pseudoscalar combination of three momenta (two final products and the momentum of the original neutron). But this would violate also time reversal symmetry. The problem is solved by including spin variables to perform the experiment with polarized neutrons and measuring the fragment asymmetry $(\mathbf{s}_n \cdot \mathbf{p}_f)$ with respect to the neutron polarization.

It is impressive to see a considerable asymmetry of motion of hundreds of nucleons depending on the polarization (helicity) of one starting neutron. Figure 15.11 shows the fragment asymmetry with respect to the neutron polarization in fission events with different final mass or kinetic energy distributions in the Grenoble experiments [136]. The asymmetry is on the level 4×10^{-4} being indeed enhanced by several orders of magnitude compared to the "natural" order of weak effects. The fact that the resulting asymmetry essentially does not depend on the special characteristics of a given decay (masses and kinetic energies) confirms that the parity violation took place in the compound phase of chaotic internal dynamics before the stage when the final characteristics of the fragments were decided. The magnitude of the effect agrees with the estimates based on the chaotic mixing (here there is no "kinematic" enhancement present for the mixing of p- and s-neutron resonances).

Equally spectacular are the results of the experiments [127,137] measuring just the total cross sections of interaction of longitudinally polarized neutrons with heavy nuclei where both sources of enhancement are present. The effect of parity violation, visible at the neutron resonances, leads to difference of cross sections $\sigma_{\pm}$ for neutrons of opposite helicity. At energies of various resonances, one can see the effect by the naked eye, Figure 15.12 for ^{238}U. Here the difference

$$\mathcal{P} = \frac{\sigma_+ - \sigma_-}{\sigma_+ + \sigma_-}, \tag{15.59}$$

is of the order 1–2% and has usually random signs; it reaches 10% in lanthanum. That the effect is noticeable at neutron resonances is natural as the chaotic mechanism requires a relatively long lifetime of the intermediate compound state. In an exceptional case of ^{232}Th,

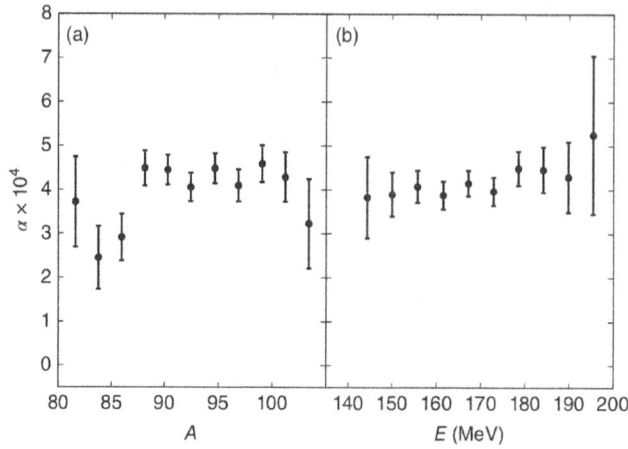

FIGURE 15.11
The fragment asymmetry in various fission events with different mass (a) or kinetic energy (b) distributions from experiment [136].

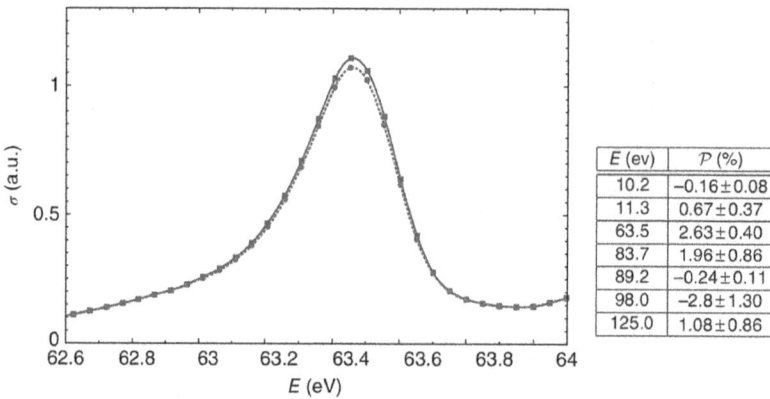

E (ev)	P (%)
10.2	−0.16±0.08
11.3	0.67±0.37
63.5	2.63±0.40
83.7	1.96±0.86
89.2	−0.24±0.11
98.0	−2.8±1.30
125.0	1.08±0.86

FIGURE 15.12
The absorption peak around 63.5 eV resonance in the transmission of polarized neutrons for two helicity states in ^{238}U. The table on the right shows asymmetries for various resonances [140,141].

the cross-section difference (15.59) has the same sign at many resonances [138]; the corresponding physics may be connected with effects of octupole deformation in thorium isotopes [139].

15.5.4   The Doorway State Model

In references [131,132], a doorway state model was introduced. Among the various approaches to parity violation in nuclei, the single-particle parity-violating force was first considered by Michel [142]. Such a one-body force results from the convolution of a two-body parity-violating force with the single-particle nuclear density. Since the one-body potential has to be a pseudoscalar, invariant under time reversal, the possibilities of simple forms for such a potential are limited. The simplest possibility is [131,133]

$$V_{\text{PV}} = g(r)(\vec{\sigma} \cdot \mathbf{p}), \tag{15.60}$$

where $g(r)$ is a function of the radial coordinate only, and $\vec{\sigma}$ and $\mathbf{p}$ are the vectors of spin and momentum of the nucleon, respectively. The parity-violating potential might be an isoscalar as well as isovector, and therefore,

$$g(r) = g_0(r) + g_l(r)t_3, \tag{15.61}$$

where t_3 is the third component of the nucleon isospin.

Consider a simplified nuclear Hamiltonian:

$$H = \sum_{i=1}^{A} \left[\frac{p_i^2}{2M} + U(r_i) \right] + \sum_{i=1}^{A} g_i(\vec{\sigma}_i \cdot \mathbf{p}_i) \equiv H_0 + \sum_i g_i(\vec{\sigma}_i \cdot \mathbf{p}_i), \tag{15.62}$$

with g_i containing an isoscalar and isovector part. Defining the nuclear eigenstate of a parity conserving part,

$$H_0|\Phi\rangle = E|\Phi\rangle, \tag{15.63}$$

it can be shown [131] that the solution to the equation with the full H can be written in the first order of g in the form:

$$|\Psi\rangle = \left(1 + iM \sum_i g_i(\vec{\sigma}_i \cdot \mathbf{r}_i) \right) |\Phi\rangle. \tag{15.64}$$

The second term in this wave function is the parity-violating component. We note that

$$\sqrt{\frac{4\pi}{3}} \sum_i (\vec{\sigma}_i \cdot \mathbf{r}_i) = \sum_i r_i [\sigma_i \otimes Y_1(\hat{r}_i)]^{0^-}, \tag{15.65}$$

where the expression on the right-hand side denotes the coupling of the spin vector $\vec{\sigma}$ viewed as a tensor of rank 1, to the spherical harmonics of rank 1, to form a tensor of rank 0 and negative parity, i.e., a pseudoscalar. This is, of course, the $J^\Pi = 0^-$ component of the spin-dipole operator. This operator when acting on $|\Phi\rangle$ will produce particle-hole excitations of the spin-dipole type ($S = 1, L = 1$) [144]. Multiplying this operator by g_i will produce isoscalar and isovector spin-dipole configurations.

One can generalize this approach by postulating that in the case when the strong interaction nuclear Hamiltonian is more complete and contains two-body parts, then also the one-body parity-violating potential in Eq. (15.60) and the $J = 0^-$ spin-dipole state are responsible for parity violation in compound nuclear states. We will now refer to the case when H_0 in Eq. (15.63) is replaced by

$$\tilde{H}_0 = H_0 + \sum_{i>j} V_{ij}. \tag{15.66}$$

In this general case, we should expect that the $J = 0^-$ spin-dipole configurations will serve as doorways for parity mixing. Moreover, two-body parts in the nuclear Hamiltonian $\tilde{H}_0$ will correlate the $1p - 1h$ states and produce collective, giant spin-dipole states or resonances. We will use the following definitions of "ideal" giant $J = 0^-$ spin-dipole (SD) states. The isoscalar is

$$|SD; 0^-\rangle_0 = \frac{1}{N_0} \sum_i r_i [\sigma_i \times Y_l(\hat{r}_i)]^{0^-} |\tilde{\Phi}\rangle \tag{15.67}$$

and the isovector is

$$|\text{SD}; 0^-\rangle_1 = \frac{1}{N_1} \sum_i r_i [\sigma_i \times Y_1(\hat{r}_i)]^{0^-} \tau_3(i) |\tilde{\Phi}\rangle, \tag{15.68}$$

where $|\tilde{\Phi}\rangle$ is an eigenstate of $\tilde{H}_0$ and N_0, N_1 are appropriate normalization factors equal to the square root of the total $J = 0^-$ isoscalar and isovector $L = 1, S = 1$ strengths, respectively.

The one-body parity-violating potential will couple and the states in Eqs. (15.67, 15.68) give

$$|\Psi\rangle = |\tilde{\Phi}\rangle + a_0 |\text{SD}; 0^-\rangle_0 + a_1 |\text{SD}; 0^-\rangle_1, \tag{15.69}$$

$$a_k = \frac{\langle\tilde{\Phi}| \sum_i g_i \vec{\sigma}_i \cdot \mathbf{p}_i |\text{SD}; 0^-\rangle_k}{\hbar\omega_k - E}, \tag{15.70}$$

where $\hbar\omega_k$ are the energies of the spin-dipole giant resonances built on the state $\tilde{\Phi}$ and E is the energy of the state $\tilde{\Phi}$. (Note that g_i contains the isoscalar and isovector parts of the parity-violating operator, and these will couple to $k = 0$ and $k = 1$, respectively.) One then calculates the corresponding spreading width [131]:

$$\Gamma_{\text{PV}} = \frac{|\langle\tilde{\Phi}|V_{\text{PV}}|\text{SD}\rangle_0|^2}{(\hbar\omega_0 - E)^2 + \Gamma_{\text{SD},0}^2/4} \Gamma_{\text{SD},0}$$

$$+ \frac{|\langle\tilde{\Phi}|V_{\text{PV}}|\text{SD}\rangle_1|^2}{(\hbar\omega_1 - E)^2 + \Gamma_{\text{SD},1}^2/4} \Gamma_{\text{SD},1}, \tag{15.71}$$

where $\Gamma_{\text{SD},0}$ and $\Gamma_{\text{SD},1}$ are the spreading widths of isospin $T = 0$ and $T = 1, L = 1, S = 1$, $J = 0^-$ giant resonances evaluated at the energy of $|\tilde{\Phi}\rangle$.

The spin-dipole resonance, in particular of the isovector ($T = 1$) type, was studied both experimentally [51,143] and theoretically [144]. Theory indicates that the average energy of the $J = 0^-$ is several MeV higher than that of $J = 1^-$ and 2^-. Usually more than $1/3$ of the total $L = 1$, $S = 1$ ($T = 1$) strength is contained in the $J = 0^-$ component. As for the isoscalar ($T = 0$) spin-dipole, there is only scarce experimental information. One should expect $S = 1, L = 1$ ($T = 0$) strength to be found around $1\hbar\omega$.

It has become clear in the past decade that the Axel-Brink postulate, which states that every state has its giant dipole built on it, is well satisfied [145]. The mechanism of parity mixing we discuss here should be viewed as an application of this postulate. The doorways for parity mixing in the ground and excited states are the spin-dipole giant resonances built on these states. The mixing of p- and s-resonances discussed in the literature [127–129] is an example of this. The single-particle p- (or s-) resonance in an odd-even nucleus which mixes with the s- (or p-) is part of a spin-dipole built on the latter. The full $S = 1, L = 1$ strength contains in addition p-h components of the core.

Although we predict that most of the nuclear states will acquire similar parity-violating spreading widths, one should not be deceived by this fact when considering the neutron polarization experiments. In these experiments, only states having single-particle components in the neutron plus target ground-state compound system will be detected. Moreover, because of penetrability effects, only components of single-particle p-states having admixtures of s-states will be detected. The parity-violating spreading width found in the ^{239}U experiment [146] is around $\Gamma_{\text{PV}} \approx 10^{-7}$ eV. We use Eq. (15.71) to make a rough estimate for the parity mixing matrix element: replace the two doorways, the $T = 0$ and $T = 1$, by a single one located at an average energy which in a heavy nucleus is about 10 MeV. The width of a dipole resonance is roughly of the order of MeV in such a nucleus.

These numbers give a parity-violating matrix element of $\langle\Phi|V_{\rm PV}|{\rm SD}\rangle = \sqrt{2}$ eV for a state in the heavy nucleus. This is not inconsistent with the size of the matrix element found in other parity-violating experiments.

15.5.5 Anapole Moment

The spin-dipole is connected to an additional parity-violating effect, namely to the anapole moment. The existence of the anapole moment was first suggested by Zel'dovich [147] soon after parity violation was discovered. The anapole moment exists in the situation when parity is violated, but time reversal symmetry is preserved. Pioneering calculations of the anapole moment were done in Ref. [148]. It was suggested that the anapole moment could provide information about the nature of the NN parity-violating force, in particular about the π and ρ exchange contributions to the weak NN force.

The anapole operator is given by [149,150]

$$\hat{\mathbf{a}} = -\pi \int d\mathbf{r} r^2 \mathbf{j}(\mathbf{r}), \tag{15.72}$$

where $\mathbf{j}(\mathbf{r})$ is the nuclear electromagnetic current density. It has been found [148,150] that the dominant part of the anapole operator stems from the spin part of $\mathbf{j}(\mathbf{r})$ and is given by

$$\hat{\mathbf{a}}_s = \frac{\pi e \mu}{m}[\mathbf{r} \times \vec{\sigma}], \tag{15.73}$$

where μ is the nucleon magnetic moment, m is the nucleon mass, $\vec{\sigma}$ is the nucleon spin operator, and $\mathbf{r}$ is its coordinate.

For the nucleus, we write this part of the anapole operator as a sum over all nucleons:

$$\hat{\mathbf{a}}_s = \frac{\pi e}{m} \sum_{i=1}^{A} [\mu + (\mu_p - \mu_n)t_3(i)](\vec{r}_i \times \vec{\sigma}_i), \tag{15.74}$$

where $\mu = (\mu_p + \mu_n)/2$, and μ_p and μ_n are the proton and neutron magnetic moments in units of nuclear magnetons, respectively. The operator written in vector spherical harmonics as a tensor couple product is

$$\vec{r}_i \times \vec{\sigma}_i = -i\sqrt{2}\sqrt{\frac{4\pi}{3}} r_i [Y_{L=1}(\hat{r}_i) \otimes \vec{\sigma}]^{\Delta J=1}, \tag{15.75}$$

which is the $\Delta J = 1, L = 1, S = 1$ spin-dipole operator. The anapole operator in Eq. (15.74) involves, therefore, the isoscalar and isovector $J = 1^-$ spin-dipole operators. The distribution of isovector spin-dipole strength was studied extensively both experimentally and theoretically [144,152], and there is a considerable amount of information about the isovector spin-dipole strength distribution. For the isoscalar spin-dipole, there is little information.

The anapole moment of the nucleus is defined as the expectation value

$$a = \langle\psi|\hat{a}_z|\psi\rangle_{(J_z=J)}, \tag{15.76}$$

where ψ is the ground-state wave function of the nucleus. It is clear that since the operator $\hat{\mathbf{a}}$ is odd under parity and time reversal operations ($\mathcal{P}$-odd, $\mathcal{T}$-odd), its expectation value can be nonzero only if parity mixing occurs in the wave function ψ. Note that one does not need time reversal violation in this case because the operator $\hat{\mathbf{a}}$ contains, unlike, for example, the electric dipole, the spin operator $\vec{\sigma}$. In this sense, it is similar to the magnetic moment operator which, however, does not require parity violation. This becomes clear from the vector model that shows that the matrix elements of any vector operator, including $\hat{\mathbf{a}}$, inside a multiplet of states $|JM\rangle$ are given by

$$\langle JM|\hat{\mathbf{a}}|JM'\rangle = \frac{\langle\langle(\hat{\mathbf{a}}\cdot\hat{\mathbf{J}})\rangle\rangle}{J(J+1)}\langle JM|\hat{\mathbf{J}}|JM'\rangle, \tag{15.77}$$

so that the vector operator has to be $\mathcal{T}$-odd.

In 1997, the first measurement of an anapole moment was made in ^{133}Cs. This was the first observation of a *static* moment that is due to violation of reflection symmetry. This work prompted a number of theorists to calculate the values of anapole moments in heavy nuclei, see, for example, [150,151] and [152]. More experiments are needed in order to understand and describe better this interesting phenomenon of anapole moments.

15.5.6   Parity Violation in Neutral Currents

Here we say few words about ongoing experiments at the Jefferson Laboratory (JLab) where there are electron accelerators with the beam energy in the GeV region. Using the longitudinally polarized electron beams, one can look for the scattering from the nuclear target of interest due to the *weak neutral currents*. Here, there is no neutrino as in the process of exchange by the neutral Z-boson the charge of the electron projectile does not change. The parity-violating asymmetry in the electron scattering from a nucleus tells us that the process was weak and therefore induced by the exchange of the Z^0-boson rather than of the charged W-boson.

In the electroweak theory, the electron scattering from the nuclear target interacts mainly with the neutrons. Indeed, collecting the neutrons and protons from quarks (ddu and uud, respectively, with electric quark charges $q_u = +2/3$ and $q_u = -1/3$), we see [39] that the nuclear weak charge ratio is

$$\frac{f_n}{f_p} = -\frac{1}{1 - 4\sin^2\theta_W}, \tag{15.78}$$

where the *Weinberg angle* θ_W is the parameter of the electroweak theory close to $30°$ as established by the parity-violating scattering of polarized electrons from a proton target [153]. The polarized electron scattering at small angles in principle should give a direct information on the mean radius of the neutron distribution in the nucleus, while the proton distribution (mean charge radius) can be determined from the electromagnetic scattering. The measured quantity is the difference of cross sections for electrons with right and left helicity.

The existence of the *neutron skin* in heavy nuclei with $N > Z$ seems to be natural even if not precisely extracted from experiments until now. There are two experiments of this type under way at the JLab, PREX for ^{208}Pb and CREX for ^{48}Ca. A detailed knowledge of neutron and proton densities in heavy nuclei is important for the properties of neutron matter in stars and for physics of heavy-ion collisions. The PREX results from the run of 2010 for the difference of the mean square radii for neutrons and protons were [154]

$$R_n - R_p = 0.33^{+0.16}_{-0.18} \text{ fm}. \tag{15.79}$$

15.6 Electric Dipole Moment

15.6.1 Possible Nuclear Enhancement of EDM

Almost a decade after the discovery of parity ($\mathcal{P}$)-violation in weak interactions [6], experiments involving the decay of kaons found that also $\mathcal{CP}$ is not conserved [9] ($\mathcal{C}$ means charge conjugation). Later, $\mathcal{CP}$-violation was also found in the study of B-mesons [10,155,156] and recently also in the D-mesons study [157,158]. Through the $\mathcal{CPT}$ theorem, violation of $\mathcal{CP}$ is equivalent to time reversal ($\mathcal{T}$) non-conservation [159] and explained by the corresponding phase in the CKM matrix (see Section 15.3) which sets limits on the deviations from the standard model. For this mechanism of quark mixing in weak interactions, the predictions for the EDM of elementary particles, atoms, or nuclei are below the limits of possible measurements. One of the famous Sakharov conditions for baryogenesis asymmetry [11] is $\mathcal{CP}$-violation. In this case, the $\mathcal{CP}$-violating parameter in the CKM matrix is too small to render an explanation for the baryogenesis asymmetry. Various extensions of the standard model were suggested providing other sources for $\mathcal{CP}$-violation, see the review articles [160–163].

Since the 1960s, many attempts have been made to observe $\mathcal{T}$-violation in systems different from the kaons. Time reversal violation has not been observed so far in such experiments, but upper limits for $\mathcal{T}$-conservation have been established. The search for time reversal violation encompasses a large variety of physical systems and involves many methods. One of the more widely used approaches involves the search for static $\mathcal{T}-$, $\mathcal{P}$-odd electromagnetic moments that would be absent if the Hamiltonian of the system would be even under time reversal and reflection. Such moments include the electric dipole moment (EDM), the electric octupole, the magnetic quadrupole, etc. Early on, attempts were made to measure the EDM of the neutron, and at present, significant upper limits on the existence of such moment were established [164,165].

The neutron was not the only system in which attempts were made to find a static EDM. Experiments with atoms and molecules were performed where upper limits for EDM of the respective systems were established. In fact, the measurements of dipole moments of mercury and xenon atoms [166,167] and TlF molecule [168] have established upper limits for time reversal violating NN and quark-quark interactions that are of the same order (or maybe even exceed) the limits obtained in the measurement of the neutron EDM. The existence of a static atomic EDM may be due to the following three reasons:

1. the possible existence of a dipole moment of the electron (the best limits on the electron EDM were obtained in tallium atom measurements in Ref. [169]);

2. time reversal violation in the electron-nucleon interaction, thus in the lepton-hadron interactions;

3. the possible existence of a static $\mathcal{T}$-odd, $\mathcal{P}$-odd moment of the nucleus arising from the time reversal violating component of the hadron-hadron interactions. The experiments with mercury gave the best limit on this interaction [170]. It is thought that the dominant contribution to the atomic EDM comes from $\mathcal{P}$- and $\mathcal{T}$-violating components in NN forces. These components of the interaction induce $\mathcal{P}$-,$\mathcal{T}$-violating electromagnetic moments in the nucleus. In turn, these forbidden nuclear moments produce $\mathcal{P}$- and $\mathcal{T}$-violation in the electromagnetic field, which can induce static EDM in the atom. This is the working assumption in the study of the atomic dipole moment [170–172].

The measurements of the atomic dipole moment are presently very advanced and have reached a very high degree of precision [166,167,170]. These experiments are performed with neutral atoms. It has been shown [174] that in such cases, the field of the EDM of the nucleus is screened by the field of the electrons, and there is no effect when the atom is placed in an external electric field. This situation applies only when one considers point-like particles. When the finite size of the nucleus is taken into account, one can go to higher orders in the expansion of the electromagnetic field and find that in the next order in the expansion, it is the Schiff moment that can induce an EDM in the atom. The upper limits for the existence of dipole moments obtained in the atomic measurements cannot provide directly limits on the presence of $\mathcal{T}$- and $\mathcal{P}$-violation in the nuclear Hamiltonian. One needs theoretical input. To make the connection, one must have precise calculations of moments in nuclei produced by the $\mathcal{T}\mathcal{P}$ non-conserving part of the NN interaction.

A suggestion was put forward about 25 years ago [171,172] that rotating nuclei that have static octupole deformations when viewed in their intrinsic (body) frame of reference will have enhanced $\mathcal{T}$-odd, $\mathcal{P}$-odd moments if a time reversal and parity-violating interaction is present in the nuclear Hamiltonian. In the intrinsic frame, the nucleus with an octupole deformation has large octupole, Schiff, and dipole moments. Orientation of these moments is connected to a nuclear axis $\mathbf{n}$ (e.g., the dipole moment is $\mathbf{d} = d\mathbf{n}$). In a stationary rotational state, the mean orientation of the axis vanishes ($\langle \mathbf{n} \rangle = 0$) since the only possible correlation $\langle \mathbf{n} \rangle \sim \mathbf{I}$ violates time reversal invariance and parity (here $\mathbf{I}$ is the total angular momentum of the system). Therefore, the mean values of electric dipole, octupole, and Schiff moments vanish in the laboratory frame if there is no $\mathcal{T}$-, $\mathcal{P}$-violation. In the nuclei with the octupole deformation and nonzero intrinsic angular momentum, there are doublets of rotational states of opposite parity with the same angular momentum $\mathbf{I}$ (in molecular physics, this phenomenon is called parity doubling). A $\mathcal{T}$-, $\mathcal{P}$-odd interaction mixes these rotational levels. As a result, the nuclear axis becomes oriented along the total angular momentum, $\langle \mathbf{n} \rangle \sim \alpha \mathbf{I}$ where α is the mixing coefficient. Due to this orientation of the nuclear axis by a $\mathcal{T}$-, $\mathcal{P}$-odd interaction, the mean values of the $\mathcal{T}$-, $\mathcal{P}$-odd moments are not zero in the laboratory frame, e.g., $\langle \mathbf{d} \rangle = d\langle \mathbf{n} \rangle \sim d\mathbf{I}$.

One finds two basic enhancement factors in this mechanism: first, in the intrinsic frame, the nucleus with an octupole deformation will have large octupole, Schiff, or dipole electric moments because a large number of nucleons will contribute to the moments, and second, due to the appearance of closely spaced parity doublets in the spectrum of the nucleus with octupole deformation. It is not only that the spacing between the members of the doublets is small but also ($\mathcal{T}$-,$\mathcal{P}$-odd) interaction will mix well two such states. The enhanced nuclear Schiff moments that result in such nuclei with a reflection asymmetric shape will induce 1000 times enhanced atomic electric dipoles, and measurements performed with such atoms may improve upper limits for time reversal violation.

15.6.2   Schiff Moment

Let us start our consideration with the expression for the electrostatic potential of a nucleus screened by the electrons of the atom. If we consider only the dipole $\mathcal{T}$-, $\mathcal{P}$-odd part of screening (Purcell-Ramsey-Schiff theorem [175]), one finds

$$\varphi(\mathbf{R}) = \int \frac{e\rho(\mathbf{r})}{|\mathbf{R} - \mathbf{r}|}\, d^3r + \frac{1}{Z}(\mathbf{d} \cdot \nabla) \int \frac{\rho_s(r)}{|\mathbf{R} - \mathbf{r}|}\, d^3r. \qquad (15.80)$$

Here, $\rho(\mathbf{r})$ is the nuclear charge density, $\int \rho(\mathbf{r})\, d^3r = Z$ includes the spherically symmetric part of $\rho(s)$, and $\mathbf{d} = \int e\mathbf{r}\rho(\mathbf{r})\, d^3r$ is the EDM of the nucleus. The multipole expansion of the potential contains both $\mathcal{T}$-, $\mathcal{P}$-even and $\mathcal{T}$-, $\mathcal{P}$-odd terms. The dipole part in (15.80) is canceled out by the second term in this equation:

$$+\frac{1}{2Z}(\mathbf{d}\cdot\nabla)\nabla_\alpha\nabla_\beta\frac{1}{R}\int\rho_s(r)r_\alpha r_\beta\,d^3r \tag{15.81}$$

$$-\int e\left(\mathbf{r}\cdot\nabla\frac{1}{R}\right)\rho(\mathbf{r})\,d^3r+\frac{1}{Z}(\mathbf{d}\cdot\nabla)\frac{1}{R}\int\rho_s(r)\,d^3r=0. \tag{15.82}$$

The next term is the electric quadrupole which is $\mathcal{T}$-,$\mathcal{P}$-even; thus, the first nonzero $\mathcal{T}$-,$\mathcal{P}$-odd term is

$$\varphi^{(3)}=-\frac{1}{6}\int e\rho(\mathbf{r})\,r_\alpha r_\beta r_\gamma\,d^3r\,\nabla_\alpha\nabla_\beta\nabla_\gamma\frac{1}{R} \tag{15.83}$$

$$+\frac{1}{2Z}(\mathbf{d}\cdot\nabla)\nabla_\alpha\nabla_\beta\frac{1}{R}\int\rho_s(r)r_\alpha r_\beta\,d^3r. \tag{15.84}$$

Here, $r_\alpha r_\beta r_\gamma$ is a reducible tensor. After separation of the trace, there will be terms that contain a vector $\mathbf{S}$ (Schiff moment) and a rank 3, $Q_{\alpha\beta\gamma}$ (electric octupole) moments:

$$\varphi^{(3)}=\varphi^{(3)}_{\text{Schiff}}+\varphi^{(3)}_{\text{octupole}}, \tag{15.85}$$

$$\varphi^{(3)}_{\text{Schiff}}=-(\mathbf{S}\cdot\nabla)\Delta\frac{1}{R}=4\pi(\mathbf{S}\cdot\nabla)\delta(R), \tag{15.86}$$

$$\varphi^{(3)}_{\text{octupole}}=-\frac{1}{6}Q_{\alpha\beta\gamma}\nabla_\alpha\nabla_\beta\nabla_\gamma\frac{1}{R} \tag{15.87}$$

where

$$\mathbf{S}=\frac{1}{10}\left(\int e\rho(\mathbf{r})r^2\mathbf{r}\,d^3r-\frac{5}{3}\mathbf{d}\frac{1}{Z}\int\rho_s(r)r^2\,d^3r\right) \tag{15.88}$$

is the Schiff moment and

$$Q_{\alpha\beta\gamma}=\int e\rho(\mathbf{r})[r_\alpha r_\beta r_\gamma-\frac{1}{5}(\delta_{\alpha\beta}r_\gamma+\delta_{\beta\gamma}r_\alpha+\delta_{\alpha\gamma}r_\beta)]\,d^3r, \tag{15.89}$$

$$Q_{zzz}=\frac{2}{5}Q_3=\frac{2}{5}\sqrt{\frac{4\pi}{7}}\int e\rho(\mathbf{r})r^3Y_{30}\,d^3r, \tag{15.90}$$

is the tensor octupole moment.

In the absence of $\mathcal{T}$- and $\mathcal{P}$-violating interaction, the EDM of an atom vanishes. The interaction between atomic electrons and the $\mathcal{T}$- and $\mathcal{P}$-odd part of the electrostatic nuclear potential in Eq. (15.83) will mix atomic states of opposite parity and thus generate an atomic EDM:

$$D_z=-e\langle\tilde{\psi}|r_z|\tilde{\psi}\rangle=-2e\sum_{|k_2\rangle}\frac{\langle k_1|r_z|k_2\rangle\langle k_1|-e\varphi^{(3)}|k_2\rangle}{E_{k_1}-E_{k_2}}, \tag{15.91}$$

where $\tilde{\psi}$ denotes the perturbed atomic wave function, $|k_1\rangle=|k_1,J_1,J_{1z}\rangle$ is the unperturbed electron ground state, and $|k_2\rangle$ is the set of opposite parity states with which $|k_1\rangle$ is mixed due to the perturbation $-e\varphi^{(3)}$.

15.6.3   The EDM of ^{199}Hg

The most accurate upper limit on an atomic EDM has been obtained from the measurement of the atomic EDM in the mercury atom [170]. The present upper limit for the EDM in this atom is $|d| < 7.4 \times 10^{-30}$ e cm. The nucleus ($Z = 80$) has an odd number of neutrons. The lowest-order contribution to the Schiff moment comes from a $2p - 1h$ configuration (a proton $1p - 1h$ added to the neutron) admixed into the ground state [176]. Because of the importance of this nucleus for the present experimental studies, it was useful to examine in more detail the influence of core polarization effects on the Schiff moment [177,178]. The main conclusion was that the core-polarization contributions to the Schiff moments are sizable. These corrections reduce the lowest-order values obtained in the past by factors of 2–3. This has important consequences when trying to interpret the implications of the EDM measurements. The reductions of the Schiff moments (compared to the values obtained in the past) mean that using the experimental upper limits for the atomic dipole, one obtains now a higher upper limit for the strength parameters in the weak $\mathcal{P}$ and $\mathcal{T}$ non-conserving component of the NN interaction. Time reversal violation was studied in other atoms with spherical nuclei, such as ^{126}Xe and ^{133}Cs [179].

A very different idea of evaluating the $\mathcal{T}$-violating nuclear forces was also tested the statistical distribution of level spacings s between the closest stationary states with the same values of spin and parity behaves linearly, $\propto s$, in time-reversal invariant cases and quadratically, $\propto s^2$, in the presence of $\mathcal{T}$-violation. However, due to the poor statistics of such close encounters of measured levels, it is hard here to extract a useful estimate.

15.6.4   Rotational Doublets and $\mathcal{T}$-, $\mathcal{P}$-Odd Moments

If a deformed nucleus in the intrinsic (body-fixed) reference frame is reflection asymmetric, it can have collective $\mathcal{T}$-,$\mathcal{P}$-odd moments without any $\mathcal{T}$, $\mathcal{P}$-violation. As a consequence of this reflection asymmetry, rotational doublets appear in the laboratory frame. Without $\mathcal{T}$-,$\mathcal{P}$-violating forces, a $\mathcal{T}$-,$\mathcal{P}$-odd moment vanishes exactly in the laboratory. A $\mathcal{T}$-,$\mathcal{P}$-odd interaction, however, may reveal such intrinsic $\mathcal{T}$-,$\mathcal{P}$-odd moments in the laboratory frame.

Consider a nearly degenerate rotational parity doublet in the case of an axially symmetric nucleus. The wave functions of the members of the doublet are written as [180]

$$\Psi^{\pm} = \frac{1}{\sqrt{2}} \left(|IMK\rangle \pm |IM - K\rangle \right). \tag{15.92}$$

Here, I is the nuclear spin, $M = I_z$ and $K = (\mathbf{In})$, where $\mathbf{n}$ is a unit vector along the nuclear axis. The intrinsic dipole and Schiff moments are directed along $\mathbf{n}$:

$$\boldsymbol{d}_{intr} = d_{intr}\,\boldsymbol{n}, \tag{15.93}$$

$$\boldsymbol{S}_{intr} = S_{intr}\,\boldsymbol{n}. \tag{15.94}$$

For these good parity states, $\langle \Psi^{\pm}|(\mathbf{I}\cdot\mathbf{n}|\Psi^{\pm}\rangle = 0$ because K and $-K$ have equal probabilities which means that there is no average orientation of the nuclear axis $\langle \Psi^{\pm}|\mathbf{n}|\Psi^{\pm}\rangle$ in the laboratory frame. This is a consequence of $\mathcal{T}$- and $\mathcal{P}$-conservation since the correlation $(\mathbf{I} \cdot \mathbf{n})$ is $\mathcal{T}$-,$\mathcal{P}$-odd. As a result of $\langle \Psi^{\pm}|\mathbf{n}|\Psi^{\pm}\rangle = 0$, the mean value of the $\mathcal{T}$-,$\mathcal{P}$-odd moments (whose orientation is determined by the direction of the nuclear axis) is zero in the laboratory frame. A $\mathcal{T}$-,$\mathcal{P}$-odd interaction V^{PT} will mix the members of the doublet. The admixed wave function of the predominantly positive parity member of the doublet will be $\Psi = \Psi^+ + \alpha\Psi^-$ or:

$$\Psi = \frac{1}{\sqrt{2}}[(1 + \alpha)|IMK\rangle + (1 - \alpha)|IM - K\rangle], \tag{15.95}$$

where α is the $\mathcal{T}$-,$\mathcal{P}$-odd admixture,

$$\alpha = \frac{\langle \Psi^- | V^{PT} | \Psi^+ \rangle}{E^+ - E^-}, \tag{15.96}$$

and $E^+ - E^-$ is the energy splitting between the members of the parity doublet. In the $\mathcal{T}$-,$\mathcal{P}$-admixed state,

$$\langle \Psi | (\mathbf{I} \cdot \mathbf{n}) | \Psi \rangle = \langle \Psi | \hat{K} | \Psi \rangle = 2\alpha K, \tag{15.97}$$

i.e., the nuclear axis $\mathbf{n}$ is oriented along the nuclear spin $\mathbf{I}$:

$$\langle \Psi | n_z | \Psi \rangle = 2\alpha \frac{KM}{(I+1)I}. \tag{15.98}$$

Therefore, in the laboratory system, the electric dipole and Schiff moments obtain nonzero average values. For example, in the ground state of a well-deformed nucleus, usually $M = K = I$ and

$$\langle \Psi | S_z | \Psi \rangle = 2\alpha \frac{I}{I+1} S_{\text{intr}}. \tag{15.99}$$

15.6.5 Enhancement Factors in Pear-Shaped Nuclei

The idea of enhancements of $\mathcal{T}$-violation was first suggested in the papers [171,172] and [173]. The initial calculations of the quadrupole plus octupole deformed (termed "pear-shaped") neutron-rich actinide nuclei were described using a two-liquid drop model (Figure 15.13). The surface of an axially symmetric deformed nucleus is

$$R = c_V(\beta) R_0 \left(1 + \sum_{l=1} \beta_l Y_{l0} \right), \tag{15.100}$$

c_V is of order unity and $R_0 = r_0 A^{1/3}$. We present here the result for the calculated intrinsic Schiff moment leaving out the details of the calculation, which can be found in Ref. [172]:

$$S_{\text{intr}} \simeq eZR_0^3 \frac{3}{20\pi} \sum_{l=2} \frac{(l+1)\beta_l \beta_{l+1}}{\sqrt{(2l+1)(2l+3)}}. \tag{15.101}$$

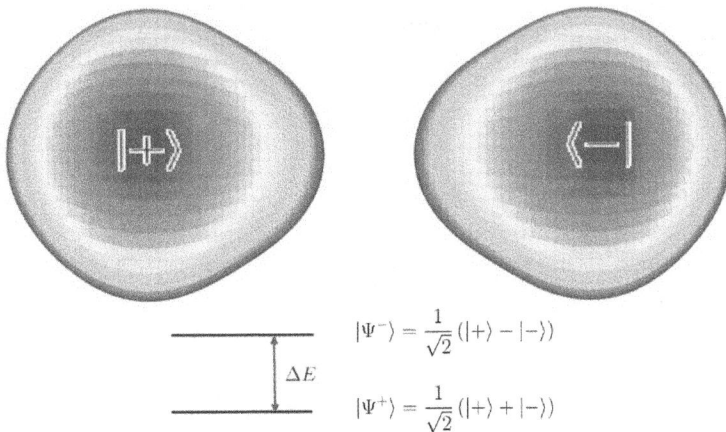

$$|\Psi^-\rangle = \frac{1}{\sqrt{2}} (|+\rangle - |-\rangle)$$

$$|\Psi^+\rangle = \frac{1}{\sqrt{2}} (|+\rangle + |-\rangle)$$

FIGURE 15.13
A schematic illustration of the doublets in the "pear-shape nuclei."

In order to calculate the laboratory Schiff moment, we use in Eqs. (15.100) and (15.101) a simple form for the one-body $\mathcal{T}$-,$\mathcal{P}$-odd potential:

$$V^{PT} = \frac{G}{\sqrt{2}} \frac{\eta}{2m} \rho_0 \sum_i (\vec{\sigma}_i \cdot [\nabla_i f(r_i)]), \qquad (15.102)$$

where $\rho_0 f(r)$ is the nuclear density. The coefficient η is the unknown strength of the potential above. For a detailed description of the method of calculation of Schiff moments, see Ref. [172]. The calculations were performed for a number of neutron-rich actinide atoms. The coefficients β_2 and β_3 are empirically known and shown in Table 15.1.

The nuclear Schiff moments for the octupole-deformed nuclei calculated here are two to three orders of magnitude larger than those obtained in Refs. [175–179] for isotopes of ^{129,131}Xe, ^{199,201}Hg, and ^{203,205}Tl (see Table 15.3). The reasons for these huge enhancements can be summarized as follows:

1. The quadrupole + octupole deformation leads to large intrinsic Schiff moments.

2. The parity doublets in the reflection asymmetric nuclei are very close in energy.

3. At the same time, the parity and time reversal violating matrix element between the members of the doublet is relatively large.

4. The nuclei that are quadrupole + octupole deformed are also heavy, having large Z.

5. Atomic physics in these atoms is of help.

Calculations of microscopic nature have been performed [181,182] confirming the large enhancements found in Refs. [171,172].

In several papers [183,184], it was indicated that the possibility of enhancement of the Schiff moment might exist also in the odd-even nuclei with the neighboring even-even nucleus being deformed and having a low-lying octupole state. The odd nucleon will couple to the core states producing a similar effect as found in the pear-shaped nuclei. Even a more extreme case was considered [184], when the odd nucleon couples to a spherical core with low-lying quadrupole and octupole excitations.

15.6.6   Experiments with Pear-Shaped Nuclei

In order to perform experiments that search for dipole moments in atoms with nuclei that are octupole deformed in addition to the quadrupole deformation, one has to study the properties of these nuclei. One has to confirm that indeed they exhibit parity doublets, determine how large is the spacing between the doublets, what is the lifetime of such

TABLE 15.1

Intrinsic g.s. Deformations and Energy Splittings between Opposite Parity Core States

	^{223}Ra	^{225}Ra	^{223}Rn	^{221}Fr	^{223}Fr	^{225}Ac	^{229}Pa
β_2	0.125	0.143	0.129	0.106	0.122	0.138	0.176
β_3	0.100	0.099	0.081	0.100	0.090	0.104	0.082
β_4	0.076	0.082	0.078	0.069	0.076	0.078	0.093
β_5	0.042	0.035	0.024	0.045	0.033	0.038	0.020
β_6	0.018	0.016	0.023	0.020	0.022	0.013	0.015
E_c (keV)	212	221	213	305	212	206	333

nuclei, etc. The experimental data regarding the nuclei shown in Table 15.2 are discussed in the reviews [185–188] and in a more recent publication [189]. The ^{223}Ra and ^{225}Ra nuclei were considered in detail in Refs. [190,191]. The case of francium is especially interesting in light of the experiments involving trapping of atoms [192]. The experimental and theoretical studies of ^{221}Fr [193] and ^{223}Fr [194,195] provide strong evidence of intrinsic reflection asymmetry in the ground state.

The result for the Schiff moment of ^{221}Fr is smaller than that of ^{223}Fr because of the following factors. Firstly, the factor $M/[I/(I+1)]$ in Eq. (15.98) is $1/7$ for ^{221}Fr, whereas for $I = 3/2, K = 3/2$ ground state of ^{223}Fr, it is $3/5$. Secondly, the calculation gives the admixture coefficient α for ^{221}Fr about three times less than for ^{223}Fr. Because of these factors, the Schiff moment of ^{221}Fr in the laboratory frame is about 12 times less than that of ^{223}Fr, although in the intrinsic frame, the values of S_{intr} are roughly equal.

There is a controversy regarding the ground-state spin and parity doublet in ^{229}Pa, which is on the border of the region of octupole-deformed nuclei. Two assignments, (a) $K = 5/2, I = 5/2$ ground state and 220 eV energy splitting within the parity doublet [187,188] and (b) $K = 1/2, I = 3/2^-$ ground state and unidentified parity partner level [188], were made. In case (a), the calculations using the experimental value of the energy splitting give the admixture coefficient $640 \times 10^7 \eta$ and the Schiff moment in the laboratory frame $S = 230000 \times 10^8 \eta e$ cm. Note, however, that in Table 15.2, the results for ^{229}Pa are an order of magnitude smaller because a theoretical value of 5 keV was used for the energy splitting of the doublet.

Apart from the measurements in ^{199}Hg, which we discussed previously, there are now several experiments performed or planned, that attempt to look for the EDM in some of the actinide nuclei that are suggested in [171] and [172], see Table 15.3. At KVI in Groningen and at TRIUMF, Vancouver, experiments with ^{223}Rn are planned. The most advanced research of an atomic EDM in the actinide atoms is the ^{225}Ra experiment at the Argonne National Laboratory. The first results of this experiment were published in 2015 [196]. A year later, an improved upper limit for the EDM was published [197]. The corresponding upper limit for $\mathcal{T}$-violation is still a few orders of magnitude away from the ^{199}Hg result, in spite of the large enhancement of the Schiff moment in ^{225}Ra. However, the potential for improving considerably the upper limit using this method is quite feasible [196,197].

TABLE 15.2

Admixture Coefficients α (Absolute Values) and Theoretical Energy Splitting between the g.s. Doublet Levels $\Delta E = E^- - E^+$, Parities of the Intrinsic (Reflection-Asymmetric) Single-Particle g.s. Calculated Using the Woods-Saxon (WS) and Nilsson (Nl) Potentials, Experimental Energy Splitting, *Intrinsic* Schiff Moments and Schiff Moments Calculated with the Woods-Saxon Potential, and Induced Atomic Dipole Moments

	^{223}Ra	^{225}Ra	^{223}Rn	^{221}Fr	^{223}Fr	^{225}Ac	^{229}Pa
α(WS) $(10^7 \eta)$	1	2	4	0.7	2	3	34
ΔE(WS) (keV)	170	47	37	216	75	49	5
π_p(WS)	0.81	-0.02	0.17	-0.55	-0.34	-0.35	0.01
α(Nl)$(10^7 \eta)$	2	5	2				
ΔE(Nl) (keV)	171	55	137				
ΔE_{expt} (keV)	50.2	55.2		234	160.5	40.1	0.22
$S_{intr}(e\ fm^3)$	24	24	15	21	20	28	25
$S(10^8\ \eta e\ fm^3)$	400	300	1000	43	500	900	1.2×10^4
d(at) $(10^{25}\ \eta e\ cm)$	2700	2100	2000	240	2800		

The values are given for pear-shaped nuclei.

TABLE 15.3

The Calculated Schiff Moments and Atomic Dipole
Moments for Spherical Nuclei

	^{199}Hg	^{129}Xe	^{133}Cs
$S(10^8 \ \eta e \ fm^3)$	-1.4	1.75	3
$d(\text{at}) \ (10^{25} \ \eta e \ cm)$	5.6	0.47	2.2

15.7   Neutrino-Nucleus Reactions

I have done a terrible thing. I have invented a particle that cannot be detected.
(W. Pauli)

Neutrinos they are very small
They have no charge and mass
And do not interact at all
The earth is just a silly ball
To them, through which they simply pass
Like dust maids down a drafty hall.
(John Updike, 1960)

(In 2019, one can say that the only statement that is correct is *They have no charge*, the rest is proven to be incorrect).

15.7.1   Introduction

Historically processes induced by weak interaction on nuclei such as beta-decay have been exploited to study the properties of the weak interaction and also investigate nuclear structure. Maybe one of the best examples were the experiments that determined the non-conservation of reflection symmetry in the weak interaction [6,7]. On the other hand, many such experiments studying beta decay and muon capture (Eq. (15.110)) have provided essential information about nuclear structure [57,198]. This activity was enlarged and intensified with the appearance of meson factories in the 1980s [199]. The study of weak interactions and nuclei was extended to reactions of scattered neutrinos on nuclei [199–201]. Such studies revealed new features of nuclear structure and lead to a better understanding of the reaction mechanisms of weakly interacting particles. Moreover, these studies had significant practical importance in the exploration of neutrino properties.

Nuclei are used quite often as neutrino detectors so that understanding the reactions induced by neutrinos on nuclei becomes an important ingredient in the study of neutrino oscillations and neutrino masses, and in the search of additional types of neutrinos beyond what the standard model includes [202]. Neutrino studies also are instrumental in astrophysics. The detection and analysis of neutrinos emitted by stars and in particular by supernovae during their collapse provide information about nuclear reactions occurring in stellar environments [203,204]. To summarize the importance of neutrino-nucleus interactions:

1. Study of many aspects of the nuclear structure.

2. Study of neutrino oscillations and neutrino masses via neutrino-nucleus interactions.

3. Neutrinos as "messengers" bringing information about the interiors of stars.

4. Nucleosynthesis. Supernova physics.

In the present review of neutrino reactions on nuclei, we will deal only with the low-energy (neutrinos with $E_\nu \leq 300$ MeV) part of such studies. At such energies, nuclear structure is revealed in great detail. There is presently a large number of experimental and theoretical studies of neutrino reactions on nuclei at much higher energies [205], but we will not deal with this here.

15.7.2 Experimental Studies

The meson factories produced intense beams of pions, which then decay into leptons to provide intense fluxes of neutrinos and antineutrinos. The neutrinos obtained in these decays were both electron neutrinos and muon neutrinos. There are two types of neutrino beams: DIF (decay in flight) and DAR (decay at rest). In the 1980s and 1990s, the main two facilities that were active and produced pioneering work on neutrino-nucleus scattering were the LSND (Los Alamos) and the KARMEN at Karlsruhe collaborations [199–201]. The experiments performed involved charge currents, with the aim of detecting neutrino oscillations. While this was the main aim, nuclear structure information was also obtained. The nuclear target used in these experiments was $^{12}_{6}$C, a nucleus whose properties are well known.

The types of charged current reactions explored:

- $_Z A(\nu_e, e^-)_{Z+1} A^*$ at LSND, KARMEN
- $_Z A(\nu_\mu, \mu^-)_{Z+1} A^*$ at LSND

with the target being $^{12}_{6}$C. In addition, these data are supplemented by μ^- capture rates: $_Z A(\mu^-, \nu_\mu)_{Z+1} A^*$ inclusive and exclusive. These are charge-exchange transitions and therefore of isovector character. Two classes of neutrino scattering experiments were performed, inclusive and exclusive to the ground state.

The neutrino (or antineutrino) beams are not monochromatic. The energy distribution of the fluxes can be determined experimentally. As an example, we show in Figure 15.14 the fluxes used in the LSND collaboration. Two sources of neutrinos are used in the experiments. Electron neutrinos originated from the DAR of muons and muon neutrinos were obtained from the DIF of pions.

15.7.3 Theoretical Framework

The theory of nuclear response to weak probes has been formulated years ago, and there is extensive literature that discusses it. A good place to look it up is the reference [206]. The general expression for the cross section of the reaction $\nu_l + {}^{12}C \to l + {}^{12}N$ is

$$\sigma = (2\pi)^4 \sum_f \int d^3 p_l \delta(E_l + E_f - E_\nu - E_i) |\langle l(p_l); f | H_{\text{eff}} | \nu_l(p_\nu); i \rangle|^2, \tag{15.103}$$

where $E_f(E_i)$ is the energy of the final (initial) nuclear state, $E_\nu(p_\nu)$ is the incident neutrino energy (momentum), and $E_l(p_l)$ is the outgoing lepton energy (momentum). The effective single-particle Hamiltonian H_{eff} is derived by carrying out the Foldy-Wouthuysen transformation and retaining terms up to $\mathcal{O}(|\mathbf{q}|/M)^3$ (M is the nucleon mass, and $\mathbf{q}$ is the momentum transfer) since the momentum transfer involved in the nuclear scattering of neutrinos produced by accelerator can be large. The detailed expressions can be found in [207]. If the nuclear recoil effects are neglected, we have

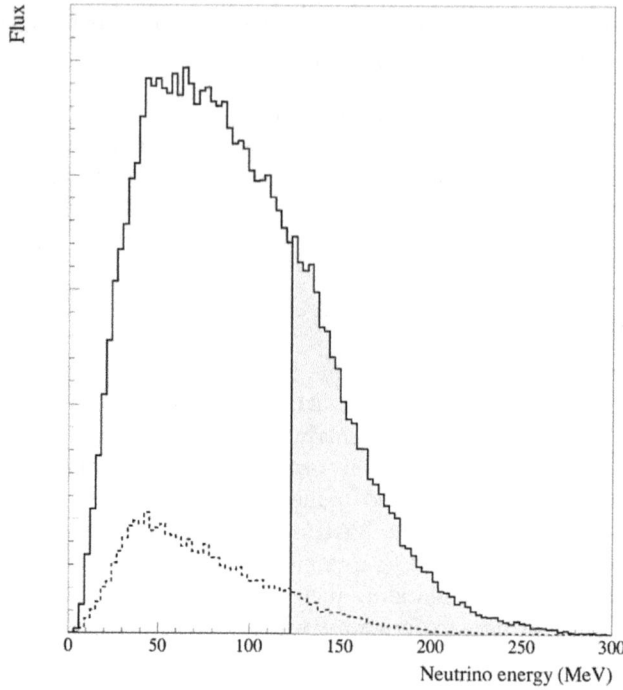

FIGURE 15.14
The solid line shows the flux shape of ν_μ from π^+ decay in flight. The region above the muon production threshold is shaded. The dashed line shows the $\bar{\nu}_\mu$ flux from π^- decay-in-flight for the same integrated proton beam.

$$\sigma = \frac{G^2}{2\pi}\cos^2\theta_C \sum_f p_l E_l \int_{-1}^{1} d(\cos\theta) M_\beta, \qquad (15.104)$$

where $G\cos\theta_C$ is the weak coupling constant, θ is the angle between the momenta of the incident neutrino and outgoing lepton, and M_β is given by

$$M_\beta \equiv M_F |\langle f|\tilde{1}|i\rangle|^2 + M_{G0}\frac{1}{3}|\langle f|\tilde{\sigma}|i\rangle|^2 + M_{G2}\Lambda. \qquad (15.105)$$

The squared nuclear matrix elements are

$$|\langle f|\tilde{\sigma}|i\rangle|^2 = \frac{4\pi}{(2J_i+1)}\sum_{l,K}|\langle J_f||\sum_k t_+(k)j_l(qr_k)[Y_l(\hat{\mathbf{r}}_k)\sigma]^{(K)}||J_i\rangle|^2, \qquad (15.106)$$

$$\Lambda \equiv \left(\frac{5}{6}\right)^{\frac{1}{2}}\sum_{l,l',K}(-1)^{l/2-l'/2+K}\sqrt{2l+1}\sqrt{2l'+1}\begin{pmatrix} l & l' & 2 \\ 0 & 0 & 0 \end{pmatrix}$$

$$\times \begin{pmatrix} 1 & 1 & 2 \\ l & l' & K \end{pmatrix}\frac{4\pi}{(2J_i+1)}\langle J_f||\sum_k t_+(k)j_l(qr_k)[Y_l(\hat{\mathbf{r}}_k)\sigma]^{(K)}||J_i\rangle$$

$$\times \langle J_f||\sum_{k'} t_+(k')j_{l'}(qr_{k'})[Y_{l'}(\hat{\mathbf{r}}_{k'})\sigma]^{(K)}||J_i\rangle^*. \qquad (15.107)$$

In Eq. (15.106), k labels the space and spin-isospin coordinates of the kth nucleon, l and l' are the orbital angular momenta, and K is the total angular momentum of the transition operators. The coefficients M_F, M_{G0} and M_{G2} [207] appearing in Eq. (15.105) depend on the momentum transferred to the nucleus [$q = (q, iq_0) = p_l - p_\nu$] and the nucleon form-factors $f_V(q^2), f_A(q^2), f_W(q^2), f_P(q^2)$. Second-class current form-factors are ignored.

Various corrections resulting from the Coulomb distortions of the outgoing lepton wave functions are described in the literature [208]. As there is no monochromatic beam of neutrinos, one must introduce a flux-averaged cross section $\langle \sigma_f \rangle$ that can be compared to experiment

$$\langle \sigma \rangle_f = \int dE_\nu \sigma(E_\nu) \tilde{f}(E_\nu), \qquad (15.108)$$

with

$$\tilde{f}(E_\nu) = \frac{f(E_\nu)}{\int_{E_0}^\infty dE'_\nu f(E'_\nu)}, \qquad (15.109)$$

$f(E_\nu)$ being the neutrino flux and E_0 the threshold energy.

There is another neutrino reaction closely related to the muon-neutrino scattering, namely the capture of a negatively charged muon from an atomic orbit:

$$\mu^- + (A, Z) \to (A, Z - 1) + \nu_\mu. \qquad (15.110)$$

When the muon is captured from the $1s$ atomic orbit, the inclusive rate is:

$$\Lambda_c = \frac{m_\mu^2}{2\pi} |\phi_{1s}|^2 [G_V^2 M_V^2 + G_A^2 M_A^2 + (G_P^2 - 2G_P G_A) M_P^2], \qquad (15.111)$$

if we neglect the recoil term that represents a correction of a few percents [209]. Here, ϕ_{1s} is the muon $1s$-bound state wave function evaluated at the origin, i.e., $|\phi_{1s}|^2 = R(Z)^3/\pi$, R being a reduction factor accounting for the finite size of the nuclear charge distribution ($R = 0.86$ for ^{12}C) [206], and m_μ being the muon reduced mass. The constants G_V, G_A, and G_P [209] are the "effective coupling constants" which depend only slightly on the neutrino momentum p_ν. This can be simply obtained from the energy and momentum conservation, $p_\nu = m_\mu - (m_n - m_p) - |E_\mu^B| - E_{fi}$, where m_n and m_p are the neutron and proton masses, respectively, and $|E_\mu^B|$ is the nuclear excitation energy measured with respect to the parent nucleus ground state. The capture rate can be factorized as in Eq. (15.111) if we neglect the dependence of the coupling constants on p_ν. The squares of the vector, axial-vector, and pseudoscalar matrix elements are

$$M_V^2 = 4\pi \sum_l (2l + 1) \sum_f \left(\frac{p_\nu}{m_\mu}\right)^2 |\langle J_f| \sum_k t_+(k) j_l(p_\nu r_k) Y_{l0}(\hat{\mathbf{r}}_k)|J_i\rangle|^2, \qquad (15.112)$$

$$M_A^2 = 4\pi \sum_{l,K} (2K + 1) \sum_f \left(\frac{p_\nu}{m_\mu}\right)^2 |\langle J_f| \sum_k t_+(k) j_l(p_\nu r_k) [Y_l(\hat{\mathbf{r}}_k) \times \sigma]^{K0}|J_i\rangle|^2, \qquad (15.113)$$

$$M_P^2 = 4\pi \sum_{l,l',K} (2K + 1) \sum_f \left(\frac{p_\nu}{m_\mu}\right)^2 |\langle J_f| \sum_k t_+(k) \theta_P(K))|J_i\rangle|^2, \qquad (15.114)$$

$$\theta_P(K) = \sqrt{\frac{K}{(2K + 1)}} j_l(p_\nu r_k) [Y_l(\hat{\mathbf{r}}_k) \times \sigma]^{K0}$$

$$+ \sqrt{\frac{K + 1}{(2K + 1)}} 2j_{l'}(p_\nu r_k) [Y_{l'}(\hat{\mathbf{r}}_k) \times \sigma]^{K0}, \qquad (15.115)$$

with $l = K - 1$ and $l' = K + 1$. The decay corresponds to the limit of zero momentum transfers of the transition probabilities in Eq. (15.105), that is

$$M_\beta^0 \equiv f_V^2(0)|\langle f|1|i\rangle|^2 + f_A^2(0)\frac{1}{3}|\langle f|\sigma|i\rangle|^2, \tag{15.116}$$

where

$$\langle f|1|i\rangle|^2 = \frac{1}{2J_i + 1}|\langle J_f||\sum_k t_+(k)||J_i\rangle|^2, \tag{15.117}$$

$$|\langle f|\sigma|i\rangle|^2 = \frac{1}{2J_i + 1}|\langle J_f||\sum_k t_+(k)\sigma(k)||J_i\rangle|^2, \tag{15.118}$$

and the transition operators are of the usual F or GT type. The ft value is given by

$$ft = \frac{2\pi^3 \ln 2}{(G^2 \cos^2 \theta_C m_e^5)} \frac{1}{M_\beta^0}. \tag{15.119}$$

The neutrino reaction cross section, the muon capture rate, and the ft value for the β decay depend on the wave functions of the initial and final nuclear states involved in these processes.

The experimental muon scattering cross sections from LSND were of two types: transitions to the ground state of the final nucleus, which we term exclusive, and transitions to all available final states, which are called inclusive. The experimental results that have been obtained in the scattering experiments are shown below: $^{12}C(\nu_\mu, \mu^-)^{12}N_{\text{g.s.}}$ exclusive; $^{12}C(\nu_\mu, \mu^-)^{12}X$ inclusive (these are DIF neutrinos), and the source of ν_μ neutrino is π^+, and $^{12}C(\nu_e, e^-)^{12}N_{\text{g.s.}}$ exclusive, $^{12}C(\nu_e, e^-)^{12}X$ inclusive (these are DAR neutrinos), the source of neutrinos is μ^+. The numerical values found in experiments by LSND [199] are

- Inclusive $\sigma(\nu_\mu, \mu^-) = (10.6 \pm 0.3 \pm 1.8) \times 10^{-40}$ cm^2
- Exclusive to g.s. $\sigma(\nu_\mu, \mu^-) = (0.56 \pm 0.08 \pm 0.1) \times 10^{-40}$ cm^2.

The corresponding experimental electron scattering cross sections by LSND [199] are

- Inclusive $\sigma(\nu_e, e^-) = (14.1 \pm 1.2) \times 10^{-42}$ cm^2.
- Exclusive $\sigma(\nu_e, e^-) = (9.6 \pm 0.4 \pm 0.9) \times 10^{-42}$ cm^2.

The experimental muon capture rates for ^{12}C [210] are

- Inclusive $\Lambda = (3.79 \pm 0.05) \times 10^4$ s^{-1}.
- Exclusive $\Lambda = (0.06 \pm 0.01) \times 10^4$ s^{-1}.

15.7.4 Random Phase Approximation and the Shell Model

The transitions involved in the processes described above are of isovector type (see the transitions depicted in Figure 15.15). The framework that is suited to describe such excitations is the charge-exchange RPA (or QRPA) [208,212,213] first introduced in Ref. [214]. In this approach, one determines the nuclear states in the form

$$|\lambda\rangle = \left(\sum_{I,a} X_{Ia}^{(\lambda)} p_I^\dagger n_a + \sum_{i,a} Y_{iA}^{(\lambda)} p_i^\dagger n_A\right)|\tilde{0}\rangle, \tag{15.120}$$

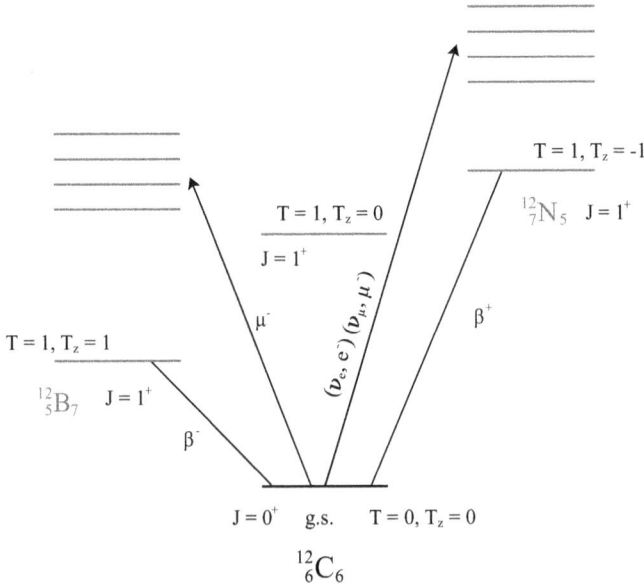

FIGURE 15.15
Various weak interaction transitions for the $A = 12$ isospin triplet are indicated in this figure and discussed in the text.

where $\tilde{0}$ is the RPA ground state. The amplitudes $X^{(\lambda)}$ and $Y^{(\lambda)}$ are the solution of the charge-exchange RPA equations. For a one-body charge-exchange operator

$$\mathcal{O} = \sum_{\alpha,\beta} \mathcal{O}_{\alpha\beta} p_\alpha^\dagger n_\beta, \tag{15.121}$$

the transition amplitude can be expressed simply as

$$\langle \lambda | \mathcal{O} | \tilde{0} \rangle = \sum_{Ia} X_{Ia}^{(\lambda)*} \tilde{O}_{Ia} - \sum_{i,A} Y_{iA}^{(\lambda)*} \mathcal{O}_{iA}, \tag{15.122}$$

with the notations $i, j, \ldots (a, b, \ldots)$ proton (neutron) occupied states and $I, J, \ldots (A, B, \ldots)$ proton (neutron) unoccupied states.

The basis of the RPA calculations in the neutrino studies was particle-hole states obtained from the self-consistent Hartree-Fock theory usually using the Skyrme-type interactions. The results of the RPA (QRPA) of Ref. [208] are shown in Tables 15.4–15.7.

These calculations reproduce the experimental findings except for the muon-neutrino cross sections [208,212] when the experimental cross sections are considerably smaller than the RPA results. This is particularly striking in the case of the inclusive muon-neutrino cross section where the discrepancy is almost a factor of two. This appears in all various RPA calculations.

Another approach to the nuclear structure relevant to the neutrino reactions uses an extended shell-model framework [208,215], see Tables 15.4–15.7. In Ref. [215], configurations were composed, in terms of oscillator frequency, of excitations up to $1\hbar\omega$ for negative parity states and $2\hbar\omega$ for positive parity states. In Ref. [208], the space was larger, up to $3\hbar\omega$ excitation for negative parity states and up to $2\hbar\omega$ for positive parity states. Examining these tables, one sees that the results are strongly dependent on the choice of single-particle wave functions and on the size of the shell-model space. The muon inclusive cross sections in

TABLE 15.4

Flux-Averaged Inclusive Cross Section $\langle\sigma\rangle_f$ within the Different Approaches Used, Namely the Shell Model (SM), the Random Phase Approximation (RPA), and Charge-Exchange RPA (QRPA)

	(ν_μ, μ^-) **DIF**	(ν_e^-, e^-) **DAR**
	$\langle\sigma\rangle_f$ (10^{-40} cm^2)	$\langle\sigma\rangle_f$ (10^{-42} cm^2)
SM (HO wf) $(0+1+2)\hbar\omega$	18.71	14.21
SM (HF wf) $(0+1+2)\hbar\omega$	13.33	13.94
SM (WS wf) $(0+1+2)\hbar\omega$	11.1	12.1
SM (HO wf) $(0+1+2+3)\hbar\omega$	21.08	16.70
SM (HF wf) $(0+1+2+3)\hbar\omega$	15.18	16.42
SM (WS wf) $(0+1+2+3)\hbar\omega$	(13.2)	(12.3)
RPA	19.23	55.10
QRPA	20.29	52.0
CRPA [212]	18.18(17.80)	19.28(18.15)
Expt.	$12.4 \pm 0.3 \pm 1.8$ [199]	$14.1 \pm 1.6 \pm 1.9$ [201]
		$14.8 \pm 0.7 \pm 1.4$ [199]
		14.0 ± 1.2 [201]

Different SM results are given according to various choices of radial wave functions, i.e., oscillator functions with the length parameter $b = 1.64$ fm (HO wf) and Hartree-Fock wave functions (HF wf), and for either $(0+1+2)\hbar\omega$ or $(0+1+2+3)\hbar\omega$ model space. A comparison of continuum RPA (CRPA) with fractional occupancies [212] and a shell model calculation [214] is made. In the former case, the results are obtained with the finite-range G-matrix derived from the Bonn NN potential (BP) and with the Landau-Midgal (LM) force (in brackets). In the latter case, Woods-Saxon wave functions (WS wf) have been used and the results within the $(0+1+2+3)\hbar\omega$ model space (in brackets) are obtained by extrapolation.

TABLE 15.5

The Same as Table 15.4, but for Flux-Averaged Exclusive Cross Sections

	(ν_μ, μ^-) **DIF**	(ν_e^-, e^-) **DAR**
	$\langle\sigma\rangle_f$ (10^{-40} cm^2)	$\langle\sigma\rangle_f$ (10^{-42} cm^2)
SM (HO wf) $(0+1+2)\hbar\omega$	0.70	8.42
SM (HF wf) $(0+1+2)\hbar\omega$	0.65	8.11
SM (WS wf) $(0+1+2)\hbar\omega$	0.58	8.4
RPA	2.09	49.47
QRPA	1.97	42.92
CRPA	1.06 (1.03)	13.88 (12.55)
Expt.	$0.66 \pm 1.0 \pm 1.0$ [199]	$10.5 \pm 1.0 \pm 1.0$ [201]
		$9.1 \pm 0.4 \pm 0.9$ [199]
		$9.1 \pm 0.5 \pm 0.8$ [199]

both these shell-model calculations are reduced compared to the RPA results, bringing them closer to the experimental value. Still, this might be an "illusion" because of the reduced space in the shell-model calculations as compared to the RPA, where the space is infinite. Figure 15.16 illustrates this point for $J^\pi = 2^+$. The strength of the $J^\pi = 2^+$ calculated in the RPA continues to higher energies after the strength calculated in the shell model has diminished to zero (Tables 15.8–15.10). In Table 15.11, the RPA strength is calculated in the same model space as in the shell model ($3\hbar\omega$). In this case, the results of the two calculations agree. One can summarize the advantages and drawbacks of the two methods as follows:

TABLE 15.6
The Same as Table 15.4, but for the Inclusive Muon Capture Rates Λ_c

	$\mu^-(^{12}\mathrm{C},^{12}\mathrm{B})\nu_\mu$
	$(10^4\mathrm{s}^{-1})$
SM (HF wf) $(0+1+2+3)\hbar\omega$	3.32
SM (WS wf) $(0+1+2+3)\hbar\omega$ [215]	(4.06)
RPA	5.12
CRPA	5.79 (5.76)
Expt.	3.8 ± 0.1 [210]

TABLE 15.7
The Same as Table 15.6, but for the Exclusive Muon Capture Rates Λ_c

	$\mu^-(^{12}\mathrm{C},^{12}\mathrm{B_{g.s}})\nu_\mu$
	$(10^4\mathrm{s}^{-1})$
SM (HO wf) $(0+1+2+3)\hbar\omega$	0.50
SM (HF wf) $(0+1+2+3)\hbar\omega$	0.48
SM (WS wf) $(0+1+2+3)\hbar\omega$ [215]	0.66
RPA	2.54
CRPA	2.37 (2.43)
Expt.	0.62 ± 0.03 [210]

1. Exclusive cross sections (to the ground state):
   Easy in the extended shell model, but difficult in the RPA (involves configuration mixing).

2. Inclusive cross sections:
   Easy in the RPA, but difficult in the shell model (involves high-lying states, many major shells).

In Ref. [216], yet another shell-model approach was taken to describe various weak interaction processes involving the $A = 12$ isotriplet. The calculations are based on the ground states connected to $0, 1, 2\hbar\omega$ final states. The strategy is to assume that the role of ground-state correlations beyond $0\hbar\omega$ can be taken into account by effective operators. The simplest possible choice of an overall renormalization accounted for most of the experimental data, see Table 15.12 for the results. The thing to notice here is that all quantities measured, except the muon neutrino reaction, agree with experiment after a reduction factor of 0.64 is introduced.

It is not clear what is the origin of the reduction factor. We list several possibilities:

1. The RPA-type correlations reduce the strength due to the repulsive nature of the isovector particle-hole interaction;

2. Other correlations due to the short-range part of the interaction are also responsible for a reduction in the absolute spectroscopic factors;

3. Removal of strength in the spin channel due to the coupling to the resonance.

In all the approaches taken, it seems that there is usually an agreement between theory and experiment. The only outstanding cases are the charge-exchange reactions with muon neutrinos measured by the LSND group, where the experimental cross sections are considerably smaller than the calculated ones. We believe this controversy still exists until now.

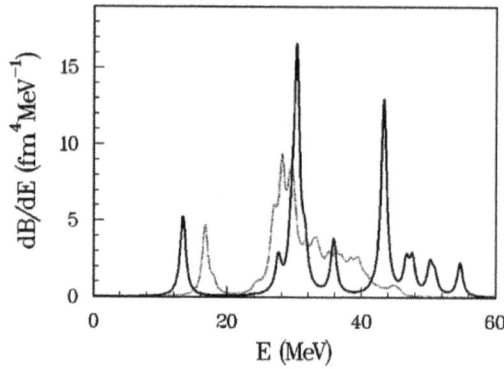

FIGURE 15.16

Strength distributions for the $J^\pi = 2^+$ obtained for the Fermi-type operator $\hat{O} = \sum_k r_k^2 Y_2(r_k) t_+(k)$ with the RPA approach (full line) and SM (dotted line). The lines are the results of the folding with a Lorentzian of 1 MeV width.

TABLE 15.8

ft Value for the β^+ Decay from $^{12}N_{g.s.} \to {}^{12}C_{g.s.}$

	ft (s)
SM (HO wf) $(0+1+2)\hbar\omega$	17008
SM (HF wf) $(0+1+2)\hbar\omega$	16425
RPA	3032
Expt.	$13182 \pm 1.$ [211]

TABLE 15.9

Contribution of the Most Important Multipolarities J^π to the Inclusive (ν_μ, μ^-) DIF Cross Sections $\langle\sigma\rangle_f (10^{-40})$ cm^2, within the RPA, QRPA, and SM Approaches

J^π	RPA	QRPA	SM	SM
			HF wf	HO wf
0^-	0.96	0.95	0.82	1.35
1^+	4.42	4.48	2.47	4.11
1^-	3.53	3.48	3.11	4.06
2^+	2.04	1.94	1.38	2.29
2^-	3.78	4.21	3.87	4.53
3^+	1.71	2.25	1.58	2.33
3^-	0.70	0.80	0.47	0.58
4^-	1.36	1.35	1.11	1.19

For the later, the results obtained with two different choices of the wave functions, Hartree-Fock (HF wf) and harmonic oscillator (HO wf), are shown.

Some charge current calculations for electron and muon neutrinos on a few lead isotopes were performed [217,218]. The main purpose of these calculations was to assess the possibility of using lead for detectors of supernova neutrinos. Recently such calculations were extended to the neutrino neutral current [220].

TABLE 15.10

Contribution of the Most Important Multipolarities J^π to the Inclusive (ν_e, e^-) DAR Cross Sections $\langle\sigma\rangle_f(10^{-42}\text{cm}^2)$, within the RPA, QRPA, and SM Approaches

J^π	RPA	QRPA	SM	SM
			HF wf	HO wf
0^-	0.3	0.5	0.6	0.6
1^+	50.0	45.6	9.6	9.8
1^-	1.7	2.0	2.1	2.2
2^+	0.1	0.1	0.1	0.1
2^-	3.0	3.7	4.0	3.9
3^+	0.0	0.1	0.0	0.1

For the later, results obtained with two different choices of the wave functions, Hartree-Fock (HF wf) and harmonic oscillator (HO wf), are shown.

TABLE 15.11

Contribution of the Inclusive (ν_μ, μ^-) DIF Cross Sections $\langle\sigma\rangle_f(10^{-42}\text{ cm}^2)$, within the RPA, RPA in the Same Model Space as the SM, and SM with HF Wave Functions.

J^π	RPA	RPA $(3\hbar\omega)$	SM (HF wf)
0^+	0.15	0.14	0.15
0^-	0.96	0.81	0.82
1^+	4.42	3.76	2.47
1^-	5.53	3.14	3.11
2^+	2.04	1.78	1.38
2^-	3.78	3.09	3.87
3^+	1.71	1.23	1.58
3^-	0.70	0.40	0.47
4^+	0.18	0.07	0.07
4^-	1.36	1.13	1.11
5^+	0.39	0.24	0.15
6^+	0.01	0.00	0.00
Total	19.23	15.79	15.18

TABLE 15.12

Results for Cross Sections, Rates, and $B(GT)$ Values

Process	Final states	Expt.	$0\hbar\omega$ g.s.	$0\hbar\omega$ g.s. $\times 0.64$
$\langle\bar{\sigma}\rangle$ (DAR)	1^+ g.s.	$9.1 \pm 0.4 \pm 0.9$ [199,201]	14.6	9.3
			23.7	15.1
$^{12}C(\nu_e, e^-)^{12}N$		14.1 ± 1.2 [199,201]		
$\langle\bar{\sigma}\rangle$ (DIF)	1^+g.s.	0.66 ± 0.14 [199]	1.4	0.9
			30.0	19.2
$^{12}C(\nu_\mu, \mu^-)^{12}N$		$12.4 \pm 0.3 \pm 1.8$ [199]		
$\Lambda\mu^-$ capture ^{12}C to ^{12}B	1^+g.s.	6.0 ± 0.4 [210]	9.4	6.0
	Total	37.9 ± 0.5 [210]	52.4	33.5
$B(GT)$ for ^{12}C to ^{12}N	1^+g.s.	0.99 ± 0.01 [219]	1.45	0.93
$B(GT)$ for ^{12}C to ^{12}B	1^+g.s.	0.87 ± 0.01 [219]	1.45	0.93

The units for $\langle\bar{\sigma}\rangle$ (DAR) are 10^{-42} cm^2, the units for $\langle\bar{\sigma}\rangle$ (DIF) are 10^{-40} cm^2, and the units for Λ are 10^3/s. The experimental values are compared to the calculated values based on the $0\hbar\omega$ g.s. wave function.

Acknowledgments

The work on various aspects of weak interactions in nuclei was repeatedly supported by the grants from the National Science Foundation (USA) and Binational Science Foundation US-Israel. We are grateful to Alex Brown, Vladimir Dmitriev, Victor Flambaum, Nguyen Van Giai, Mihai Horoi, Roman Sen'kov, Vladimir Spevak, and Alexander Volya for fruitful collaboration during many years of common work.

References

[1] C.S. Wu and S.A. Moszkowski, *Beta Decay,* Wiley Interscience, N.Y. (1966).

[2] S. Bilenky, *Introduction to the Physics of Massive and Mixed Neutrinos,* Springer, Berlin (2010); the book contains the text of the famous Pauli letter to a Tübingen conference with the neutrino hypothesis.

[3] M.C. Goodman, *Am. J. Phys.* **84**, 907 (2016).

[4] C.L. Kowan, Jr., F. Reines, F.B. Harrison, H.W. Kruse, and A.D. MxGuire, *Science* **24**, 103 (1956).

[5] F. Reines *et al.*, *Phys. Rev.* **117**, 159 (1960).

[6] T.D. Lee and C.N. Yang, *Phys. Rev.* **104**, 254 (1956).

[7] C.S. Wu, E. Ambler, R.W. Hayward, D.D. Hoppes, and R. P. Hudson, *Phys. Rev.* **105**, 1413 (1957).

[8] L.D. Landau, *ZhETF* **32**, 405 (1957).

[9] J.H. Christenson, J.W. Cronin, V.L. Fitch, and R. Turlay, *Phys. Rev. Lett.* **13**, 138 (1964).

[10] B. Aubert *et al.*, *Phys. Rev. Lett.* **86**, 2515 (2001).

[11] A.D. Sakharov, *ZhETF Pis'ma.* **5**, 32 (1967); reprinted in *Sov. Phys. Usp.* **34**, 392 (1991).

[12] L. Wolfenstein, *CP violation,* NorthHolland, Amsterdam (1989).

[13] M.S. Sozzi, *Discrete symmetries and CP violation,* Oxford University Press (2008).

[14] W. Greiner and B. Müller, *Gauge Theory of Weak Interactions,* Springer (2000).

[15] K. Zuber, *Neutrino physics,* 2nd ed., CRC Press, Boca Raton (2012).

[16] IceCube Collaboration, *Phys. Rev. D* **91**, 072004 (2015).

[17] C. Iliadis, *Nuclear Physics of Stars,* Weinheim, Germany: Wiley-VCH (2007).

[18] T.P. Cheng and L.F. Li, *Gauge theory of elementary particle physics,* Oxford University Press (2006).

[19] A.T. Yue, M.S. Dewey, D.M. Gilliam, G.L. Greene, A.B. Laptev, J.S. Nico, W.M. Snow, and F.E. Wietfeldt, *Phys. Rev. Lett.* **111**, 222501 (2013).

[20] A.P. Serebrov *et al.*, *Phys. Rev. C* **78**, 035505 (2008).

[21] P. Kienle *at al.*, *Phys. Lett. B* **726**, 638 (2013).

[22] E. April *et al.* (XENON collaboration), *Nature* **568**, 532 (2019).

[23] A.S. Barabash, arXiv:1209.3743.

[24] S.R. Elliott, A.A. Hahn, and M.K. Moe. *Phys. Rev. Lett.* **59**, 2020 (1987).

[25] S.M. Bilenky and C. Giunti, *Int. J. Mod. Phys. A* **30**, 1530001 (2015).

[26] J. Schechter and J.W.F. Valle, *Phys. Rev. D* **25**, 2951 (1982).

[27] J.B. Albert *et al.* (EXO-200 Collaboration), *Phys. Rev. Lett.* **120**, 072701 (2018).

[28] M. Jung *et al.*, *Phys. Rev. Lett.* **69**, 2164 (1992).

[29] F. Bosch *et al.*, *Phys. Rev. Lett.* **77**, 5190 (1996).

[30] M.P. Silverman, *EPL* **114**, 62001 (2016).

[31] K. Riisager *et al.*, *Phys. Lett. B* **732**, 305 (2014).

[32] D. Anthony *et al.*, *Phys. Rev. C* **65**, 034310 (2002).

[33] R. Raabe *et al.*, *Phys. Rev. Lett.* **101**, 212501 (2008).

[34] N. Auerbach, *Phys. Rep.* **98**, 273 (1983).

[35] A.L. Cole, T.S. Anderson, R.G.T. Zegers, S.M. Austin, B.A. Brown, L. Valdez, S. Gupta, G.W. Hitt, and O. Fawwaz, *Phys. Rev. C* **86**, 015809 (2012).

[36] K. Amos, A. Faessler, and V. Rodin, *Phys. Rev. C* **76**, 014604 (2007).

[37] K. Ikeda, *Progr. Theor. Phys.* **31**, 434 (1964); J-I. Fujita and K. Ikeda, *Nucl. Phys.* **67**, 143 (1965).

[38] N.B. Gove and M.J. Martin, *At. Data Nucl. Data Tables* **10**, 205 (1971).

[39] V. Zelevinsky and A. Volya, *Physics of Atomic Nuclei*, Wiley-VCH, Weinheim, 2017.

[40] J.R. Armstrong, A.A. Sakharuk, and V.G. Zelevinsky, in *Nuclei and Mesoscopic Physics, WNMP 2007*, Eds. P. Danielewicz, P. Piecuch, and V. Zelevinsky, AIP Conf. Proc. **995**, 12 (2008).

[41] P.G. Hansen, H.L. Nielsen, K. Wilksy, and J.G. Cuninghame, *Phys. Lett. B* **24**, 95 (1967).

[42] F.N.D. Kurie, J.R. Richardson, and H.C. Paxton, *Phys. Rev.* **49**, 368 (1936).

[43] J.R. Arnold and W.F. Libby, *Science* **110**, 678 (1949).

[44] C. H. Johnson, F. Pleasonton, and T. A. Carlson, *Phys. Rev.* **132**, 1149 (1963).

[45] N. Auerbach and V. Zelevinsky, *Phys. Rev. C* **90**, 034315 (2014).

[46] V. Rodin and A. Faessler, *Phys. Rev. C* **66**, 051303 (2002).

[47] R.J. Blin-Stoyle and M. Tint, *Phys. Rev.* **160**, 803 (1967).

[48] D.H. Wilkinson, *Phys. Rev. C* **7**, 930 (1973).

[49] I.S. Towner, *Phys. Rep.* **155**, 263 (1987).

[50] R.R. Doering, A. Galonsky, D.M. Patterson, and G.F. Bertsch, *Phys. Rev. Lett.* **35**, 1691 (1975).

[51] C. Gaarde *et al.*, *Nucl. Phys.* **A369**, 258 (1981).

[52] G.F. Bertsch and I. Hamamoto, *Phys. Rev. C* **26**, 1323 (1982).

[53] A. Klein, W.G. Love, and N. Auerbach, *Phys. Rev. C* **31**, 710 (1985).

[54] F. Osterfeld, *Rev. Mod. Phys.* **64**, 491 (1992).

[55] T. Wakasa *et al.*, *Phys. Rev. C* **55**, 2909 (1997).

[56] M. Ericson, A. Figureau, and C. Thevenet, *Phys. Lett.* **45B**, 19 (1973).

[57] A. Bohr and B. Mottelson, *Nuclear Structure*, vol. 1 (Benjamin, N.Y. 1975).

[58] M. Toki and W. Weise, *Phys. Lett.* **B97**, 12 (1980).

[59] G.E. Brown and M. Rho, *Nucl. Phys.* **A372**, 397 (1981).

[60] A. Bohr and B. Mottelson, *Phys. Lett.* **100B**, 10 (1981).

[61] L. Zamick and N. Auerbach, *Phys. Rev. C* **26**, 2185 (1982).

[62] N. Auerbach, *Phys. Rev. C* **27**, 1346 (1983).

[63] S. Pittel, J. Engel, J. Dukelsky, and P. Ring, *Phys. Lett. B* **247**, 185 (1990).

[64] K. Yako *et al.*, *Phys. Lett. B* **615**, 193 (2005).

[65] V. Zelevinsky, N. Auerbach, and B. M. Loc, *Phys. Rev. C* **96**, 044319 (2017).

[66] N. Auerbach, D.C. Zheng, L. Zamick, and B.A. Brown, *Phys. Lett. B* **304**, 17 (1993).

[67] N. Frazier, B.A. Brown, D.J. Millener, and V. Zelevinsky, *Phys. Lett. B* **414**, 7 (1997).

[68] P. Gysberg *et al.*, *Nature Phys.* **15**, 428 (2019).

[69] N. Cabibbo, *Phys. Rev. Lett.* **10**, 531 (1963).

[70] M. Kobayashi and T. Maskawa, *Progr. Theor. Phys.* **49**, 652 (1973).

[71] L.L. Chau and W.-Y. Keung, *Phys. Rev. Lett.* **53**, 1802 (1984).

[72] I.S. Towner and J.C. Hardy, *Rep. Prog. Phys.* **73**, 046301 (2010).

[73] J.C. Hardy and I.S. Towner, *Phys. Rev. C* **91**, 025501 (2015).

[74] N. Severijns and O. Naviliat-Cuncic, *Phys. Scr.* **T152**, 014018 (2013).

[75] M. Conzalez-Alonso, O. Naviliat-Cuncic, and N. Severijns, *Prog. Part. Nucl. Phys.* **104**, 165 (2019).

[76] S.S. Gershtein and J.B. Zeldovich, *Zh. Eksperim. Teor. Fis.* **29**, 698 (1955) [English translation Sov. Phys. JETP **2**, 576 (1955).]

[77] R. Feynman and M. Gell-Mann, *Phys. Rev.* **109**, 193 (1958).

[78] T.D. Lee and C.N. Yang, *Nuovo Cim.* **3**, 749 (1956).

[79] J.C. Hardy and I.S. Towner, *Phys. Rev. C* **71**, 055501 (2005).

[80] I.S. Towner and J.C. Hardy, *Phys. Rev. C* **77**, 025501 (2008).

[81] W.E. Ormand and B.A. Brown, *Phys. Rev. C* **52**, 2455 (1995).

[82] N. Auerbach, *Phys. Rev. C* **79**, 035502 (2009).

[83] V. Rodin, *Phys. Rev. C* **88**, 064318 (2013).

[84] A.M. Lane, *Nucl. Phys.* **35**, 676 (1962).

[85] N. Auerbach and A. Klein, *Nucl. Phys.* **A395**, 77 (1983).

[86] S.Y. Igashov, V. Rodin, A. Faessler, and M.H. Urin, *Phys. Rev. C* **83**, 044301 (2011).

[87] R.G.T. Zegers *et al.*, *Phys. Rev. Lett.* **90**, 202501 (2003).

[88] W.J. Marciano and A. Sirlin, *Phys. Rev. Lett.* **96**, 032002 (2006).

[89] O. Naviliat-Cuncic and N. Severijns, *Phys. Rev. Lett.* **102**, 142302 (2009).

[90] L.S. Brown and G. Gabrielse, *Rev. Mod. Phys.* **58**, 233 (1986).

[91] J.C. Hardy and I.S. Towner, *Phys. Rev. Lett.* **88**, 252501 (2002).

[92] C.-Y. Seng, M. Gorchtein, H.H. Patel, and M.J. Ramsey-Musolf, *Phys. Rev. Lett.* **121**, 241804 (2018).

[93] L. Xayavong and N.A. Smirnova, *Phys. Rev. C* **97**, 024324 (2018).

[94] K. Langanke and M. Wiescher, *Rep. Prog. Phys.* **64**, 1657 (2001).

[95] H. Grawe, K. Langanke, and G. Martinez-Pinedo, *Rep. Prog. Phys.* **70**, 1525 (2007).

[96] K. Langanke and H. Schatz, *Phys. Scr.* **T152**, 014011 (2013).

[97] E.M. Burbidge, G.R. Burbidge, W.A. Fowler, and F. Hoyle, *Rev. Mod. Phys.* **29**, 547 (1957).

[98] R.A. Alpher, H.A. Bethe, and G. Gamow, *Phys. Rev.* **73**, 803 (1948).

[99] R.A. Alpher, J.W. Follin, Jr., and R.C. Herman, *Phys. Rev.* **92**, 1347 (1953).

[100] Ya.B. Zel'dovich, *Adv. Astron. Astrophys.* **3**, 241 (1965).

[101] P.J.E. Peebles, *Astron. J.* **146**, 542 (1966).

[102] S. Weinberg, *Cosmology*, Oxford University Press, 2008.

[103] G. Gamow, *Phys. Rev.* **74**, 505 (1948).

[104] E.G. Adelberger *et al.*, *Rev. Mod. Phys.* **70**, 1265 (1998).

[105] Q.R. Ahnad *et al.* (SNO Collaboration), *Phys. Rev. Lett.* **89**, 011301 (2002).

[106] B. Meyer, *Annu. Rev. Astron. Astrophys.* **32**, 153 (1994).

[107] A. Aprahamian, K. Langanke, and M. Wiescher, *Prog. Part. Nucl. Phys.* **54**, 535 (2005).

[108] K. Wisshak *et al.*, *Phys. Rev. Lett.* **87**, 251102 (2001).

[109] G.M. Fuller, W.A. Fowler, and M.J. Newman, *Astrophys. J.* **252**, 715 (1982).

[110] E. Caurier, K. Langanke,G. Martinez-Pinedo, and F. Novacki, *Nucl. Phys.* **A653**, 439 (1999).

[111] J. Cooperstein and J. Wambach, *Nucl. Phys.* **A420**, 591 (1984).

[112] K. Langanke and G. Martinez-Pinedo, *Rev. Mod. Phys.* **75**, 819 (2003).

[113] J.-J. Liu and D.-M. Liu, *Astrophys. Space Sci.* **361**, 246 (2016).

[114] D.G. Yakovlev, A.D. Kaminker, O.Y. Gnedin, and P. Haensel, *Phys. Rep.* **354**, 1 (2001).

[115] J.J. Liu and W.M. Gu, *Astrophys. J. Suppl.* **224**, 29 (2016).

[116] L.G. Mann, D.C. Camp, J.A. Miskel, and R.J. Nagle, *Phys. Rev.* **137**, B1 (1965).

[117] S.K. Bhattacherjee, S.K. Mitra, and H.C. Padhi, *Nucl. Phys.* **A96**, 81 (1967).

[118] V. Avetisov and V. Goldanskii, *Proc. Natl. Acad, Sci. USA* **93**, 11435 (1996).

[119] I.M. Ternov, V.N. Rodionov, O.F. Dorofeev, and A.E. Lobanov, *Izv. Vuzov Fizika* No. 3, 82 (1986).

[120] R.A. Harris and L. Stodolsky, *Phys. Lett. B* **78**, 313 (1978).

[121] I.B. Khriplovich, *Parity Nonconservation in Atomic Phenomena*, Nauka, Moscow, 1981.

[122] M. Cattani, *JQSRT* **52**, 831 (1994).

[123] R. Berger and J. Stohner, Wiley Interdisciplinary Reviews, *https://doi.org/10.1002/wcms.1396*.

[124] E.G. Adelberger and W.C. Haxton, *Annu. Rev. Nucl. Part. Sci.* **35**, 501 (1985).

[125] B. Desplanques, J.F. Donoghue, and B.R. Holstein, *Ann. Phys. (N.Y.)* **124**, 449 (1980).

[126] D. Blyth *et al.* (NPDGamma Collaboration), *Phys. Rev. Lett.* **121**, 242002 (2018).

[127] P. Alfimenkov *et al.*, *Nucl. Phys.* **A398**, 93 (1983).

[128] J.D. Bowman *et al.*, *Phys. Rev. Lett.* **65**, 1192 (1990).

[129] V.V. Flambaum and O.P. Sushkov, *Nucl. Phys.* **A412**, 13 (1984).

[130] V.E. Bunakov, E.D. Davis, and H.A. Weidenmueller, *Phys. Rev. C* **42**, 1718 (1990).

[131] N. Auerbach, *Phys. Rev. C* **45**, R514 (1992).

[132] N. Auerbach and J.D. Bowman, *Phys. Rev. C* **46**, 2582 (1992).

[133] N. Auerbach and B.A. Brown, *Phys. Lett. B* **340**, 6 (1994).

[134] G.V. Danilyan, *Sov. Phys. Usp.* **23**, 323 (1980).

[135] O.P. Sushkov and V.V. Flambaum, *Sov. Phys. Usp.* **25**, 1 (1982).

[136] A. Koetzle, P. Jesinger, F. Goennenwein, G.A. Petrov, V.I. Petrova, A.M. Gagarski, G. Danilyan, O. Zimmer, and V. Nesvizhevsky, *Nucl. Instrum. Methods A* **440**, 750 (2000).

[137] J.D. Bowman, G.T. Garvey, M.B. Johnson, and G.E. Mitchell, *Annu. Rev. Nucl. Sci.* **43**, 829 (1993).

[138] S.L. Stephenson *et al.*, *Phys. Rev. C* **58**, 1236 (1998).

[139] V.V. Flambaum and V. Zelevinsky, *Phys. Lett. B* **350**, 8 (1995).

[140] B.E. Crawford *et al.*, *Phys. Rev. C* **58**, 1225 (1998).

[141] G.E. Mitchell, J.D. Bowman, and H.A. Weidenmueller, *Rev. Mod. Phys.* **71**, 445 (1999).

[142] F.C. Michel, *Phys. Rev.* **133**, B329 (1964).

[143] D.J. Horen *et al.*, *Phys. Lett. B* **99**, 383 (1981).

[144] N. Auerbach and A. Klein, *Phys. Rev. C* **30**, 1032 (1984).

[145] D.M. Brink, 2009, available online at *http://tid.uio.no/workshop09/talks/Brink.pdf*.

[146] C.R. Gould *et al.*, AIP Conf. Proc. **238**, 747 (1991).

[147] Ya.B. Zel'dovich, *Sov. Phys. JETP* **6**, 1184 (1958).

[148] V.V. Flambaum and I.B. Khriplovich, *Sov. Phys. JETP* **52**, 835 (1980).

[149] C.S. Wood, S.C. Bennett, D. Cho, B.P. Masterson, J.L. Roberts, C.E. Tanner, and C.E. Wieman, *Science* **275**,1759 (1997).

[150] V.F. Dmitriev and V.B. Telitsin, *Nucl. Phys.* **A613**, 237 (1997).

[151] V.V. Flambaum and D.W. Murray, *Phys. Rev. C* **56**, 1641 (1997).

[152] N. Auerbach and B. A. Brown, *Phys. Rev. C* **60**, 025501 (1999).

[153] D. Androic *et al.* (Q_{weak} Collaboration) *Phys. Rev. Lett.* **111**, 141803 (2013).

[154] S. Abrahamyan *et al.* (PREX Collaboration) *Phys. Rev. Lett.* **108**, 112502 (2012).

[155] K. Abe *et al.* (Belle Collaboration), *Phys. Rev. Lett.* **87**, 091802 (2001).

[156] R. Aaij *et al.* (LHCb collaboration), *Phys. Rev. Lett.* **110**, 221601 (2013).

[157] R. Aaij *et al.*, (LHCb collaboration), *Phys. Rev. Lett.* **108**, 11160 (2012).

[158] R. Aaij *et al.*, *J. High Energy Phys.* **02**, 126 (2019),

[159] J. Schwinger, *Phys. Rev.* **82**, 914 (1951).

[160] J. Engel, M.J. Ramsey-Musolf, and U. van Kolck, *Prog. Part. Nucl. Phys.* **71**, 21 (2013).

[161] T.E. Chupp, P. Fierlinger, M.J. Ramsey-Musolf, and J.T. Singh, *Rev. Mod. Phys.* **91**, 015001 (2019).

[162] J.S.M. Ginges and V.V. Flambaum, *Phys. Rep.* **397**, 63 (2004).

[163] N. Auerbach and V. Zelevinsky, *J. Phys. G* **35**, 093101 (2008).

[164] K.F. Smith *et al.*, *Phys. Lett. B* **234**, 191 (1990).

[165] J. N. Pendlebury *et al.*, *Phys. Rev. D* **92**, 092003 (2015).

[166] T.G. Vold, F.J. Raab, B. Heckel, and E.N. Fortson, *Phys. Rev. Lett.* **52**, 2229 (1984).

[167] J.P. Jackobs, W.M. Klipstein, S.K. Lamoreaux, B.R. Heckel, and E.N. Fortson, *Phys. Rev. A* 52, 3521 (1995).

[168] D. Cho, K. Sangster, and E.A. Hinds, *Phys. Rev. Lett.* **63**, 2559 (1989).

[169] V. Andreev *et al.*, *Nature* **562**, 355 (2018).

[170] B. Graner, Y. Chen, E.G. Lindahl, and B.R. Heckel, Phys. Rev. Lett. **116**, 161601 (2016) [Erratum ibid. **119**, 119901 (2017)].

[171] N. Auerbach, V.V. Flambaum, and V. Spevak, *Phys. Rev. Lett.* **76**, 4316 (1996).

[172] V. Spevak, N. Auerbach, and V.V. Flambaum, *Phys. Rev. C* **56**, 1357 (1997).

[173] V. Spevak, N. Auerbach, *Phys. Lett. B* **359**, 254 (1995).

[174] L.I. Schiff, *Phys. Rev.* **132**, 2194 (1963).

[175] O. P. Sushkov, V. V. Flambaum, and I. B. Khriplovich, *Sov. Phys. JETP* **60**, 873 (1984).

[176] V. V. Flambaum, I. B. Khriplovich, and O. P. Sushkov, *Nucl. Phys.* **A449**, 750 (1986).

[177] V. F. Dmitriev and R. A. Sen'kov, *Yad. Fiz.* **66**, 1988 (2003).

[178] J. Engel, M. Bender, J. Dobaczewski, J. H. de Jesus, and P. Olbratowski, *Phys. Rev. C* **68**, 025501 (2003).

[179] V.F. Dmitriev, R.A. Sen'kov, and N. Auerbach, *Phys. Rev. C* **71**, 035501 (2005).

[180] A. Bohr and B. Mottelson, *Nuclear Structure*, Vol. 2, Benjamin, N.Y. 1975.

[181] J. Dobaczewski and J. Engel, *Phys. Rev. Lett.* **94**, 232502 (2005).

[182] S. Ban, J. Dobaczewski, J. Engel, and A. Shukla, *Phys. Rev. C* **82**, 015501 (2010).

[183] J. Engel, J.L. Friar, and A.C. Hayes, *Phys. Rev. C* **61**, 035502 (2000).

[184] N. Auerbach, V.F. Dmitriev, V.V. Flambaum, A. Listskiy, R.A. Sen'kov, and V.G. Zelevinsky, *Phys. Rev. C* **74**, 025502 (2006).

[185] G.A. Leander and Y.S. Chen, *Phys. Rev. C* **37**, 2744 (1988).

[186] A.K. Jain, R.K. Sheline, P.C. Sood, and K. Jain, *Rev. Mod. Phys.* **62**, 393 (1990).

[187] I. Ahmad and P.A. Butler, *Annu. Rev. Nucl. Part. Sci.* **43**, 71 (1993).

[188] P.A. Butler and W. Nazarewicz, *Rev. Mod. Phys.* **68**, 349 (1996).

[189] L.P. Gaffney *et al.*, *Nature* **497**, 199 (2013).

[190] R.K. Sheline, Y.S. Chen, and G.A. Leander, *Nucl. Phys.* **A486**, 306 (1988).

[191] R.K. Sheline, A.K. Jain, K. Jain, and I. Ragnarsson, *Phys. Lett. B* **219**, 47 (1989).

[192] J. E. Simsarian, A. Ghosh, G. Gwinner, L. A. Orozco, G. D. Sprouse, and P. A. Voytas, *Phys. Rev. Lett.* **76**, 3522 (1996).

[193] R.K. Sheline, C.F. Liang, and P. Paris, *Int. J. Mod. Phys. A* **5**, 2821 (1990).

[194] W. Kurcewicz *et al.*, *Nucl. Phys.* **A539**, 451 (1992).

[195] R.K. Sheline, C.F. Liang, P. Paris, J. Kvasil, and D. Nosek, *Phys. Rev. C* **51**, 1708 (1995).

[196] R.H. Parker *et al.*, *Phys. Rev. Lett.* **114**, 233002 (2015).

[197] M. Bishof *et al.*, *Phys. Rev. C* **94**, 025501 (2016).

[198] H. Ejiri, J. Suhonen, and K. Zuber, *Phys. Rep.* **797**, 1 (2019).

[199] C. Athanassopoulos *et al.* (LSND collaboration), *Phys. Rev. C* **56**, 2806 (1997).

[200] R.C. Allen *et al.*, *Phys. Rev. Lett.* **64**, 1871 (1990).

[201] D.A. Krakauer *et al.*, *Phys. Rev. C* **45**, 2450 (1992).

[202] V. Barger, D. Marfatia, and K. Whisnant, *The physics of neutrinos*, Princeton Univercity Press, Princeton, 2012.

[203] J.N. Bahcall, *Neutrino astrophysics*, Cambridge University Press, 1989.

[204] F. Halzen, *Nature Phys.* **13**, 232 (2017).

[205] L. Alvarez-Ruso *et al.*, *Prog. Part. Nucl. Phys.* **100**, 1 (2018).

[206] J.D. Walecka, Section 4 - *Semileptonic weak interactions in nuclei*, in V.W. Hughes, and C.S. Wu, editors, *Muon Physics*, Academic Press, 1975, p. 113.

[207] T. Kuramoto, M. Fukugita, Y. Kohyama, and K. Kubodera, *Nucl. Phys.* **A512**, 711 (1990).

[208] C. Volpe, N. Auerbach, G. Colò, T. Suzuki, and N. Van Giai, *Phys. Rev. C* **62**, 015501 (2000).

[209] N. Auerbach and A. Klein, *Nucl. Phys.* **A422**, 480 (1984).

[210] D.F. Measday, *Phys. Rep.* **354**, 243 (2001).

[211] F. Ajzenberg-Selove, *Nucl. Phys.* **A433**, 1 (1985).

[212] E. Kolbe, K. Langanke, F.-K. Thielemann, and P. Vogel, *Phys. Rev. C* **52**, 3437 (1995).

[213] N. Auerbach, N. Van Giai, and O.K. Vorov, O. K., *Phys. Rev. C* **56**, R2368 (1997).

[214] N. Auerbach, A. Klein, and N. Van Giai, *Phys. Lett. B* **106**, 347 (1981).

[215] A.C. Hayes and I.S. Towner, *Phys. Rev. C* **61**, 044603 (2000).

[216] N. Auerbach and B.A. Brown, *Phys. Rev. C* **65**, 024322 (2002).

[217] C. Volpe, N. Auerbach, G. Colò, and N. Van Giai, *Phys. Rev. C* **65**, 044603 (2002).

[218] W. Almosly, B.G. Carlsson, J. Suhonen, J. Toivanen, and E. Ydrefors, *Phys. Rev. C* **94**, 044614 (2016).

[219] E.K. Warburton and B.A. Brown, *Phys. Rev. C* **46**, 923 (1992).

[220] W. Almosly, B.G. Carlsson, J. Suhonen, and E. Ydrefors, *Phys. Rev. C* **99**, 055801 (2019).

Index

Note: **Bold** page numbers refer to tables and *italic* page numbers refer to figures.

For Product Safety Concerns and Information please contact our EU
representative GPSR@taylorandfrancis.com
Taylor & Francis Verlag GmbH, Kaufingerstraße 24, 80331 München, Germany

www.ingramcontent.com/pod-product-compliance
Lightning Source LLC
Chambersburg PA
CBHW080653220326
41598CB00033B/5195